Electromyography for Experimentalists

Electromyography for Experimentalists

Gerald E. Loeb and Carl Gans

The University of Chicago Press
Chicago and London

Gerald E. Loeb is medical officer for research at the Laboratory of Neural Control, National Institute of Neurological and Communicative Disorders and Stroke, National Institutes of Health. He is also president of Biomed Concepts, Inc. and adjunct associate professor of bioengineering at the University of Utah.

Carl Gans is professor of biological sciences at the University of Michigan. He is the coauthor with Thomas S. Parsons of *A Photographic Atlas of Shark Anatomy,* also published by the University of Chicago Press, the author of *Reptiles in Color* and *Biomechanics*, and the editor of the *Biology of the Reptilia.*

The University of Chicago Press, Chicago 60637
The University of Chicago Press, Ltd., London

Printed in the United States of America

95 94 93 92 91 90 89 88 87 86 5 4 3 2 1

Library of Congress Cataloging-in-Publication Data

Loeb, Gerald E.
Electromyography for experimentalists.

Bibliography: p.
Includes index.
1. Electromyography. 2. Muscles—Research—Methodology. I. Gans, Carl, 1923– . II. Title.
QP321.L64 1986 599′.01852 85–28934
ISBN 0–226–49014–9
ISBN 0–226–49015–7 (pbk.)

For Sandy and Mabel

Contents

SUMMARY OF CONTENTS

DETAILED CONTENTS

Introduction

As muscles are activated, they generate action currents that flow through the resistive medium of the tissues. The voltage gradients so produced may be recorded as the myoelectrical signal, which has three meanings to biologists. To biophysicists it reflects a fundamental process whereby the contractile mechanism is triggered and synchronized, thus permitting an important look into the biophysics of membrane and synapses. To physiologists the signal is the end product of chains of command and reflexes that they hope to elucidate by quantifying amplitude and timing. To functional morphologists and kinesiologists the signal is a useful epiphenomenon by which the mechanical contributions of muscles to gross anatomical movements may be estimated. This book is intended for the latter two groups of investigators, who share the problems of trying to use electromyography to make inferences about processes that are complexly and nonlinearly related to EMG (we use the abbreviation EMG for electromyography, electromyogram, and electromyographical).

Most kinesiologists and functional anatomists would prefer to use signals that are directly proportional to the tension and displacements generated by single muscles rather than this token of the mechanical output. However, recording of EMG signals requires a minimum of interference with the movement in progress, and EMG signals can be obtained from muscles that cannot be instrumented with strain gauges. Similarly, most physiologists would prefer to have an intracellular record of the transmembrane potentials of the motoneurons instead of the inconsistent bursts of synaptically generated EMG signal. However, unlike more sophisticated (and direct) measurements, the EMG can be obtained simultaneously from multiple muscles without significantly affecting the behavior of the system.

For any type of inquiry the use of an indirect measurement obligates the investigator to support the inferences to be drawn. The investigator can fulfill this obligation in one of two ways: by performing controls that directly demonstrate the relationship or by using techniques that, according to theoretical considerations, involve no significant chance of error.

In fact, most experimentalists use a third method—namely to do what everyone else has been doing. Once techniques become widely accepted, they cease to be questioned, particularly by those who practice them and, in turn, referee the submissions of others. This collective suspension of caution risks the obvious calamity—suppose everyone has been wrong? While not unheard of in the history of science, the problem usually comes to light early through the cumulative self-checking process of research. A more pernicious aspect is that no two experiments are the same in biological properties. Techniques that seemed reliable in one animal, one limb, or one muscle may not be appropriate at another, seemingly similar, recording site. Apparently minor variations on the selection of equipment, the fabrication of electrodes, and their attachment to recording sites may not be minor at all. Although it is unlikely that everyone has been wrong, this is scant consolation to the individual investigator who needs to be certain that he or she is not wrong.

Actually, there is rather little in the way of valid historical precedent for many approaches to the EMG of experimental animals. That methodology has been adapted from the voluminous clinical experience in humans. However, the justification for a clinical technique is not that it allows study of a fundamental biological process, but that it correlates well with a differential clinical diagnosis.

In experimental studies the diversity of species involved and the range of experimental conditions that may be encountered are vast and, for the most part, uncharted. Paradoxically, the biophysical and electrical processes that produce EMG signals are quite well understood and appear generally applicable to most muscles and species. Therein lies the purpose of this book: to provide the experimentalist with bases for the design and conduct of experiments that facilitate the solution of problems and minimize the chance of serious error.

This is also a practical book. There are, to be sure, many ambitious attempts in the literature to derive rigorous mathematical descriptions of the relationships between the signal sources and the recorded EMG. We acknowledge our indebtedness to the insights provided by such models, at the same time that we point out their general lack of utility for the average experimentalist. We hope to bridge this gap by translating the differential equations into subjective understanding. We make brutally simplifying assumptions, but we try to justify them on anatomical and physiological grounds. We introduce analogies and examples to avoid camouflaging our ignorance with equations. We provide alternatives whenever ideal techniques and controls are impractical.

The art of EMG can be found in the countless recipes for electrode designs, insertion techniques, connection schemes, and signal processors, some of which we describe here. The science lies in knowledge of how to select the theoretically correct approach. The engineering lies in development of suitable and cost-effective techniques for a given application, as well as convincing and efficient demonstration that one has chosen wisely.

We have subdivided this book into two parts. The first part deals with general theoretical matters, including the nature of muscle, the nature of electrical signals, concepts of antenna and electrode theory, and electrode placement and verification. The second part is more practical, dealing in detail with the several steps that must be considered in the design, construction, and operation of an EMG facility and the performance of successful experiments. In the first chapter we provide an introduction to the subject, with an overview and references to specific topics taken up in more detail later in the book—theory first, and practicality second. Toward this end, in this chapter we briefly review major steps in successful myographical analysis. Throughout it we refer the reader to the particular sections in which each of the topics alluded to is discussed in more detail. In the final chapter we present a most simple and inexpensive experimental arrangement that allows the experimentalist to carry out pilot experiments. Principles tend to become obscured with time. As problems arise during the process of investigation, and the experiments shift from the simple to the complex, the reader may wish to refer repeatedly to the first and last chapters.

We have desisted from organizing this work along historical lines or as a review of the literature. In most cases, the general theoretical points we need to make derive from a general consideration of biophysical theory as it has evolved over the past half-century rather than from any explicit references. Even the specific methods discussed here usually represent an amalgamation and abstraction of technical details. Their presentation in the original references tends to obscure and constrain the general utility of and operative principles behind them. Therefore, we have relied on internally consistent logic rather than appeals to external authority to justify each statement. For the reader interested in consulting the relevant literature, we provide an annotated bibliography that indicates the particular significance of articles we have found useful in developing these logical arguments as well as a number of general reviews of the various topics. While these reviews are obviously not encyclopedic, they do at least represent and provide citations for important ideas and methods that the reader may need to cite or review.

Any book on methods for scientific study faces the serious problem of obsolescence in a rapidly changing world of electronic data processing. Sources of supply, particularly for specialized items, are prone to appear and disappear without warning. In these matters, a network of active colleagues and adherence to the tradition of listing critical materials and their sources in published papers will always be essential. Much attention is being paid to the microcomputer explosion, which has made sophisticated data processing available in even modestly funded laboratories. Whereas this may change the appearance of published figures, it does not change the basic electronic principles required to generate a reliable analog signal from a biological preparation. In this respect, the electronics of EMG is somewhat like the current state of high-fidelity audio products; new boxes with new bells and whistles will continue to appear, but actual performance has not changed significantly nor is it likely to in this highly evolved field. Furthermore, many of the tried-and-true forms of signal processing (e.g., rectification and integration) are so simple and effective as electronic techniques that there may be little need and less reason to replicate or improve on them with cumbersome digital techniques. When considering anything new, we should make sure it is also better; innovation is not synonymous with improvement.

We thank Martin Bak for technical advice on specific issues, Fay Hansen-Smith and Frances J. Richmond for sharing their expertise about the many histochemical techniques incorporated in appendix 1, and Coen Ballintijn, Jeannine de Vree, Ellengene Petersen, and Gerhard Roth for advice on specific methods. Coen Ballintijn, Henricus J. de Jongh, Abbot S. and Sandra Gaunt, Gerard C. Gorniak, and Sven-Erik Widmalm commented on all or most of the manuscript, an early version of which was also discussed by students in Zoology 435 at the University of Michigan during the fall of 1983 and by the Neurokinesiology Group of the Laboratory of Neural Control (NINCDS, NIH) in the summer of 1984. Susan M. Konchal, Katherine Vernon, and Kyoko A. Gans typed portions of the manuscript. Brent A. Bauer translated our rough sketches into the final drawings, and John Lacy and George Dold performed some of the photography. We thank vendors for use of the illustrations identified in the captions and those colleagues who assisted with illustrations. Carl Gans wishes to acknowledge grants of the United States National Science Foundation and National Institutes of Health and the contributions of students and colleagues, which supported the studies underlying the development of some of the approaches reported here. Gerald Loeb wishes to acknowledge the support of the Intramural

Research Program of the National Institutes of Neurological and Communicative Disorders and Stroke and the contributions of many colleagues in the Laboratory of Neural Control over the years during which many of the ideas and techniques described here were developed.

Part 1 THEORY

1 Orientation—This End Up

A. PHILOSOPHY

The striated muscles of vertebrates are assemblages of variously shaped cells containing elongate contractile systems. Whenever these cells are stimulated, they attempt to shorten; in the process, they stiffen and exert a pull that tends to bring their ends together. Most larger animals use such muscles to exert forces and achieve displacements, to position and move their component parts. The muscular function appears most simple when stated this way, but it differs among muscles and animals, and the actual performance of particular muscles involves many interesting complications.

Most of the joints in animal bodies are bridged by multiple muscles, some arranged in parallel and others in series. Several such muscles may well produce equivalent actions when individually stimulated; however, they may be used at different times during locomotive, respiratory, and masticatory cycles. In contrast, some other muscles may produce different effects when individually stimulated; however, such muscles may then act in concert with quite distinct muscles so that their effects become equivalent. Ensemble patterns may differ not only for different groups of muscles but also for different activities. Thus, similarity of position does not imply similarity of action. Not only do the action patterns of entire muscles differ, but so do those of their layers and fascicles. The larger and more complex the anatomy of a muscle is, the more probable it is that its action is complex as well. However, one cannot simply predict function only from structure.

Zoologists, physiologists, and other students of organisms have numerous reasons for wanting to know how and why organisms produce or avoid movements. Other biologists attempt to determine when, under which conditions, how strongly, and for how long particular muscles will act while the animals are resting, biting, mating, jumping, and flying. In this book we attempt to introduce you to ways of studying the performance of such muscles. As experimentalists cannot extrapolate this information from the structure of the animal, they need an alternate technique. For many reasons, the technique should not cause the animal any pain; one reason is that pain will affect an animal's behavior so that it may avoid using precisely those muscles being studied. Consequently, the technique should have minimal effect on the behavior. Electromyography is a relatively painless technique that lets us discover when and how the animal is using its muscles. It involves a briefly invasive approach, which should cause only minimal and transient trauma. Thereafter, the EMGs should indicate when and how strongly the animal uses particular muscles during a variety of behaviors.

Unfortunately EMG by itself has limited utility. One important message of this book is that the technique must be part of a series of anatomical and behavioral observations. Also, it must proceed in parallel with a study of the movements performed and the forces exerted. Consequently, the kind of EMG technique ultimately utilized, the kind of mechanical indicators selected, the way EMGs are analyzed, and the way data are ultimately presented must obviously depend on what we are trying to find out. This chapter is intended to help with that decision, to define some of the terms referring to different experimental approaches, and to provide a guide to the rest of the book. We suggest that it be read first.

B. PERSPECTIVE

To many of us, the initial view of an EMG is a bit confusing. Assuming that everything is working, the more or less constant baseline, seen as a sweep signal on an oscilloscope screen, suddenly displays a series of large wiggles or the pen of a chart recorder becomes temporarily excited. The instrument is detecting some kind of an event. What should we have done and what must we know in order to interpret what may be happening within the animal?

We should start by recognizing that an EMG at best represents the major changes in currents and voltages that occur whenever muscle fibers are being activated by their motoneurons. To date electromyography has proved to be of limited utility in detecting other, slower events—for instance, various kinds of tonus and even the excitation-contraction threshold of slow fibers. The EMG will indicate the responses of faster fibers with variable effectiveness depending on the skill with which the electrode has been formed and placed and the appropriateness with which the resulting signals have been filtered, amplified, and otherwise treated. Lacking these conditions, the signal is likely to indicate only that the electrode moved within the tissue mass or that the signal derives from some muscle other than that in which the electrode was originally implanted.

The problem of unambiguously identifying an EMG as originating entirely from the activity of one particular muscle will always be the most difficult and controversial aspect of this method. This is the fundamental problem alluded to by the term "cross-talk." The burden of proof remains with the investigator, and in chapter 8 we discuss a variety of approaches to discharging this obligation.

Few experimentalists are interested only in the electrical events within the muscle. Most are likely to study muscular activity in order to understand the movement, its neural control, or even the behavior of an animal. This means that, rather than only obtaining a reproducible EMG, they wish to use it as a substitute for a direct indication of the forces exerted by the muscle and of the displacements it facilitates. In short, the EMG provides only one datum of a desirable set.

Electromyography can indicate not only the start and end of muscular activity but also something about the number of active motor units and the frequency at which they fire. As discussed in chapter 3, which deals with the organization of muscle, the EMG only provides an indirect indication of the force output or of the excursion that the muscle produces; therefore, a separate discussion is devoted to correlation of EMG level and force (chapter 16). Any quantitative comparison among EMG records requires prior standardization of electrodes and signal treatment (chapter 17). Standardization involves a number of steps that must be carefully planned and executed. Without these one does not obtain results that are repeatable and suitable for analysis—results that represent more than random wigglings on a screen.

The disciplines of kinesiology (study of

movement), muscle mechanics (study of the forces causing movement), muscle physiology (study of the structure and properties of the biological motors), and motor neurophysiology (study of the control systems for motor behavior) must all come together to provide an understanding of these indirect messages of the EMG signal. Figure 1.1 shows one way of ordering the various sources of data and the transformations from one parameter to another both in reductionistic and synthetic directions. It is important to note the many levels of anatomical structure and their potential interactions. In order to understand them, we must relate the molecular processes in a muscle fiber to the electrical signals they generate, on the one hand, and to the output forces and motion we can observe, on the other.

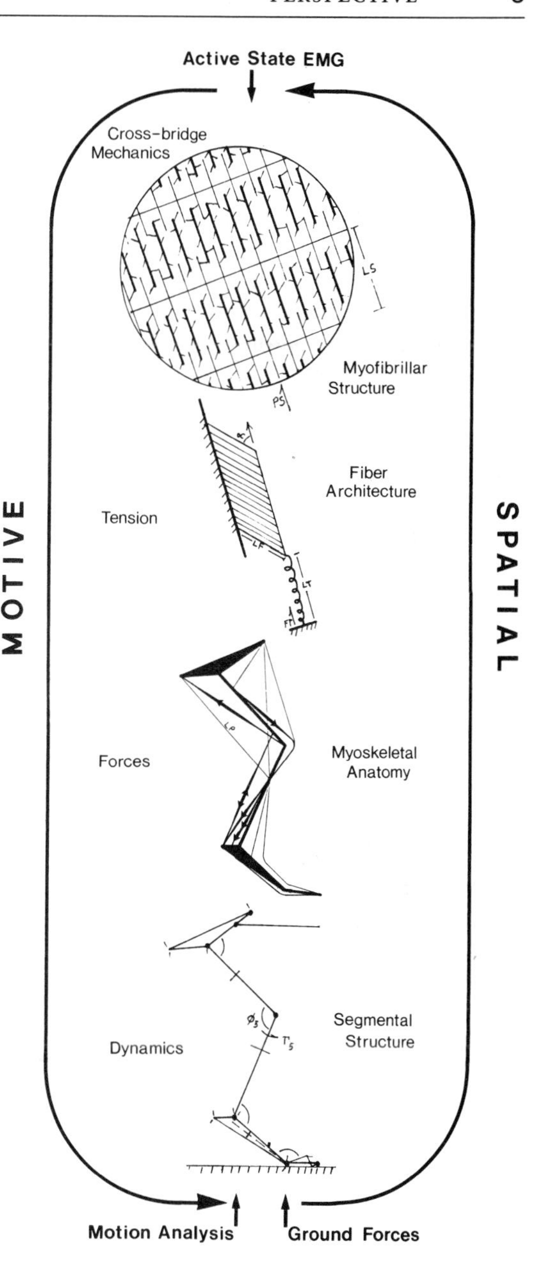

1.1. Relationships among the various kinesiological (movement in space) terms and kinematic (motive force) terms describe a closed circle of logically related terms. The apparent motion of the whole structure (as seen on videotape or movie film) can be converted into a stick figure (*bottom*). From a knowledge of the gross muscular anatomy (such as origins, insertions, pulleys, and lever arms), one can generate the time course of length and velocity for each whole muscle (*second from bottom*). A knowledge of muscle fiber architecture lets one derive the time course of motion in muscle fibers (*second from top*) and sarcomeres (*top*); this defines the force-generating capabilities of the muscle for a given level of activation, usually indicated by the EMG recording (*top*). The time course of force output by the sarcomere may be transformed by the architectural relationships (in the opposite direction) to calculate tension output in the tendon and individual muscle torques applied to limb segments. The internal torques should cross-check against the net torque that must be applied at each joint of the segmental structure to account for the ground forces and accelerations noted in the motion analysis.

C. FUNDAMENTALS

The form of an EMG signal is determined by both anatomical structure and electrical function. Experimentalists must understand at least the basic principles of each before they can appreciate how these factors combine to produce the EMG signal. In chapters 2 and 3 we present the terminology and basic concepts of electronics and anatomy, respectively, in preparation for their fusion in chapter 4, in which we describe the generation of the EMG signal. In the following chapters we deal with factors contributing to the specific details of the recorded signals (chapter 5) and considerations underlying the general design of recording electrodes (chapter 6).

As a science develops and as its students must master more complex and advanced techniques, there is a danger that the new generation will lack the opportunity to understand from first hand experience the simple and elegant principles upon which the science is based. Mastery of state-of-the-art principles is necessary. However, a science as young as experimental biology still yields important revelations from the reexamination of processes and methods at fundamental levels. Too often, the apparent sophistication of a highly evolved experimental science serves to mask fundamental weakness and even error. The opportunity to detect and correct these failings tends to come to those who understand their experiments on the basis of first principles rather than performing them by rote from manuals. A textbook can go only so far in conveying a deep, intuitive understanding of either anatomy or electronics. The student is well advised to follow in the footsteps of the great experimentalists of both fields, who spent a great deal of time observing with an open mind and tinkering in the laboratory.

An appreciation of functional anatomy begins with dissection. Most of the features of muscle fiber architecture and myoskeletal anatomy are best discovered with a good eye and a skilled hand. By careful dissection one can establish a list of the candidate muscles that might contribute to the behavior of interest and those that might generate mechanical actions and electrical signals that might confuse the interpretation of kinesiological and EMG data. Careful notes are invaluable in planning surgical approaches that will probably minimize trauma and recovery time after implantation of electrodes and transducers. Comparison of the observations with the available literature may suggest the need for more quantitative functional measurements (e.g., measurement of the contractile properties of living muscles) or more detailed structural studies (e.g., histochemical profiles of fiber types); these are just more refined types of dissection. We direct the reader to appendix 1 for some commonly used techniques in structural anatomy and histology.

An appreciation of electronics begins with a working familiarity with the equipment for the indirect display of the invisible ebb and flow of electrons in conductors. In chapter 2 we draw on the analogy of water flowing in hoses to remind the reader of the basic simplicity that can be obscured by confusing knobs and buttons on the front panel of modern electronic devices. We strongly encourage the reader to sit down with the particular display equipment (usually an oscilloscope or chart recorder) and see how it responds to simple adjustments of signals from a battery, from a sine-wave generator, and from simple attenuator, filter, and rectifier circuits.

An understanding of these fundamentals will enable the experimentalist to develop ideas in the form of testable hypotheses, rather than simply presenting anecdotal observations. Once an experimental configuration can be defined in terms of all possible alternatives and their consequences, it becomes possible to apply rules of logic to experimental design and therefore to draw definitive conclusions. We must avoid confusing the collection of data with the pursuit of knowledge. As equipment becomes more sophisticated and readily available, it becomes easy to collect data without asking why. Whereas most of this book deals with such equipment and with data collection procedures, we would remind the reader that the responsibility for experimental design and strategy rests entirely with the experimentalist. The selection of the particular methods from the extensive catalog described herein depends on the consideration of both the principles underlying each method and the real agenda of the experimentalist.

D. PRACTICALITIES

1. Scale and Time Course

Many animal experiments involve surgical implantations; all involve maintenance of the experimental subjects. Consequently, we have included a brief discussion of animal maintenance and surgery (chapter 19) and information concerning applicable policies (appendix 3).

Animal experiments can be divided into acute, semichronic (or semiacute), and chronic, depending on their time course. "Acute" experiments are made entirely in one sitting, usually on an anesthetized, immobilized animal or surgically reduced preparation. "Chronic" experiments are made over extended periods using subjects chronically implanted or fitted with measurement apparatus and perhaps trained to perform reliably. In between is a gray area of experiments performed briefly, but usually after some procedure for attaching instrumentation.

The selection of one of these options obviously depends on the nature of the animal, the facilities available, and the complexity of measurements to be obtained. It must also involve careful consideration of the number of repetitions required to obtain a statistically valid description. The performance of motor tasks is notoriously variable, even for apparently regular, periodic behavior, such as chewing, walking, and breathing. Statistics are needed to test whether most of the variation is interindividual and whether records differ among seasons, days, or times of day. Quasi-periodic events with reproducible form but random occurrence may require even longer sequences of continuously recorded data as well as patient and painstaking selection and analysis. Truly random events pose the most serious problems, in part because they may easily be confused with purely artifactual events and because of the interaction between naturally occurring periodic and quasi-periodic events.

2. Systems Complexity

Many of the techniques described in part two (application) require or enable the investigator to construct highly complex preparations and data structures. Such complexity presents both the opportunity to obtain a comprehensive understanding of the motor behavior under study and the danger of overwhelming the study (and the subject) with problems of yield, reliability, and artifact (see chapters 13 and 14). It is always easy to suggest that cross-talk can be ruled out by recording simultaneously from all muscles (chapter 8) or that muscle force and length should be monitored directly via implanted transducers (chapter 16). Although these may be the most direct solutions to such problems, the experimentalist must impose a certain discipline in deciding whether it is justified to incorporate them and to carry them out.

3. Gadgetry

As most of this book is about gadgetry, it will be helpful to keep in mind a hierarchy of gadgets; this lets you know what to expect from any given instrument and where to look for solutions to a particular need (fig. 1.2). We have divided most of the electronic and electromechanical gadgets to be found in the laboratory into those that actually involve the animal and aid collection of the data, those that are used to preserve a reasonably faithful copy of the source signal, and those that help to reduce and analyze the data so that the pertinent findings emerge or can be presented in a photogenic and compelling way. A list of sources of supply for commonly used gadgets is provided in appendix 2.

Most of this book deals with the gadgets known as electrodes—their fundamental categories (chapter 6), their materials (chapter 7),

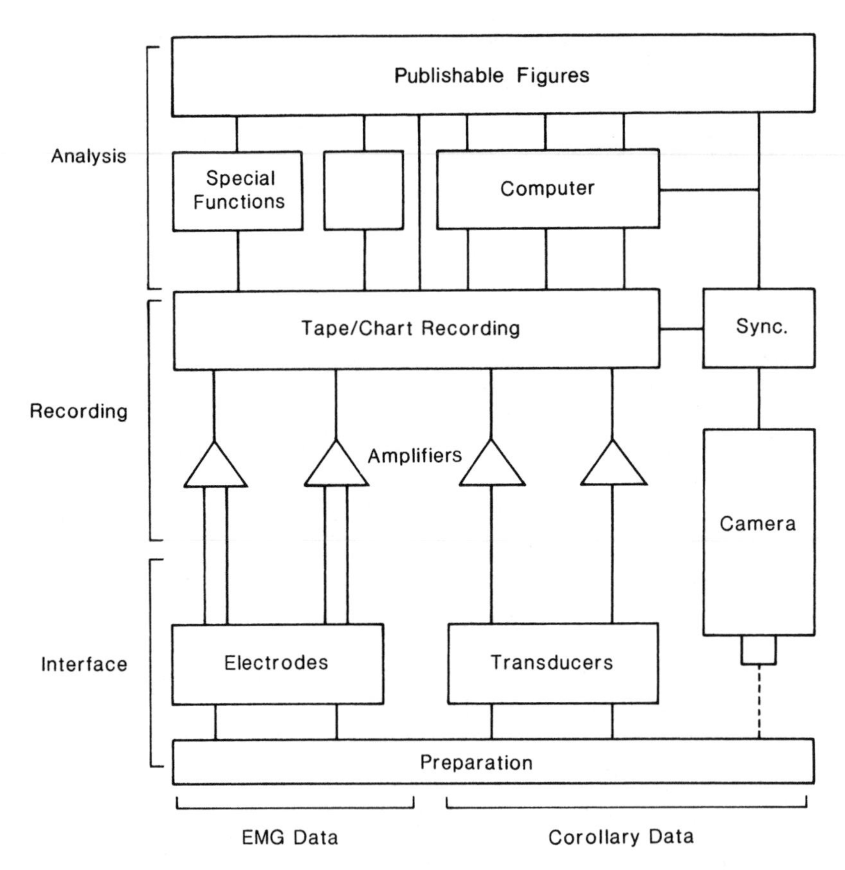

1.2. Typical functions of an EMG laboratory, from the collection of data to the generation of publishable data. Most equipment needs can be divided into three discrete levels; interfaces with the preparation itself, electronics for conditioning and recording raw data, and electronic and graphic techniques for analyzing, reducing, and summarizing pertinent results.

their specific designs (chapter 10), and their methods of construction (chapter 9). Other primary sources of signals include implanted and external transducers of physical quantities, such as force and position (chapter 16), and image-gathering devices, such as video and cinephotographical cameras (chapter 15).

A variety of electronic components is needed to condition and translate the events detected by these devices into records for storage and analysis. These include such diverse elements as cables and connectors (chapter 11) and amplifers, telemeters, and tape and chart recorders (chapter 12). All of these elements share the goal of preserving the original signals with as little distortion as possible.

Finally, there is the function of analysis—an open-ended set of techniques ranging from simple filtering and rectification to Fourier analysis of spectral content and inverse dynamic analysis of stick figures. In chapters 17 and 18 we survey the more common techniques.

This stage of analysis brings us full circle to the goals and hypotheses that presumably instructed the original experimental design. Graphic methods may aid the presentation (chapter 18). However, two other factors will determine whether the results will be considered significant: the quality of the experimental design and the reliability of its execution. In this book we hope to contribute ideas useful for both factors.

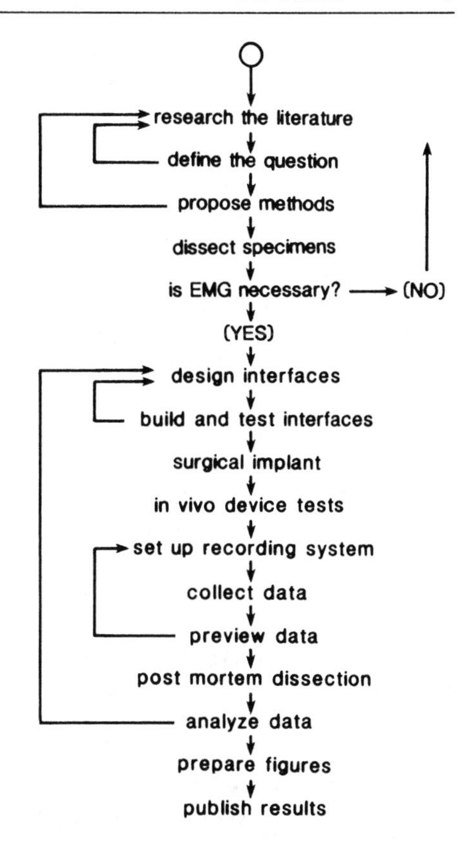

1.3. Flow chart of experimental design. Note that the selection of any particular method such as EMG should never be a foregone conclusion; rather, it must result from a careful analysis of the biological questions to be answered and review of the pertinent data from the literature. Even after EMG has been selected as at least part of the experimental approach, the particular techniques for electrode interfaces and electronic processing and recording should be designed incrementally and iteratively as the results of pilot experiments indicate the nature of the signals available and disclose technical or scientific problems.

E. TO START

The considerations outlined in this chapter should be addressed during the planning of each experiment; early decisions tend later to incorporate irreversible limitations (fig. 1.3). EMG provides a powerful set of techniques, but like all complex techniques, it demands discipline of the experimentalist. Thus, in chapter 21 we suggest efficient ways of conducting pilot experiments that may aid this planning.

One reason for caution is that full-scale studies involve the multichannel recording of multiple simultaneous signals, usually including several EMG sources plus kinematic, timing, and annotation information; the experiments are intrinsically complex. There are three important aspects of complexity that affect system reliability. The first is the probability that all pieces of equipment are individually functional at any given moment. This is the product of the probabilities that each item is functioning correctly. An individual 95% probability of correct function for a device may sound acceptable, but a laboratory system composed of 20 such items will only be completely functional about 36% of the time—that is, only every third experiment will work. As laboratory systems grow and as equipment is added, attention to systematic calibration, preventive maintenance, and routine service become increasingly important (see chapter 14). Finicky, homemade gadgets and systems loom ever more menacing as their number increases. Just keeping track of and repairing or discarding defective equipment and cables require disciplined attention to established procedures.

The second aspect of complexity is that the number of potential states of any system increases in proportion to the square of the number of devices. Each additional device (even an additional cable) increases the number of possibilities for interconnecting all devices and thus increases the risk of making electrically incompatible interconnections. The addition of one more connection may seem innocuous, but it may cause a critically loaded driver to break into oscillation. Unsophisticated users of electronic equipment often respond to the frustration of dealing with complex and poorly documented systems and procedures by trying things at random until either a useful signal or a puff of smoke emerges. Important goals of systems design are the reduction of the level of potential frustration and the limitation of such random behaviors to those that neither threaten life nor invalidate months of careful work (see chapter 11).

The third factor is memory. If you have been using the same, simple experimental procedures for a decade or if you can completely analyze your data at the end of each experiment, then your memory may cover for marginal note taking. However, if you plan to spend months or years developing new electrodes, new surgical procedures, and new experimental protocols and collecting miles of multichannel chart paper, tape recordings, and movie film, then you will need to develop your personality to new levels of compulsiveness. A tape recorder (perhaps voice actuated) lets you expand on items, such as filter settings, tape counter and film readings, recording speeds, and amplification, that are best written out before the experiment starts. It helps not only to note the times at which procedures were changed during a run but also to retain your incidental (and often ephemeral) insights into animal response.

The best insurance for a successful experiment may be to run through the whole experimental sequence once prior to including the experimental animal. Ultimately, the experimentalist needs to be certain that all of the apparatus is properly set up and documented and that readable data are being recorded on film and tape; then he or she can make decisions regarding the activities of the animal's muscles. Freedom from worry about techniques for data collection will allow the experimentalist to concentrate on the organisms being studied as well as on biological observations and their interpretation.

2 Coping with Ohm's Law

A. VIEWPOINT

There are two kinds of engineering: theoretical and practical. The theoretical engineer takes each problem and meticulously reduces it to its components, selects the appropriate mathematical relationships from an arsenal of physical laws, and then rigorously derives the optimal solution (or grows old and dies first). The practical engineer sizes up the general nature of the problem, recalls some appropriate analogies or generalizations from experience, and starts to tinker with the available parts until something useful emerges (or the parts start to melt). The biologist designing a new experiment needs to apply just enough of both kinds of engineering to get on with the science.

If you cannot remember that $E = IR$, there are probably some things we should get straight first (fig. 2.1). Electricity flows like water through a hose. The pressure (voltage, E) can push a small flow (current, I) through a long, skinny hose (high resistance, R, or impedance, Z) or a large flow through a short, fat one (low resistance or impedance). Hoses (resistors) in parallel obviously let more total water through than if they are hooked end to end (in series). If you want to deduce the pressure at the source of the hose from a measurement of the pressure at its outlet, you need to cap the outlet as tightly as possible (high input impedance on the measuring device); otherwise, the pressure is dissipated as the water flows through the hose. The dissipation is especially troublesome if the hose leads from a valve that is almost closed (high source impedance). If the hose has a leak (shunt) someplace along its length, you cannot measure all the pressure (source signal voltage) that is really present. If the hose crosses an elephant trail, the outlet pressure will show a lot of extraneous fluctuations (noise) every time an elephant steps on the hose.

The above-mentioned relationships cover almost everything in electrophysiology, from the generation of the signal at the source to the design of electrodes and amplifiers for its recording. The situation for AC signals, such as EMGs, is slightly more complicated by the fact that the signal frequency enters into the calculations of the degree of impediment to flow in electronic circuits (hence the term "impedance"). But the simple fact remains that electricity is a flow through a medium caused by something analogous to a pump. Some of this flow must be diverted through a measuring device and then returned to the circuit if we want to know what the pump is doing without unduly disturbing it. The complexity arises from the fact that, in biological circuits, there may be many pumps (generators) operating simultaneously, sharing the same pool of water (charged particles). The pumps are often spatially separated and may propel the fluid in directions that are not always obvious or in agreement. In fact, a rigorous treatment of a typical situation is likely to be completely intractable unless one makes simplifying assumptions.

Biologists looking to use EMG as a fairly gross indicator of what is going on in a whole muscle are entitled to make rather drastic simplifications. They are not interested in the meaning of every little squiggle of the trace but in its overall trends. In fact, they are usu-

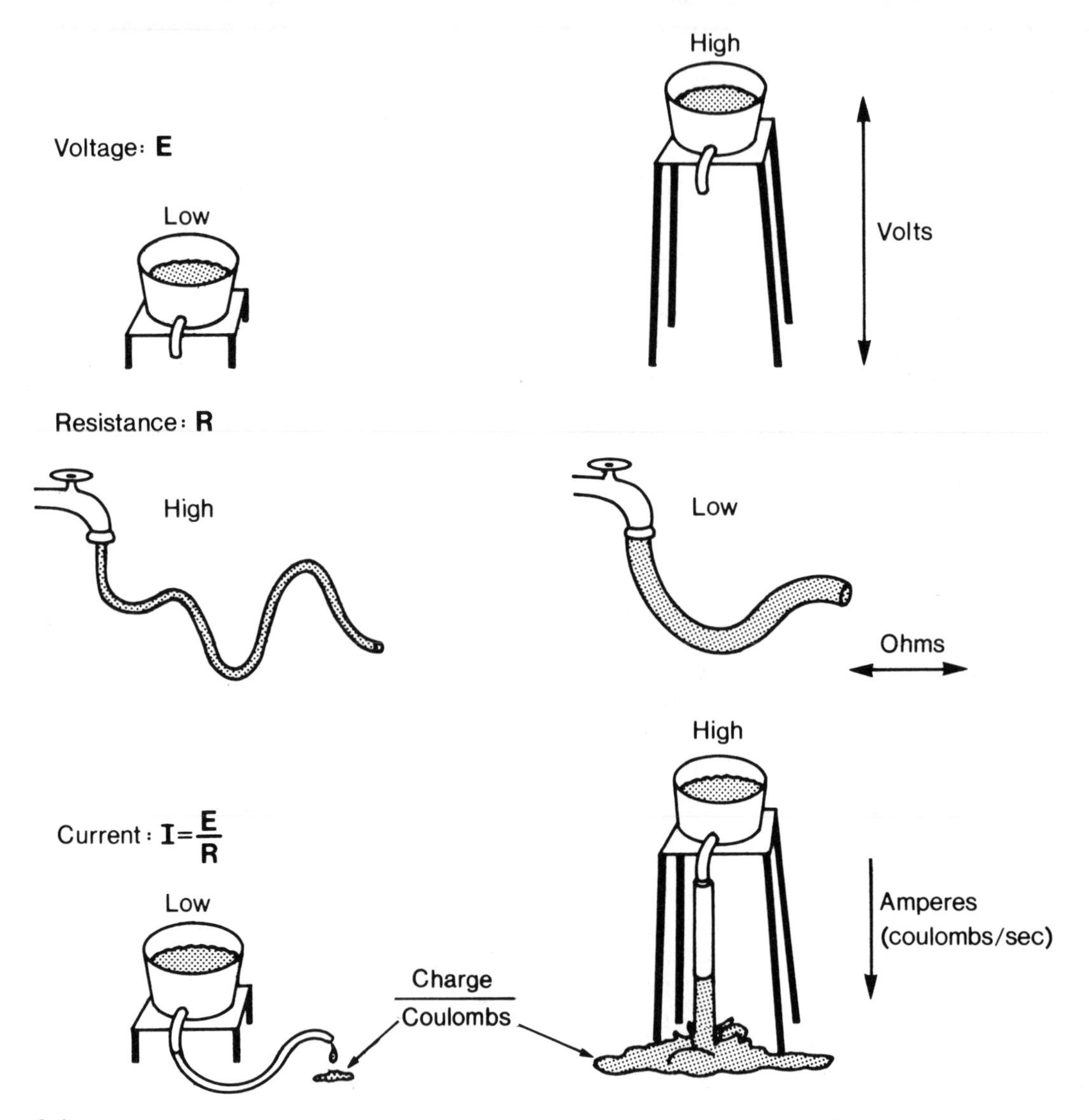

2.1. Ohm's law revisited. One can appreciate most of the important relationships in electrical circuits intuitively by thinking of the analogy of water flowing through a hose. Voltage is analogous to the pressure head available for pushing the water through the hose, and resistance to flow comes from the dimension of the hose. Obviously, the flow of water (or electrical current) will be greatest whenever the pressure is highest and the resistance to flow is lowest.

ally more concerned (or should be) with the physical location of the source of the recorded waveforms than with the waveforms themselves. These problems are quite addressable with the simple multiplicative relationship $E = IR$ of Ohm's law, but only after the terms are properly defined. It is far more important to understand the following terms subjectively than to remember the equations that relate them; problems usually arise when terms are used inappropriately rather than when equations are miscalculated.

B. BASIC DEFINITIONS

1. Voltage

Abbreviation: E (for electromotive force)
Synonym: Potential difference
Unit: Volt (V)
Usual range:
Millivolt (mV) = 10^{-3} V
Microvolt (μV) = 10^{-6} V

Sometimes voltage is measured for its own sake, as in the resting potential of a cell, which describes the battery-like behavior of the cell membrane. More often, in electrophysiology the voltage differential is measured to indicate the amount of current flowing through some piece of tissue as a result of some distant biological generator of electrical current. In this case, it is not so much a cause but a result of the product of the current flow times the resistance to flow of the tissue. Of course, the same voltage can be viewed as a cause, for example, of the current flow through an amplifier. It is important to remember that a voltage only exists with respect to a point of reference, hence the term "potential difference." The voltage is the amount of force it takes to move that current between two such points, through the resistance between them. In electrical circuits, one often picks a place as the reference (ground) and all voltages are measured in terms of the voltage it would take to move the current through this place. In biological circuits, the selection of the reference point is not self-evident and must be specifically mentioned so that a reported voltage will be meaningful.

2. Current

Abbreviation: I
Unit: Ampere (A)
Usual range:
Milliampere (mA) = 10^{-3} amp
Microampere (μA) = 10^{-6} amp

Current is the measure of flow through a circuit. A flow of water could be described in molecular terms as the passage of a certain number of molecules per second (fig. 2.2). The flow of electrical current may be described similarly as the passage of a certain number of charged particles per second. The unit of flow chosen is the ampere. An ampere is 1 coulomb (coul) of charge per second; the coulomb is defined electrochemically as the number of electrons necessary to plate out 1.1180 mg of silver (monovalent cation from a solution of silver nitrate). There are 96,500 coulombs in 1 mole (gram weight) of silver; thus, there are $6.02 \times 10^{23}/96500$ or 6.24×10^{18} electrons per coulomb. This odd definition reflects an attempt to reconcile the electrochemical roots of electricity with the physics of electrostatic force (1 coul of charge will be stored in a 1 farad (F) capacitor at 1 V).

In aqueous environments, the charged particles are positively or negatively charged ions. Ions with multiple charges are counted multiple times. In electrical circuits, the charge carriers are electrons; these are negatively charged. For simplicity of calculation, electronic circuits are often treated as if they were

2.2. Electrochemical terminology. Electrical current is defined as the flow of positively charged particles (or the equivalent opposite flow of negatively charged particles such as the free electrons in a wire). Positive charge accumulated at an anode exerts a positive voltage with respect to the negative charge (or absence of positive charge) at a cathode. In an aqueous solution containing anions (negatively charged ions such as chloride) and cations (positively charged ions such as sodium), the anions will be attracted to the anode and repelled at the cathode, whereas the cations will be attracted to the cathode and repelled at the anode. The magnitude of the difference in electrical potential between the two electrodes (anode and cathode) and their physical separation in space give rise to a dipole source of electrical current flow.

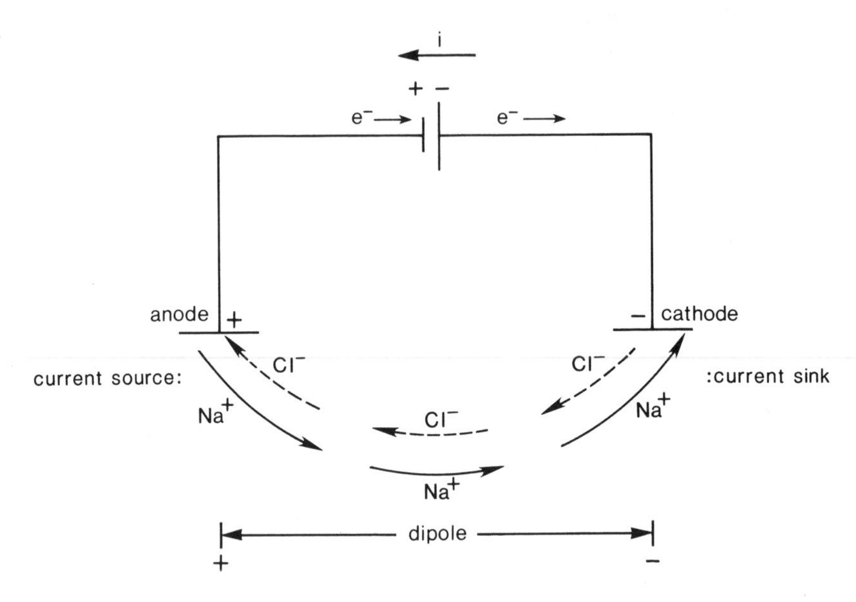

currents of positively charged particles. We can thank Ben Franklin for the resultant confusion, but then it was just an unlucky guess, because there is really no simple way to distinguish between a flow of electrons from left to right or a flow of positrons from right to left. This is actually fortunate, because most of the pumped charge carriers in active biological membranes are, in fact, the positively charged cations such as sodium, potassium, and calcium. By convention, the current flow is considered positive between a point of higher positive voltage and a point of lower voltage, and it is usually shown symbolically in a circuit diagram as an arrow pointing from higher to lower voltage.

3. Resistance

Abbreviation: R
  Unit: Ohm (Ω [omega])
  Usual range:
    Kilohm (kΩ) = 10^3 Ω (also written kilohm, kohm, or just k in circuits)
    Megohm (MΩ) = 10^6 Ω (also written megohms, megs, or just M in circuits)

Resistance refers to the difficulty of passing a given amount of current through an element of a circuit or a whole circuit. All materials from the best conductors, such as copper and gold, to the best insulators, such as glass and air, can be described by their resistance, which always is greater than zero and less than infinity. Of course, one cannot define the resistance of anything until one knows its shape and dimensions, because this defines the path that the current must take. A large cross-sectional area means many paths may be available; great distance means the path will be long. There are then two independent factors: the shape of the conductor and the inherent properties of its material. The latter is called resistivity. For a homogeneous path, the resistance is the product of this abstract resistivity (Ro) times the length (l) and divided by the area (A):

$$R = Ro \times l/A.$$

If we are using cgs metric units, the resistivity must be expressed in units of ohms times centimeters for the answer to be expressed in ohms. The resistivity of body fluids is 65 Ω cm in mammals and 80 Ω cm in amphibians. It may be up to one order of magnitude higher in dense body tissues, such as bone and fascia. However, the tissue resistivity falls pretty much in a midrange. It is high enough to keep the active cells that float around in it from shorting each other out, low enough to be used as a current pathway where confined in tubes, and constant enough to simplify our task.

4. Conductance

Abbreviation: G
Unit: mho (ohm spelled backward)
Siemen (new term for mho)

Conductance is simply the reciprocal of resistance. It is sometimes convenient to think that a channel may open, thus permitting current to flow through it. Therefore, rather than saying that its resistance is decreasing, we say its conductance is increasing, but they are physically the same process.

5. Capacitance

Abbreviation: C
Unit: Farad (F)
Usual range: Microfarad (μF) = 10^{-6} F
Picofarad (pF) = 10^{-12} F

Positive charges attract negative charges. When positive ions meet negative ions in a solution, they may combine into an uncharged entity and precipitate as a salt. If they are held apart by some nonconductive barrier, they will tend to line up on each side of the barrier to get as close to the oppositely charged ions as possible. However, like charges repel each other. As all the charges of one sign start elbowing each other out of the way to get near the barrier, they create a repelling force, which, at equilibrium, is equal and opposite to the attractional force. The number of charged particles that one can gather on each side of the barrier depends on the surface area, the amount of force (voltage) available to squeeze them in, and the thickness of the barrier (because small gaps allow the force to be concentrated across a small distance). These concepts are combined into a single entity called capacitance, which describes the amount of charge that can be stored on either side of such a gap for a given amount of applied force.

The unit of capacitance is the farad (after Michael Faraday), which is one coul of charge per applied volt. This is an enormous quantity, equal to all the electrons that would flow in a current of 1 A for 1 second. Thus, the usual units are microfarads (μF, large capacitors for power supplies and filters) and picofarads (pF, the capacitance of biological membranes and electrode interfaces). Capacitance is important in electrophysiology because the electrical potentials are the result of local concentrations of ions in the body. They are usually detected by the capacitive changes induced in the flow of electrons in metal conductors, such as electrodes. The capacitive gap measures only a few angstroms, defined by the mean distance of approach of ions colliding thermally with an exposed metal surface. The experimentalist has little control over this gap except to make certain that the metal surface is truly clean (removing oils, cellular debris, or metal oxides). However, the surface area is easily manipulated, either by enlargement of the overall geometric "apparent surface area" or roughening of the electrode to increase the "real surface area."

6. Impedance

Abbreviation: Z
Unit: *see* Resistance

Impedance is the term for resistance in circuits in which the current is alternating rather than constantly applied or direct. Operationally, impedance is handled just like resistance and, for many circuit elements, it is identical to resistance. In biological circuits, the main difference between resistance and impedance arises through the effects of capacitance. Imagine a circuit consisting of a battery, a switch, and two metal plates facing each other but separated by a perfect insulator (infinite resistance). Obviously, no continuous loop of current can flow even after the switch is closed because the circuit is incomplete. However, for an instant after the switch is closed, electrons will move. The action of the battery will cause them to line up on the negatively charged plate of the pair and be sucked away from the positive plate. During the instant in which this happens, there is a movement of electrical charge, or a current; this means the impedance at that instant must not have been infinite. If we now turn the battery around, leaving the switch closed, the current will surge in the other direction, again for an instant.

Each time the supply voltage changes sign or even amplitude, the capacitor acts like a conductor of electricity, although no particle actually crosses the gap. The smaller the plates, the less charge they can store and the higher their impedance. However, the impedance is determined also by the rate at which the voltage changes (the number of battery reversals per second), as it is effectively infinite for a direct current or zero rate of change. Biological tissue and the surfaces of metal electrodes act like combinations of both pure resistors (no frequency dependency) and capacitors. Impedance is a blanket term for the effective obstruction to current flow and must be specified for a specific frequency of oscillation. By convention, we specify it for a sinusoidal change of voltage.

7. Power

Abbreviation: P
Unit: Watts (W)
Usual range: Milliwatts (mW) = 10^{-3} W
Microwatts (μW) = 10^{-6} W

Although physiologists have little cause actually to describe their measurements in terms of power, we will frequently invoke this concept in considering the processes of amplification and the susceptibility of transmitted signals to noise pickup. Power is the product of current and voltage, although Ohm's law provides the more useful relationship of voltage squared divided by resistance.

8. Frequency

Abbreviation: f
Unit: Hertz (Hz) (cycles per second)
Usual range: Kilohertz (kHz) = 10^3 Hz

Frequency in a strict electronic sense refers to the number of cycles per second of a pure sinusoidal oscillation and can be used directly in calculations such as that for impedance. However, biological waveforms usually are not purely sinusoidal but show an irregular shape. One of the basic tenets of electronics is that any waveform can be considered to be made up of the linear addition of purely sinusoidal waveforms of various frequencies. This is the basis for the commonly used procedure of Fourier transformation, in which a signal is described by the weighting factors needed to synthesize it from a series of sinusoids that are harmonically related to each other. For signals such as EMG, it is often important to know the range of frequencies in which most of the energy of the signal lies—that is, those with the largest weighting factors. This provides information about the design and selection of electrodes, amplifiers, and filters; obviously, the impedance must be adjusted to avoid distortion of the real signal and also to reject unwanted signals (noise) with differing frequencies. The term "frequency" is often used in a looser sense to include the time course of intermittent processes, such as nerve impulses; however, it is better to discuss the "rate" of such phenomena.

9. Rate

Unit: Pulses per second (pps)

Discontinuous events, such as action potentials, are usually quantified by counting. Even when they are asynchronous or irregular, they are usually described in terms of their mean rate, in pulses per second. Occasionally, an instantaneous rate is defined as the inverse of the period between successive pulses, again in pulses per second. A gradually changing rate of firing can be described by plotting as a stepped graph called a frequency-gram (fig. 20.4).

C. CONVENTIONAL DEFINITIONS

The following additional terms are often seen in the literature. They do not describe fundamental properties; rather, they are terms of tradition or convenience for devices and processes.

1. Anode

An anode is an attractor of anions, or negatively charged particles, in an electronic circuit. It thus carries a voltage that is positive with reference to the source of such particles. The term "anode" is frequently used to identify the positive electrode of a stimulating pair or other voltage or current source.

2. Cathode

A cathode is an attractor of cations, or positively charged particles, in an electronic circuit. A cathode thus carries a voltage that is negative with reference to the source of such particles. The term "cathode" is frequently used to identify the negative electrode of a stimulating pair or other voltage or current source.

3. Dipole

A dipole is a pair, consisting of an anode and a cathode, separated by some specified physical distance (their moment) across which current flows. This is an important concept in electrophysiology, because any biological signal source can be thought of as such an idealized signal source. Of course, the currents produced by biological generators do not emerge from pointlike places. More typically, biological generators consist of constantly shifting regions of membranes across which electrical charge flows inward or outward or shows zero net flow. We shall see that the calculation of an equivalent dipole moving in time and space provides powerful insight into the proper design and orientation of EMG electrodes.

4. Current Source

In an electronic circuit the current is always defined as the movement of positive charge. The source of current is thus the positive electrode (anode). Frequently this term is used to refer to the positive end of an idealized dipole.

5. Current Sink

Current sink is the opposite of source, in which the positive current returns to the signal generator, and thus the negative electrode (cathode) or negative end of a dipole.

6. Volume Conductor

An electronic circuit made up of components and wire differs from an electrophysiological circuit in the body in that the nature of the connections in the body is generally unclear. All the signal sources, as well as the recording electrodes, are floating in the same pool of weakly conductive saline. The situation is somewhat analogous to the transmission and reception of radio waves through air. The term "volume conductor" connotes such a situation. Under these circumstances, the flow of electrical current is better described as a gradient of continuous distribution rather than a particular pathway.

7. Gradient

The flow of current through a volume conductor does not take specific, discrete paths. Therefore, any quantitative measure must include the notion of a smoothly varying gradient of intensity from one place to another. This is particularly useful in the consideration of bioelectrical measurements, because what is being measured often is related to the steepness of the gradient rather than to absolute values of voltage or current with respect to an indifferent reference point. This, in fact, is the basis for differential amplification, which measures *differences*.

One can speak of gradients of either current or voltage. In the case of current, one is usually speaking of current density, or the rate of flow of charged units per second through a given unit cross-section of tissue. A point source of current in a uniformly conductive medium will thus produce a set of isocurrent density shells. These shells will be arranged as concentric spheres around the source and their spacing will define the intensity of the current density gradient. In the case of potentials, the gradient is in units of volts per unit of distance. In the case of the point source in a uniform volume conductor, the concentric spherical shells would be isopotential surfaces as well, and their spacing would reflect the steepness of the gradient.

8. Differential Recording

Signals are detected by comparison of (differential amplification of the difference between the potential at) two points that are separated by this gradient. Sources with gradients oriented perpendicular to the line connecting a pair of recording points thus produce no detectable signal, because both electrodes see the same signal and the difference is zero. Pieces of nonconductor can be inserted into the volume conductor to change the gradients by diverting the current flow. The basis for the design of recording electrodes depends on making reasonable estimates of the location of the gradients. Electrodes should be arranged to span maximally the gradient desired and avoid spanning the gradients of extraneous signal sources to be rejected.

9. Amplifier

We usually think of an amplifier as a device that makes small signals into big signals. But how do we define small? Obviously, a low voltage is often turned into a higher voltage by an amplifier. However, amplifiers can also turn small currents into larger currents or turn currents into voltages. Bioelectrical signals are usually small in two respects: they have very low voltage (millionths to thousandths of a volt) and they have very little capability to drive current. The latter problem arises from our use of relatively high impedance electrodes, which act like the partially closed water valve described earlier. As it passes from the body fluids to the measuring circuit, electrical current must change from the flow of ions to the flow of electrons. It does so reluctantly, so the amplifier must be designed to make do with as little current flow as possible.

The source impedance problem is a sneaky one because it is not intuitively as clear as the problem of the low-voltage signal. Often we inadvertently make it worse by building small electrodes to improve selectivity or extending their operation to low-frequency signals for which impedance can rise drastically. Low-current signals coming from high source impedances are particularly sensitive to noise, both noise picked up from external, extraneous signals and internally generated noise resulting from the random motion of electrons (described below). Conceptually, these sources of noise are similar to the instability of water flow through a valve that is almost closed; slight temperature changes and little taps on the handle can drastically change the flow and be perceived as random fluctuations or noise.

An amplifier contributes both voltage gain, which can usually be controlled by the user, and current gain, which is determined by the fixed ratio of the input impedance to the output impedance of the amplifier. The actual power gain that could be obtained from an amplifier (but only if the rest of the system is designed very specifically) is the product of these two forms of gain. A rule of thumb is that the input impedance of an amplifier should be at least 10 times higher than the source impedance (electrodes); its output impedance should be at least 10 times lower than the impedance of all the loads connected to its outputs (which may be lower than imagined as they are in parallel and may include long lengths of coaxial cable having high capacitance). Although this sacrifices some current gain, it ensures a reasonably stable calibration of the voltage gain, which is usually the parameter being measured.

10. Bandwidth

As we have described, a biological signal is usually an irregular waveform that can be mathematically decomposed into a set of pure sinusoids with different weighting coefficients. For any particular signal source, the nonzero coefficients will be a clustered group of frequencies. For EMG signals, the cluster may span the frequencies from 50 to 500 Hz for certain muscles and recording techniques, or 300 to 3000 Hz for other muscles and techniques. It is important to have some idea of the bandwidth of the source signal, because amplifiers generally have some restricted range over which their gain is constant. This range is the bandwidth of the amplifier.

If the bandwidth of an amplifier does not include all of the frequencies in which the source signal is produced, the signal will be distorted; if its bandwidth is much greater than this range, it will introduce unnecessary noise. Most amplifiers have adjustable filters (see below) that allow their bandwidth to be changed. The usual way to estimate the bandwidth of biological signals is to open up these filters completely (in all stages) and then gradually start closing in from above and below until there is some noticeable change in the amplitude or shape of the signals. There are many places in a signal recording system where bandwidth is important and where signals may inadvertently be seriously degraded. Chart recorders constitute one place where the bandwidth is critical. It is important that you have some seat-of-the-pants idea of the bandwidth of your particular signal before you select equipment for coping with your problems. Simply setting your bandwidth as wide as possible or in agreement with someone else's published values can lead to problems.

11. Filters

What do the settings on filters really mean? There are two basic filters—those that remove high frequencies, called high-cutoff filters or low-pass filters, and those that remove low frequencies, called low-cutoff filters or high-pass filters (fig. 2.3). For either type, the frequency usually specified is one at which only 70.7% of the input signal amplitude gets through, the so-called 3 decibel (dB) power point. For a low-pass filter, frequency components higher than this specified frequency are increasingly more attenuated for frequencies that are further above the cutoff point. The rate at which the attenuation increases for

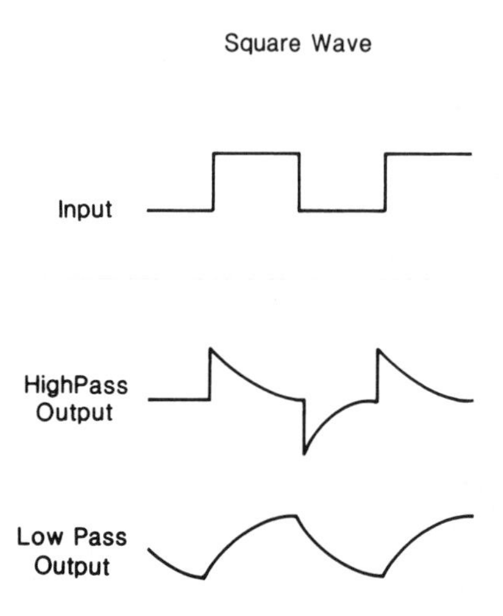

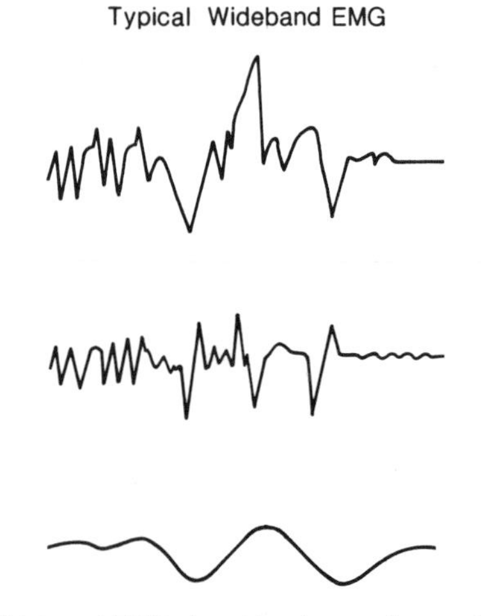

2.3. Effects of filters. A square-wave test signal will be distorted into different shapes by a high-pass filter (which removes all of its low-frequency components, leaving just the rising and falling edges) and by a low-pass filter (which removes the high-frequency components, leaving slowly rising and falling oscillations). A wideband EMG signal is also made up of a wide spectrum of frequencies, although not in so orderly a pattern as a square-wave. The filter circuits cause profound changes in the shape and overall amplitude of different parts of the input signal, depending on the instantaneous mix of frequency components.

progressively higher frequencies is called the rolloff (see also the discussion of bandpass and filters in chapter 12). Most simple filters, such as those found on biological amplifiers and oscilloscopes, have a rolloff of 6 dB per octave. What this obscure-sounding set of units means is that for each doubling of frequency, the attenuation doubles. For a high-pass filter, which attenuates lower frequencies, you simply reverse the terms to attenuate frequencies progressively lower than the specified cutoff frequency. The 6 dB per octave filtering action is modest but useful and inexpensive; you can make such a filter with only two components: a resistor and a capacitor. In fact, you may inadvertently make such a low-pass filter by poorly designing highly resistive electrodes and excessively capacitive leadout cables, another reason for knowing what your signal should look like. We will deal later with special filtering techniques for getting rid of particular frequencies and for purposely changing the appearance of the source signal, the better to reveal features such as the shape of the envelope.

12. Noise

The term "noise" generally refers to any signal that is not wanted; it is the bane of the electrophysiologist's existence. It is said that Eskimos have 50 words for snow, which helps them to describe and understand the implications of something they must learn to live with. Electrophysiologists similarly recognize multiple kinds of noise. If you never learn to differentiate amongst the various sources and kinds of noise, you had better hope you never encounter them! As we have seen many different kinds of noise, we have devoted a whole chapter to its nuances. Briefly, though, be aware that there are three general kinds of noise: wide band, narrow band, and biological noise.

Wide band noise, also called thermal noise, Johnson noise, or white noise, is the fundamental noise made by "hot," vibrating electrons crashing into each other. Its amplitude can be calculated as $E_n = \sqrt{(4kTBZ)}$, where k is the Boltzman constant, T is temperature in degrees Kelvin, B is bandwidth in hertz, and Z is source impedance in ohms. The obvious lessons are that electrode impedances should be kept low, amplifier filters set tightly to the proper bandwidth, and the preparation kept frozen in liquid helium (two out of three is acceptable).

Narrow-band noise is caused by AC line hum, fluorescent lights, motor brushes, car ignition systems, radio and television stations, centrifuges, frost-free freezers, and switching of power supplies. This is the most frustrating noise because it may be radiated through the air or conducted through the power lines, from right under the table or from the next building. Proper electrode and amplifier design will minimize the tendency to pick up this noise, but getting rid of it often means finding the source, which means learning to recognize characteristic signatures.

Biological noise includes internally generated signals such as electrocardiographical signals, respiratory artifact, and cross-talk. All three noise signals might be music to someone else's ear, and they have just as much right to be there as the signals you want to record. Much of this book is devoted to the design of electrodes that selectively detect the voltage gradient generated by the signal dipole under study instead of these "noise" sources.

There is one general question to ask whenever the output "looks noisy"—namely, how noisy? These days, we have multiple gain stages and oscilloscope sensitivities that change with the flick of a wrist over six orders of magnitude. This makes it easy for the neophyte simply to turn up the gain until the signal fills the screen, with no idea what this corresponds to in actual millivolts presumably present on the electrode. Several hours may go by before the "poor quality" of the signal is correctly attributed to the fact that the electrodes have fallen out of the preparation or that the first stage battery is dead.

Noise can be your friend if you know what to expect. If the noise level is too high, perhaps the source impedance is too high—that is, electrodes are sitting in air instead of tissue. If the noise level is too low, perhaps the source impedance is too low—that is, the electrodes have shorted out or the first stage amplifier is grounded or turned off. The magical signal-to-noise ratio we seek to optimize is meaningful only when one factor is held to some known value as the other is improved.

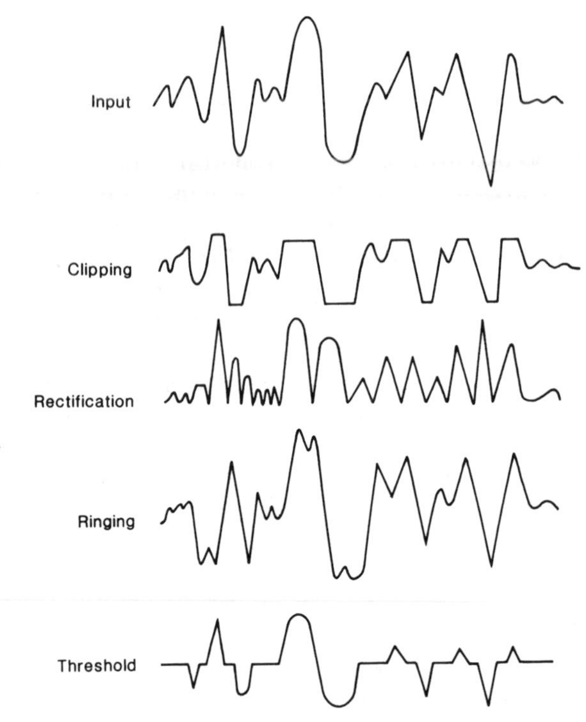

2.4. Nonlinear distortion effects. Electronic circuits intentionally or inadvertently can distort an imput signal by clipping (inability to follow inputs over a certain amplitude in one or both polarities), rectification (elimination of one polarity of input or its reversal to the opposite polarity, as shown here), ringing (inability to settle at a stable output after a rapidly changing input), and threshold (insensitivity to input transitions under a certain threshold level). Certain electronic components regularly give rise to a combination of effects. Thus, if a diode is placed in series with a signal, it passes that portion of the signal greater than its threshold voltage; if it is placed in parallel to ground, the diode can be used to clip signals that exceed its threshold voltage.

13. Distortion

As it passes through various pieces of electronic equipment, the waveform of a signal can be changed in many complex ways (fig. 2.4). Distortion is a blanket term for such changes, which range from simple attenuation of certain frequency components as a result of limited frequency response (usually a linear process) to various nonlinear changes, such as rectification (elimination of either the positive or the negative portion of the signal), clipping (cutting the tops of some spikes), and ringing (causing the signal to oscillate).

14. Cross-talk

Electrical signals are usually imagined as confined to the components and conductors that are drawn on schematic circuit diagrams. Unfortunately, there are unspecified ways in which electrical signals can pass from one component to another supposedly independent one. This is called cross-talk. There are two levels of such cross-talk. The first level occurs in the animal. Electrical currents distribute throughout a volume-conductive substance, such as the tissues of the body. Their confinement to one place, such as the muscle tissue in which they are generated, is only a relative matter. Thus, electrical signals generated by one source tend to show up in another. The second level of cross-talk occurs in the recording system as a result of stray capacitances and inductances arising in the physical proximity of components. It can be avoided or markedly reduced by good circuit design and equipment configuration.

D. HOARY MISCONCEPTIONS

There is something strange about the interaction of electrical engineering and biology. Even people who get the terms right and know all the equations often persist in believing things that are simply untrue. Vitalism lives on, either because it seems sacrilegious to be so objective about nature or because fully rigorous models of complex tissues are rarely available. This leads one to rely on a common sense that may be less common than one might first guess. We here present and attempt to dispense with three of the most common, pernicious misconceptions that get in the way of objective experimental design.

Relative Conductivity of Tissues. It is obvious that subcutaneous fat pads and solid bone have lower conductivity to electrical currents than does a beaker of saline solution. However, even a cursory histological examination of any biological tissue reveals the myriad of cellular, extracellular, and vascular fluid spaces that are required to sustain life. Even if tightly joined cells form seemingly impermeable sheets of lipid membranes, these membranes will be found to be studded with ionically conductive channels. Furthermore, alternating electrical currents pass quite readily across these membranes by capacitive conductance, so that ionic movement on one side of the barrier produces ionic movement on the other as if there were no barrier at all.

Thus, it is quite unreasonable to expect that any of these structures (and especially thin ones such as sheets of fascia) will have the dielectrical properties of Teflon. In fact, even the most resistant biological structures have bulk resistivities that are rarely more than one order of magnitude larger than free saline fluid. Now a factor of 10 may seem significant, but in the volume-conductive world of the body, such factors tend to disappear. This is because these structures represent such a small (and often spatially distributed) percentage of the total volume of any complex structure such as a limb. So a little more of the current flows around them rather than through them? So what? For almost all of the macroscopic recording configurations considered in this book, the heterogeneity of the tissue and the volumetrically distributed flow of current reduce any local feature to an insignificant secondary or tertiary effect. So you need not worry that muscle fascia or scar tissue may isolate your electrode from the biopotentials to be measured, and you should not think that these same layers will keep signals from adjacent structures from being recorded by your electrodes. The first-order considerations of electrode spacing and orientation (chapters 5 and 6) are where the action is.

High-impedance (or small) electrodes are more selective than low-impedance (large) ones. There is a type of recording electrode for which the microenvironment of current flow, very near the surface of single, active muscle fibers or neurons, provides the critical signal. Such so-called monopolar microelectrode recording does require electrode surfaces that are small enough to get the entire exposed contact very close (see chapter 20), and these contacts tend to have rather high tip impedances (this is not in itself desirable but derives from metal-electrolyte interface properties).

However, a great deal of EMG applications

(those covered in all but chapter 20 of this book) represent recordings of diffuse field potentials in bulk tissue. Only closely spaced bipolar electrodes with differential amplification can make such recordings spatially selective. For these, the bipolar spacing of the two contacts is the critical factor, not their individual size or surface area. Now, obviously a pair of closely spaced contacts cannot each be much larger than the spacing between them, but there is no reason to make them artificially smaller either. Small contacts have higher impedance than larger ones (causing more thermal noise, chapters 7 and 13) and may be inordinately influenced by microenvironment processes of no interest here, such as motion artifact and damaged fibers. The religious zeal for high impedance comes from the fact that electrical insulators (on electrodes, leads, and connectors) usually fail by having too little impedance rather than too much. That is, the only way for the expected electrode impedance to go is down, usually signifying a leak in the insulation somewhere (see chapter 14). Obviously one wants the electrode impedances to stay at the high end of the usual range, but there is nothing intrinsically good about high impedance.

Ground Electrodes can prevent artifacts and cross-talk. Suppose you have gotten bad currents from stimulation pulses and adjacent muscles and other biological sources flowing about in the muscle in which you have put your recording electrodes. Why not prevent them by grounding them out with "guard" or "flanker" electrodes on either side of the recording electrode? Well, if you wanted to ground part of an electrical circuit, you would do so with very low-resistance pathways connected directly to ground. The criterion for "low resistance" is one that imposes much lower resistances than those through which the currents normally would have to pass in the remaining part of the circuit.

The problem in biological circuits is that the interfacial resistance between any metal contact and the body fluids tends to be rather high. For almost all of the dimensions and materials likely to be encountered, this interfacial impedance into the ground electrode tends to be greater than the access impedance of the volume-conductive tissues between the spurious current source and the recording electrode. Consequently, ground electrodes are unlikely to have any significant effect on the disturbing current flows, which reach the recording electrode virtually unattenuated.

3 The Organization of Muscle

A. GENERAL PROBLEM

Muscles are tissues that tend to shorten upon chemical and electrical stimulation. Any muscle can produce displacement and force; however, the displacements and the forces actually produced depend on the nature of the imposed resistance. This, in turn, depends on the ways in which the parts of the muscle are formed, placed, and attached. EMGs cannot be interpreted in terms of muscle work until the nature and mechanical arrangement of the muscles have been examined. Their nature involves factors such as the kinds of actin and myosin organized into the sarcomeres (which will form the bridges and result in shortening), the sarcoplasmic reticulum (which will establish the temporal course of force generation), and the mitochondria and myoglobin pigments (which supply energy and hence affect fatigability). The mechanical arrangement involves factors such as the number of contractile units between origin and insertion, the angle of the fibers to the axis of contraction, and the stiffness or elasticity of the connecting tissues.

The electromyographer is not likely to study all of these factors. For that matter, EMG is generally restricted to the study of phasic (twitch-type) rather than slow tonic (graded) muscles; we refer to conditions in slow muscles only in passing. However, all of the above-mentioned factors affect the potential behavior of muscle and we must recognize them as variables of the system that might otherwise be unsuspected. Even more important is the fact that both the nature and mechanical arrangements of muscle fibers and their organization into motor units are prime factors in determining the amplitude, frequency, and spatial distribution of the EMG signals that can be recorded from any particular muscle (chapters 4 and 5). We conclude this chapter with a brief account of methods available for analyzing each such level of organization.

Although there are a series of elegant pilot studies, the organization of muscles in whole animals represents a field at the very beginning of useful analysis. Many of the plausible generalizations that have crept into textbooks are based on examination of a few muscles in a limited number of species performing very few actions. Many studies published during the past decade have generated "surprising" results. Such unexpected results are the hallmark of a young field. They also suggest that caution is desirable and that generalizations should be checked before they are accepted.

B. LEVELS OF ORGANIZATION

1. Principles

The shortening and force-generating properties of muscle fibers reside in the sarcomeres—regular arrays of parallel, overlapping, thin and thick filaments that can slide by each other (fig. 3.1). The organization of the contractile filaments into sarcomeres, repetition into strings called myofibrils, and coordination among the parallel myofibrils of a muscle fiber produce the characteristic banded or striated pattern of (striated) skeletal muscle. The transverse spacing of the sliding filaments is maintained in a regular lattice by transverse planes of connective proteins and by the dynamics of cross-bridge formation. The cross-bridges provide the structure wherein energy is transformed from the chemical to the mechanical state. The various filaments are described from their appearance in long section; their transverse connections are called lines. The thick filaments are interconnected by fine structures near their middle (M line); the thin filaments are interconnected by filamentous platelike structures called Z lines (after Zwischen-Scheiben). They emerge from both sides of these plates, like the parallel spikes from a floral trivet. As the Z lines are most obvious, the sarcomeres have traditionally been defined as running from one Z line to the next and to consist of two (half-) sets of interconnected thin filaments that slide among a single set of interconnected thick filaments.

The sliding of the filaments causes the sarcomeres to change length, by changing the overlap of the thick and thin filaments. Thus, the potential number of cross-bridge linkages and the potential force output may be determined from an accurate measurement of the length between the Z lines. The microscopists of the nineteenth century noted that the sarcomeres of particular muscles showed regional

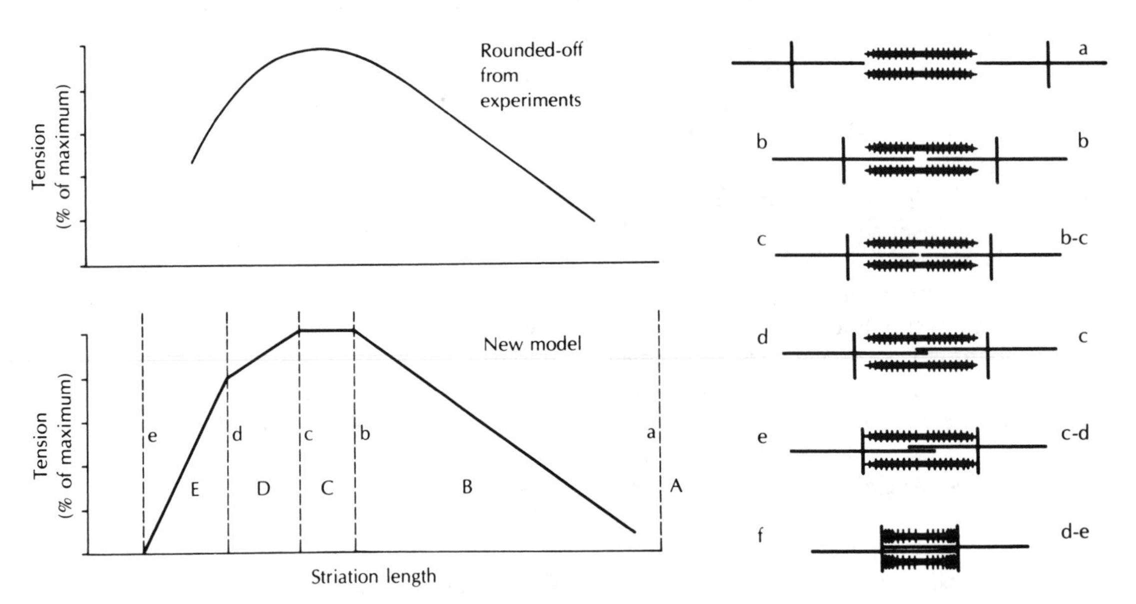

3.1. Patterns of overlap of the myofilaments that underlie the nature of force production with the variable length of muscle fibers. *Top,* The length/tension curve represents a smoothed-out generalization based on the actual data obtained for isometric twitches of a muscle. It can be compared with the series of straight lines that describe the actual behavior of the fibers. *Bottom,* The model is divided into five regions that correspond to the degrees of overlap between the thick and thin filaments. It predicts that the force output will be proportional to the number of cross-bridges that can form between adjacent thick and thin lines. (From Gans, 1980; modified from Gordon, Huxley, and Julian, 1966.)

subdivisions when viewed in dark-field microscopy, an effect more recently tested with polarized light microscopy or laser diffraction. These subdivisions were referred to as the central A (anisotropic or "dark") and the peripheral I (isotropic or "light") bands. Also described was a central light zone (H for Helle Scheibe). In general, these lines (cylinders) match the regions of filamentous overlap or lack of it. Recent research findings suggest that there are other filamentous proteins present that may be significant contributors to the elastic properties of relaxed muscles.

Laterally to the fibrils lie cisternae, a system of reticular membranes that form longitudinal vacuoles parallel to the filaments (fig. 3.2). Also, there are the transverse tubular spaces (T system) of the sarcoplasmic reticulum, which is physically continuous with the sarcolemma. The transverse tubules make regularly spaced mechanical and electrical connections with the cysternae. These vacuoles

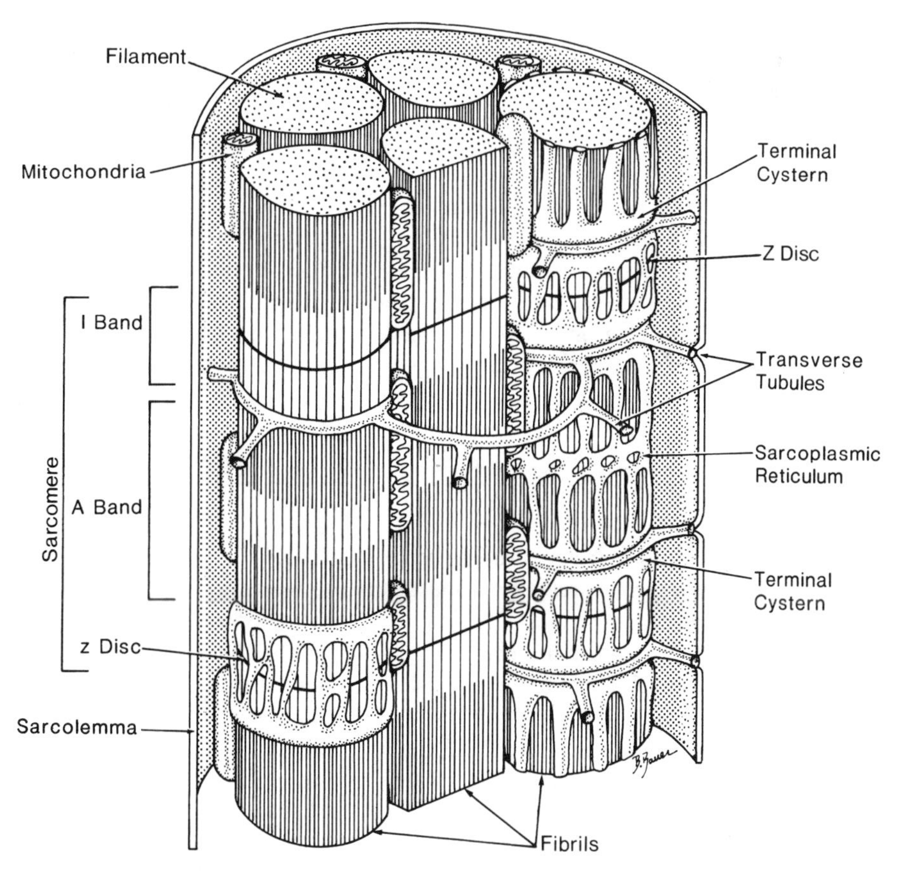

3.2. Three-dimensional reconstruction of portion of a muscle fiber. Note that the muscle fibrils are composed of filaments arranged in an array of sarcomeres. The drawing shows the sheathing cisternae, the transverse tubules, the scattered mitochondria, and the sarcolemma (which is surrounded by a meshwork of collagen fibrils). See text for details.

mediate calcium release and resorption, thereby controlling contraction (chapter 4). Calcium released from these vacuoles into the sarcoplasm induces the formation of the interfilamentous bridges; these must break and reform, cycling as the muscle shortens (or is lengthened). The formation of bridges uses energy provided by the various kinds of respiratory pigments and enzymes distributed through the cytoplasm; the mitochondria stacked among fibrils thus determine the number of times a muscle may contract before it fatigues and consequently determine the power output of the muscle.

The fibrils, their sarcoplasm, mitochondria, and multiple nuclei are enclosed by a membranous sarcolemma, which defines the muscle fiber or cell. Single or multiple nerve terminations called motor end plates or neuromuscular junctions lie directly on the sarcolemma. Overlying these is a meshwork of collagen filaments in a matrix. The sarcolemma is an electrically excitable membrane very much like that of an unmyelinated nerve fiber. Each end plate has both voltage-dependent sodium and potassium channels capable of generating a propagating action potential and acetylcholine-sensitive membrane channels; in concert they function as a chemical synapse.

The contractile and force-generating behavior of single muscles, then, reflects properties at several levels of organization. The first level, obviously, involves the molecular structure of the thick and thin filaments. The second level concerns their organization into sarcomeres. It includes consideration of the T system of the sarcoplasmic reticulum, as well as of the pigments, mitochondria, and lipid inclusions. The third level deals with the way the sarcomeres are arranged in series to form the fibrils and these in parallel to form the fibers. The sarcolemma and its adjacent coatings of connective and elastic materials are also treated here.

The contractile and force-generating properties of whole muscles depend on another organizational level. Study of this level deals with the architectural interconnection of muscle fibers and their suspension among the collagenous tensile supports, which attach them to each other and ultimately to skeletal elements. Finally, there are the electrical interconnections via nerve fibers as well as functional relationships with intrinsic and extrinsic sense organs such as muscle spindles and tendon organs, which establish the control characteristics of the muscle. The striated muscle fibers of vertebrates may differ on each of these levels, within single muscles, among the muscles of an animal, and among those of different species.

Each of these levels tends to be studied by distinct methods; although they are discussed independently here, they are synergistic. A proper understanding of the pattern of the action of a particular group of muscles and its modulation demands that these levels be considered in parallel. The levels ignored in any particular study will contribute to the remaining uncertainty.

2. Biochemical Components of the Contractile System

The contractile mechanism of the cell resides in the thick and thin filaments sometimes referred to as threads of myosin and actin, although each is composed of several molecular components including tropomyosin and troponin. An ongoing series of studies has shown that actin, myosin, tropomyosin, and troponin are involved directly in force production, whereas other components have noncontractile functions. Each thick filament is seen in cross-section to be flanked by six thin ones.

Cross-bridges extend between them at angles near 45°. The formation and cycling of these bridges ultimately furnish the movements and forces that the muscle can produce. In short, the formation of cross-bridges represents the key component of the sliding filament model of muscular action proposed by A. F. Huxley (fig. 3.1).

The past few decades of research in the physiology and biochemistry of muscle have revealed most of the surprisingly complex sequence of chemical events whereby the energy of the phosphate bonds in adenosine triphosphate (ATP) is utilized, although the nature of the actual mechanical force generator remains a mystery. Whereas there is no need here to detail all of these events, certain features have important consequences for the ability of muscles to generate force and motion efficiently and rapidly. In particular, it now appears that the unattached cross-bridges reside in a "cocked" state in which the adenosine triphosphate has already been converted into a tightly bound adenosine diphosphate molecule. The energy stored in this bound state permits the attached cross-bridge to exert its pulling force between the filaments, but the adenosine diphosphate cannot be released unless and until the filaments slide past each other to allow the cross-bridge to complete its power stroke. On the other hand, the amount of force generated by the cross-bridge decreases as the rate of shortening motion between the filaments increases. Furthermore, the force output required to pull apart an attached cross-bridge in opposition to its direction of pull is actually greater than its output during isometric tension. These large changes in force with motion applied to the muscle result in a high degree of intrinsic "stiffness" (ratio of force change to length change in an elastic structure), even in the absence of reflexes mediated by proprioceptors.

Analysis of muscle performance reveals the somewhat paradoxical situation that a muscle produces the least tension and consumes energy at the highest rate when it is allowed to shorten at its maximum velocity. Conversely, an activated muscle that is being pulled apart against its attempt to contract generates the greatest force and consumes the least metabolic energy. Because of these strongly nonlinear relationships between force output and metabolic rate near zero velocity, the characteristics of muscle in the "isometric state" tend to depend on how much elasticity and squirming motion occurs within the muscle and its matrix of connective tissue. Also, it should be apparent that even if the EMG can be taken as a good approximation of the number of activated cross-bridge binding sites, it has only an indirect relationship to the actual number of bound cross-bridges and their force output.

This brief discussion of contractile mechanisms leads to the simplistic impression that the other components are less important to the force-generating process. We have already noted the effects of the sarcoplasmic reticulum and of the mitochondria. Also important is the occurrence of networks of elastic proteins that may connect the thick filaments to the Z lines, exerting significant passive elastic force in inactive but stretched muscles (Magid et al., 1984).

3. Properties of the Sarcomeres

The functional unit of contraction is the sarcomere. All of the series and parallel sarcomeric units of a muscle fiber are generally presumed to behave more or less homogeneously, although significant heterogeneity has been observed or inferred in many muscles. Three processes operating at this level affect the force output of the muscle.

First, the length of the sarcomere is directly related to the overlap of the thick and thin filaments, thereby placing an upper limit on the number of cross-bridges located within binding range of activated binding sites (fig. 3.1). The degree of overlap gives rise to the classic length/tension curve, which tends to be measured under isometric conditions. As the overlap increases, more bridges can form and the potential tension increases linearly. Once sliding of the filaments proceeds to the point at which the ends of the active portion of the thin filaments overlap the central inactive portion of the thick ones, every bond permitted at one end of the active portion of the thick filaments will be matched by the loss of a bond on the other; the tension reaches a plateau. Further sliding leads to central overlap of the ends of the thin molecules so that they interfere with each other and, as this overlap increases, the number of bridges decreases, as does the contractile force. Finally, the sliding causes the ends of the thick filaments to touch the Z lines; further shortening starts to deform them, and the force rapidly drops to zero.

Second, the force generated by a bound cross-bridge changes dynamically if there is relative motion between the thin and thick filaments—that is, the length of the fiber is not fixed or isometric. If the fiber is shortening, there are rate-limiting processes in the motion of the cross-bridge that reduce the force output, up to the limit of zero force defined by A. V. Hill at maximum velocity (V_{max}). If the total extent of the motion is greater than the range of motion of a single cross-bridge, then the rate at which cross-bridges can find and attach to new binding sites limits the force output as well. If the fiber is being lengthened against its attempt to contract, the force output can be greater than the usual isometrically measured tension. This is because the cross-bridges act like stiff springs that must be broken away from their binding sites, requiring about twice as much force as they put out at the normal beginning of their stroke motion. These uncycled cross-bridges can quickly rebind to the next potential binding site that passes by, so the extra tension output of active but lengthened muscle persists even at relatively high velocities of lengthening. Thus, the force/velocity relationship is highly nonlinear, virtually discontinuous through zero velocity, and quite asymmetrical for the lengthening versus the shortening parts of the curve.

Third, the kinetics of calcium release, diffusion, and resorption accompanying each action potential constitute the limiting factor in the number and time course of activated binding sites (assuming that the muscle is not fatigued and has abundant adenosine triphosphate to keep all cross-bridges in the cocked state). These kinetics are extremely complex, having hysteretic effects related to the temporal pattern of activation experienced over many seconds and even minutes. Such effects include short-term fluctuations in the amount of calcium available in the cysternae of the sarcoplasmic reticulum, temporary saturation of resorption sites for active transport back into the sarcoplasmic reticulum, and rate-limiting diffusion and bound transport of calcium within both the sarcoplasm and the tubular systems. Such effects probably contribute to the short- and long-term changes in tension development noted by muscle physiologists and identified under controlled conditions of study by names such as potentiation, sag, and catch. It is probable that all of these effects contribute significantly to modulation of force under at least some naturally occurring conditions; however, they have not been sufficiently well modeled to permit quantitative predictions for diverse conditions.

4. Fibers and Fiber Properties

The cellular unit of striated muscle, the muscle fiber, is multinuclear, reflecting its origin from the fusion of multiple myoblasts, to form the myotubes, which in turn develop into the fibers. In most fibers the nuclei lie peripherally, strung along the sarcolemma; however, they are occasionally central. Fibers of higher vertebrates are cylindrical with more or less conical ends, their shape reflecting the position of the fiber in the fascicle or bundle. However, in some fishes the fibers are platelike structures. Other exceptional architectures such as spiral wrappings have been described.

In the cylindrical fiber the myofibrils are packed adjacent to each other and extend the full length of the fiber. Toward the fiber end there are various minor invaginations of the sarcolemma and passage of collagen filaments toward the fibrillar terminations. However, this is not obvious for all fibrils, and there is some evidence that the tension generated by sarcomeres is transmitted laterally, as well as via the Z bands, so that it passes to and along the sarcolemma and its overlying coatings. This envelope is continuous with the tendinous fibers that leave its ends. More complex patterns are seen whenever fibers are attached to the sides of other fibers.

The fibrils are surrounded and interconnected by the sarcoplasm, which provides the pathway for action current flow from the sarcolemma, for transport of products associated with gas exchange and synthesis, and for force transmission. The basic architecture is distinct in insects; they use tracheoli to shift the gas exchange site into the cells adjacent to the myofibrils. The tracheolar system of insects may permit some volumetric change of the muscle during contraction; however, the volume of the vertebrate muscle fibers is essentially constant (water being almost incompressible). Shortening (and lengthening) must be compensated for by changes in girth. As the sarcolemmal collagen might resist circumferential expansion, one may expect the hydrostatic pressure within the fiber to increase slightly, even though the surface area required to cover a fiber decreases as it shortens.

The number of mammalian muscle fibers apparently does not change significantly after maturity. In general, mitosis does not occur, although the so-called satellite cells may retain mitotic capacity. During longitudinal growth, new sarcomeres appear to be added at the ends of the fibrils. Circumferential increase appears to involve synthesis of additional filaments in parallel to existing ones. Several studies indicate that this occurs as a muscle is exercised and that the fibrils split thereafter. However, the fixity of muscle cell number during life is much less clear in species from other classes of vertebrates; thus, various amphibians and fishes definitely add muscle fibers during life.

Consequently, it is necessary to consider both the taxonomical position of the organism and its life history. Some amphibian muscle fibers, for instance, may be completely reformed at the time of metamorphosis, and the pectoral muscles of some male frogs both hypertrophy and change their histochemical profile during the mating season (when these muscles are used to clasp the females). Other seasonal changes occur in the axial muscles of fishes and in the flight muscles of some migratory birds. The hatching muscles of birds (those that power the cracking of the eggshell) regress after emergence from the egg. The penial retractor muscles of turtles (beloved of pharmacologists) are sex specific and their expression is under hormonal control. Even less obviously sex-associated muscles may show sex differences in some species and this may need to be taken into account when experiments are planned.

It is generally assumed that fiber stimulation is an all-or-none event and that all sarcomeres are activated simultaneously; however, this applies mainly to higher vertebrates in which the muscles generate propagated action potentials. Even in these animals experiments do show local and poorly explained differences in sarcomere length along particular fibrils. This matter is discussed further in the account of innervation pattern.

The force/velocity relationships between sarcomeres and entire fibers deserve a brief comment here. The force produced by a fibril will theoretically be equal to that of any one of its component sarcomeres. In contrast, the displacements, and hence the velocity of the end of a fibril, will be the sum of those of all of the serially arranged sarcomeres into which it is divided.

This permits application of the force relation, derived from the sliding filament model, to that for an entire fiber, indeed an entire muscle. Assuming that the length/tension state of the component sarcomeres has been established, the force produced by a fiber will be a function of the number of thick filaments lying in parallel across the cross-section of the fiber—a concept that represents the essence of the so-called physiological cross-section of a muscle.

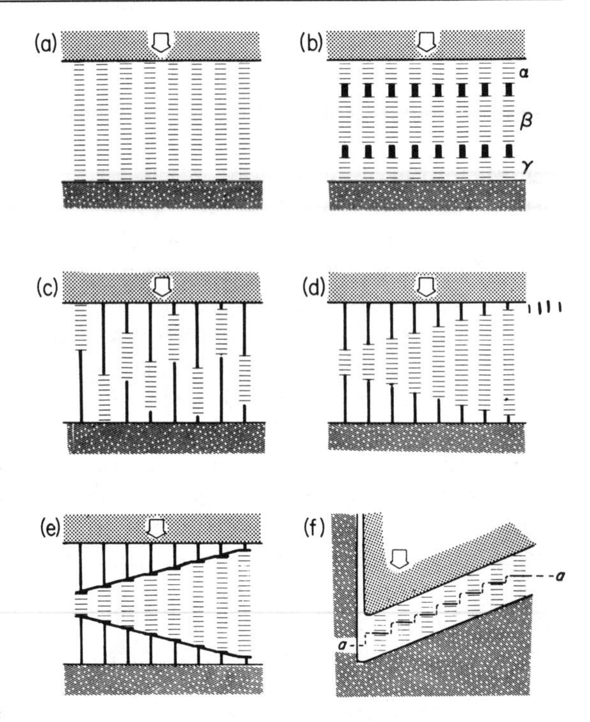

3.3. Permutations of the way muscle fibers (or muscles) may be arranged in parallel to the line of force generation. *Hatched areas* represent origin and insertion sites and are fibers that tend to approach each other whenever the muscle is stimulated. *a,* Fibers run end to end between origin and insertion. *b,* Fibers are interrupted by segments of connective tissue (inscriptions); whereas these are shown separate, they may be interconnected, joining blocks of fibers (or muscles). *c,* Short fibers of approximately equal length are attached to origin and insertion sites by filaments of connective tissue that interdigitate among the adjacent fibers. *d,* Lengths of the fibers differ (here regularly) so that equal shortening would impose quite different effects on the individual sarcomeres or muscle fibers. *e,* The pattern is as in *d,* but the adjacent fibers are cross-linked, generating a complex and possibly indeterminate arrangement of force transmission. *f,* The array of fibers is staggered. However, because the relative excursion of the two surfaces is constrained, the fibers continue to act in parallel to the line of motion. (From Gans, 1982a.)

5. Muscle Architecture

In a relatively small muscle all of the fibers may lie in parallel and be bound by a connective tissue envelope (the epimysium) into a homogeneous bundle. However, sheets of connective tissue (the perimysium) subdivide most muscles into fascicles; these are surrounded by endomysium. The perimysia also provide the pathways for the blood vessels and nerves. Domestic cattle deposit fat in these sheets; this phenomenon produces marbling in beef. Mere inspection of the three-dimensionally complex external shape of many muscles suggests that their internal arrangement of fascicles and fibers is also complex (figs. 3.3 and 3.4).

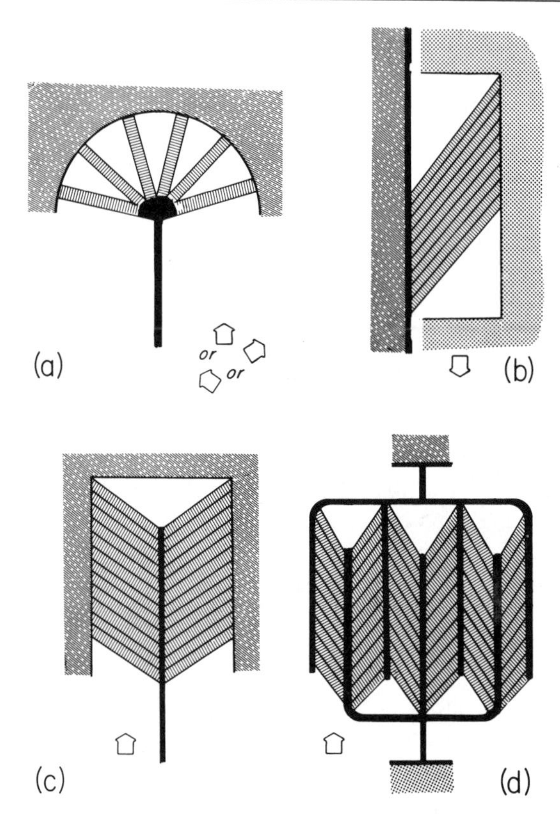

3.4. Permutations of muscle fiber arrangements. In each case, the force transmitted by the tendon lies at an angle to the force produced by the individual muscle fibers. *a,* Radial arrangement of fibers. *b,* Singly pinnate muscle. *c,* Doubly pinnate muscle. *d.* Multiply pinnate muscle. In *a,* activation of the different fibers would not only generate differential force, but also shift the site at which the muscles attach to the tendon. The angle of the fibers will change as the fibers shorten. In *b,* the surfaces of origin and insertion are maintained in parallel by mechanical stops; in *c* and *d* they are maintained in parallel by the action of fibers inserting on their opposing sides. (From Gans, 1982a.)

The fibers constituting these fascicles may be of equal length and arranged in parallel so that the length of any fiber is that of the entire muscle; most often these conditions are not met. Fibers may lie curved or straight, in parallel or pinnately (feathered); their origins and insertions may be arranged along single lines and surfaces or may be staggered. Generally, fibers are significantly shorter than the muscle from which they derive. To make some sense of these arrangements, let us consider them in terms of muscle function.

Each fiber is likely to exert its main force along its axis. We may then speak of parallel arrays, in which the axis of the force produced by the fibers parallels that of the muscle as a whole, and angled arrays, for which the force generated by any fiber lies at an angle to that of the muscle as a whole. A small, but significant, third category includes muscles with fibers that remain curved so that each end is pulling in a different direction. Obviously, this analysis considers muscles while they are in tension; the fibers of many soft tissue systems change in orientation during activation (as do the muscles of skeletal elements while these shift about joints).

As collagen is more than 50 times as resistant to deformation than is muscle tissue, the areas of attachment of tendons are always smaller than is the cross-sectional area of the belly of the muscle. This causes some slight divergence among the lines of force of the component fibers, even for strap muscles, which are theoretically parallel fibered. However, it is incorrect to assume from their highly specific stiffness that tendons and fascicles are indistensible elements that can be neglected. Tendons have complex and nonlinearly springlike properties that contribute substantially to elastic energy storage and shock absorbance. Their stiffness is matched to that of their muscles by virtue of their proportionately smaller cross-sectional area.

Only that component of the force that acts along the line of the muscle as a whole generally contributes to its action. With any divergence comes a loss of effective force; the component at right angles is often counteracted by that of a fiber placed opposite to it, and both represent a loss. This virtual loss increases with the angle to which the fibers diverge from the main line of action and exceeds 30% as the angle reaches 45°. Geometric considerations show that the length of angled fibers tends to be shorter than that of the muscle as a whole and that this difference increases with greater complexity of architecture (e.g., multipinnate muscle). The higher the angle of pinnation and the thinner the muscle in this plane, the shorter the muscle fibers.

What benefit does an animal derive from these complex packing patterns? The main advantage is that complexity permits the muscle to be designed to provide the force, displacement, and velocities required by the system. For instance, it is possible to place far more fibers in parallel (between the origin and insertion) in an angled (pinnate) muscle than in a parallel one (see fig. 3.3). This permits the animal to increase the effective overall force generated by a given volume of muscle, although each fiber loses a small fraction of the force it produces and the total displacement (and velocity) achievable at the tendon via these shorter fibers may be much less (although each pinnately arranged fiber may move the tendon slightly more than it shortens because of the levering action if the long section maintains a parallelogramic shape). Such architectural relationships are important, because they permit the animal to match fiber lengths to the demand for excursion and velocity. Also, they let the flat (maximal force) portion of the length/tension curve fall at any desired portion of the range through which a muscle may move.

Once the desired relations of fiber length and pinnation angle have been established, various angled arrangements and their multipinnate variants permit the effective use of the available space by means of selection of the most suitable from a number of functionally equivalent arrangements (fig. 3.3). Angled arrangements have the further advantage that the fibers and their tendons may radiate from an attachment site, rather than leaving it in only a single direction (fig. 3.4). Differential activation of such a system may allow the insertion to be shifted at an angle to the main line of action, thus producing complex movements.

6. Motor Units and Motor Control

Striated muscle fibers receive their stimulation via the motor end plates, which are variously shaped neuronal terminations that lie on the sarcolemma usually near the midpoint of each fiber. The architecture and the foldings of the underlying membranes are related to the stimulation pattern and the temporal contraction pattern (fast, slow). Each motoneuron communicates with from fewer than 10 to 1000 muscle fibers at such motor end plates; the assemblage is called a motor unit. Hence, the axon of the single process arising from a cell body in the ventral horn of the spinal cord or brain stem must bifurcate many times before reaching the terminals within the muscle.

The fibers of multiple motor units tend to be interspersed so that each motor unit is widely spread (exceptions being seen in certain diseases—i.e., when a motor cell reinnervates a muscle following earlier denervation by trauma or disease). One observes all degrees of overlap from conditions in which the centers of fiber distribution cluster closely to those in which the units overlap widely, including patterns in which groups of motor units are restricted to particular fascicles permitting differential activation of parts of the muscle. In some muscles one notes longitudinal subdivisions with each group of muscle fibers in series supplied by a distinct set of motor units or by branching of each motoneuron to multiple subdivisions. Sometimes these bands are defined by intermediate tendinous inscriptions; in other cases the ends of the fibers interdigitate so that only histochemical stains for connective tissues and motor end plates disclose the architectural pattern. These patterns may relate to functional specialization, such as separation of motor units relating to postural stabilization from those generating rapid, forceful movement. They are common and often overlooked throughout the animal kingdom, including mammals.

During the early stages of ontogeny, vertebrate muscle fibers show electrotonic (communicating or gap) junctions; multiple fibers are electrically interconnected and often show spontaneous contraction patterns. With the arrival at this muscle of the outsprouting axon terminals of the motoneuron, the electrotonic junctions close and the fibers become independent. The axons of several motoneurons apparently compete for these end plate sites. This competition results in a phase during which individual muscle fibers show multiple innervation, and the ultimate architectural pattern of overlapping motor units is established. Various reports suggest that innervation of muscle fibers by more than one motoneuron persists as an adult condition in some species (particularly in nonmammalian forms such as teleost fishes and amphibians). Whenever this is suspected, it is necessary to establish the actual pattern of innervation. More than one motoneuron may contribute to a single motor end plate; multiple motoneurons, perhaps deriving from different nerves, may each provide a separate motor end plate, and (as in arthropods and teleosts) a single motoneuron may supply several motor end plates that are serially arranged along a muscle fiber.

The systems described so far represent only the efferent (outflowing) limb of the motor loop. The firing pattern observed in the motor units is controlled by neural circuits utilizing the information provided by a series of afferent (sensory) inputs that are centrally modulated by various interneuronal and descending circuits. Afferent information available is most varied. It derives from integument and major sense organs, as well as from a pair of elegant proprioceptors, the tendon organ (placed in series with small groups of the muscle fibers and involved mainly in the detection of active muscular force), and the muscle spindle (in parallel with muscle fibers and sensitive to muscle length and velocity). Muscle spindles incorporate intrafusal muscle fibers (hence the muscle fibers of major interest to the electromyographer are called extrafusal), which are apparently activated in complex patterns that, in mammals, are only sometimes simultaneous with those of the adjacent extrafusal fiber units. Such patterns modulate the sensitivity of the afferent endings and may be parts of reflex loops (fig. 3.5).

The distribution and the mechanical relationships of proprioceptors to extrafusal muscle fibers are not well described even for simple mammalian muscles. In the few complex muscles of mammals and nonmammals that have been studied, the distributions have been nonuniform and the mechanical relationships appear to be surprisingly complex. Both the nature of the proprioceptive information and its ultimate use in motor control circuits are likely to be at least as specialized as the muscles served.

7. Muscle Types and Fiber Types

The electromyographer tends generally to study only phasic or spike-producing (twitch) striated muscles; in these the membrane potential changes propagate rapidly during activation (hence the terms "action potential" and "spike"). In contrast, the changes of membrane potential of smooth muscle fibers are relatively slow. Similarly, the so-called

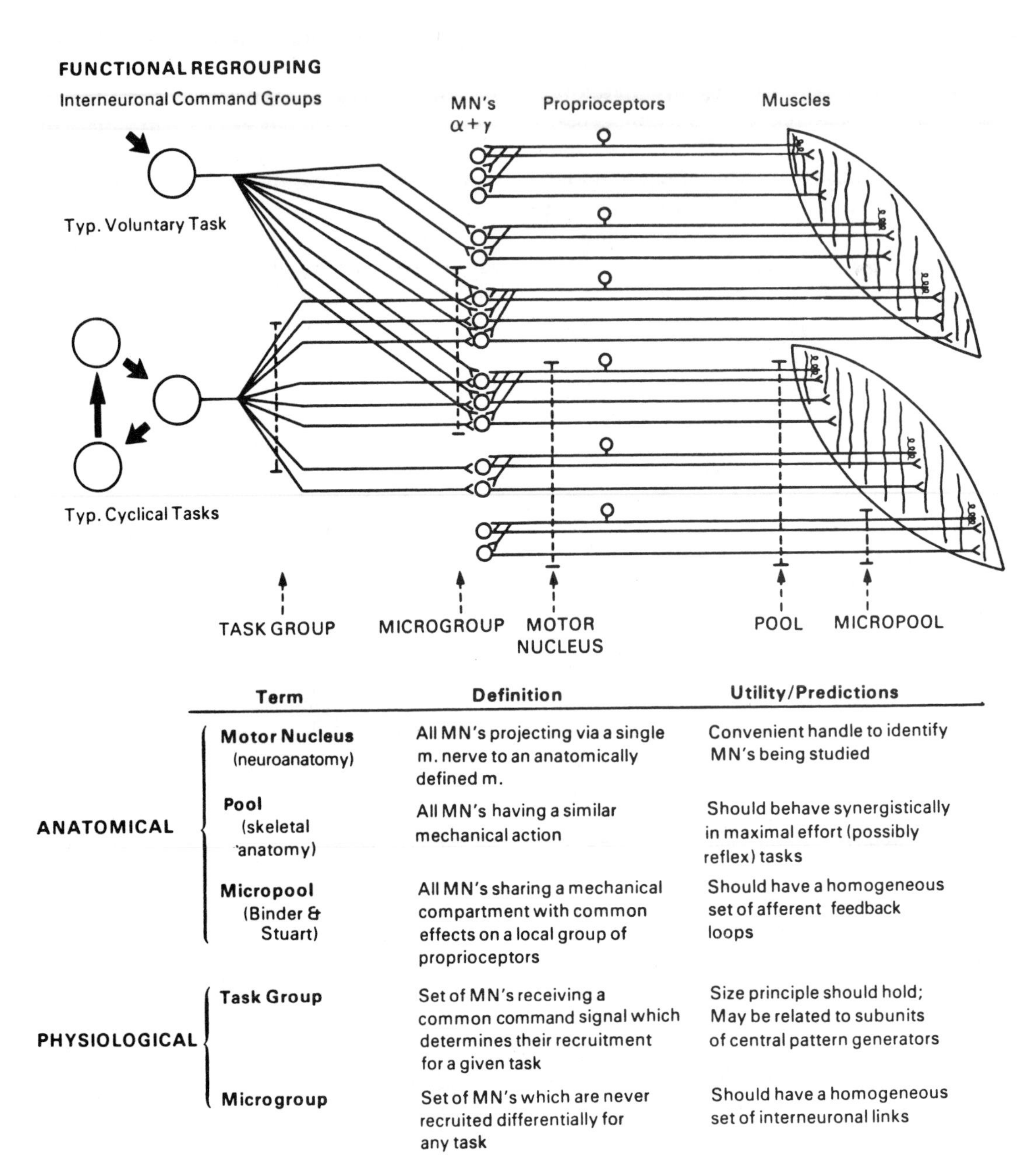

	Term	Definition	Utility/Predictions
ANATOMICAL	**Motor Nucleus** (neuroanatomy)	All MN's projecting via a single m. nerve to an anatomically defined m.	Convenient handle to identify MN's being studied
	Pool (skeletal anatomy)	All MN's having a similar mechanical action	Should behave synergistically in maximal effort (possibly reflex) tasks
	Micropool (Binder & Stuart)	All MN's sharing a mechanical compartment with common effects on a local group of proprioceptors	Should have a homogeneous set of afferent feedback loops
PHYSIOLOGICAL	**Task Group**	Set of MN's receiving a common command signal which determines their recruitment for a given task	Size principle should hold; May be related to subunits of central pattern generators
	Microgroup	Set of MN's which are never recruited differentially for any task	Should have a homogeneous set of interneuronal links

3.5. Innervation of muscles imposes several additional organizational levels, both anatomical and functional. The neural anatomy gives rise to motor nuclei (spinal cord and brain stem) that are operationally defined as all the motoneurons projecting to a single muscle nerve (usually but not always a single contiguous clustering of cell bodies). The mesenchymal anatomy of the muscle gives rise to motoneuron pools and micropools that are characterized by the homogeneity of their mechanical contribution to skeletal movement and of their effect on local proprioceptors within the muscles. Physiological recruitment of at least some muscles involves subsets of the motoneurons in the nucleus. These functional elements may be recruited in some orderly (usually size-related) manner to the exclusion of other motoneurons projecting to the same muscle. Such sets have been called task groups. Their organization may relate to the control circuitry required to program and make effective use of the proprioceptive feedback from muscle spindles under specialized kinematic conditions of the task. (From Loeb, 1984, copyright 1984, Macmillan Publishing Company. Reprinted by permission of the publisher.)

slow striated fibers also produce slowly rising transients; they are sometimes referred to as tonic. The signals generated by smooth and tonic muscles fall well within the frequency range of those artifacts generated by the movement of loose metallic electrodes in contracting muscle and by various slow electrical field effects. In standard EMG practice, these frequency bands are generally considered to contain mainly noise; they tend to be filtered out explicitly or severely attenuated by ordinary recording procedures. Furthermore, the signals from slow muscle may be masked by signals from other sources in the organism (see chapter 13). Attempts to record from slow muscles often involve the use of anesthetized animals in which the kind and magnitude of extraneous signals are likely to be reduced. The fibers of cardiac muscle, of course, produce relatively slow spikes, but they do so in synchrony so that the resulting signals have much larger magnitude than slow EMG. This (and the obvious practical applications) makes recording and analysis of the electrocardiograms a discipline of its own.

Consumers of chickens and some other domestic fowl know that the various striated muscles of these vary in color. It becomes clear that the more frequently used muscles (e.g., leg versus wing muscles) tend to have a darker color. Some 20 years ago the old finding that the kinds of fibers seen in a single muscle might fall into distinct morphological categories was rediscovered. Furthermore, it became clear that various anatomical, histological, and biochemical correlates were regularly associated with the physiological properties of particular muscle fibers and of their innervating motoneurons (fig. 3.6). The main difficulty has been that the patterns disclosed differ substantially among the several classes of vertebrates—for instance, among bony fishes, amphibians, reptiles, birds, and mammals. This has led to several conflicting nomenclatures of the fiber types presumably recognizable in different animals and to some uncertainty as to whether any of them are analogous.

In mammals three main types of phasic muscle fibers have been recognized and named in accordance with their contractile properties—namely fibers that are fast twitch/fatiguing, fast twitch/fatigue resistant, and slow twitch (also fatigue resistant). A different terminology refers to fast glycolytic, fast oxidative/glycolytic, and slow oxidative fibers based on the histochemical demonstration of the concentration of various metabolic enzymes. Intermediate types of fibers have been characterized in some muscles, such as the primate masseter and temporalis muscles. The pattern differs in birds; indeed, we will see that each class of vertebrates has its own unique categories of fiber properties.

In some muscles one sees differences in fiber diameter associated with fiber type. However, this breaks down in many cases and may reflect functional differences—for instance, between the muscles of locomotion and those of mastication. Similar doubts apply to such generalizations as the placement of the oxidative fibers in the mechanically most advantageous position relative to the joint system.

An important generalization is Henneman's size principle. According to this, the smallest motoneurons (of the spinal cord) are likely to be recruited first in response to depolarizing activity in the premotoneural command circuits. Because the size of each motoneuron correlates with the number of fibers it innervates and their glycolytic capacity, the least strong but most fatigue-resistant muscle fibers will be recruited first.

The difficulty with such appealing concepts is that muscles play multiple and sometimes conflicting roles in different kinds of behavior. Some exceptions have already been noted and the principles cannot be assumed to pertain to every motor task.

8. Interspecies Variation

Although we have mentioned some exceptions, much of the preceding discussion pertains to many mammalian muscles, some of which have been intensively studied at multiple levels. However, we know enough about the muscles of other classes of vertebrates to confirm that these differ substantially so that it remains essential to consider the differences when evaluating experimental results. A very important generalization is that mammalian muscle appears relatively stereotyped at the cellular and ultrastructural levels; the various body muscles of single individuals of many other vertebrates show much greater morphological and physiological specialization down to the levels of the sarcomeres and myofilaments.

Some differences occur on the gross structural level; thus, the nuclei may be central or peripheral, and certain fishes have muscles

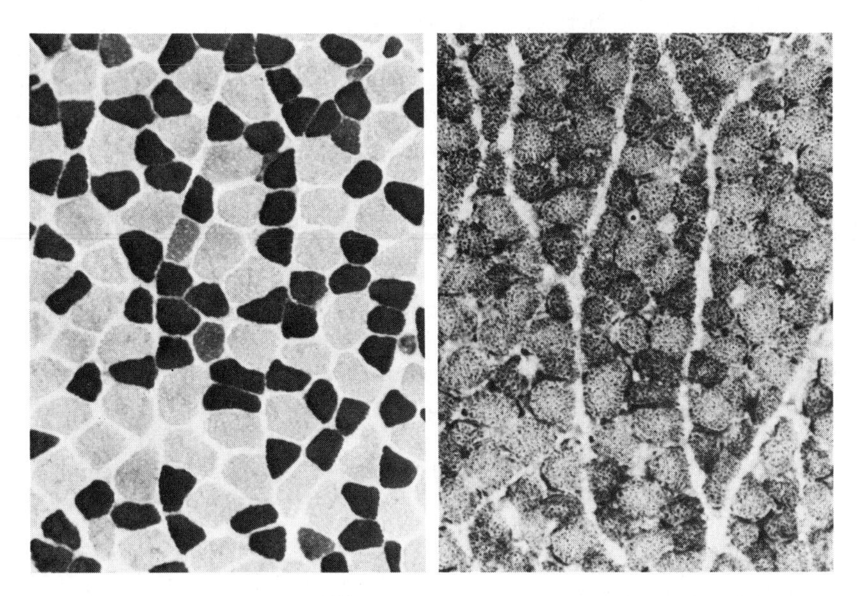

3.6. Parallel cross-sections of muscle after histochemical treatment show the different responses of equivalent fibers and the way the fibers are arranged in an overlapping mosaic. Gastrocnemius muscle of juvenile rat stained for myofibrillar ATPase (*left*) and succinic dehydrogenase (*right*). (Courtesy Fay Hanson-Smith.)

that are platelike rather than spindle shaped. The biochemistry differs so that the actins and myosins may be distinct, even to the extent of having filaments of different length for different muscles of a single species. The architecture of such fundamental aspects as the T system differs in its relation to the Z lines. The nature and distribution of the motor end plates have certain unique features in amphibians.

The physiology of muscle has other intergroup aspects; thus, the relatively slow properties of reptilian muscle have attracted much attention. These differences also extend to the nature of the control and coordinating systems; there are regional and interspecific differences in the nature and presence of muscle spindles and in the kinds and wiring of sensory receptors that these animals use to control muscular activity.

C. COMPLEXLY ARRANGED MUSCLES

1. General

Textbooks stress that muscles run from origin to insertion and that they tend to approximate these sites whenever they are stimulated. We have already noted the two aspects that complicate this situation. First, the tension generated by sarcomeres is not transmitted just at the fiber ends but passes down the sarcolemma. Second, muscle fibers are attached to each other in complex arrays so that the activation of one motor unit may deform all or part of adjacent ones.

Under these circumstances displacement, force, and control aspects may not be obvious. Of course, there is no theoretical scheme that takes all variants into account. Hence, we will review briefly some special cases as examples in the hope that these will sharpen the experimentalist's vision and widen expectation of nonstandard circumstances.

2. Multijoint Muscles

If a muscle bridges a single joint, the degrees of freedom for their placement are limited. The further the insertion from the fulcrum, the greater the velocity in the muscle for a particular velocity of skeletal displacement and the less force required to establish particular joint torques. In short, the required fiber properties are established by the position of the required muscle, and the number of sarcomeres must be sufficient to accommodate the full excursion at the joint.

In contrast, a two-joint muscle achieves some independence from the position of either joint. If the joints can be made to move in opposite directions (so that motion at one lengthens the muscle whenever the other shortens it), a situation can be obtained in which a muscle needs to act over a relatively small excursion range. This facilitates the use of muscles with short, angled fibers that generate substantial force per mass. This may not only have direct energetic benefit but may also reduce the mass of muscle that contributes to the inertia of swinging elements.

Multijointed muscles paradoxically may simplify the problem of control. Thus, the flexors and extensors of the human digit are not arranged in antagonistic pairs for each joint. Rather their tendons generally bridge several joints. With this system, the distal finger pads may be positioned accurately, but the position of the intermediate joint (knuckles) is established secondarily and is not under voluntary control.

3. Muscles in Series

Various muscle units consist of blocks of fibers, or fascicles, that are serially arranged. Examples are the semitendinosus muscles of many mammals, which have two sets of motor units in series, the rectus abdominis muscles of many species, in which multiple bundles are subdivided by tendinous inscriptions; and the digastric system of mammals, in which there may be two discrete muscles in series, a connected pair, or a superficially single mass. More complex situations appear in the axial masses of fishes, in which stacks of roughly parallel fibers (myomeres) run between the zigzagging connective tissues (myocommas), and in the axial muscles of snakes, in which three or more muscles, each belonging to a different vertebral segment, are arranged in triradiate arrays separated by stretches of tendon that are many times the length of the active muscular components.

These serial arrangements show no single common denominator. In some cases we see remnants of evolution. The digastric apparently derives from two distinct muscles in the pharynx of fishes; this explains its innervation from both the facial and trigeminal nerves. In other cases a recent role (biologically significant function) has been more obviously influential; witness the elongation of snake muscles, clearly associated with their role in forming smooth, long-radius body curves.

4. Generation of Fluid Pressure

a. Contraction around Fluid-filled Spaces. Contraction around fluid-filled spaces is one of two ways that muscles may generate fluid pressures; the other is contraction within fluid-filled spaces and muscular skeletons (discussed below). Arrangements of straight fibers may compress a lumen only if the fibers are linked to skeletal structures such as the rib cage. Such straight fibers may activate levers (ribs being rotated about joints with the vertebrae). However, curved slings of more or less straight-fibered muscles may contract to reduce the curvature and affect the contained space (as in the transverse mandibular muscles of frogs and the diaphragm of mammals).

Peripheral muscle fibers may be wrapped completely around a central cavity so that the lumen narrows upon contraction. The pressure such fibers can exert on the lumen is given by the ratio of the thickness of the fiber layer to the radius of the enclosed space times the force per unit area produced by the muscle. This relationship suggests that substantial wall thickness is needed if the system is to generate great pressures and that the shortening of each sarcomere must be limited if the fibers are to operate in the desirable portion of their length/tension curve. However, such systems often incorporate the need for a substantial reduction of the luminal diameter requiring significant shortening of the sarcomeres (approximately 30% shortening to halve the volume). Hence, they require different muscle types, and one rarely sees striated muscle in such a role among the vertebrates.

b. Contraction within Fluid-filled Spaces and Muscular Skeletons. Placement of muscle fibers within the fluid volume to be compressed has the advantage that the ratio of fibers to fluid may be much greater than for a walled system in which the structural rearrangements required during contraction may impose viscosity problems limiting velocity.

In general, such fiber-in-fluid systems occur in tongues. In these systems the lymphatic fluid is shifted from one portion of a chamber to another by means of pulling in selected portions of the periphery. There is a regularly distributed attachment of the muscle fibers across the portions of the limiting membrane that are to be pulled inward. Only if the membrane is to be dimpled does one see a local concentration of attaching fibers.

Although contracting muscle fibers increase their diameter, this increase in contraction ordinarily does not produce significant internal pressure. However, whenever fibers are constrained from swelling, such as by very tight packing or wrappings of connective tissues, substantial internal pressures may be generated within fibers and muscles. This phenomenon underlies the swelling of the biceps (used by men of certain ages to impress some women and by circus strong men to break chains). Some frogs use this mechanism to transform normally flaccid tissues in their tongue into temporarily rigid elements so that the tongue can be catapulted at prey.

It should be noted that the pressurization need only occur in the portion of the fiber that is constrained. The free portion of such a locally constrained fiber can perhaps contract further than the constrained part. Furthermore, local constraint would permit locally segregated, multiple innervations to produce differential shortening along portions of a single fiber.

D. HOW TO STUDY MUSCLE ARCHITECTURE

The experimentalist faces the complexity described above at the start of any analysis. Presumably, there is some question about the way some bones are moved or some motor act is performed. How does one approach the problem? Is EMG an appropriate source of data and are there other perhaps better techniques for understanding the myoskeletal system being investigated?

Clearly, one must start with an understanding of gross anatomy, with the placement of major muscle masses, tendons, and skeletal elements. First, one needs to know something about the arrangement of the major muscle masses. In higher vertebrates this is often easy, as the groupings of fibers are well defined and fascial planes separate adjacent muscles; whenever muscles tend to move relative to each other during contraction, one often sees well-defined collagenous coatings with loose connective tissue spacings. However, under other circumstances two muscles attaching onto an element close to one another may share a single sheet of collagen and thus be more difficult to separate. Not only the zones of origin and insertion but also the zones of lateral attachment are important.

Dissection is necessarily destructive. Although we attempt to derive maximal benefit from each specimen, several specimens must usually be used, not only to understand variation but also because approaches by serial section or from different angles permit a more complete conceptualization first of topography and later of function.

The configuration of the myoskeletal system must be characterized at the various physiological positions of each joint, with attention to those features that seem related to behavioral kinesiology, such as the position of the hindlimbs as a horse kicks and gallops and the position of its jaws when it bites in a fight or grinds alfalfa. If specimens are to be preserved for later dissection, some should be fixed at each extreme (wide open, fully closed) and others at the position occupied whenever the animal performs biologically important actions. The overall mass of each muscle is of interest in only the most general terms, as it tells nothing about either fiber architecture or fiber type. It is only one of several factors used to determine physiological cross-section. At best it offers some indication of the range of differences among sexes, populations, and species, but the absence of difference in any series does not guarantee that the muscles of two animals are indeed similar. Anomalous structures in individual specimens may confuse the surgeon or physiologist.

The major groupings of muscles may be established by dissection, which should be well documented by means of careful notes and sketches and, preferably, by a standard series of photographs. If a high-resolution cine photographical setup or video tape recorder is being used for behavior studies, individual frames may be used instead of still photography; however, the camera position must be locked down and the specimen maintained in a fixed position so that the sequences appear from a common vantage point (see chapter 15). As films may be projected forward and backward, the way that superficial and deeper layers relate can be illustrated. Nerves and blood vessels (see chapter 8) should be noted, both to ensure correct electrode placement and muscle identification and to avoid modification, through damage, of the output of the muscles when the electrodes are put in place. Particular attention should be given to major nerve branches and to multiple innervations; either of these may indicate the potential for functional segregation or architectural specialization.

At this level it is often useful to generate primary hypotheses about the workings of the system. If adequate numbers of living subjects are available, it is useful to attempt stimulation of individual muscle masses with the animals under deep anesthesia. It may be worthwhile to film such tests. It may also be useful to install strain and pressure gauges (see chapter 16) to monitor the effects of such stimulation. Certainly it is important in such studies to match the plateau level of the length/tension curve to a particular joint position.

The next step in analysis is to determine the fiber architecture of the component muscles. A relatively quick check of the architecture of either fresh or preserved materials is facilitated by a combination of stains: toluidine blue is used to mark surface texture such as collagen fibrils and Lugol's solution (tincture of iodine) to stain sarcoplasm. Intermittent drying of the surface (compressed air jet or modified hair dryer) and application of concentrated sodium hypochlorite solution (sold as laundry bleach) selectively break down the collagen, permitting an approximate check of fascicle length and angles as one passes down through the layers of the muscle.

A more adequate indication of the length and placement of fibers is achieved by the use of nitric acid (appendix 1). This dissolves the connective tissues so that the fibers fall apart; the partly or entirely dissolved tissues can then be stored in glycerin. It is essential to check the fiber ends under a compound microscope to ensure that the full fiber rather than a broken segment is being measured. An even better approach utilizes silver or gold staining to permit visualization of the connective tissues and motor end plates (appendix 1). This is particularly useful for cases in which there are very fine muscle fibers interdigitated among those of larger diameter, but this technique can only be used on fresh material.

Mapping of the area of single motor units is a field in itself and tends not to be combined with simple EMG approaches. It involves stimulation of the cell body or single fiber of the motoneuron and assay of the effect either by shifting of electrodes through its field in the muscle or by staining and histological reconstruction. In the latter case the repetition rate is set sufficiently high to deplete the glycogen stores of the muscle fibers so that the motor unit stands out after staining by periodic acid–Schiff techniques (appendix 1). Unfortunately, this approach is costly as it requires one specimen per motor unit, although motor units of multiple muscles could theoretically be analyzed in parallel. Care must be taken to avoid damaging the muscle fibers, as these would then appear glycogen depleted without having been stimulated. The recent development of modifications of the horseradish peroxidase techniques, coupled with methods for clearing the muscle so that the path of individual axons may be followed in space, offers the potential for powerful and easier approaches.

The mapping of EMG activity from various points on the surface of a complex muscle may be a relatively simple way to sort out complex innervation patterns and multiple or branched motor nerves. Local stimulation of grossly visible nerve bundles can give a direct indication of the distribution of groups of motor units, because the action potentials so evoked in any branch can spread through the entire arborization of the motoneuron. Closely spaced differential recording is often a much more sensitive indicator of local activity than visible twitch motion, which can be conveyed mechanically to passive regions of muscle. One should use the surface ball electrodes to record and stimulate (neurophysiologists make them by melting the ends of fine platinum wires).

The characterization of fiber types and their distribution usually requires a combined approach based on both histochemistry and contractile physiology. The various histological approaches are detailed in appendix 1.

Not all of the listed techniques are applicable or practical for use in a particular case. The choice of techniques used and aspects examined (or omitted) ultimately determines the level of certainty for functional hypotheses and should always be kept in mind. Although a careful search of the literature may often turn up much of value, this should never be used as a substitute for a personal "hands on" inquiry geared to the issues of interest. Also, apparently "homologous" muscles in similar species may actually share little more than their phylogeny. The high degree of trophic interaction observed in developing and mature muscles leaves their actual form and function uncertain in each species and even in individuals.

4 How Muscles Generate Electricity

A. ACTION POTENTIALS AND THE CONTRACTILE MECHANISM

EMG involves the measurement of electrical events in muscle. How are these generated? Let us start by noting why an organ with a strictly mechanical function generates electricity at all.

The contractile apparatus of striated skeletal muscle consists of longitudinally arranged, interdigitated sets of the filamentous proteins actin and myosin. In the presence of calcium ions, small extension arms spaced out along each myosin filament respond by bonding to receptor sites on adjacent actin filaments. Having done so, they exert a sliding force like little lever arms. The force is provided by high-energy bonds in locally bound adenosine triphosphate molecules. At each cross-bridge, this force falls to zero when the filaments slide far enough past each other to relax the strain in the bonds. The muscle fiber controls the amount of force being generated at any time by the release of calcium ions into the general intracellular space and its re-uptake into a compartmentalized space called the sarcoplasmic reticulum. Under normal resting conditions, almost all of the intracellular calcium is held in the sarcoplasmic reticulum, so that the muscle has few or no active cross-bridges and is relaxed, generating little or no tension.

In most muscles, the fibers are actually syncytia, consisting of many fused cells that are some tens of microns in diameter. They may run the entire length of the muscle, often for many centimeters. Obviously, if calcium ions were to be released among the sliding filaments in one region of the fiber but not in others, this region would contract at the expense of simply stretching out the inactive regions. No force would then be conveyed to the bone or tendons at either end of the muscle. Some means must be devised for coordinating the release of calcium ions, so that calcium is released almost simultaneously over the entire muscle fiber. This is the role of the myoelectrical action potentials that make up the EMG.

In terms of its electrical properties, the muscle fiber can be thought of as a large-diameter, unmyelinated nerve axon, somewhat like the squid's giant axon of neurophysiological renown. The axon of the motoneuron terminates on a motor end plate usually located near the longitudinal midpoint of the muscle fiber. When a nerve impulse arrives, the electrical disturbance in the nerve terminal leads to the release of a neurotransmitter (acetylcholine in vertebrates). This quickly diffuses across the synaptic cleft, where it binds to receptors in the muscle fiber membrane and opens channels in that membrane (fig. 4.1). The muscle fiber, like any neuron, actively maintains its intracellular environment at a potential of about 80 mV negative with respect to its surroundings. The fiber collects potassium ions and evicts sodium ions, thereby creating the concentration gradients that produce the resting potential (as quantified by the Nernst equation). Whenever the acetylcholine-controlled channels are opened, this resting potential drops momentarily; this, in turn, leads to the opening of voltage-sensitive channels that admit only sodium ions. Still more electrical current rushes into the cell at these places, in turn depolarizing even more remote sections of the cell membrane of the muscle fiber.

Left unchecked, the muscle fiber would soon lose all of its membrane potential. However, the opened channels quickly close. Only a tiny amount of ionic flow is actually required to discharge the membrane capacitance, thus changing the transmembrane potential. The enzyme acetylcholinesterase rapidly breaks down the acetylcholine that started the process. The channels that opened to admit sodium ions are soon joined by channels that pass only potassium ions. As these ions are concentrated inside the cell, their outward flow down their concentration gradient soon restores the resting membrane potential, permitting both the sodium and potassium channels to close. At any given patch of membrane, the sequence of events includes (1) passive outward flow of current from adjacent actively depolarizing regions (discharging the membrane capacitance of the patch), (2) active inward flow of current through opened sodium gates of the patch, (3) active outward flow of current through opened potassium gates, and (4) stable resting mem-

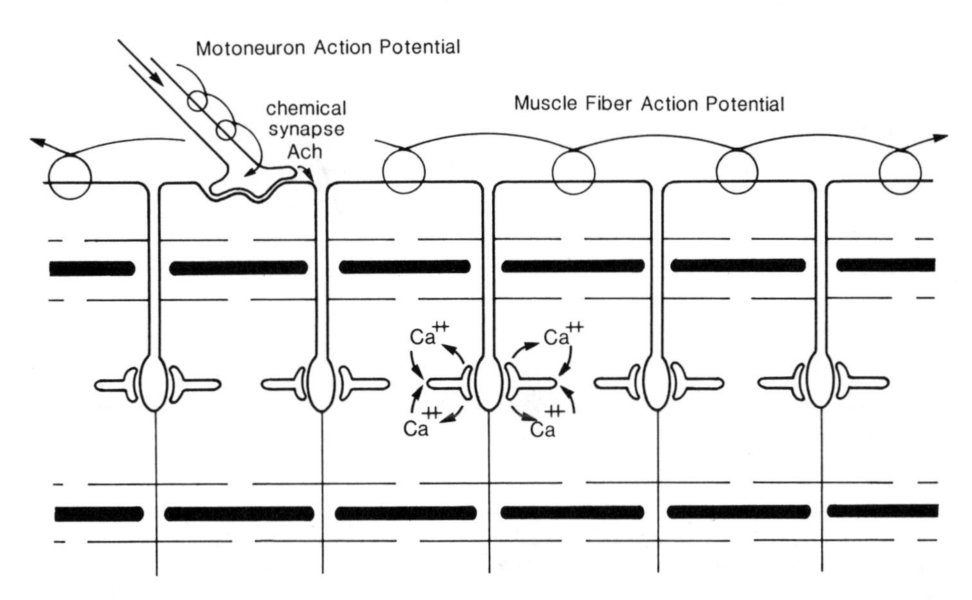

4.1. Electrochemical events in neuromuscular transmission. An action potential arrives at the end plate of a motoneuron on the muscle fiber and there releases acetylcholine (*ACh*) into the synaptic cleft. The acetylcholine locally depolarizes the sarcolemma, leading to a new propagated action potential that passes along the sarcolemma toward the ends of the muscle fiber and downward from the surface into the transverse tubular system. Here the disturbance couples electrotonically into the longitudinal tubular system, stimulating the release of calcium ions into the sarcoplasmic space. The calcium ions bind to troponin molecules on the actin filaments, presumably causing a stereochemical shift that gives rise to favorable conditions for binding to the cross-bridges that arise from the myosin filaments.

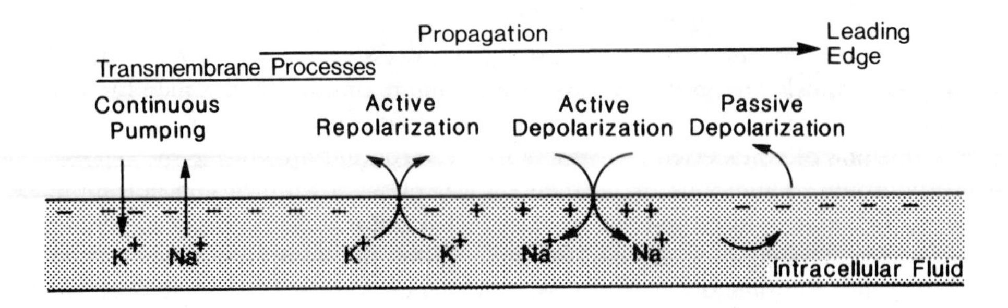

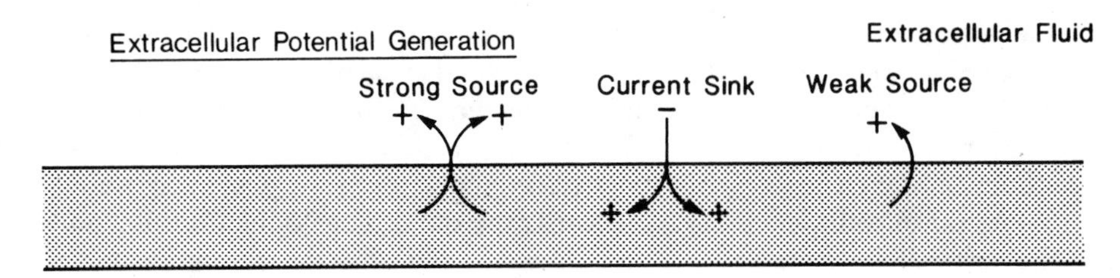

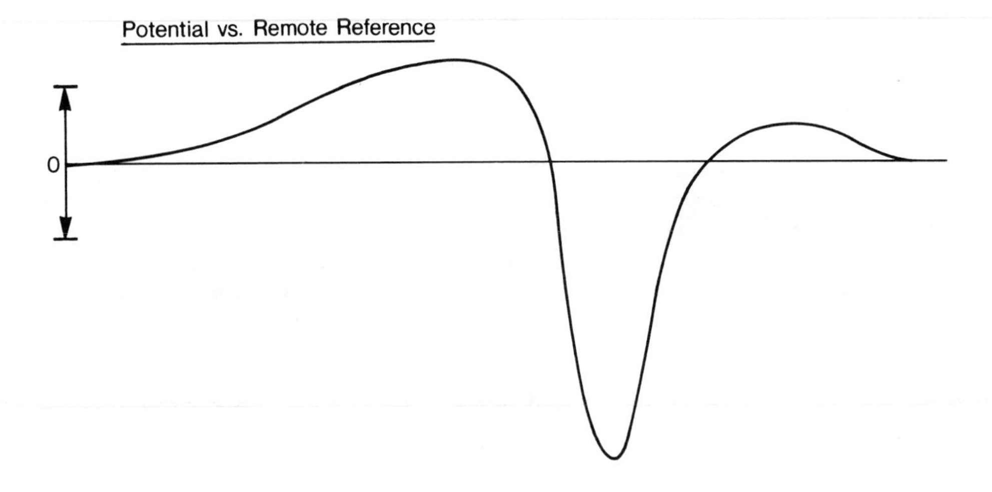

4.2. Propagation of the action potential along a nerve or muscle fiber consists of a series of ion-specific changes in the permeability of the cell membrane to sodium and potassium ions. Along the leading edge of the action potential, the normally high membrane resistance to the flow of sodium ions breaks down as a result of local depolarization of the membrane. Positively charged sodium ions rush into the negatively charged cytoplasm and further depolarize both this actively depolarizing region of the cell and a passively depolarizing front ahead (*right*). This will break down similarly once the depolarization reaches threshold. A similar decrease of resistance to flow for potassium ions occurs with a slight time lag. As the potassium ions are highly concentrated in the cell, their outward diffusion (active repolarization phase) quickly restores the inside-negative polarization of the membrane, allowing the sodium channels to return to their highly resistive state. The ebb and flow of these two positively charged ion elements give rise to extracellularly recordable potential gradients (with respect to a remote, passive reference point). These potential gradients, moving in both time and space, constitute the EMG signal as recorded from active muscle fibers.

brane potential with no net flow of current.

The whole chain of events moves physically down the muscle fiber at about 2 to 5 m per second, so every part of a muscle fiber several centimeters long will experience the action potential within a few milliseconds (fig. 4.2). The electrical signal reaches the sarcomeres deeply within the muscle fiber via the T tubules. The conducted disturbance in the resting membrane potential is the trigger that opens voltage-sensitive channels in the cisternae of the sarcoplasmic reticulum. Because the time course for the diffusion and re-uptake of the calcium lasts tens of milliseconds, the start of the resulting contraction is, for all practical purposes, simultaneous along the length of the muscle fiber. To achieve this condition, some long, nonpinnate muscles divide their long, parallel fascicles into interdigitated, short muscle fibers in series with each other. Branched motoneurons assure simultaneous and mechanically balanced tension over the length of the fascicles.

B. FLOW OF EXTRACELLULAR CURRENT AND ELECTRODES

We can divide the cell membrane of the muscle fiber into several regions that, at any given instant, are experiencing different fluxes of ions. At the instant when the action potential starts to propagate from the point of innervation (near the middle), the major flux is occurring only in this central region and consists entirely of inward flow of positive sodium ions. This makes this region look like a current sink, but where is the source? Electrically the circuit from outside to inside must be completed by a complementary source of current. This current arises in the regions adjacent to the sink, which are having their cations stripped from the outer surface and their anions stripped from the inner surface of the membrane. This capacitive current flow is discharging the resting membrane potential, which normally consists of an accumulation of internal negative charge with respect to outside. An instant later, these adjacent, passively depolarizing regions themselves become sinks for sodium ions, with the passive source lying still further out to the ends of the muscle fiber. There is also an active current source consisting of the outward flow of potassium ions, which is reestablishing the resting membrane potential in the central region that was first to depolarize.

If we position an electrode in the extracellular fluid around such a discharging muscle fiber some distance from the innervation point, it will alternately find itself near a current source (cathode), then a current sink (anode), then a current source again (fig. 4.3). What voltage waveform will be recorded? Obviously this depends on the reference point of the measurement. If the voltage is measured as a comparison between the above electrode and another one in the same position on the other side of the fiber, then the resulting measurement of all this current surge will be zero (see definition of gradients in chapter 2). If the comparison electrode were so far away that none of the action current could flow through it, then there would still be no measurable potential because at least some current must flow through the amplifier to be detected. Of course, the body is a volume conductor, so that some potential will be recorded anywhere, representing that fraction of the action current flow that spreads throughout the volume conductor rather than flowing directly between the source and sink. Suppose that the two electrodes are positioned so that, at one instant, one lies right over a current sink and the other over an active current source. At that instant they will indicate a maximal potential difference, the amplitude of which is the product of the action current times the resistance of the local extracellular fluid through which the current (mostly) flows.

Ideally, we might slide the electrodes along the muscle fiber at exactly the same speed as the disturbance moves, with the pair spaced apart longitudinally by the distance between the source and sink; this would produce the greatest signal for the longest time. More practically, we can set the spacings and orientation of our electrodes optimally and wait for the action potential to come by. What is the optimal spacing? If the current source and sink were actually single points on the membrane at any instant, then this dipole spacing would be the one to pick for our antenna spacing. We can think of each region of membrane as having a sort of center of gravity for its current flow, somewhere near the center of the region, perhaps weighted slightly off center by incompletely turned-on or off membrane channels near the leading and trailing fringes of the region (see fig. 4.2).

How do we measure the distribution of current flow through these rapidly changing regions? From intracellular recordings of the transmembrane action potential at a single point along a fiber with a diameter of 50 μ, we know that the membrane potential takes about 0.5 msec to go from the resting potential state of −80 mV to the action potential peak of about +10 mV. It then takes about another 0.5 msec to get repolarized back to near resting level. During the first phase, current must be flowing in; during the second phase, current is flowing out through the local cell membrane. If this disturbance is known to be moving in tandem down the fiber at say, 5

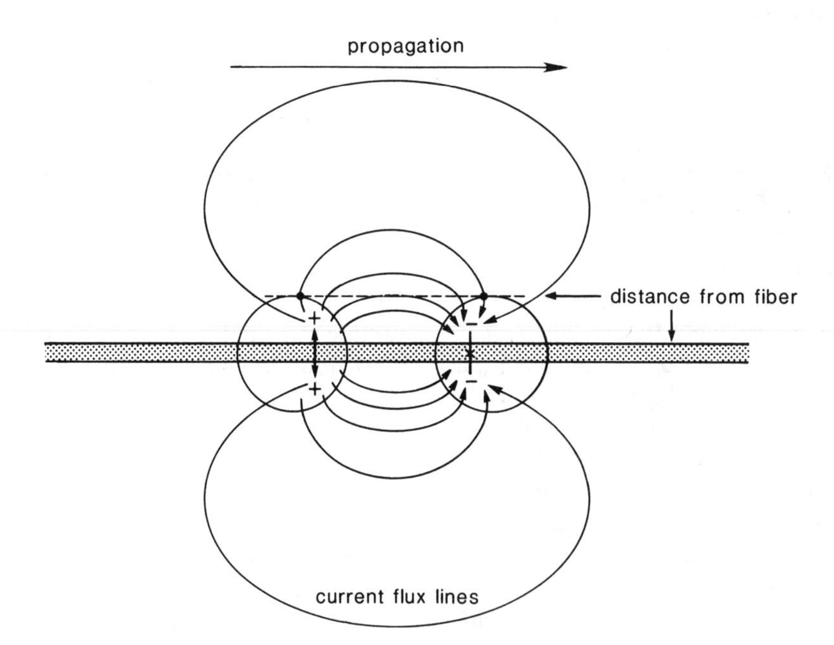

4.3. Flow of ions induced in the extracellular fluids around a muscle fiber generating an action potential decreases rapidly with distance from the surface of the fiber. The effect of distance is relative to the dipole separation of the current source and sink, shown here as a single equivalent point at the weighted center of distributed source and sink regions of the action potential coursing along the fiber.

mm per millisecond (5 m per second), then the length of the patch acting as a source or sink at any instant must be 0.5 msec × 5 mm per millisecond = 2.5 mm. The center-to-center spacing between source and sink will also be 2.5 mm since the repolarizing action follows immediately on the heels of the depolarizing action. Should the velocity be 2 m per second and the time of depolarization still 0.5 msec, the dipole spacing would be 1 mm.

The preceding analysis assumed that we can place the two electrodes right on top of the discharging muscle fiber. Even if the muscle is only 10 mm in diameter, an intramuscular electrode can lie next to only a small percentage of its relatively few fibers; even for these the sheer physical size of a macroelectrode contact will mean that much of its surface area lies some distance from the fiber surface. Action current is still flowing in these regions of the volume conductor, but the distributions make it look as if the dipole is larger, a process called spatial filtering (see Andreassen and Rosenfalck, 1981).

We can imagine the current flow at a distance as occurring because only some of the current can squeeze through the limited amount of rather resistive fluid right up next to the fiber. The remaining current will have to take the long way around, going first in the "wrong" direction and swinging around in a big arc that encompasses the more distant extracellular regions where our electrodes lie. The paths of current flow thus look like the magnetic flux lines surrounding a short bar magnet, the poles of which are separated by a distance equal to that between the centers of the source and sink of our dipole. A set of isopotential lines can then be plotted passing perpendicular to each current flow line, in a plane through the axis of the fiber. These lines will curve around the dipole spacings. The tangency points between these curves (at each end) and lines parallel to the dipole at various distances from the fiber surface (fig. 4.3) will mark the maximum voltage differences at any distance. Such pairs of tangency points tend to be located further apart longitudinally and represent lower potential differences as we get further away from the axis of the dipole.

A first approximation of a fairly reasonable electrode spacing may be obtained by taking the square of the perpendicular distance from the electrodes to the fiber and adding it to the dipole spacing for the fiber calculated above. This is a very simplistic attempt to adapt some lessons from electrical dipole antenna theory to a rather complex milieu. Precision is obviously not critical in this estimation, because there will be an unavoidably large range of distances to all the fibers that contribute to the whole muscle EMG from the various parts of the muscle. What will become critical later is an understanding of how this antenna spacing can be used to improve the selectivity for action potentials from the fibers of one muscle as opposed to those of an adjacent muscle.

The EMG recorded from large surface electrodes (so-called gross electrodes or macroelectrodes) is the result of many of these conducted action potentials in many muscle fibers. Some fibers, innervated by axonal branches of the same motoneuron, discharge more or less simultaneously as part of the motor unit. Other fibers from different motor units are likely to be activated in an overlapping but asynchronous manner. The manner in which these various potentials add and subtract from each other is the subject of the next chapter.

5 Structural and Functional Factors Influencing the EMG Signal

A. TASK OF ANALYSIS

We now have some fairly specific information about the location and timing of the extracellular electrical currents generated by the action potential of a single muscle fiber. However, a single motor unit is usually a composite of several score of such muscle fibers, which may be distributed over a relatively large portion of the muscle and will be more or less synchronized by their single innervating motor axon. Rarely does one wish to record the activity pattern of a single unit by itself (as described in chapter 20). Most studies examine gross EMG signals that tend to be the composite of multiple motor units, usually firing asynchronously. The action currents produced by these tens or hundreds or thousands of muscle fibers all flow within the same volume conductor. As their directions differ, some currents are additive; others cancel each other. Obviously, we can expect the waveform of the EMG signal to be highly complex and random in its details. Contrast this with the situation in cardiac muscle, in which the sequence of action potentials in the entire array of muscle cells spreads smoothly through space from a single point of origin. This produces a large, regular, low-frequency waveform that thus may be recorded as the electrocardiogram.

It may be instructive to build up the EMG waveform from its component parts. Whereas a rigorous theoretical approach to this problem is extremely complex, we can at least consider some of the structural and functional factors that would affect the resulting signal.

First, how do the action potentials of a single motor unit sum? The simplest case would be if the various muscle fibers constituting the unit were all located adjacent to each other, were all the same size, and were all innervated at the same point along their length. In such a case the action potentials of the several fibers would be completely in phase, and the whole would act simply as the geometrical sum of the parts—that is, the extracellular motor unit potential would be a multiple of each single fiber potential (resembling the situation in cardiac muscle). To what extent are these simplifying conditions met in real muscles?

B. DISTRIBUTION OF MUSCLE FIBERS IN SPACE

1. Sampling Problems

The fiber distribution of single motor units has been mapped in a few, mainly mammalian, muscles. It appears that the fibers of such units are loosely distributed over a considerable proportion, perhaps one third to one half, but not all of the cross-sectional area of the muscle. There may be small aggregates here and there in the distribution, but it appears generally random in a normal muscle. This does not apply at all in muscles that have been denervated and then reinnervated by regenerating motor axons. Such muscles usually have many fewer motor units, each of which consists of a much larger number of fibers often densely packed into one or more irregularly shaped areas of distribution. It is important to consider both of these patterns, as an EMG electrode inserted in or placed on the surface of a muscle will necessarily be closer to some of the muscle fibers of a motor unit than to others (fig. 5.1). Dual wire bipolar electrodes tend to record selectively from an elliptical region the major axis of which is related to the electrode separation. If the motor units are each distributed randomly throughout the muscle, then the sampling of any small fraction of its volume would give an unbiased estimate of the whole. However, if the distributions are patchy or show local clumping, the locally common muscle fibers will obviously have a much larger effect on the recorded waveform than their true representation in the muscle as a whole.

2. Segregation by Fiber Type

If all the motor units of a muscle were identical in their mechanical performance and physiological recruitment, it really would not matter whether the EMG signals were to be recorded from only a few local units. This would still represent a fair sampling of the total motor pool. However, muscle fibers come in several distinct physiological types, differing and usually characterized by their cross-sectional area, force output, speed of contraction, fatigability, and physiological order of recruitment. Each motor unit is composed entirely of muscle fibers of one such type. Thus,

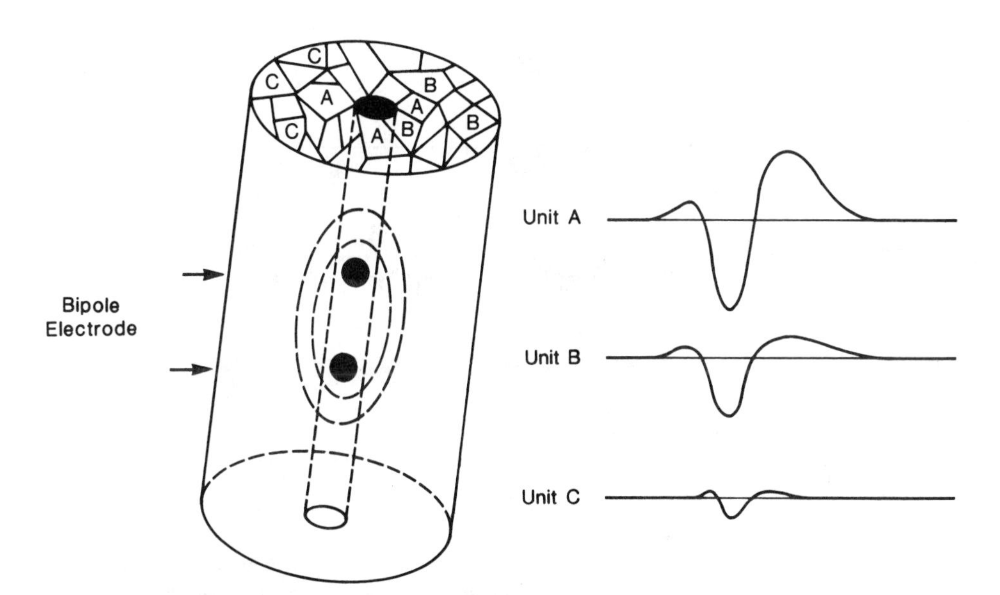

5.1. Effect of fiber distribution on recorded potential. Motor units *A, B,* and *C* each consist of a similar number of similarly sized muscle fibers that are simultaneously activated by the firing of their parent motoneurons. However, the bipole electrode is right in the middle of the *A* fibers, close to one edge of the *B* fiber distribution in the muscle, and remote from the *C* fibers near one surface of the muscle. As a result, unit *A* gives rise to a large EMG with high-frequency components, *B* gives rise to a somewhat smaller EMG signature, and *C* generates very little recordable activity.

a recording from any single unit may give a somewhat distorted picture of the time at which the muscle as a whole is first recruited or reaches peak activation (fig. 5.2). A local recording will not be thus biased if the different types of units are distributed evenly and if their muscle fibers are intermingled. However, the various types of fibers tend to be segregated, even in muscles with simple fusiform shapes. One common pattern in mammals is that the inner core of the muscle tends to be comprised of slow-twitch, low-force, fatigue-resistant fibers, while the more superficial parts have a higher proportion of fast-twitch, high-force, easily fatigued fibers. In the pharyngeal muscles of fishes the fiber types seem to stratify asymmetrically. Slow-twitch fibers may be recruited somewhat selectively for tonic tasks such as postural maintenance, whereas the fast-twitch fibers may be used

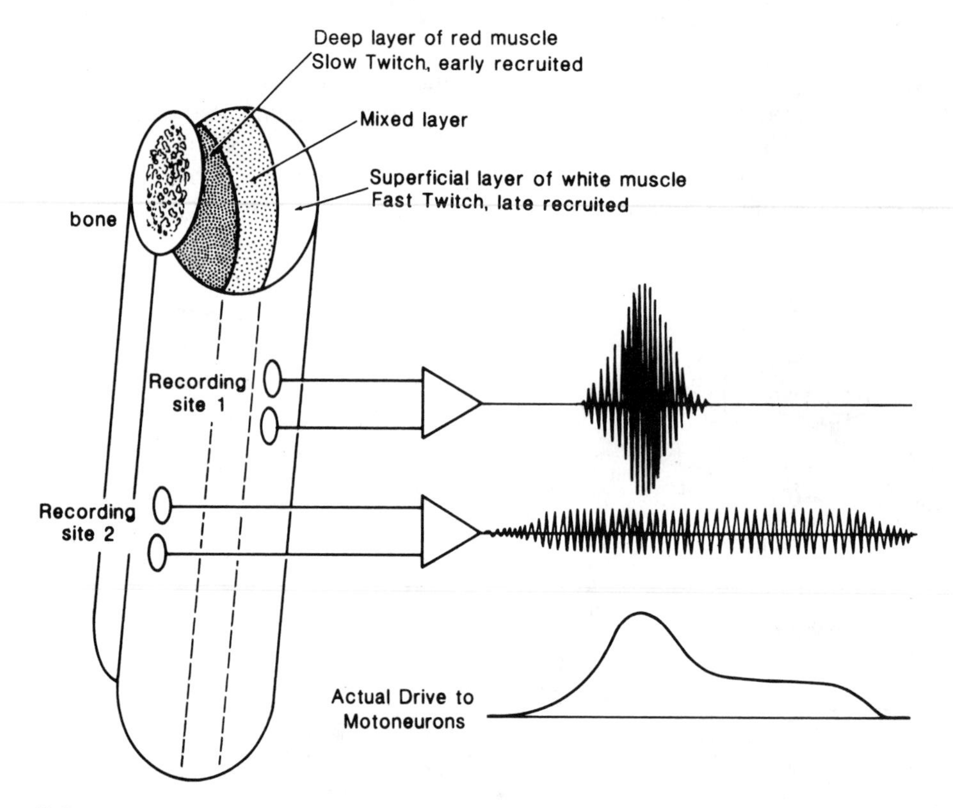

5.2. Effects of fiber type segregation and recruitment. In this hypothetical but realistic muscle, all the motoneurons receive the same neuronal command signal, but those supplying fast-twitch muscle are recruited only when this command becomes greater than their relatively high thresholds. Because these motoneurons are selectively distributed to muscle fibers in the superficial part of the muscle, their activity will result in recorded EMG signals only if these are electrodes suitably disposed to record from this part of the muscle. Conversely, electrodes deep in this muscle will pick up the first muscle activity at the beginning of the command signal, but they may not reflect the increasing command in the middle of the recruitment period should there be no late recruited units in this part of the muscle.

5.3. Normal distribution of innervation from each motoneuron in a muscle results in a highly dispersed mosaic. *Left,* Fiber distributions for four (*top*) and one (*bottom*) motor units. *Right,* Typical EMG. Any local region of the muscle consists of the intermingled fibers of perhaps 20% to 50% of the total motor units present in the muscle. Following denervation, the fastest regenerating motor axons tend to acquire control over large, clumped groups of muscle fibers. Thus, the activation of the regenerated muscle may give rise to an EMG pattern that is dominated by a small number of large, spiky events rather than the normal smooth distribution of small, overlapping electrical signals.

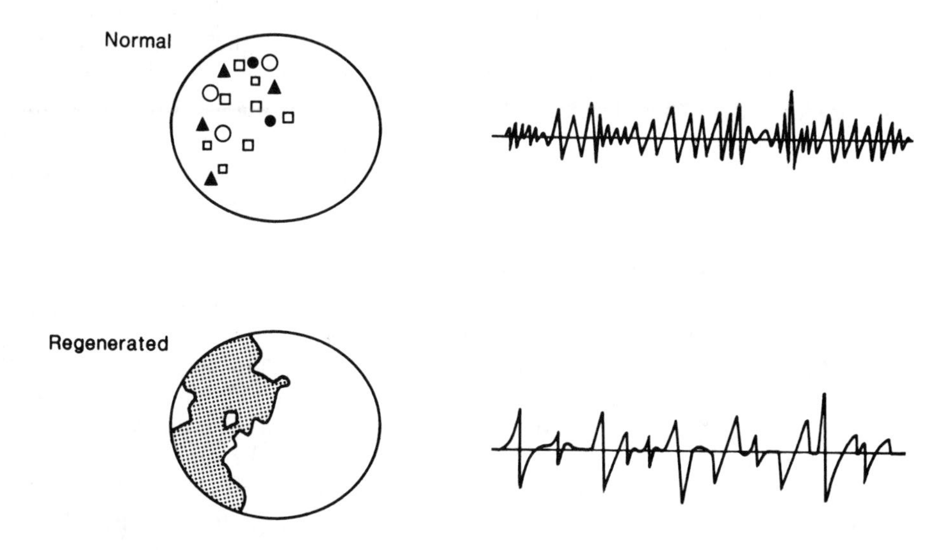

only or exclusively for tasks requiring fast, intermittent bursts of force. An EMG electrode buried in such a complexly stratified muscle might give a rather different impression of the function of a muscle than would an electrode located near or on its surface.

The study of single unit distributions of fibers in muscles is difficult and tedious (usually done by glycogen depletion and histological examination), not gladly undertaken just to improve EMG electrode design. (However, it has been done on far too few muscles and animals and may be of interest in and of itself.) From the above-mentioned arguments, we can see that the most important factors are actually the nonhomogeneity of fiber-type distributions in mixed muscles and the possibility of extremely local clumping, most often seen in reinnervated muscle. The former may be apparent from simple visual inspection of living tissue, because the slow-twitch units are the classically identified "red" muscle and the fast-twitch units are the classically identified "white" muscle. Relatively simple histochemical stains are also available for fiber typing whole muscle (appendix 1).

The local clumping problem is often apparent from the EMG records themselves. If the EMG record is examined at high time resolution (say 10 msec per division on an oscilloscope), it may not show a smoothly graded distribution of large and small wiggles, but rather one or more stereotyped waveforms repeating at fairly regular intervals (fig. 5.3). These are likely to be single-unit potentials, which should not be recordable from gross EMG electrodes in a muscle in which a large number of units are active simultaneously. If the record consists of such discrete units, then it is likely that there has been considerable denervation and perhaps subsequent reinnervation. If this was not a part of the experimental paradigm, the investigator is well advised to consider the possibility that the nerve or muscle was seriously damaged by or during the implantation of the recording electrodes. Much will be said later about the design of electrodes to avoid such problems or to compensate for them.

3. Segregation by Task Group

One additional nonhomogeneity in muscle fiber distribution must be considered; however, it should be considered as a possible finding rather than an artifact. It has recently become apparent that a given muscle may actually contain two or more rather separate pools of motor units. This may obviously be suspected in muscles with distributed attachments, where one part of the muscle may have a distinctly different mechanical action than another. However, this situation may be present even for muscles with single discrete points of origin and insertion, all muscle fibers of which have the same mechanical action. The situation seems to arise in muscles that are polyarticular or that are needed during kinematically different phases of the same behavior. Their gross EMG pattern may then show two phases of electrical activity. The electrical activity of these phases may appear similar, but different motor units may actually contribute to each phase (fig. 5.4). If the anatomical distribution of fibers from one such task group of units within the muscle does not completely overlap the distribution of the other task group, then the location of recording electrodes may influence the relative magnitudes of the recorded phases.

Whenever the distribution of fiber lengths seems unusually scattered or the physical action or EMG pattern recorded from a muscle appears complex, one must attempt to control for the possibility that the muscle is nonhomogeneous. The simplest and most direct solution is to implant pairs of electrodes in two or more different regions of the muscle and then to compare their signals in order to see whether they are similar. This will not rule out the possibility of functionally segregated motor pools; however, it will at least reveal physically segregated pools and may explain a previously noted variability in the apparent function of the muscle.

C. FIBER SIZE

The diameters of the fibers differ substantially among many muscles and are generally not random. The diameters of slow-twitch fibers tend to be somewhat smaller than those of fast-twitch fibers, which often constitute units with high force output. As muscle fibers are electrically similar to unmyelinated nerve fibers, we can use some of the well-established relationships between nerve fiber diameter and electrical properties of the action potential. Generally, conduction velocity is proportional to fiber diameter and action current is proportional to the square of fiber diameter. As discussed in the previous chapter, the wavelength or dipole separation is produced by the conduction velocity times the duration of the conductivity disturbances that make up the action potential. The timing of the openings and closings of the membrane channels is relatively constant. Therefore, a fiber twice the diameter of another will have approximately twice its dipole separation (fig. 5.5). This, in turn, means that it is going to be best recorded by an electrode with twice the interelectrode spacing.

We really want to know how this effect will bias the recording of a given electrode for units composed of fibers with different diameters. For this we must know something about the current densities being contributed by the various fibers; these currents will cause gradients of potential in the volume-conductive tissues between and around the poles of the dipole. Fibers of twice the diameter will have perhaps four times the action current of their slimmer counterparts. However, it is even more significant that the small-force, slow-twitch units tend to have many fewer muscle fibers per unit. The force output during a single twitch may be only 1% of the twitch force produced by a fast unit; this difference is al-

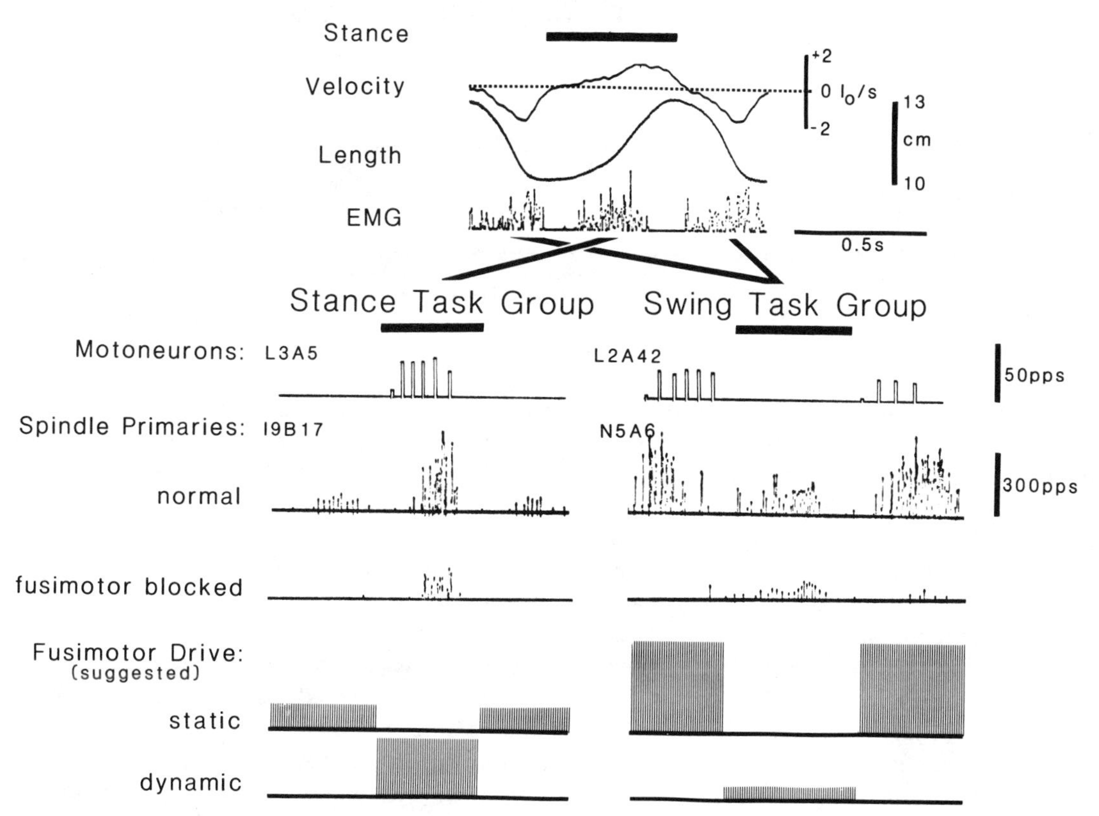

5.4. EMG of the anterior sartorius muscle of the cat shows two bursts during each walking cycle. The first occurs during stance simultaneous with lengthening of the muscle (*top, middle of each trace*) and the second during swing as the muscle is shortening(*top, each end of each trace*). However, recordings (made in the ventral spinal root) from the axons of individual motoneurons reveal that this muscle contains two independent recruitment groups, one for each EMG burst (*bottom, first trace of each group*). Recordings from muscle spindle afferents (*bottom, second trace of each group*) also suggest two different patterns of fusimotor recruitment (gamma motoneurons projecting only to the intrafusal muscle fibers that control the sensitivity of these stretch receptors). (From Loeb, 1984, copyright 1984, Macmillan Publishing Company. Reprinted by permission of the publisher.)

5.5. Large-diameter fiber conducts faster than a small-diameter fiber and has a longer wavelength and associated dipole spacing. This is because the spread of passive depolarization in front of the actively depolarizing region depends on the ratio between the amount of membrane capacitance to be depolarized (proportional to fiber diameter) and the resistance to current flow along the fiber length (inversely proportional to the square of the fiber diameter).

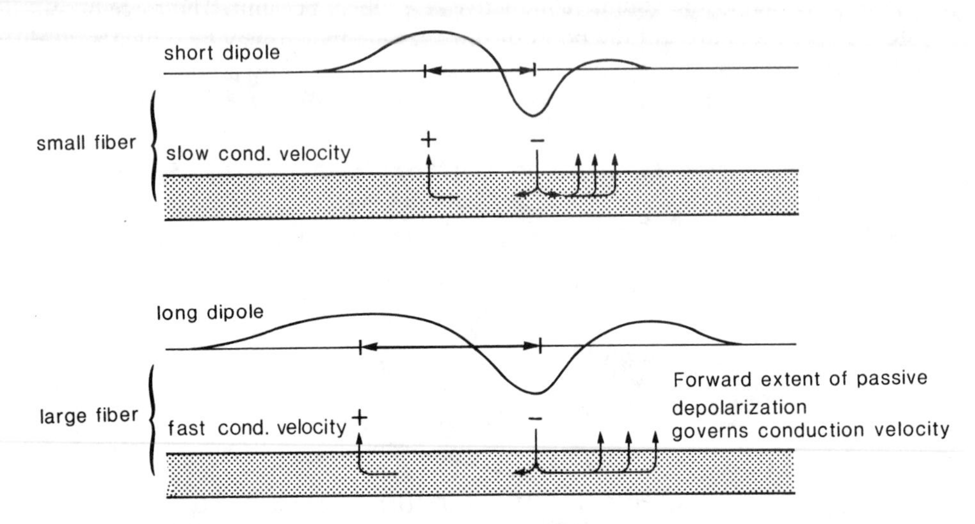

most completely due to their smaller total cross-sectional area with small numbers of smaller fibers. As cross-sectional area, force, and action current are all related to the square of fiber diameter, we have some reason to expect that the total action current contributed by a motor unit is reasonably correlated with its total force output. This is a very useful relationship that bypasses the heterogeneity in fiber diameter.

Thus far we have been talking about action currents, not about gradients in extracellular potential. We can derive the latter from the former only when we take into account the dipole separation. Now, the total extracellular voltage gradient, from its highest point to its lowest point, will be proportional to this action current, but we really want to know what percentage of this gradient will be spanned by a bipolar electrode with some fixed interelectrode spacing. This is related to the ratio between the electrode spacing and the dipole spacing and also to the perpendicular distance to the fiber(s). We know that for optimal recording the electrode spacing should be equal to the virtual dipole spacing. (We say virtual dipole because, as previously discussed, the apparent dipole spacing increases with perpendicular distance from dipole to recording site.)

Obviously, a mixed muscle with large- and small-diameter fibers having longer and shorter dipoles, respectively, will lack a single optimum. If the interelectrode spacing is on the short side, then it will be closer to optimal for the small-diameter fibers (slow units) and further from the optimum for the large-diameter fibers (which are still likely to generate larger unit records but no longer proportionately so). If the interelectrode space is relatively long, then the record may be disproportionately biased in favor of the large-diameter fibers (fast units). Another factor that comes into play is that large interelectrode spacings will be closer to optimum for more distant fibers than are small, interelectrode spacings. This further aggravates the problem of cross-talk from muscle fibers in nearby portions of adjacent muscles. Under such conditions cross-talk is particularly likely to derive from distant fibers of large diameter that may be active intermittently so that it is difficult to attribute their signal correctly. In the next chapter we shall describe ways of arriving at a reasonable compromise regarding this thorny problem.

D. INNERVATION POINT

We have still to consider the question of innervation point. We feel intuitively that the electrodes should be somewhere near the center of the muscle, but is this always true, and why should it be so? Let us examine some extreme cases. If a motor unit is made up entirely of muscle fibers of uniform size, then the extracellular action currents will be in phase and will add constructively only as long as their action potentials traverse the lengths of the fibers together. If they are innervated at different points along their lengths, then the various sources and sinks traveling up and down each fiber from the point of innervation will variously augment and cancel each other, reducing the amplitude of the extracellular potential gradient. Fortunately, muscle fibers of a given unit and even across units tend to receive their innervation in spatially ordered and restricted regions.

Now, if our electrode symmetrically straddled the innervation point, an even worse problem would occur! As the action potentials propagate symmetrically in either direction, the two poles of our bipolar electrode could, theoretically, be looking at exactly the same time course of extracellular potential, and the differential record would cancel to zero. This effect has been used to locate the motor innervation point, by sliding a bipolar electrode along the muscle. Fortunately, nature is not so neat, nor are electrode placements so precise, so there is little chance of seriously degrading the record, at least for most of the fibers in the vicinity. However, this same lack of uniformity can seriously degrade recordings from the ends of a muscle. If all the action potentials in the various fibers start off at the same instant, near the same place, then they should arrive at the ends of the fibers at the same time if the velocities are the same. However, fiber diameters, and consequently the conduction velocities, vary by at least 10%, even within a given motor unit. Now, if the disturbance of the action potential lasts only about 1 msec, the difference in travel time for a 20 mm distance at 2.0 versus 2.2 mm per millisecond is similar. Obviously, the sources and sinks in the various fibers of the unit will be completely out of phase by the time they reach the ends of the muscle (fig. 5.6).

This distance effect accounts for the long, polyphasic "signature" of single unit EMG potentials. Furthermore, the waveform tends to vary as the unit producing the signature changes its length. The volume of the fiber remains constant, so that a stretched fiber obviously will be narrower and a shortened one thicker. As the conduction velocity of the fibers reflects their diameter and narrower fibers will conduct more slowly than thicker ones, the temporal scatter of their action potentials at the recording electrode will increase further. All of this suggests that recording electrodes should be placed reasonably close to the innervation point in order to obtain signals of maximal and constant amplitude.

We have been a bit cavalier about the geometry of fibers in our hypothetical muscle. The textbook example of a symmetrical fusi-

5.6. Formation of polyphasic unit waveforms. Even if two muscle fibers are innervated by the same motoneuron at the same place along the longitudinal extent, slight differences in their conduction velocity will disperse their action potentials in time and space as they propagate down the fibers. Their action currents sum in the extracellular space, giving rise to complex, polyphasic potentials.

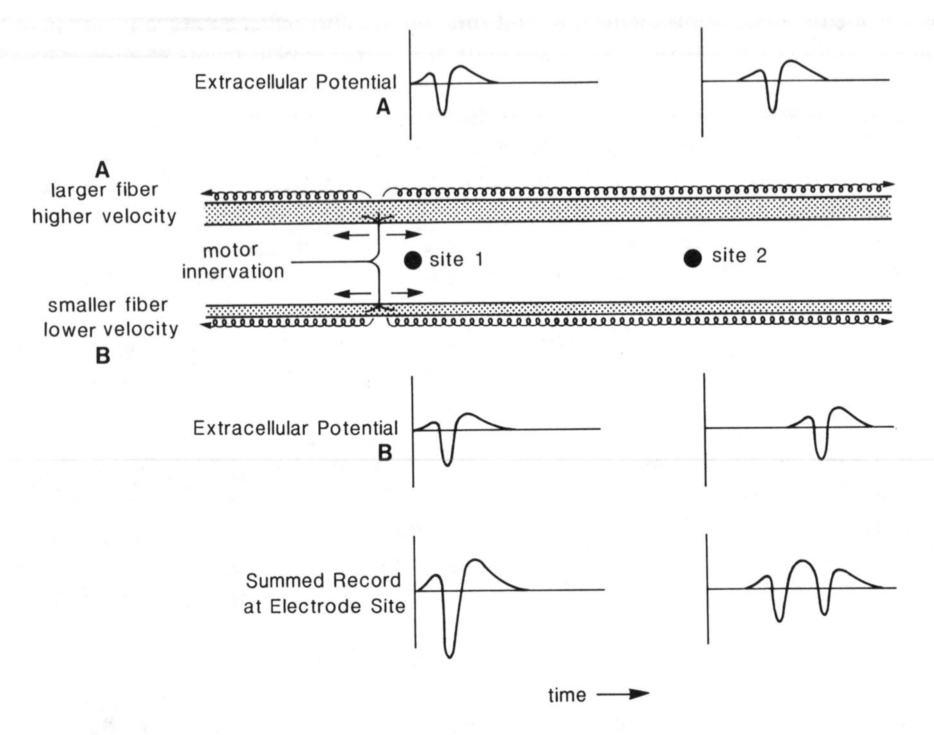

form muscle with individual muscle fibers running in parallel over the entire length is more the exception than the rule. Even muscles with an outside profile that is more or less symmetrically fusiform may have an internal arrangement of relatively short oblique or pinnate fibers. This arrangement clouds the question of what is meant by the "middle" and the "end." Usually, the innervation band crosses a muscle near the midpoint of its muscle fibers, so it may actually run in an arc or oblique line across the muscle. (See appendix 1 for methods of staining.)

As previously noted, there are even more exceptions to this rule of midpoint innervation; for instance, in the sartorius and tenuissimus muscles of cats, the innervation points are scattered in patches up and down the length of the muscle, representing branches of the same motoneurons (fig. 5.7). Or consider the semitendinosus muscle, which looks like a single muscle from the outside but which actually has two bands of innervation serving completely separate fibers that are inserted into each other at a fascial band near the anatomical midportion; hence, this is a pair of muscles in series. The mammalian digastric muscle sometimes has two bellies in series, although these may have fused centrally. However, the innervation generally continues to reach this muscle by quite distinct nerves.

Such anomalous situations must be kept in mind when the EMG patterns recorded from a muscle are suspiciously small or inconsistent. Sometimes they can be resolved by careful dissection of fresh or preserved specimens; occasionally they may warrant detailed study with specialized techniques, such as histochemical labeling for acetylcholinesterase activity. Again, such problems can be investigated and even corrected for by specialized designs of recording electrodes, as covered in the next chapter.

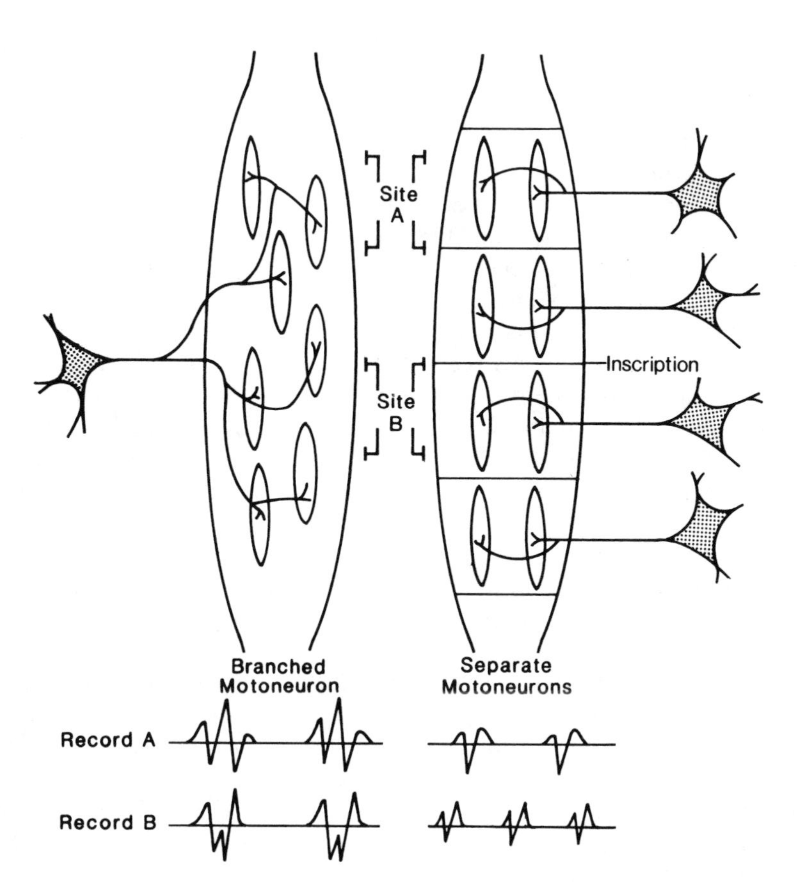

5.7. Because of the inherently slow conduction velocity of muscle fibers, long muscles must be assembled from shorter fibers in some series (or staggered) parallel (pinnate) configuration. In nonpinnate, serially arranged muscles, the innervation of the entire length of the muscle may be accomplished by highly branched motor units and/or separate populations of motor units confined to parts of the muscle separated by inscriptions. To prevent mechanical instability, the level of activation (number of active muscle fibers) must be always approximately equal at any point along the muscle, but the two different innervation patterns will differ in EMG details as shown in the sample records.

6 The Electrode as an Antenna

A. CONCEPTS

We introduced the notion of antennas when it was noted that the muscle fibers radiate voltage fields into the volume-conductive space of the fluids surrounding them. As these muscle fibers act as biological signal generators, they have been described in terms of their field strength and their dipole spacing and orientation, these being the classic engineering descriptors for transmission antennas. We have seen how much of the complexity in these fields (and of the related EMG recordings) arises from the multiplicity of signal sources and their physical proximity. We now address the problem of designing the EMG electrodes, which we may best conceptualize as receiving antennas.

Engineers may quarrel (some have) with this fast-and-loose use of the word "antenna." When not referring to the sensory structures on the heads of many invertebrates, this word is properly used specifically for electrical devices that emit or detect oscillating magnetic fields, as do radio and television transmitters. Their magnetic fields are necessarily accompanied by electrostatic fields such as the fields generated by muscle fibers causing current flow in a volume conductor. However, the magnetic fields are oriented perpendicular to the electrostatic fields. Therefore, the orientation of a radio or television antenna for optimal reception is unlike that of recording electrodes with respect to their signal sources. However, we have adopted the term "antenna" because it brings to mind all of the important concepts that govern the recording of EMG signals—namely the notions of gradients, potential differences, references, wavelength, orientation of signal sources and detectors, noise, interference, and cross-talk.

Let us briefly review the series of bioelectrical events that give rise to a recorded voltage (fig. 6.1).

First, changing conductivities in the membranes cause action currents to flow across the membranes and in the extracellular fluids around active cells.

Second, the extracellular currents cause potential gradients as they flow through the resistive extracellular fluids.

Third, the changing potential gradients give rise to electrical currents in the electrode leads by capacitive conductance across the metal/electrolyte interface of the electrode contacts (see chapter 7, section B2a). The currents actually flow through the high-impedance circuits of the amplifier input stage.

Fourth, the amplifier converts these weak currents into large output voltages.

We should next define a number of terms that are more traditionally used to describe EMG electrodes and relate these to the antenna concept (fig. 6.2).

B. CLASSES OF ELECTRODE GEOMETRY

1. Monopolar Electrodes

As we have already noted, there is no such thing as a pure voltage, only differences in voltage (potential) between two points. Thus, a monopolar electrode would seem to be a contradiction in terms. Actually, monopolar electrodes represent single recording points, the voltage of which is established relative to a diffuse and remote reference point. The remote reference point is variously referred to as a reference electrode, indifferent electrode,

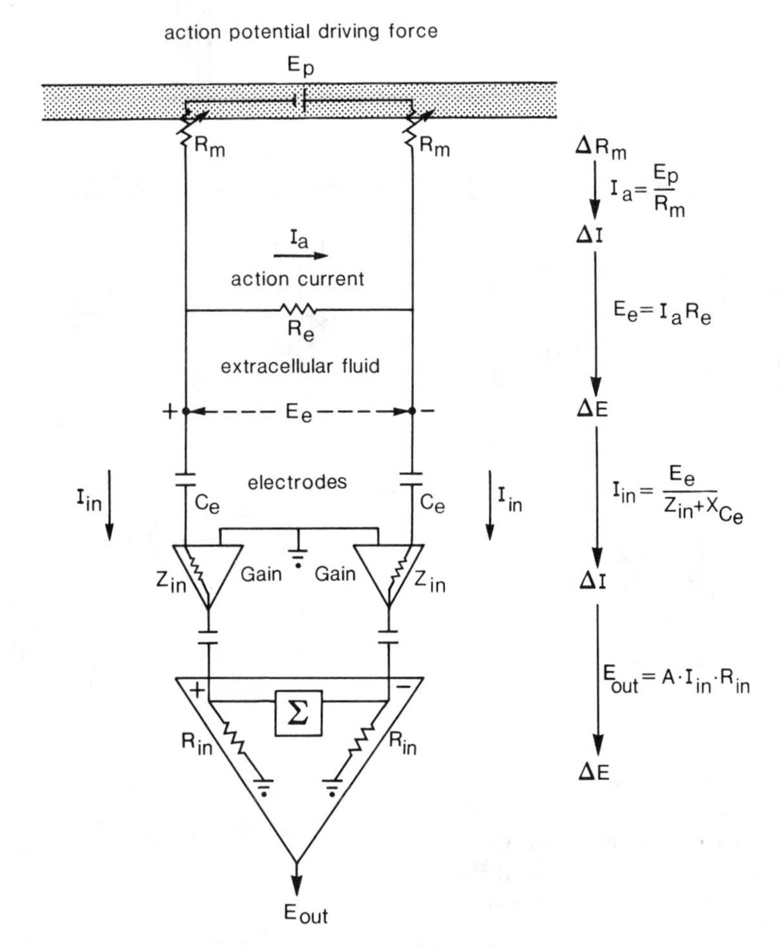

6.1. Ohm's law relationships in recording EMG. The final signal recorded as the output of an EMG amplifier is the result of a series of transformations among the parameters of voltage, impedance, and current in the simple equations known as OHM's law: E = IR. The action potential itself arises from the current flow (I_a) induced by the resting membrane potential (E_p) acting against a rapidly changing membrane resistance (R_m). The extracellular portion of this current flow (I_a) acts against a small but non-zero resistance of extracellular fluids (R_e) to produce the extracellular action potential (E_e). This potential induces a tiny but detectable current flow (I_{in}) in the series circuit made up of the reactive impedance of the electrode surface (X_{Ce}) and the input impedance of the amplifier (Z_{in}). After amplification in the first stage, the currents from each of the bipolar contacts produce voltage drops across input resistors (R_{in}) in the summing amplifier, where their difference is calculated and amplified by gain factor (A) to produce (E_{out}).

6.2. Classes of electrodes and their domains. The various electrode configurations can all be grouped into a small number of fundamentally distinct categories, each of which tends to record potentials the gradients of which are confined within the region indicated (*dotted lines*).

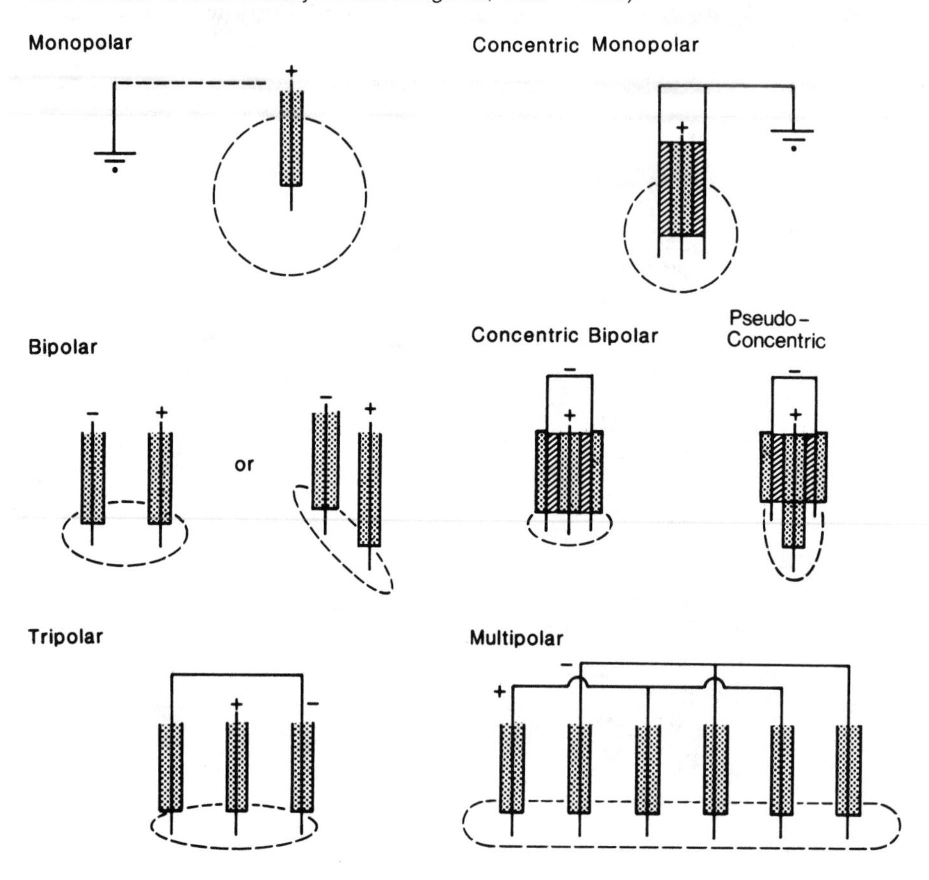

common electrode, or ground (although it may not actually be electrically connected directly to the physical ground of the earth). Thus, the animal into which a monopolar electrode is to be placed must always bear another electrode or other electrical connection. Generally, a monopolar recording should be made with a single-ended rather than a differential amplifier (after all, there is nothing except the indifferent or ground electrode to hook into the negative input of a differential amplifier).

2. Bipolar Electrodes

A bipolar electrode generally has two electrical contacts, each with an independent electrical lead. It is (or should be) designed to measure the voltage difference between two specifically defined points in space. The exact location, spacing, and orientation of those points are critical because they dictate which gradients of local potential the points will span and how completely they will span them (fig. 6.3). A bipolar electrode must be used with a differential amplifier, which treats each input equally in terms of its sensitivity and which produces as its output an accurate reading of the potential difference between them. Because such amplifiers actually preprocess each input separately before performing the subtraction, there must be a third contact, similar to the reference discussed above.

There are two commonly used general forms of bipolar electrode. In the dipole form the two contacts are similar to each other and are separated by some distance along a line that defines their dipole moment. One end of the line thus represents the positive input to the amplifier, the other the negative. If the potential gradient actually produces a positive voltage difference from the minus electrode to the plus electrode, the output of the amplifier will be a negative voltage (with respect to the ground of its power supply).

The concentric bipolar electrode employs a pointlike recording contact surrounded by a ringlike recording contact. Usually, the center contact is connected to the positive input of a differential amplifier and the circumferential ring to its negative input. Such a configuration actually measures the potential difference between the central point and an average of all the potentials that surround it at some fixed distance. This geometrical pattern is most commonly and usefully employed as a microelectrode, to detect very localized fields near single fibers. The main advantage of concentric bipolar electrodes is that even very small ones are easy to fabricate and have great physical strength. However, they are generally inferior electrically in terms both of sensitivity and of common mode rejection of extraneous electrical noise (discussed in chapter 12).

The term "concentric bipolar" is often misused. For example, it should not be used to describe electrodes in which the central contact is displaced significantly away from the plane of the concentric contact. Once the distance away from the plane becomes greater than the radius of the ring, the electrode begins to function much more like a dipole, the axis of which is along the length of this projection.

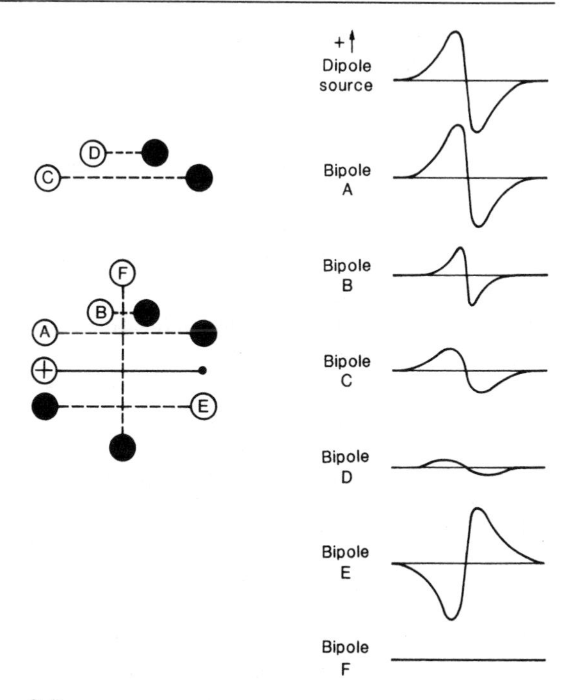

6.3. Effects of various bipolar antennas on records from dipole sources. Important factors include the spacing between electrode poles, the orientation of the bipolar electrode axis with respect to the dipole source axis, and the spacing between the electrode and the source. *Circles* refer to the corresponding negative poles of the lettered bipolar inputs.

Conversely, the term should not be used to refer to electrodes in which the concentric contact is actually an extended length of bare metal sheath, as when such electrodes are made by passing a central contact down the shaft of a hypodermic needle that is not insulated along its extended outside surface. In such an electrode the outer surface actually functions as a distributed common electrode, effectively "grounding" the side of the differential amplifier to which it is connected. Because a metal/electrolyte interface necessarily has a high impedance, it cannot significantly change the potential gradients actually present in the tissues by shorting them to ground. Thus, whereas this configuration forms a very effective electrical shield such as for the exclusion of radiated power line hum, actually it only has the recording antenna properties of a monopolar electrode.

3. Tripolar Electrodes

The tripolar configuration has become quite popular for the recording of nerve potentials and has some utility in the exclusion of certain kinds of biological noise. A typical tripolar configuration consists of three electrode contacts equally spaced along a straight line. The central contact of such an electrode is best connected to one active input (usually the positive one) of a differential amplifier, whereas both of the flanking contacts are connected to the other active input (usually the negative one). This configuration again requires a remote reference electrode.

Tripolar electrodes are specifically designed to reject voltage gradients that have a constant slope along the electrode axis. The negative input to the amplifier will be the average of the potentials at the two flanking electrodes—an average that should be exactly equal to the potential that the constant gradient produces at the central contact. If the source of the voltage gradient is sufficiently remote from all of the contacts, the gradient will have the requisite constant slope in the vicinity of the tripolar contacts. However, should a source lie at or beneath any of the contacts, particularly the central one, then its gradients will be maximally different in the two implicit dipole segments on either side of the center (in fact, they will have opposite sign). Thus, they will "subtract constructively."

Tripolar electrodes are most commonly applied to record action potentials from nerve fibers confined in a tubular, nonconductive sheath. This sheath can be a rubber cuff that is implanted around the nerve in the body, or it may be generated by elevation of a section of the nerve into a pool of mineral oil or other nonconductive liquid. Assuming that the fluid column in the sheath has a uniform diameter and uniform resistivity along its length, all of the currents generated by electrical signal sources outside of the nerve (and outside of the sheath) will produce a voltage gradient of constant slope within the sheath. Although it is rarely possible to surround an entire muscle by such a sheath (e.g., tenuissimus muscle of cats), the voltage gradients from distant sources such as the heart or electrical stimulators may well have nearly constant slope in the vicinity of the muscle, particularly if the muscle lies deep in the body, away from the surface of the skin. Of course, if one knows the direction of the voltage gradient, one can cancel it out by simply orienting the bipolar electrode with its axis perpendicular to the gradient. However, this orientation may be a poor one for the desired voltage gradient one wishes to record. The tripolar configuration provides an antenna that not only is selectively sensitive to local current sources (as are all bipolar electrodes) but also actively cancels out the gradients that are constant rather than locally changing. This is mathematically equivalent to calculating the second spatial derivative of the potentials.

Some caveats are in order. First, the electrical average between the potentials picked up by the two flanking contacts will be an accurate mean only if the contact impedances of these two surfaces are very well matched. Second, the central contact will only see a mean potential if it is exactly equidistant between the flankers. Third, the common mode rejection of an amplifier, which allows these two similar signals to be precisely subtracted, depends on a similar source impedance from the two inputs. However, one input comes from a single contact, perhaps in the middle of a resistive encasing sheath, and the other input derives from two parallel contacts (each with its own impedance), which both lie near the ends of that sheath. Consequently, the asymmetry may be considerable unless special steps are taken.

4. Multipolar Electrodes

As noted earlier, dipolar electrodes have the propensity to record selectively from sources located with an ellipsoid, the major and minor radii of which reflect the physical separation of the bipolar contacts. Also, biological dipole sources produce their steepest voltage gradients close to their axes, and these gradients are most fully spanned if the sampling points, spaced along this axis, are separated by a distance equal to the effective dipole moment spacing between source and sink of the action current. However, these source dipole moments are relatively short only close to the fiber. Thus, a similarly short bipolar electrode oriented parallel and close to a muscle fiber will record very selectively from that muscle fiber and very poorly from muscle fibers and other signal sources located further away. This makes it possible to obtain a very clean picture of the electrical activity of a very small volume of muscle and to eliminate cross-talk almost completely.

However, whenever just a few motor units are being sampled, such high-impedance electrodes will give a very noisy picture of the activity envelope of the muscle, because the recorded EMG signal will consist of only a few brief action potentials occurring at relatively low frequency. In contrast, scores of motor units may be active asynchronously in the whole muscle, and these combine to produce a smooth force output. What is worse, the few motor units being recorded may not be a valid sampling of the diverse types of muscle fibers within the muscle, either because of the statistics of small samples or because of local concentrations of distinct fiber types (as discussed in the previous chapter).

One solution would be to distribute lots of small bipolar electrodes throughout the muscle and to sum their recorded signals to provide a mean activity profile. Of course, this kind of multiple bipolar electrode would require two electrical leads for each electrode site, plus a differential amplifier for each, plus a summing amplifier with a large number of inputs. However, our discussion of the tripolar electrode indicated that we can convert the potentials recorded at two separate places to a mean by simply connecting the two recording contacts before connecting them to the input of an amplifier. The multipolar electrode extends this trick to include a large number of contacts for each input. The simplest way to achieve such a configuration is to cut multiple exposure windows into the insulation along a pair of parallel insulated wires (see chapter 10, section C6).

Caveats apply, similar to those discussed for the tripolar electrode only multiplied in their importance. The electrical impedances of the exposed contact areas (essentially the surface area) must be quite similar so that the signals picked up at each local bipole will be equally weighted and so that the differential amplifier can produce a high common mode rejection.

One advantage of having a separate lead for each of many contacts is that various combinations can be tried selectively to enhance the recording of certain units, based on idiosyncracies in their fiber locations. The efficacy of this approach would reflect the heterogeneity of the mosaic distribution of fibers making up each unit and is usually severely limited by the volume-conductive properties of the tissue.

5. Barrier or "Patch" Electrodes

All previously mentioned electrodes involve different antenna schemes for the numbers and orientations of the electrode contacts. Patch electrodes are usually typical bipolar electrodes that are intimately connected to a dielectrical (nonconductive) barrier that redirects the flow of current through the tissue in its vicinity (fig. 6.4).

Most of the electrode configurations deal with the extracellular potential gradients as a given, to be detected passively by the antenna (see chapter 6, section A, step 3). The patch electrode intervenes (chapter 6, section A, step 2), by redirecting some of the flow of extracellular current. Consequently, desirable action currents generate larger potential gradients in the vicinity of the electrode contacts, whereas the gradients generated by undesirable action currents become smaller. This is possible because electrical currents, like flowing water, take the path of least resistance.

Suppose that two thin muscles lie adjacent to each other (e.g., the mammalian anterior digastric muscle between the platysma and mylohyoid muscles) and that it is of interest

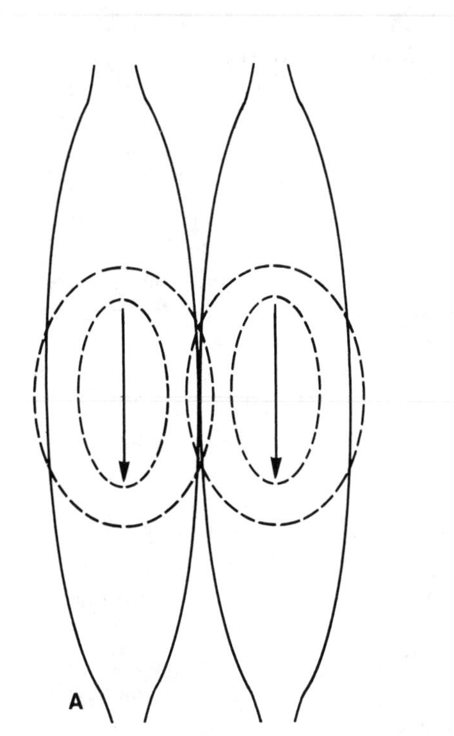

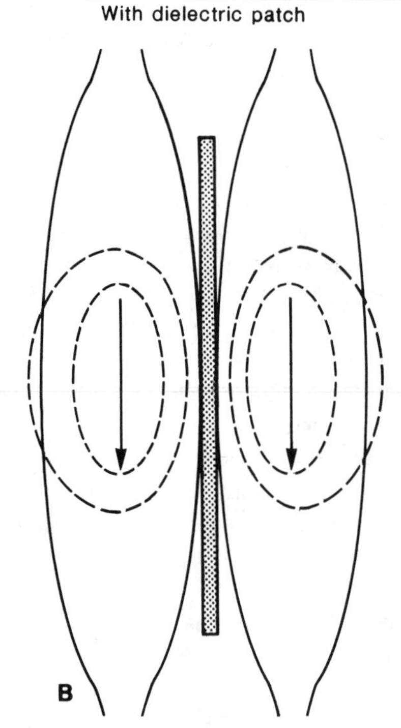

6.4. Mixing of extracellular fields in two adjacent muscles. The insertion of a large, nonconductive barrier such as the patch electrode substrate can cause a flow redistribution of extracellular action current and the resulting recordable fields.

to record a clean EMG signal from one without undue cross-talk from the other. If a bipolar electrode with an intercontact spacing of, say, 5 mm is placed longitudinally within or alongside one muscle, it will record this muscle's EMG well. As it is less than 5 mm from fibers of the other muscle, it will also record the activity of those fibers (as discussed later in this chapter). Whereas we cannot change the physical distance between the fibers of the two muscles, we can change the relative length of the electrical path from the fibers of each to the recording electrode by placing a flexible patch of a nonconductive material such as silicone rubber between the muscles. This patch confines the action currents generated by the fibers of each muscle and keeps them from spreading into the adjacent one. If the electrodes are placed on one side of the patch, most of the flux of current within the extracellular media on the contact side will be generated by activity in the muscle fibers on that side.

Conversely, the action currents arising from the muscle fibers on the opposite side of such a patch can only cause potential gradients in the vicinity of the recording electrode by long, circuitous, resistive paths. As the current flowing through parallel circuits is inversely proportional to their resistance, the opposite fibers will contribute only weak potential gradients at the contacts and will be recorded only as small signals, if at all. Obviously, if the patch were infinite in length and breadth, the isolation would be complete. In the practical case, the patch cannot and need not be more than about three times the electrode spacing to produce surprisingly good results.

One fringe benefit of the patch electrode, which will be discussed at greater length in chapter 10, is that it may achieve this highly selective performance without invasion of the muscle. The recording contacts actually need not be located in either of the muscles, but rather in the fascial plane between them, at a site that usually is easy to dissect. The barrier effectively causes the patch electrode to operate as a unidirectional surface antenna. Such an arrangement can be used to great advantage whenever one needs to record from thin and superficial sheets of muscle that overlie bulky muscles. The patch is then oriented under the sheet and the electrode is pointed toward the skin, where there are no sources of cross-talk. However, a patch electrode would be useful for recording from, say, a thin muscle sandwiched between two sources of cross-talk only if one placed a second patch on its opposite side. This would control the spread of action currents from the muscle on the far side of the recording electrode as well. Another version of patch electrode incorporates two sets of contacts, one on each side of a single patch. Not only will this minimize trauma and foreign bodies by providing two channels of data from a single implanted device, but it can provide a very valuable control for the actual magnitude of the cross-talk problem (see chapter 8).

6. Belly Tendon Electrodes

The belly tendon electrode is a curiosity from the field of clinical EMG—an interesting hybrid of the monopolar and bipolar approaches to electrode geometry (fig. 6.5). It deserves special mention, as it is frequently referred to in the literature and has useful, if limited, properties. The name refers to the fact that one relatively small and well-localized contact is placed in or over the midpoint of the belly of the muscle, near the presumed innervation point. The second electrode is placed over the tendon of the same muscle, usually just distal to the termination of the contractile elements. Signals from the two contacts are amplified differentially with respect to a remote ground.

Obviously, the large spacing between the

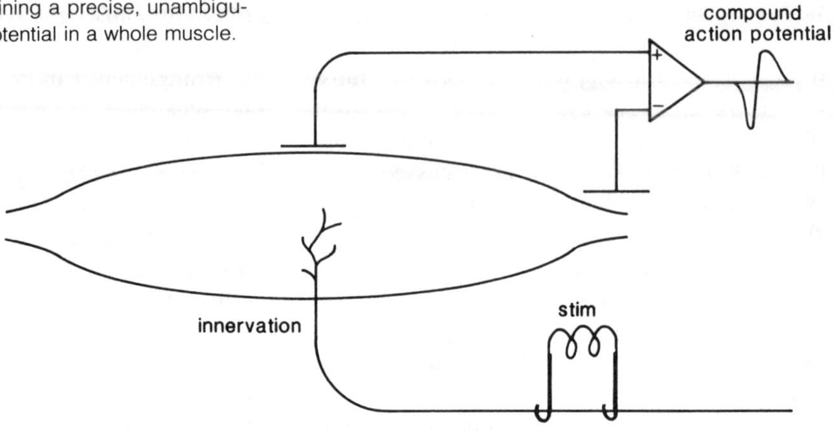

6.5. Belly tendon electrode for latency determination. The stimulator is hooked over the nerve. This represents a useful technique for obtaining a precise, unambiguous onset of the evoked potential in a whole muscle.

recording contacts and the location of one of them over electrically inactive tissue make the recording monopolar with respect to the muscle. The major advantage of the technique is that the arrival of a motoneuron volley at the end plate zone will result in a well-defined onset of a leading negative potential on the belly electrode, with virtually no active contribution from the tendon electrode to complicate the waveform. The reason for using a second contact with differential amplification is that it provides at least some common mode rejection of distant sources of electrical noise and artifact, such as electrocardiogram signals (ECG) and power line hum (see chapter 13).

Of course, the belly tendon electrode provides virtually no selectivity to separate the EMG signal of the target muscle from that of any other muscle mass located in the vicinity of either the belly or the tendon contact. This matters little for the clinical applications of this approach in which an electrical stimulus is employed to cause synchronous activation of a nerve projecting to the muscle; the conducting latency then is measured to the instant of first deflection of the baseline. The idea is simply to provide a clean onset with minimal delay caused by slow and variable propagation of a signal through muscle fibers. As with any specialized technique, the belly tendon electrode works well for its intended application and poorly for other measurements; it is not suitable for selective recording of the EMG of a single muscle during physiological activity.

C. RULES OF THUMB FOR ELECTRODE DIMENSIONS

All of the electrode geometries discussed above can be thought of as dipole antennas, either explicit in their design or implicit in their electrical behavior. The monopolar configuration is often used as an ill-defined and highly suboptimal dipole that has little place in scientific EMG. However, for very small muscles or very local measurements of muscular activity, monopolar semimicroelectrodes may be the best or only feasible approach for sampling the large potentials found near the surfaces of active single fibers (see chapter 20 for relevant biophysical factors).

The various tripolar and multipolar configurations used to sample some large muscles can be thought of as sets of dipoles, each of which should conform to the basic principles outlined below. For clarity, we will use the term "bipole separation" to refer to the center-to-center spacings of the electrodes and reserve the term "dipole separation" (or moment) to refer to the apparent spacing between the centers of the signal current sources and sinks.

The several considerations discussed above now allow us to make five recommendations regarding electrode dimensions (fig. 6.6).

1. *The bipolar electrode should be oriented parallel to the voltage gradient to be measured.* For nerve and muscle fibers, this means that the two electrode contacts should lie on a line parallel to the fibers, not across them. Bipolar contacts lying equidistant to either side of a muscle fiber passing perpendicular to the line between them will detect no signal, whatever the activity of the fiber.

2. *The bipole separation between the contacts (center to center) should be about equal to the dipole separation of the sources and sinks to be recorded.* The dipole separation is the wavelength of the action potential, given by the product of the duration of the events and the velocity of conduction. This value is about 2 to 10 mm for most mammalian mus-

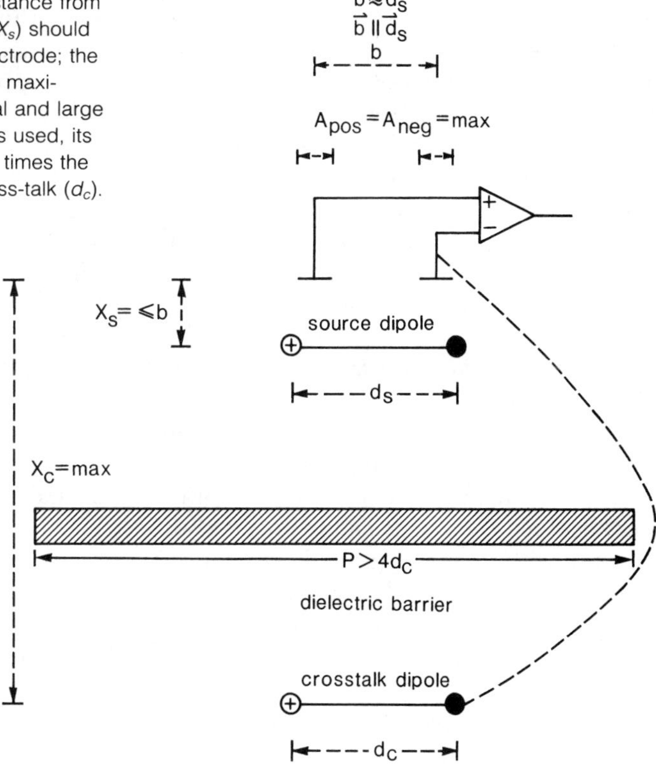

6.6 Rules of thumb for bipolar electrodes. The bipole separation of the electrode contacts (*b*) should be equivalent to the wavelength or dipole spacing of the EMG source (d_s) and lie parallel to it. The distance from the electrode axis to the dipole source axis (X_s) should be less than the bipolar separation of the electrode; the distance to a crosstalk source (X_c) should be maximized. The contact areas (*A*) should be equal and large for both poles. If a dielectrical patch barrier is used, its longitudinal extent (*P*) should be at least four times the dipole spacing of any potential source of cross-talk (d_c).

cle fibers. The dipole separation of distant fibers appears to be greater than that of closer fibers. Thus, one can achieve some degree of local selectivity by keeping the spacing of the two contacts on the short side.

3. *Each recording contact should be as large as feasible.* Usually this means at least one linear dimension equal to about half the distance between the pair. The notion that small electrode contacts provide greater selectivity permeates much of the EMG literature, and it is basically wrong for bipolar electrodes. The bipole separation defines the selectivity of the bipolar electrode. Reducing contact dimensions below half the distance between their centers adds only to noise and unduly biases the recordings by concentrating on the signals from a few fibers that may be lying right on top of one small contact. Once the physical extent of each contact surface becomes significantly smaller than the distance between them, a bipolar electrode approaches the electrical properties of two monopolar electrodes, the signals of which happen to be subtracted from each other. Each such monopolar electrode will then record large potentials from its microenvironment; these may be the only things apparent above the high levels of thermal noise and electrical interference inherent in high-impedance electrodes.

4. *The several recording contacts should be as similar in size, electrical impedance, and physical environment as possible.* The whole basis of the bipolar electrode rests on the precise subtraction of the two potentials at the differential amplifier to obtain only the gradient between them, eliminating potentials in common with respect to ground. The common mode rejection of differential amplifiers is highly sensitive to the source impedance, both its total value and its electrical phase (resistive versus capacitive conductance). The use of dissimilar contacts will defeat this precise subtraction; it may even give rise to highly undesirable effects such as electrode polarization and battery potentials.

5a. *To record selectively, the effective conductive path from dipole source to bipolar electrode should be equal to or less than the bipole separation of the electrode.*

5b. *To reject selectively, the electrical path from dipole source to bipolar electrode should be greater than four times the dipole moment of the source.* Notice that the distance to which we refer is that of the effective conductive path through the volume-conductive tissues, not the physical distance from electrode contact to signal source (fig. 6.4). As we have seen, the barrier of the patch electrode can influence the former without changing the latter. Also, notice that rule of thumb 5a refers to the bipole separation of the electrode, while rule 5b refers to the dipole moment of the signal source. Rule of thumb 2 suggests that these should be about equal. However, to improve local selectivity, the bipole separation may be made suboptimally small (and the resultant sampling problems overcome by using many such bipolar electrodes in tandem as in the multipolar electrode). This works because the amplitude of the potentials coming from sources lying closer to the electrodes than the bipole separation only decreases linearly as the separation is made shorter than their dipole moments. For potentials that originate further than four times the bipole separation from the electrode, the amplitude decreases as the square of the distance. It should be noted that closely spaced bipole contacts perform a spatial differentiation of the myoelectrical signal. Thus, they tend to record waveforms that are like time derivatives of the actual time course of the action current generated by the muscle fibers. This can shift the apparent bandwidth considerably higher than anticipated, calling for a change in signal filters.

7 Materials Science and Electrochemistry

A. FUNDAMENTAL PROBLEMS

An implantable device must be fabricated from biocompatible materials. These materials must neither cause toxic damage to the tissues in which they will lie nor be degradable or otherwise capable of being damaged by the complex chemical and physical environment in which they must function in the long term. The selection of materials and their incorporation into implantable devices demand an appreciation of the properties of this "hostile environment" (table 7.1).

We can, mnemonically and somewhat whimsically, liken the hazards to those of a voyage at sea: salt water, seasickness, and sharks.

1. Salt water. The extracellular fluid is principally an aqueous solution of table salt (sodium chloride)—an electrically conductive and corrosive agent that is aided and abetted in its damaging properties by a vast number of trace additives including enzymes, lipids, and small "solvent" molecules including alcohols and ketones.

Water and many of the dissolved small molecules and gases can permeate polymers, often passing through them by diffusion with surprising speed. Thereafter, implanted parts

Table 7.1 The Body as a Hostile Environment

Salt Water	Seasickness	Sharks
Permeation	Stress fatigue	Macrophages
Distortion	Migration	Animal subject
Shunting	Pressure necrosis	Surgeons
Corrosion		

How Living Tissue Survives	Why Implanted Devices Fail
Constant resynthesis	Constant deterioration
High elasticity and flexibility	Relative stiffness and brittleness
Buoyancy	Density
Proneness to crushing (use sharp instruments)	Proneness to puncture (use smooth instruments)
Vascular—infection resistance	Avascular—protected niches
Smooth phase transitions	Discontinuous boundaries

may swell or suffer other mechanical distortion or stress. When diffusants such as water meet phase boundaries such as those between metals and their insulating sheaths, they may condense out of the vapor phase and collect in layers that move rapidly along hydrophilic surfaces by capillary action. If this pure water encounters soluble surface contaminants, such as fingerprints, it will dissolve them and become electrically conductive. This will also result in an osmotic pressure gradient between the new solutions on the surface and the pure water in the vapor phase; this gradient further accelerates the accretion of water and the tendency to split the implant along the phase boundary. Sooner or later, any implanted device that is not hermetically sealed will have water in either the vapor or the liquid phase throughout every nook and cranny. Unfortunately, hermetic sealing techniques require bulky and rigid materials, such as metals and ceramics. Water penetration is generally inevitable, and the designer of any device intended to function unobtrusively around moving muscles must learn how to live with the stuff.

2. *Seasickness.* Those constantly moving limbs and muscles bring up another nemesis of implanted devices. Nothing, neither alloy nor polymer nor connective tissue, will tolerate mechanical distortion indefinitely. The body solves this problem by constantly renewing all parts by resynthesis and by orienting the new materials along the lines of physical stress. The experimentalist must buy time rather less elegantly with careful device design and surgical technique. In addition to the obvious problem that the metallic electrical leads will soon work-harden to the point of fracture, the electromyographer faces the tougher problem of keeping the devices accurately positioned in a body in which most parts are constantly being remodeled. If an anchoring point comes under tension, perhaps because the electrical cable connecting it to an external device lacks slack, then the body will begin to release this tension by rearranging the connective tissues to which the cable is anchored. Soon the device may migrate to an entirely new location. If the device is a locally sensitive EMG electrode, the investigator may receive a wonderful signal from a structure he or she had no intention of studying.

A more dramatic and very disconcerting problem often arises at or near the exit points of percutaneous devices. It takes very little pressure to cut off the blood supply to a local region of skin. Chronic pressure results in clinical medical problems, such as bedsores in persons confined to a bed for a long period. The tugging on the skin of a percutaneous cable or the stress concentration induced by a protruding corner can easily compromise blood supply, preventing wound healing and even causing dramatic breakdown of apparently stable surgical sites. The resulting hypoxic, devascularized, or frankly necrotic tissues make great culture media for bacteria, which can track rapidly along implanted cables and into crevices in the devices, from which they will be impossible to eradicate. As we discuss in chapter 19, wound infections are almost always the fault of the device design or of the surgeon rather than of the ubiquitous microbes upon which they are blamed.

3. *Sharks.* Under this rubric, we collect three oft forgotten culprits in the demise of implantable devices: the cellular part of the immune system, the hostile behavior of the whole animal, and the ham-handed surgeon. All can mechanically disrupt the best-laid plans.

Macrophages are in the business of breaking down foreign bodies into little pieces to be spirited away. We may not think of our carefully crafted electrodes as foreign bodies; however, the macrophage sees any implanted device and its cables, sutures, and connectors as simply another enemy. Like the roots of trees invading a foundation wall, its pseudopodia will find every surface scratch in a device and insert themselves like little levers. If the material is prone to fracturing along stress lines, as are cellophane and Mylar tape (which continue to tear once a rip has started), it may well be carried off in tiny pieces (fig. 7.1). If the split occurs between the electrical insulation layer and the metal of a wire electrode, such macrophage penetration obviously does not facilitate the recording process.

Animals tend to condition their skin, hair, and feathers and this often involves licking, chewing, rubbing, and other grooming practices as well as saliva, sweat, waxes, and oils.

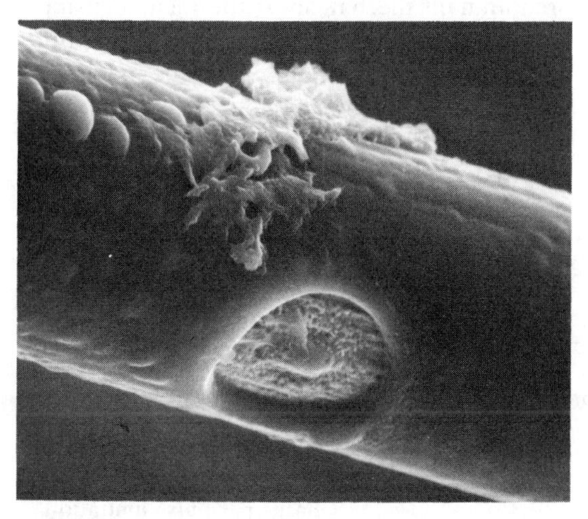

7.1. Scanning electron micrograph of a 25 μ diameter, iridium wire coated with 3 μ of Parylene-C (Union Carbide Corp.) that had been implanted for several months. Note that the biocompatible polymer is intact over the entire surface except for a round, sharp-rimmed crater. Apparently, some inclusion or other imperfection in the vapor-deposition coating provided a weak point for invasion by macrophages. A piece of the insulation coating was then detached, causing an electrical shunt. (Courtesy Dr. E. Schmidt, Laboratory of Neural Control, NINCDS, NIH.)

All of these are likely to be destructive to percutaneous devices. Individual animals may react quite idiosyncratically upon discovering such devices when awakening from anesthesia.

The physiological researcher will often have to design, build, and implant the device. Some researchers find themselves working in a team with divided responsibilities. All should be aware of a pernicious streak that, sooner or later, shows up in many surgeons confronted with a meticulously fabricated device. Successful surgeons condition themselves to have the utmost patience with and care for the delicate tissues that must be invaded. Surgeons may take out the inevitable tensions and frustrations on inanimate objects—for example, by throwing defective instruments on the floor in little fits of temper. Unfortunately, this distinction between the sacred living subject and the expendable inanimate object may prevail at just the wrong time, particularly if flaws in the design of an implant manifest themselves during attempts to position and secure it. It never ceases to amaze us how often experiments that have taken weeks to prepare and weeks to execute will be defeated in a few minutes of intraoperative carelessness. Device design must somehow allow for this. Suffice it to say that the device should be designed by, or at least with, the surgeon. Also, the person responsible for its fabrication should be present at the moment of truth in the operating room, both as a deterrent and as a consultant, and should always have a kit of tools and spare parts handy.

B. SELECTION OF ELECTRODE MATERIALS

1. Categories

Many elegant criteria can be assembled regarding the optimal materials for a device, ranging from their physical and mechanical characteristics to their cost effectiveness. Of course, most devices are built, at least the first time, from whatever happens to be on the shelf in the laboratory. Surprisingly often this works, at least for short-term devices, because the common alloys and polymers now available are actually fairly good. Stainless steel, platinum, Teflon, and silicone rubber are no longer exotic or rare in laboratories and can hardly be faulted as materials. Still, when devices malfunction or when semichronic protocols turn into long-term experiments, it becomes essential to separate two classes of problems: those of material selection and those of mechanical and electrical design. Chronic experiments simply require too much time and effort to let the experimentalist rely on trial and error debugging. A half-dozen failures might mean a whole year wasted—a tough proposition for a grant renewal or a tenure decision.

In the following discussions we divide materials into conductors and dielectrics. A passive electrical circuit, such as an electrode, employs these two basic elements to move electrical currents from one place to another. However, it must never be forgotten that the two must work in concert, both mechanically and electrically, in a device that must be feasible to construct, practical to implant, and capable of producing adequate results as it inexorably deteriorates.

2. Conductors

a. Chemistry and Surface Properties. We have had several occasions to refer to the process of electrical conduction at the metal/electrolyte interface. There are many complex reactions available by which electrical current composed of ions in water can lead to the movement of electrons in a metal conductor (table 7.2). Fortunately, for the very low voltages generated by bioelectrical signals, the predominant mechanism is the relatively simple capacitive charging of the so-called double layer (the sandwich made up of the metal surface on one side and the ions in the fluid on the other, separated by a thermodynamically required void).

Table 7.2 Causes of Metal Failure

Problem	Solution
Electrode corrosion	
H^+ Adsorption/desorption	(Solution not needed)
Surface oxide redox	(Solution not needed)
Na/Cl Double-Layer Formation	(Solution not needed)
H_2O electrolysis	Charge per pulse limitation
pH shifts in tissue	Balanced biphasic stimulation
Unstable contact impedance	Constant current driving
Metal dissolution	Alloy selection
Heavy metal toxicity	Surface pretreatment?
Electrode disappearance	
Lead breakage	
Tensile strain	Surgical routing
Discontinuity	Tapered strain relief
Flexion fatigue	Coiling, stranding
Work-hardening	Alloy selection

At low imposed voltages, the water molecules and the dissolved ions do not actually make direct electronic contact with the metal atoms and their free electrons. The thermal motion of the fluid gives rise to a condition in which its molecules and ions are constantly bouncing against the metal surface, with a mean approach distance called the Debye-Huckel radius. The gap between the fluid surface and the metal surface may be thought of as a nonconductive vacuum between two layers of conductors—that is, a capacitor. As the gap is very small, on the order of 1 to 2 angstroms, the capacitance is fairly large even for a small surface area (fig. 7.2). We recall that alternating currents will pass through a capacitor by causing positive and negative charges to accumulate alternately on the apposing plates. In the case of the metal/electrolyte interface, the electrical signal on the tissue side is the small change in the distribution of positive and negative ions coursing through the extracellular environment under the influence of the bioelectrical fields. These cause corresponding accumulations and rarefactions of the electrons sitting across the gap in the metal conductor. The motions of these electrons are the currents that the amplifier detects and enhances.

It would thus seem that all metals should be equally effective as electrodes, given comparable surface areas. However, many of the so-called corrosion-resistant metals do not actually have bare metal surfaces exposed in the presence of oxygen. Metals such as aluminum, chromium, and titanium immediately form a surface skin of their metal oxides, which are highly stable and electrically nonconductive ceramics. As the gap caused by the thermal motion is only about 1 angstrom, even the thinnest such oxide layer can greatly reduce the capacitance available to such an electrode

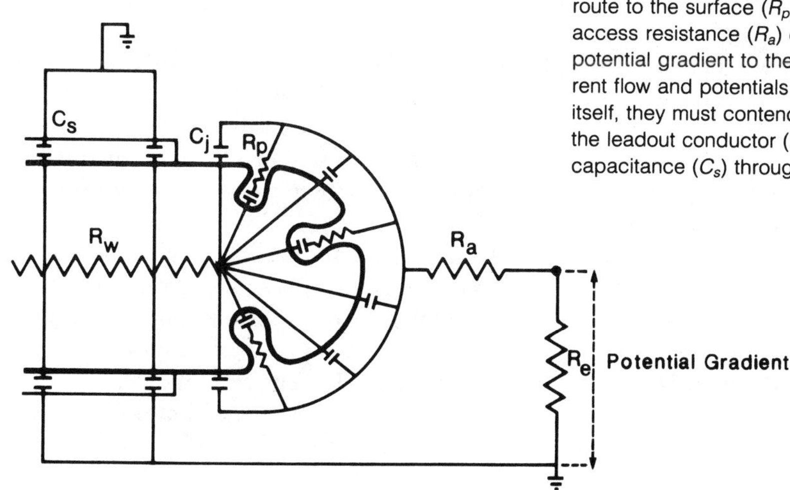

7.2. Equivalent circuit of an electrode. The rough metal surface of an electrode conducts small potentials such as EMG primarily by the capacitance of the metal/electrolyte interface (C_j). The parts of the surface that are not immediately at the apparent surface of the contact as a whole (e.g., buried between wire strands or in crevices and pits) encounter a series resistance en route to the surface (R_p). All current flow encounters an access resistance (R_a) en route from the source of the potential gradient to the electrode surface. Once current flow and potentials are induced in the electrode tip itself, they must contend with resistance to flow along the leadout conductor (R_w, usually negligible) and shunt capacitance (C_s) through the wire insulation.

by greatly increasing the effective gap between ions and electrons. The only common metal that forms no surface oxide is gold. Materials such as stainless steel and the platinum family of noble metals generally have very thin, stable oxide layers that do not degrade their electrode performance significantly. Still other metals, such as silver and copper, have excellent capacitance properties but are slowly dissolved by the chelating action of chloride ions. The process releases highly toxic heavy metal ions into the surrounding tissue.

The charge-carrying surface area that we are discussing is the "real" surface area rather than the more easily determined and physically apparent "geometrical" surface area. The difference is obvious when a supposedly smooth metal surface is examined with a scanning electron microscope. Depending on the material and its manner of working and finishing, the surface will appear to be a complex and convoluted terrain of scratches, pits, rills, and other details that greatly increase the real surface. Deliberate mechanical or electrical roughening of an electrode surface can further enhance this surface area (see fig. 7.2). However, this process has limits. First, a highly roughened surface may be mechanically fragile and subject to degradation; thus, cracks may start, accelerating fatigue failure. Second, the electrical currents may encounter significant resistance in the pockets of fluids that they must traverse to reach a real surface much deeper than the geometrical one. It is this access resistance that causes measured electrode impedances to have a resistive component, regardless of surface geometry. This series resistance term may be quite large and subject to changes as the surface, pores, and crevices become clogged with less conductive scar tissue. The phenomenon depends on very local circumstances, so that the impedance of the two wires of a bipolar electrode may be affected differently. Such imbalanced electrode impedances, of course, lead to a degraded common mode rejection in differential amplifiers.

b. Mechanical Aspects. Most of the abovementioned considerations involve relatively subtle changes in the signal-to-noise performance of electrodes, rather than fundamental changes in the recorded signals. For all intents and purposes, they offer little reason to prefer gold over stainless steel or over platinum or its various alloys. The choice among these metals is usually much more related to the mechanical problems of fabricating the contact surfaces in the required geometrical area and maintaining reliable electrical leads and connections in the presence of salt water and motion.

One of the major problems in electrical leads is work-hardening. A single strand of relatively ductile material, such as gold or pure platinum, will survive less than 1 day of normal motion in a vertebrate limb. As a bar with a finite diameter is flexed, the extreme sides in the plane of movement will be alternately compressed and stretched, resulting in migration of metal atoms (fig. 7.3). The greater the diameter of the bar and the shorter the distance of flexing (i.e., greater flexion moment), the greater the force required to flex the material near its extreme surface and the greater the level of stress at any discontinuity. Eventually a weak point develops; the imposed stresses concentrate here, and the wire soon fractures. Hence, such wire should be used only in experiments of brief duration.

Springy materials, such as stainless steel, that have high tensile strength and a high elastic modulus provide more resistance to bending, thus distributing the flexion movement more uniformly. By definition, they are also less ductile, because their metal atoms simply change their lattice spacings elastically with-

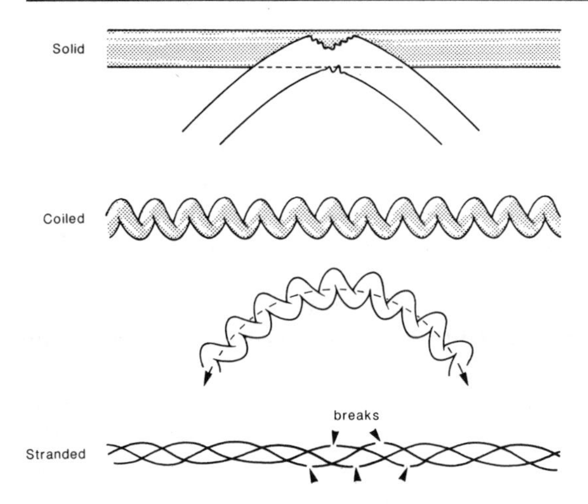

7.3. Flexion modes of various wires. Solid wires work-harden and break as a result of compressive and tensile stresses at opposite sides of a region of flexion. Such motion is redistributed as a longitudinal torsion if the wire is coiled. A stranded wire of the same current-carrying capacity (effective cross-sectional area) is also less prone to failure resulting from bending stresses that do not concentrate in any one place (although the overall tensile strength would decrease). Side-to-side conductance among the strands avoids open circuits even if each of the strands has one or more breaks along its length.

out actually changing lattice positions. Of course, this resistance to bending means that they transmit forces to the tissues surrounding them, rather than simply being bent out of the way. These forces may stress anchoring points or compress blood supplies, so stiffness alone is hardly a solution to the failure problem.

There are two well-recognized solutions to accommodating motion without excessive lead breakage. Coiling is the simplest conceptually and the most effective, but it results in bulky leads that are difficult to fabricate. However, short sections of coil can effectively reduce flexion moments. A tightly wound spring of even the stiffest wire can readily be bent axially with minimal resistance, because the flexing movement of a local point along the whole spring is distributed as a small torsional movement throughout the spring. Such leads have been used in the pacemaker industry (and by some experimentalists), but they generally require that the coiled spring be entirely embedded in an elastic polymer, such as silicone rubber, to prevent connective tissue from growing into the interstices.

Multistranding is the more common solution, as is obvious from an examination of a lamp cord. The notion here is to divide the cross-sectional area necessary for electrical conductivity and tensile strength into a large number of small strands so that the individual surface stresses are lowered. When flexed, the strands are free to slide among themselves rather than experiencing the large compressions and distentions that would occur in a rod of similar total cross-sectional area. Actually, an even more powerful effect is that of redundancy. Electrical continuity will be lost only if every single strand breaks at exactly the same point. A very high number of random breaks in individual strands will be well tolerated as long as the sides of the strands continue to touch each other and permit electrical conduction as they course in parallel. Finely stranded stainless steel wire has become widely accepted in long-term EMG. It will survive almost any amount of motion, it is biocompatible, and it is relatively inexpensive.

If at all possible, EMG electrodes should be designed to avoid electrical connections inside the body. Many of the designs (described in chapter 10) achieve this by means of a bared section of the leadout wire that serves as the electrode contact itself and an unbroken length of the wire that runs out through the skin to an external connection. Such an approach finesses one of the most vexing biomaterials problems—the battery effect. If two metals of dissimilar composition are placed into contact with each other in a conductive salt solution, the difference in electromotive force at each metal/electrolyte junction will give rise to small but steady flows of direct current (fig. 7.4). These direct currents pass across the metal/electrolyte junction not by double-layer charging but rather by the direct exchange of electrons between metal atoms and ions in solution. The resulting electrochemical oxidation and reduction reactions can rapidly corrode one or the other of the metals, as well as give rise to large pH changes (via the electrolysis of water), which are quite toxic to surrounding tissue. As water is likely to permeate all parts of the electrodes and leads, it is best to restrict junctions to welds between similar metals; introduction of solders, brazes, or other intermediaries will be dangerous. Mechanical crimping might be acceptable, but the crimping sleeve has to be ductile to form a reliable connection; it is unlikely that its metal can be of the same material as the leads.

Actually, it is rare that the working contact of the electrode cannot be made out of some clever arrangement of one or more of the strands of the leadout wire. Should a connection have to be made, say between stainless steel and platinum, it is probably best to weld the two resistively (see chapter 9) and protect the joint mechanically by potting the carefully cleaned area in silicone rubber. Soldering is often an adequate short-term solution (weeks) if similar care is taken to remove flux residues before encapsulation. It should be noted that stainless steel can be tinned (solder coated) only by use of special acid fluxes, which are highly corrosive and must be thoroughly rinsed off. Conductive epoxies, which usually contain silver (a metal highly reactive in salt water), must be avoided.

Any joint or other site at which the mechanical properties along a wire will be discontinuous should be encapsulated to relieve strain at a site of preferential flexing. Most mechanical failures in electrodes arise not along the length of the leadout, where the amount of bending is maximal, but rather just at the junction between the lead and the electrode proper (or at an intermediate joint).

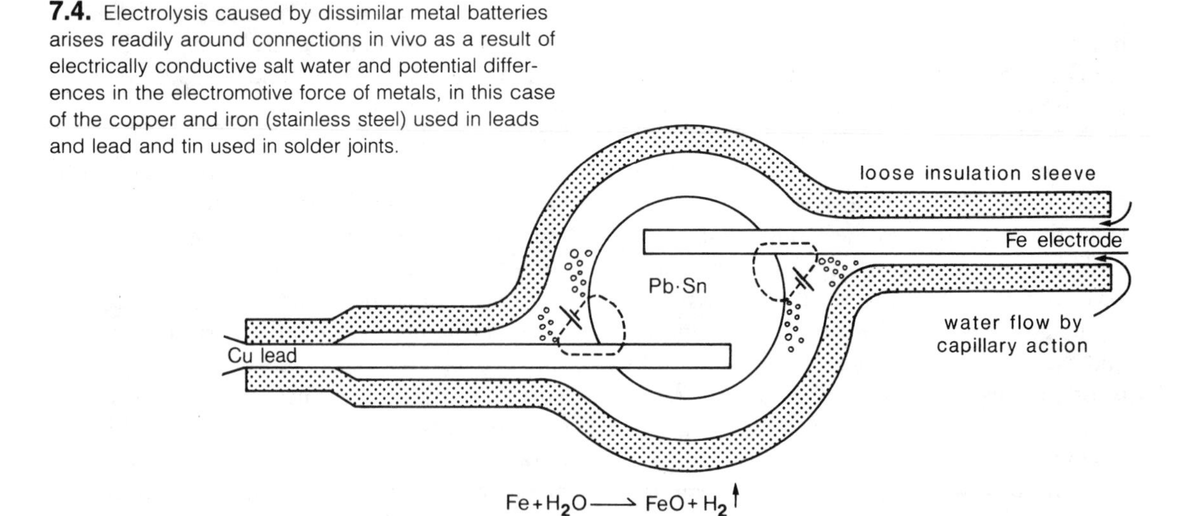

7.4. Electrolysis caused by dissimilar metal batteries arises readily around connections in vivo as a result of electrically conductive salt water and potential differences in the electromotive force of metals, in this case of the copper and iron (stainless steel) used in leads and lead and tin used in solder joints.

After all, the electrode, with its fixation apparatus and perhaps dielectric barrier (patch), constitutes a firm site; its extreme end represents a relatively large lever arm. As the electrode moves relative to the long leads, the greatest stresses will be concentrated at the attachment site of these leads. What is worse, the act of removing the insulation from the end of the stranded wire is likely to have nicked some of the strands, leaving them prone to fracture as they bend repeatedly. Design of electrodes and identification of failure modes require careful attention to such points of mechanical discontinuity. In chapter 9 we detail methods of wire stripping, joining, and strain relieving.

c. Stimulation Effects. One important caveat regarding the selection of metals for electrode contacts involves the use of electrical stimulation. Electrical pulses may be applied to EMG electrodes used otherwise for recording, either as experimental stimuli (e.g., to look for reflex responses) or as part of the protocol to confirm electrode position (see chapter 8). Once voltages in excess of about 1 V are impressed across a metal/electrolyte junction, all bets are off regarding the mechanisms of charge transfer. Any passage of net DC current, even brief monophasic electrical pulses, will certainly result in some electrolytic reaction. For a bipolar electrode, this means oxidation reactions at one contact and reduction reactions at the other, depending on the instantaneous polarity of the waveform. The usual solution to this problem involves the use of charge-balanced, biphasic stimulus waveforms, in which the change of the cathodal stimulating phase (current × duration) is immediately balanced in the second, anodal phase with an equal and opposite charge.

Even strictly charge-balanced AC pulses can lead to oxidation/reduction reactions if the total charge flowing in each phase exceeds the amount of charge storage available in the metal/electrolyte double layer. The amount of charge in a pulse phase is the product of the current (in amperes) times the duration (in seconds) and is expressed in coulombs (1 amp sec). The charge storage limit of a particular metal/electrolyte interface is given as charge per phase density, usually microcoulombs per phase per real square centimeter. This is a complex subject that is only just beginning to be investigated and understood.

For the electromyographer, a few simplifying rules of thumb may suffice. If only an occasional stimulus pulse is required, probably any electrode metal is reasonably safe, provided that the pulse is as brief as possible (usually 0.1 msec will suffice) and is absolutely charge balanced (easily accomplished by capacitive coupling to the stimulator; see chapter 12, section E). If frequent stimuli or trains of pulses will be delivered, the electrode contacts should probably be made only of noble metals and their alloys, such as platinum, iridium, and rhodium. Whereas even these can be eroded eventually, the process is extremely slow and fairly benign for charge-balanced pulses, particularly if the contact areas are kept as large as possible. Stainless steel poses serious problems with rustlike corrosion, particularly near the exit points from insulators, although special pulsing techniques have been developed to overcome at least some of this. Gold is quite unsuitable, as it is rapidly dissolved in the presence of chelating chloride ions. The same is true of copper, nickel, and especially silver, all of which can rapidly release large quantities of toxic heavy metal ions and may even corrode away entirely when used as stimulating contacts.

3. Dielectrics

The dielectrics in an electrode are generally the insulation on the leadout wires and any mechanical carriers, encapsulants, fixation points, or other structures designed to orient, retain, and protect the electrode. Of course, they must be neither biodegradable nor toxic. Most modern polymers are quite stable and biocompatible in the body by themselves. However, pure long-chain polymers are relatively hard to come by, because commercial products often contain a myriad of additives, such as catalysts, uncured monomers, plasticizers, mold release agents, fillers, and dyes. Medical grades of polymers are often limited in their selections of sizes and consistencies and tend to be very expensive. As with the conductors, though, a few reliable staples and some clever design will see most applications through without causing undue worry about the use of new and untested materials. When new materials are employed, simple gross pathological observation of a test implant site will often indicate its acceptability. Because a postmortem examination should be made to confirm the position and status of all implanted devices (see chapter 8), such data are not hard to come by.

The electrical requirement of a dielectric is simply that it block the passage of electrical currents (fig. 7.5). We have already noted that AC currents will cross even a vacuum by capacitive effects, but even the thinnest applica-

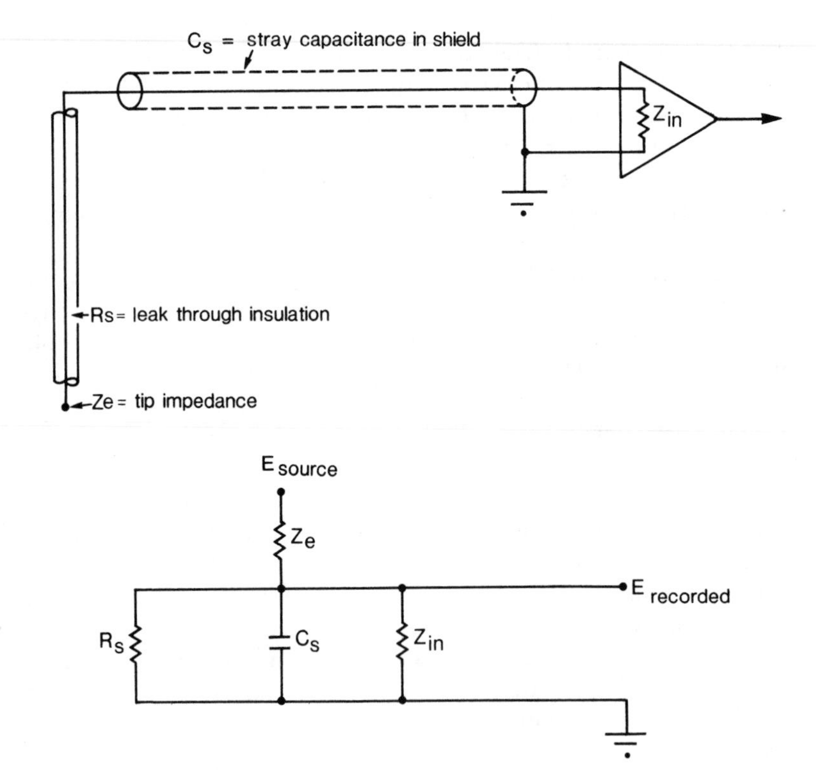

7.5. Leakage through defective insulation (R_s) and excessive stray capacitance (C_s) in connecting cables both act similarly in parallel with the amplifier input impedance (Z_{in}). They form a voltage divider that can attenuate or distort the recorded signal ($E_{recorded}$) from the value present at the source electrode tip (E_{source}).

ble coating of most polymers results in a low enough capacitative shunt vis-à-vis a gross EMG electrode that this can be ignored. However, microelectrodes tend to have small exposed contact areas and relatively high impedances (greater than 100 kΩ). In such electrodes the capacitative shunting, even through an intact coating of insulation, cannot be ignored (see chapter 20).

The biggest enemy of the dielectric is the pinhole. The surface to be insulated usually is coated with a liquid uncured polymer that often has a high surface tension. The polymer is usually cured in place, perhaps by heating. The expectation is that the liquid polymer will completely coat materials of different surface geometry and chemistry. Any speck of dust that lands in the coating will create a weak point or even a discontinuity. Also, the base materials are commonly "wetted" incompletely. Sharp corners of the underlying structure may concentrate stresses in the coating and cause local breaks long before the remainder of the coating fails. Even if the coating is perfect, it may be perforated when grasped carelessly with a metal forceps; jaw pressures may rise to astronomical levels in pounds per square inch even when the forceps are wielded by a 97-pound weakling. (This is one more reason to watch the implantation procedure.)

As we have already noted, a crack or scratch of the surface may not actually constitute an electrical leak at first, but represent a weak point that macrophages will quickly exploit. The electrical detection and effects of pinholes are discussed in chapter 14.

We have noted that water in the vapor phase permeates all polymers; however, only in the condensed liquid phase will water conduct electricity and dissect phase boundaries by capillary action (table 7.3). This suggests two, alas conflicting, approaches. If the encapsulant is hydrophobic, like Teflon, it will have a low permeability and will not encourage the condensation of water. Thus, there will be little or no capillary action, but such coatings tend to be poorly adherent once a disruption begins. On the other hand, if the encapsulant is adherent to metal surfaces (i.e., is hydrophilic), it will prevent the spread of water by adhesive seal. To make matters worse, hydrophobic polymers are likely to form poor adhesive seams with hydrophilic ones.

Actually, the saving grace in all of this is the sheer mechanical simplicity and low electrical impedance of a well-designed EMG electrode. If the electrical contacts have been made as large as appropriate for the dipole spacing, then relatively small pinholes or seam leaks will have little effect, unless they are progressively enlarging. If the pinhole has 1% of the surface area of the electrode, then the electrode will lose only 1% of its signal. A potentially more severe problem is that the undesired exposure point will itself act like a recording contact, but it will lose 99% of its signal to the effective shunt represented by the rest of the electrode. The take-home lesson is that the dielectric problem is much more easily and reliably addressed by electrode design rather than materials selection. The same goes for mechanical damage to the electrode caused during implantation or damage caused by chronic stress.

The most common biomedical polymer is silicone rubber. Well tolerated by the body, it is commercially available in uncured liquids of various consistencies and in cured sheets, tubes, and blocks. The material is naturally clear to slightly milky or translucent. Opaque and colored silicones are filled with materials such as diatomaceous earth to improve their mechanical resiliency or to make them radio-opaque. The principal disadvantages of silicone rubber are its mechanical friability and the tendency to absorb substances onto its surface, where they cause brittleness. These make it unsuitable for very thin sheets and coatings, although thin sheets with an embedded Dacron mesh are a mainstay of the barrier (patch) electrodes. There are several different catalyst systems, but most silicones will stick quite well to each other and to most hydrophilic and hydrophobic surfaces, even in the presence of moisture.

The most convenient uncured form of silicone rubber for making custom devices is the acetoxy cure system, known as a single-component RTV (for room-temperature vulcanizing). This material is even better known as bathtub caulk—the stuff that smells like vinegar as it cures overnight at room temperature. No EMG laboratory should be without it, either in the expensive medical grade as Silastic Medical Adhesive Type A (No. 902, Dow Corning Corp.) or as plain, clear bathtub caulk (local hardware store). The latter seems to have slightly inferior mechanical properties, but it is otherwise apparently biocompatible and considerably less expensive. The acetoxy curing reaction may be accelerated by warming but *only in the presence of moisture*. Putting this material in a dry-air oven with a relative humidity close to zero is not advisable. Also, because of the inward diffusion of water vapor and outward diffusion of acetic acid during the catalysis, the cure can be very slow for thick layers or those between sheets of impermeable materials, such as Teflon (in contrast, silicone rubber itself is almost transparent to water vapor, even when fully cured). Because of the high viscosity and rapid formation of a surface "skin," it may be difficult to apply small amounts precisely by hand. An air pressure–operated syringe dispenser can be

Table 7.3 Causes of Dielectric Failure

Problem	Solution
Water entry	
Diffusion through polymers	Impermeable materials
Capillary flow along hydrophilic surfaces	Adhesion between layers
Gross defects, stress points	Multilayering, contouring
Internal generation of batteries, electrolytics, curing reactions	Component selection
Condensation at phase boundaries	Adhesion to substrates
Dissolution of base materials, additives, surface contaminants	Cleanliness and purity
Osmotic dissection	Mechanical occlusion
Electrical shunting	Long shunt paths Low impedance circuits
Corrosion	
Operating voltages	Noble metal conductors
Battery voltages	Nonbimetallic circuits
Electronic failure	
Detuning of radio frequency circuits	Selective hermetic packaging
Semiconductor junction Na poisoning	Selective hermetic packaging

very handy for working on small parts (see chapter 9 and appendix 2). Even though the material will cure in aqueous solutions, it should not be implanted until fully cured; the acetic acid liberated during the cure does not help wound healing!

The other common dielectric is Teflon, a trade name for two different polymers of fluorinated hydrocarbons; TFE (tetrafluoroethylene) and FEP (fluoroethylpropylene). The Teflons are notably different from the silicone rubbers in that they are so inert that it is almost impossible to fabricate them in the laboratory. Their chemical, mechanical, and thermal refractoriness makes them ideal as insulating layers on wires, where they are usually applied as extrusions. Teflon insulation has become a standard on stainless steel wires for subcutaneous electrical leads, because it can be handled with a minimum of mechanical caution and because the body reacts minimally to its presence. The smooth, inert jacket of the wire becomes surrounded by only the thinnest layer of nonadherent scar tissue and will slide freely through this sheath months after implantation, reducing motion stress on the terminations.

The chemically inert surface of Teflon is extremely hydrophobic and very difficult to wet with adhesives. Special surface treatment agents, such as sodium naphthalene, are available to etch the surface, thus making it bondable, but they are difficult to use and may leave toxic residues. Silicone rubber adhesives will wet the regular Teflon surface, although they will not develop any appreciable bond strength. Still, this combination is acceptable for many applications, particularly those in which the joint contains mechanical strain relief and the potential pathway for seam leakage is fairly long and narrow. Teflon coating does tend to deform after steam sterilization; gas sterilization of electrodes and leads is more suitable.

C. FOR SHORT-TERM USE ONLY

The experimentalist is apt to find many other kinds of wire and insulation in the laboratory or shop and will be tempted to incorporate them in "quick and dirty" designs. Considering the amount of time invested in the typical chronic EMG study, quick and dirty techniques are rarely justified. However, human nature being what it is, the experimentalist should have at least a passing acquaintance with the properties of some of these wires.

The reason this approach works so often is that over the short term almost any electrically conductive metal, given a large enough surface area, is capable of functioning as a reasonably quiet and stable electrode for the high frequencies of EMG records. Even metals that form thick or unstable oxides (e.g., aluminum) will provide enough metal/electrolyte capacitance to admit current to a high-impedance amplification circuit. Any dielectric, given at least temporary physical integrity and sufficient thickness, will protect those parts of the conductor that are not supposed to be in electrical contact with the tissues. Of course, for small, delicate contact surface areas, both mechanical and electrical problems are magnified; the insulation must be thin, and the importance of selecting optimum materials begins to rise exponentially, even for acute studies.

1. Electromagnet and Armature Wire

Electromagnet and armature wire is often available in a tempting range of fine sizes, but it is only suitable for short-term studies. The core conductor is usually malleable copper, which is quick to work-harden and will corrode, releasing copper. What is worse, it may be pretinned, with lead and tin added to the noxious soup at the tip. In the past such wires usually were insulated with lacquer, varnish, enamel, or epoxy. All are somewhat biodegradable, especially lacquer and varnish. Lately, coatings of high temperature–tolerant polyimide (trade names Pyre-ML, Kapton, etc.) have become popular. This material is biocompatible but also brittle and easily fractured when squeezed with forceps. While nominally rated for high temperatures, this and many other coatings may be degraded by exposure to steam in an autoclave at temperatures far below its rated limit.

2. Hearing Aid and Other Ultraflexible Cables

These cables represent tempting solutions to the need for ultraflexible leads. The jackets are usually thermoplastics with low temperature tolerance. If implanted they may generate considerable fibrous tissue adhesion, perhaps defeating their mechanical innocuousness. What is worse, though, the conductors are usually silver or silver alloys formed into wire and foil. Hence, they are very fragile, being kept from failure only by the support of the dielectric jacket. Without the matrix, the foil is highly prone to fracture; usually the break occurs right at the border where the matrix stops. This imposes special tasks on electrical connectors.

3. Flexible Ribbon Cables

Flexible ribbon cables are composed either of traditional stranded wires bound side by side by fusing of their thermoplastic jackets, or of "flat" cables with copper foil conductors. In both cases the conductors are copper and likely to be pretinned, making them unsuitable for use as electrode contacts except for acute implantation. The flexibility of the flat cable comes from the fact that the thin foil is supported mechanically by the much stiffer dielectric matrix, which prevents it from flexing unevenly. Precautions cited for ultraflexible cables also apply.

4. Thermocouple Wire

Thermocouple wire represents a source of exotic alloys such as nickel-chrome (nichrome), platinum-rhodium, and iron-constantan. Actually, most alloys used are refractory or noble metals and tend to be quite inert biologically. Many form surface oxides, which may make them less than optimal electrode contacts, but with cathodic reductions (see chapter 14) and decent surface areas, they may be quite suitable. The insulation jackets are usually high temperature–tolerant polymers such as Teflon, as befits their normal applications, although one should watch for hazards such as asbestos braid. Thermocouple wires are often quite stiff and springy and may be very difficult or impossible to solder effectively without special fluxes.

8 Verification of Position

A. TASKS

In our discussion of the electrical fields generated by muscle we have emphasized the spatially oriented and localized features that let one distinguish the signals of one muscle from those of another and from other bioelectrical sources. Our discussion of electrodes has emphasized designs that are selectively sensitive to gradients of certain orientations and extents. Obviously, obtaining the optimal electrode position in the body, maintaining it throughout the experiment, and verifying that the position has been maintained are critical. Also, the electrical fields are never completely localized, and the electrodes are never affected only by local sources. Therefore, there must be some method for estimating just how good this supposedly optimal position is vis-à-vis potential sources of cross-talk (from other muscles). There is no single, best, or simplest method. Rather, the experimentalist must be prepared to select and defend a combination of techniques based on a sound analysis of the objectives of the study and the potential problems inherent in the preparation.

B. DISSECTION TECHNIQUES

If possible, it is advisable to sacrifice or at least surgically explore the experimental animals shortly after the critical data have been recorded so that the condition of the tissues near the electrodes can be accurately assessed by direct observation. Even basically sound electrode fixation schemes may be subject to bizarre migrations as a result of unforeseen mechanical stresses or the formation of scar tissue. While dissection proceeds, the electrical impedance of the electrodes should be tested, particularly if the impedance changed after implantation or if the level of cross-talk changed, suggestive of breakdown of the insulation. Suspicious-looking (i.e., apparently damaged) sections of the lead wires can be elevated out of contact with the tissue during dissection and their impedance remeasured to see if such sections are contributing a shunt (if they are, the measured impedance will then rise; see chapter 14).

The postmortem dissection may be technically much more difficult than the original operation, particularly if multiple devices have been implanted long enough so that considerable scar tissue has formed (anything longer than about one week). However, it is crucial that the actual relationship between electrode and soft tissue be examined. Shortcuts, such as x-ray examinations of whole animals, only disclose that the electrode tips did not deform during insertion; they tell little about actual position.

Dissection had best be carried out under a dissecting microscope. Careful notes from the original operation regarding the planes of surgical closure and the types and colors of sutures and anchoring materials are most helpful. The field seems easiest to dissect if the animal is sacrificed the previous day and refrigerated for about 24 hours, which allows the blood to clot and a reasonable stiffness to develop. Unfortunately, this approach interferes with subsequent histochemical or physiological assay of the particular muscle. Perfusion makes dissection of devices and wire more difficult, because the mechanical consistency of the tissue now approaches that of the implanted device. However, perfusion permits more detailed histological and histochemical studies of the electrode/tissue interface and does not significantly affect measured electrode impedances.

Histological evaluation of the area around an intramuscular electrode is quite advisable, particularly if the muscle is small or if the recordings appear unexpectedly "spikey," suggesting that only a few motor units are surviving. The amount of purely mechanical damage around an intramuscular bipolar electrode pushed or dragged into a thin muscle can be impressive; what is worse, an unfortunate choice of the path can seriously damage an innervation point, leading to widespread atrophy and electrical silencing. Of course, the metal parts of the electrode must be removed before the tissue is sectioned for histological examination, and this may compromise localization unless the block of muscle is carefully split along the track of the wire.

The use of iron staining to localize stainless steel electrode contacts is commonplace in microelectrode track reconstruction, but it appears only rarely to have been employed with EMG electrodes (see appendix 1). It is quite feasible and greatly improves one's confidence in percutaneously inserted electrode recordings or in situations in which the electrodes must be removed long before the animal becomes available for dissection. Care must be taken to distinguish electrolytic iron deposits from those left by microhematomas.

C. FUNCTIONAL VERIFICATION

A few basic measurements, easily made on a regular basis, provide vital information about potential electrode deterioration and its probable causes. First and foremost is the measurement of electrical impedance (see chapter 14). A daily record of the impedance of the implanted electrodes, commencing with an in vitro test prior to implantation, followed by intraoperative or postoperative values, can be invaluable in pinpointing the time of onset and mechanism of electrical failure (fig. 8.1). Whereas some fluctuation caused by resorption of free body fluids and development of scar tissue is normal, abrupt shifts almost always signal serious degradation of the desirable antenna properties initially designed into the electrode.

The other simple indicator of electrode performance is the absolute voltage level of the recorded signals. In this day and age of multistage amplifiers with gains continuously variable over four or more decades, many investigators have become accustomed simply to turn up the gain until the signal appears to have about the "right" amplitude. As the absolute voltage of the EMG signal may not have any physiological significance in a particular experiment, the need for rigorous calibration rarely arises. However, a reasonable idea about the true amplitude (in millivolts) referred to input and its relationship to previous experience in a given muscle can provide the first warning of a migrated electrode, denervated or reinnervated muscle, or excessive scar tissue. Also, sudden fluctuations in the electrical noise, both of wideband thermal noise and line hum, may signal breakdown of

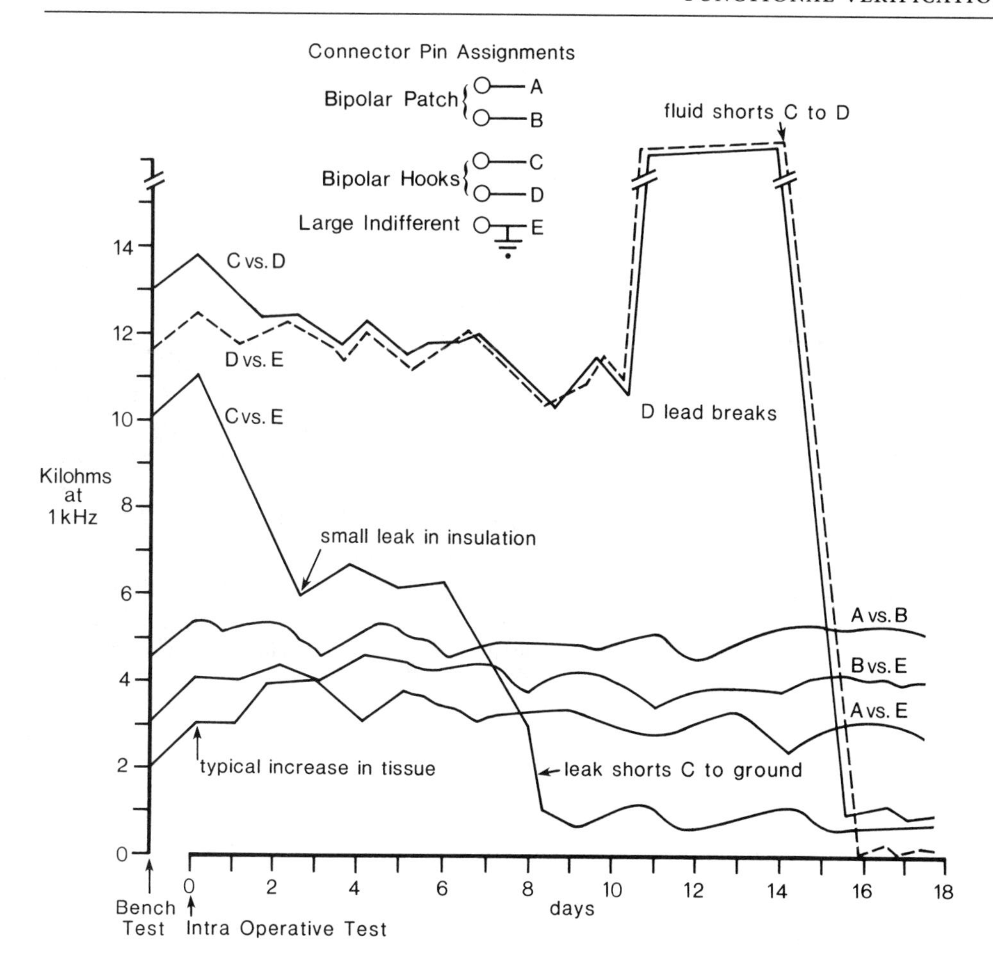

8.1. Plots of the results of electrode impedance tests both between single contacts and ground and between contacts forming bipolar pairs provide vital but different clues to problems. Hook electrodes, with a smaller surface area, start with higher impedances than patch electrodes. Once implanted in vivo, the impedance of both types rises slightly over bench test values because tissue has higher resistivity than pure isotonic extracellular fluid (saline solution). Because of their higher tip impedance, even a small leak in one lead of a hook electrode (*C*) is readily detectable as an impedance decrease to ground, although the pairwise impedance is still dominated by the intact lead. Later, the connecting lead to *D* has broken in a location in which the broken end is not touching body fluids. Both the single-ended and the pairwise impedances then become very high (although not infinite because of stray capacitive conductance in the intact part of the leads). Finally, fluid gets into the connector housing itself, shorting leads *C* to *D* at a point proximal to the wire break; this decreases all impedances to very low values (but not to zero because even the large, exposed metal/electrolyte interfaces have some junction impedance).

electrical contacts or connections with resultant open circuits, high-impedance contacts, or imbalanced source impedances.

D. TENDON STRAIN CORRELATION

The best possible verification of EMG validity is the simultaneous recording of the active tension of the muscle via an implanted strain gauge (fig. 8.2). Of course, with such a mechanical record, one might not need the EMG tracing at all. However, it may be worth the trouble to study EMG to strain gauge correlations, in a few out of a series of animals. The more easily obtained EMG may then be used more securely as the only record of muscle activity. Tendon strain correlation (or a similar direct mechanical record) often helps to clear up several possible sources of confusion.

The first potential source of confusion is due to the fact that some background cross-talk is always present anywhere; it may be very difficult to distinguish it from low-amplitude EMG signals in muscles that have small motor units and may be only weakly recruited during the task under study. Second, in certain situations, such as small, thin muscles sandwiched between large ones, it may be impossible to have any a priori confidence in the selectivity of any electrode design. Third, many muscles are very pinnate, with large numbers of short muscle fibers associated in motor units that are distributed over only a small part of the muscle. There is no guarantee that the electrical activity recorded in one part of such a muscle will be representative of that in another, and there is every reason to suspect that it may not be if the insertion of the muscle is broad, giving it a range of mechanical actions. Finally, there is no guarantee that an electrically silent muscle is not, in fact,

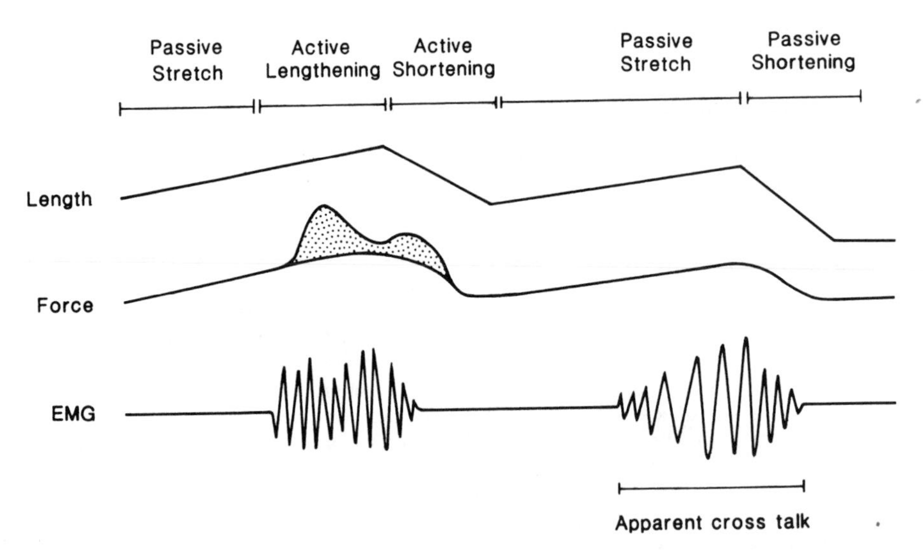

8.2. Identifying cross-talk with length and tension transducers. It should always be possible to correlate the EMG signal with the active tension indicated by the tendon strain gauge. However, the relationship will not be simple because the force output of active and passive muscle is nonlinear in relation to the amount, direction, and rate of stretch. Passive force output may not be negligible and active force output during shortening will be much less than that during isometric or lengthening conditions.

8.3. Identifying cross-talk by cutting of motor nerves. Only those parts of the EMG signal that are eliminated by cutting of the nerve supplying the muscle being studied are attributable to the muscle itself. Any other observed signals result either from cross-talk or from other artifacts.

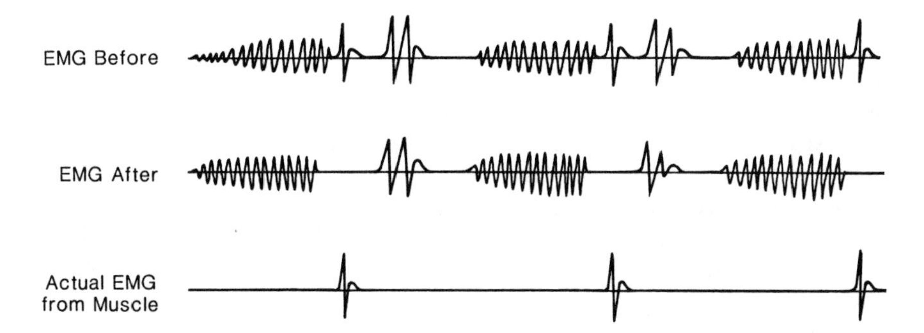

generating considerable passive tension as a result of the mechanical stiffness of its connective tissue.

One note of caution in relating EMGs to strain records: The quantitative relationship is very dependent on muscle length and velocity, particularly the direction of the velocity. Large EMG signatures may appear as only tiny forces if the muscle is shortening rapidly, whereas very low levels of recruitment (and EMG) can produce large force outputs in muscles that are lengthening (see chapter 17). Also, the EMG signature will precede force output by 5 to 100 msec depending on the species, its body temperature and size, the fiber type of the muscle, and its intrinsic speed of contraction.

E. DENERVATION CONTROLS

The best electrical verification of EMG signals is simple and often overlooked. Cutting of the nerve innervating the muscle that is presumed to be generating the recorded EMG signal should make the entire EMG signal disappear (fig. 8.3). Usually, the particular muscle is one of several synergists involved in the task, so the animal can continue to perform quite normally after this surgical denervation. Because the synergists are usually adjacent to the muscle under study, their potential contribution to cross-talk often goes unsuspected, because the data look so plausible. This little experiment can be a real eye-opener. Of course, it can only be done for one muscle at a time; a second animal must be used for the reciprocal experiment designed to check for cross-talk in the other direction. Denervation is a particularly effective procedure for distinguishing low-level recruitment from background cross-talk; the separation may be difficult to make with strain gauges because of the small forces involved.

The biggest difficulty is that of locating the nerve or nerve branch that supplies a single muscle, particularly if considerable scar tissue has formed and the animal contains several implanted devices and their leads. Consequently, it may be helpful to expose and identify the innervation point during the initial implantation and to tag it with a small, distinctive suture in order to facilitate its location during the subsequent operation. Again, careful operative notes and sketches can be invaluable.

Temporary nerve blocks by local injection of anesthetic offer a less drastic approach to the denervation control. Presumably, one would have to administer the anesthetic by percutaneous injection rather than open dissection, as the animal must be up and performing normally before it wears off. The number of such sites that one can confidently block completely, without risk of also blocking adjacent muscles, is probably quite limited. Theoretically, one could leave a small catheter implanted near the innervation point of the target muscle to serve as a conduit for the anesthetic infusion. However, after a few days, the formation of scar tissue is likely to block the catheter or may cause any significant volume of injected fluid to take diffuse and highly unpredictable courses from the tip.

F. MULTIMUSCLE CROSS-CORRELATION

One good way to establish that a particular EMG is uncontaminated by cross-talk is to record simultaneously all of the adjacent, potential cross-talk sites and show that their records do not correlate with those from the subject muscle. Even if the adjacent muscles are synergists, there should be little or no overlap in the precise timing of peaks and valleys in the raw, wideband EMG signal (fig. 8.4). Frequently, even synergists will show enough difference in the general phasing of their activity (onset and offset time) that reasonable confidence may be generated by simple inspection. Careful examination of multitrace oscillographic records (at least 5 msec per centimeter sweep) should reveal significant fine correlations whenever cross-talk occurs. If the innervation points of the two muscles are close and the electrodes are close to the innervation points in both cases, then a large spike potential in one muscle will almost simultaneously generate whatever cross-talk may be present. Of course, the signal picked up by the electrodes lying in the adjacent muscles may be considerably smaller, and the

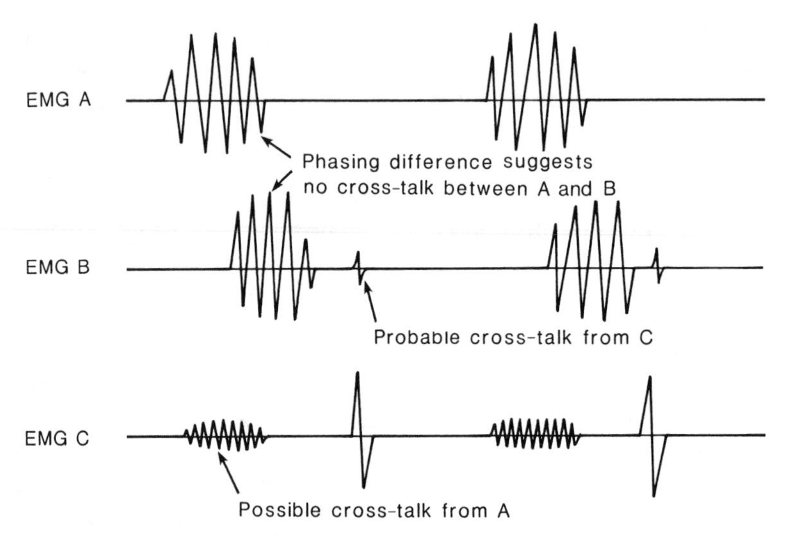

8.4. Use of gross cross-correlation to identify possible cross-talk. If simultaneous records are available from all likely sources of cross-talk, differences and similarities in the envelope of activity can be reassuring or alerting. However, if the muscles are similarly modulated synergists, cross-talk cannot be ruled out.

waveform may differ in shape and polarity. Should the innervation points lie at different longitudinal positions with respect to the electrodes, then the conduction time of the action potential down the muscle fiber may generate a temporal difference of up to several milliseconds. A consistent pattern of coincidences generally indicates significant cross-talk. The complete absence of coincidences beyond those likely to occur by chance lends some credence to the notion that the electrodes are sampling from different populations of motor units. Of course, this does not mean that they are sampling from different muscles. After all, motor units have limited domains within muscles, and a single muscle may have EMG activity in one of its compartments that is quite independent of that in another one.

The biggest weakness of this method is the hand-waving appeal to a subjective impression based on visual observation of the record. Rigorous signal-processing techniques exist for generating quantitative cross-correlations among combinations of signals. They are fairly sophisticated and time consuming in their demands on computing equipment, and there seem to be few examples of their being used previously in this manner. One obvious question that would have to be settled would be the establishment of objective criteria to judge whether a particular level of cross-correlation indicates acceptable isolation. Probably this would require that specific controls be conducted, such as deliberate implanting of electrodes intermediate between those for which cross-talk is suspected and comparison of their cross-correlation with that for putatively independent electrodes. Also, there are interneuronal circuits that may, under particular circumstances, give rise to substantial synchronization of heteronymous motoneurons, an important finding that one should not reject as necessarily artefactual.

G. ELECTRICAL STIMULATION OF MUSCLE NERVES

One of the most severe tests of cross-talk is to measure the amplitude of the signal recorded at a given electrode when its own and adjacent muscles are synchronously activated via electrical stimulation of their nerves (fig. 8.5). The synchronous activation of the muscle is, of course, very nonphysiological and will give rise to enormous EMG spikes that can be recorded by almost any design of electrode anywhere in the body. If the target muscle is small and the adjacent muscle is large, then the signals recorded with stimulation of each may not differ that much, even for rather selective recording electrodes.

Whereas this test may be unduly severe in many situations, it can be quite appropriate for evaluating potential cross-talk under conditions of physiologically synchronized activity, such as expected in reflexes. The degree of selectivity that would be acceptable for attributing reflex EMG signals to one muscle or another is much higher than that for identifying voluntary, graded recruitment. Whereas the synchronization of reflexes is not nearly as tight as that seen in electrical stimulation, the latter at least puts an upper limit on the amount of possible cross-talk. Another advantage of the electrical stimulation method is that it can be used to validate a record derived from only a single recording electrode; the various stimulations can be performed as a terminal experiment involving dissection with the animal deeply anesthetized. The multimuscle cross-correlation technique is very useful for pointing out potential sources of cross-talk during reflexes, but not for quantifying them. Nerve stimulation is probably the best technique available for validating percutaneous wire electrodes inserted for short-term use, provided they are not dislodged during the dissection necessary to expose the muscle nerves.

H. STIMULATION THROUGH RECORDING ELECTRODES

Stimulation through the recording electrode is the most commonly used electrode localization technique. It is listed distressingly far down on the list, and for good reasons. The usual technique is to apply electrical stimulation through the recording electrode to see which muscle twitches. This is most commonly done with percutaneously inserted wire and needle electrodes, and the twitch is evaluated by vision or palpation. Just deciding which muscle is twitching is nontrivial in some of the more mechanically complex parts of the musculoskeletal system. However, that is not the major weakness of this method.

Unfortunately, the structure being stimulated electrically is never the same as the structure generating the recordable potentials (fig. 8.6). Muscle fibers are biophysically like large-diameter, unmyelinated nerve fibers, whereas their innervating nerve fibers are large-diameter, myelinated nerve fibers. The electrical stimulation thresholds for unmyelinated fibers are typically a factor of 10 to 100 higher than for myelinated fibers. It may be virtually impossible to activate them electrically with stimulation pulses of brief duration. Long before any muscle fiber reaches threshold, the axons of any and all muscle nerves in the vicinity will have been activated. These in turn will transsynaptically activate all of the muscle fibers that are innervated by the moto-

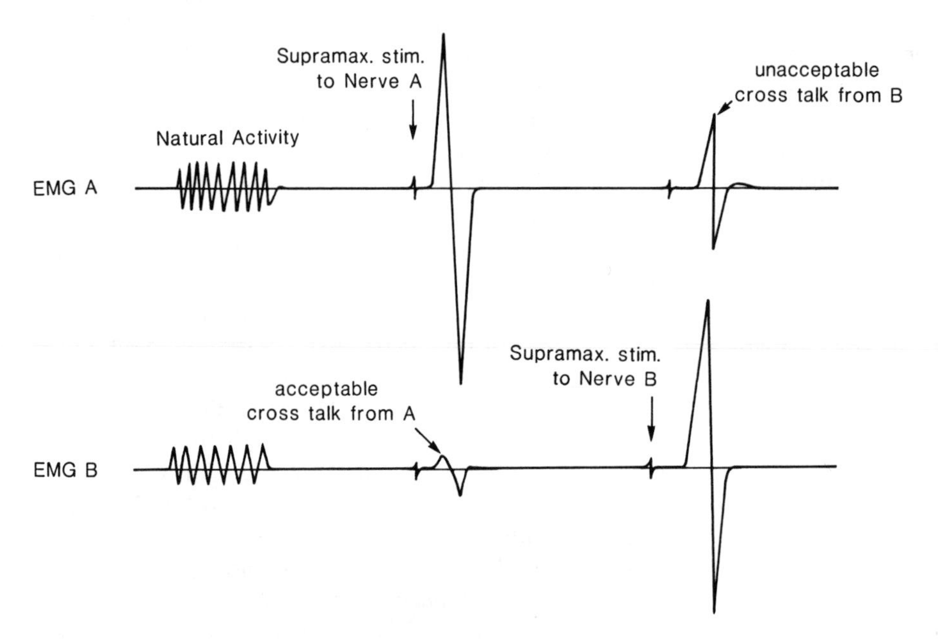

8.5. Use of electrical stimulation to identify possible cross-talk. With the recording electrodes still in situ, supramaximal electrical activation of the target muscle can be compared with the signal recorded when each adjacent muscle is similarly activated. Note that the enormous synchronous activity will almost always produce some cross-talk, so the result can only be interpreted in the context of the relative amplitudes of the naturally occurring and electrically evoked signatures in both recording sites.

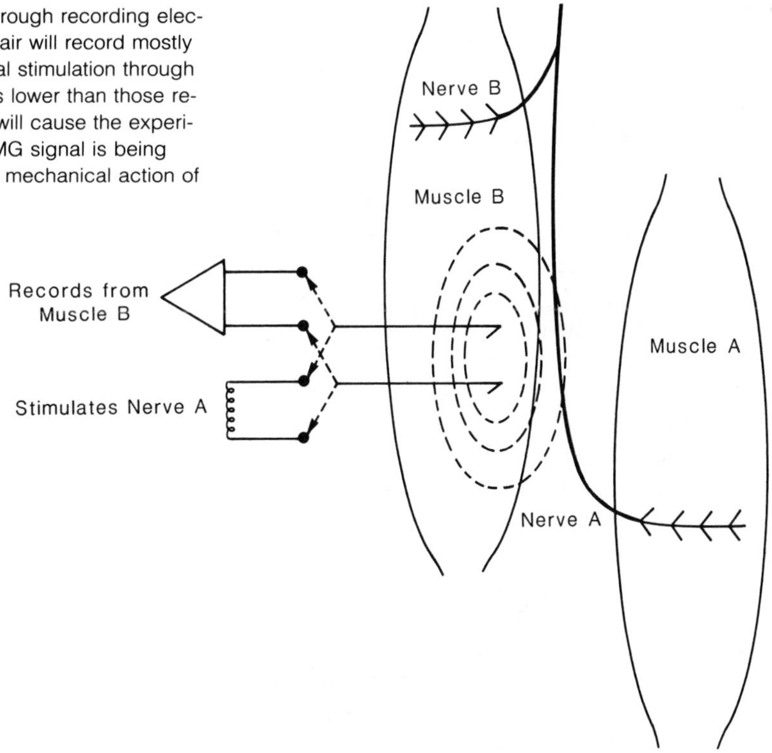

8.6. Problem with stimulating through recording electrodes. Note that the electrode pair will record mostly from *muscle B,* whereas electrical stimulation through them will recruit *nerve A* at levels lower than those required to recruit *muscle B.* This will cause the experimentalist to conclude that the EMG signal is being recorded from a muscle with the mechanical action of *muscle A.*

neurons to which the axons belong. In fact, even before the motoneurons are activated, the group 1 afferents of proprioceptors (muscle spindle primaries and Golgi tendon organs) with their large myelinated diameters will be activated, possibly giving rise to reflex twitches in some muscles.

At threshold intensity, the experimentalist will be evaluating the muscle twitches caused by the axons of whatever alpha motoneurons happen to be in the vicinity of the electrode. Now, if the electrode lies deep in the belly of the muscle near its innervation point, it is possible that these will be the motoneurons that actually activate the muscle fibers that generate the recordable EMG signal. However, if the EMG electrode has been located some distance from the innervation band of the muscle, it may lie more closely to the nerve fibers of an adjacent muscle or a passing nerve trunk. In that case the electrode will be identified spuriously as being located in the muscle that twitches and not in the muscle that generates the EMG signal detected by it.

It should also be noted that even if the correct muscle twitches first (at lowest intensity), there is no direct quantitative relationship between the selectivity of the electrode for recording and the amount of suprathreshold intensity increase that can be imposed before other muscles start to twitch. The efficacy of a bipolar stimulating electrode depends on its dipole moment orientation with respect to the target, just as that of a recording electrode depends on its orientation (see the discussion of stimulators in chapter 12). As the nerve fibers are likely to be running at nearly right angles to the muscle fibers as the nerve enters the muscle, an electrode orientation that is optimally selective for nerve stimulation may be very suboptimal for EMG recording, and vice versa.

The real question does not concern the location of the electrode but the selectivity with which it records the EMG signal of a particular muscle. Thus, electrical stimulation is particularly likely to be misleading for percutaneous electrodes. Yet these are precisely the electrodes in which distance from and orientation with respect to the innervating motoneurons are least likely to be known.

There is one potential use for this technique. If an electrode is initially implanted accurately in or on a particular muscle (e.g., a surgically implanted patch electrode), subsequent tracking of the thresholds and suprathreshold effects of electrical stimulation could provide a very sensitive indicator of the electrode's possible migration. As this test can be performed with an awake or anesthetized animal, regardless of its ability to generate motor output, it could prove most helpful in monitoring the function of electrodes during prolonged recovery from experimental manipulations (e.g., spinal transection) when normal EMG activity is unavailable. The stimulation would not be used to indicate selectivity (which must be demonstrated independently) but to provide a nonspecific indicator of changes in electrode position vis-à-vis local nerve fibers.

The above discussion assumes that the stimulation paradigm is one that causes no electrolytic damage to the electrodes or surrounding tissues (see chapter 7). Furthermore, one should rely only on constant current stimulation rather than on constant voltage stimulation, because the former compensates for changes in the contact impedance of the electrode whereas the latter does not.

Part 2 APPLICATION

9 Fabrication Equipment and Techniques

A. THE PROBLEM

As the reader is about to discover, the pursuit of EMG experiments soon leads to a proliferation of electronic and mechanical devices that must, in the main, be built and serviced by the experimentalist. Most graduate programs do not include courses in machining, soldering, electronic debugging, or the selection and use of test equipment. These less glamorous but most useful skills are left to that most demanding of institutions—the school of hard knocks.

The array of equipment commercially available to assist one in these tasks (or at least to relieve the purchaser of some grant funds) is all the more bewildering to the novice because it is necessary to buy the equipment and use it in order to determine whether the item is needed or even usable. Inevitably this leads to drawers and shelves filled with white elephants. In this chapter we review the contents of our drawers and shelves. The general message is that almost anything you are likely to need can be accommodated by a rather modest arsenal of hand tools and skills. Below we list some of these, indicating mainly those that are necessary for any laboratory but also a modicum of luxury items, which may not be absolutely necessary but can make life easier. Hobby shops and hobby supply houses are useful sources for clever tools for working on small objects.

The specialized tricks of the trade itemized below pertain particularly to the fine wires and materials used in electrode construction, but they are derived from general principles relevant to all electronic equipment. A few additional ones will be found scattered in the special sections, for instance in chapter 10.

B. EQUIPMENT

1. Soldering Gear

The great diversity of available soldering equipment gives the first clue that there is no universally satisfactory solution to soldering, which, like democracy, is the worst system except for all the others. A general purpose iron (never a soldering gun) should be light, easy to hold, and equipped with a support stand with a pilot light and a place for a damp sponge. Modest power (25 to 40 W) and temperature control are advisable, but the experimentalist should avoid those temperature-controlled irons that seem to respond very slowly to thermal loads (long warm-up time is a clue).

Small tips are getting easier to come by in this age of miniaturization. The experimentalist should make sure they are easily changed and should keep a good supply because they burn out easily, especially if the iron is left on (hence the pilot light). Most modern tips have plated surfaces, which cannot be reshaped by grinding and which eventually wear out, resulting in poor tinning. There are some very cute rechargeable battery-operated units available. These have very fine, low-power tips and limited stored energy, so they cannot substitute for a general bench iron. However, they are ideal for delicate repairs in and around the preparation, which can be a shocking experience for the animal if line-operated irons are used.

Other supplies that you will need include fluxes, solders (discussed in this chapter, section C2, Soldering) and hand tools (discussed in this chapter, section B5). Solder removers can come in very handy. There are two basic types. One is a spring-action suction plunger, which works best on stable flat surfaces such as printed circuit boards. The other is a resin-impregnated copper braid, which wicks solder away from any surface. Both are inexpensive.

Finally, a small bench edge or table-top vise is indispensable for those occasions when you run out of hands or find that your fingers are not made of asbestos. There are also several kinds of stands supporting a pair of spring-loaded clamps, each mounted on a ball bearing arm. Hobby shops have small vises designed for tying fishing flies. Some of these are inexpensive, but even cheaper models can be made from large or medium-sized alligator clips screw mounted on pieces of rod fixed to an old bench stand with a pair of machine bolts and wing nuts. Remember the specification: you want to support two or more small connectors or cables and to move them easily relative to one another, but you want them to stay in place once their position has been adjusted. Rarely need the clamps spread for more than 20 cm, but the clamp positions must be able to overlap and the angles at which they meet must be adjustable. Do not forget the need for adequate local illumination and the utility of various kinds of magnifiers.

2. Resistance Welding

Although soldering irons can do marvelous things in skilled hands, a resistance soldering/welding unit of the kind used in the manufacture of dental prostheses may have substantial advantages when one is fashioning minipin connectors or working in close quarters. Such units do not heat until after both metal tips contact the wire (hence an accidental lateral short should be avoided) and thus provide minimal lateral radiation. They are a luxury that may save much time and effort. This is the best method for joining metal parts without introducing dissimilar and corrodable intermediaries. However, they should not be used for connecting electronic components that may be damaged by current surges.

The principle of operation is the intense local heating that occurs when a large-amplitude, brief current pulse passes through the relatively high resistance between two wires that are in contact. Usually, a certain amount of pressure is applied simultaneously. The selection of material for the welding electrode is critical in terms of avoiding welding everything in the circuit into one solid mess. The adjustment of the variables of time, voltage, and pressure is both an art and a science about which the manufacturers of such equipment often have advice. Some will gladly take samples of your welding problem and try them out to find the correct equipment and settings before you make a purchase.

3. Connector Assembly Tools

Many connector types require or can be assembled with special techniques such as crimping and pin insertion. Some of the required or recommended specialized hand tools seem unconscionably expensive. However, this is no place to economize. Connector systems are fragile enough without relying on suboptimal versions.

Any laboratory employing BNC cables (they are ubiquitous as equipment jacks) should have access to hand crimping tools and crimp-type connectors sized for the coaxial cable in use (see fig. 11.9). This type of connector is very rugged and reliable when crimped correctly with the proper stripping and crimping tools; it is particularly treacherous under less than ideal assembly conditions, being prone to intermittent loss of signal, loss of ground, and short circuits between center conductor and shield. Although the crimping tools tend to be outrageously priced, they quickly pay for themselves by minimizing the number of damaged and defective connectors. Many new types of ribbon cable connectors are impossible to assemble correctly without their special presses. Whenever a connector or cable appears mechanically worn or electrically questionable, it should be marked immediately and set aside to be repaired or replaced. This rule must be drilled into everyone using a complex laboratory system.

4. Test Equipment

The only really essential test item is a plain, old-fashioned volt-ohm-meter, and the cheaper the better. It should be rugged and convenient to set on a bench or hang near equipment, and it should use commonly available, easily replaced batteries. Forget about high sensitivity, high accuracy, or digital displays. You want a moving needle you can catch out of the corner of your eye to tell you whether point A is connected to point B or whether you have blown the DC power supply. The few pieces of equipment in your laboratory that require precise calibration probably need a qualified serviceman for repairs.

It is assumed that you will have an AC impedance meter suitable for in vivo testing of electrodes with a low-current, 1 kHz test signal (see chapter 14). Once you know its output characteristics, this will probably make a dandy test signal generator for checking out and calibrating tape recorders. Most AC impedance meters have constant-current output stages, and you can turn them into voltage sources by connecting them across a precision resistor. You will probably also have an oscilloscope, which should be picked for its convenience and versatility regarding your experimental design. All modern oscilloscopes, even the cheapest, have more bandwidth and stability than you are likely to need for trouble-shooting equipment. A two-trace display is handy for comparing input with output signals and stabilizing triggering on a known signal. When many signal channels must be examined, it is often better to have a convenient switching system for the input to one trace, rather than trying to keep track of multiple simultaneous traces. Trace-storing cathode ray tubes are useful for such comparisons, although they are expensive.

Many scopes, particularly the new digital ones, come with all kinds of fancy waveform-storing, measuring, and transforming accessories, which are nice toys if you can afford the expense and the distraction from what you should be doing. There are some very ingenious computer interface and software packages for simulating a scope, which is nice except that your biggest headache is likely to be monitoring and debugging the computer and its interfaces, an interesting catch 22. In the end you will still need a scope to check that the fancy system is operating correctly.

Do not forget the extenders for printed circuit boards. Most pieces of equipment, from computers to tape recorders to modular amplifiers, employ printed circuit boards with edge card connectors. These are prone to failure and also provide the only access to important internal signals and power supply voltages. Unfortunately they tend to be virtually inaccessible in situ. The only way to get at the circuitry while it is running is to insert an extender board and plug the printed circuit board into it so that its connections may be reached. It is best to obtain an extender for each particular connector type in your laboratory. But be careful! Sticking probes on and between components with the power turned on can turn a small malfunction into a very big one. A good set of small, hooded test clips and a steady hand are essential.

5. Hand Tools

It is a simple fact that the availability of hand tools when needed (often in dire moments) depends on the degree to which you have saturated their natural binding sites in the laboratories, desk drawers, home shops, and car trunks of all your staff and colleagues. You should also remember that carbon steel rusts easily in the damp and corrosive atmospheres of biological laboratories, but this process can be controlled by periodic polish with oxalic acid powder (be careful to avoid skin contact and, even more, inhalation of the dust) or treatment with rust-resistant jellies. Consider such hand tools to be expendable supplies and buy them cheaply, in quantity, and whenever supplies start to get low. You might take the following general shopping list to your local discount house.

1. *Scissors.* Nothing cuts, trims, and strips fine electrode wires better than scissors, so make sure you have plenty of inexpensive ones on hand. Remember that your precious Mayo dissecting scissors do not improve with the cutting of stainless steel.

2. *Wire cutters.* Sturdy, side-cutting wire cutters are useful for cutting component leads, heavy cables, and other such items. Buy a version made of good steel; the edges of some drugstore variants dull rapidly.

3. *Needle nose pliers.* Fine, narrow needle nose pliers with a spring opener often prove to be helpful.

4. *Hemostats and fine-tipped forceps.* Although designed for "better" things, these tools are remarkably useful for handling small wires and components. However, label them and do not interchange them with those destined for operative use.

5. *Vise grip pliers.* These pliers are handy for the inevitable occasions when you should have used a wrench.

6. *Socket drivers.* Purchase deep-throated electronic socket drivers for tightening shaft hardware, so you will not use the pliers.

7. *Wire strippers.* It is handy to have at least one simple, adjustable wire stripper with preset stops or jaw grooves for the standard small wire sizes (22 to 36), with circular rather than V cutters, plus one adjustable-depth razor blade cutter for the jacket on coaxial cable.

8. *Hex wrenches.* You will need plenty of the small sizes to use on the set of screws of knobs and dials.

9. *Screwdrivers.* You can never have too many screwdrivers. You should purchase some long, narrow-bladed ones that can reach trim pots and the set screws in the backs of banana plugs. Also buy a set of miniature (watchmaker's) screwdrivers to adjust set screws on electronic boards (they also pop up in other places when you least expect it).

10. *Files and sharpening stones.* A set of fine, high-quality metal files and a good oil sharpening stone will come in handy for removing those inevitable nicks and burrs. They will extend the life of your other tools.

These hand tools pertain to electronic equipment only. However, the production and modification of assorted testing gear, such as walkways, mirror chambers, and camera stands, make it desirable to expand the set and also to include odd items such as pipe cleaners, tie-wraps, a can of compressed air and a small vacuum cleaner. Even if a shop is available, the laboratory might well be furnished with a set of tools for evenings, week-ends, staff vacations, and the inevitable time that the animal performs well but the shop cannot provide that critical item.

We would suggest a sturdy 3/8-inch electrical drill; a saber saw with wood, metal, and plastic cutting blades; an assortment of clamps, hammers, and screw starters; a level; and of course, a carpenter's rule.

C. TECHNIQUES

1. Wire Stripping

a. General. Judging from the great variety of insulation-stripping devices available on the market, there is no universal solution to the seemingly trivial problem of removing the insulation from the wire (fig. 9.1). Mechanical cutting may nick the wire; it may cause weak points at the exit from the protective sheath, already a vulnerable point itself. Mechanical abrasion is difficult to control, and chemical dissolution normally works poorest with precisely those materials that best insulate implanted electrodes. Burning is likely to leave a residue of potentially toxic materials that will clutter up the supposedly exposed metal surface and affect the electrical response. It may also anneal or oxidize the conductor, changing its mechanical or chemical properties. In the following section we suggest tricks that work well for stripping the particular materials mentioned as well as strategies to try on new materials.

b. Teflon-jacketed, Stranded Stainless Steel. Stranded stainless steel with a Teflon coating is a favorite EMG electrode material. It is used as a good electrode contact and a flexible leadout cable, eliminating the need for joints. Fortunately, it is easily stripped by several techniques. First, the Teflon jacket is a fairly loose sleeve over the wire strands and slides easily over them. If the sleeve is nicked slightly and then pulled, the Teflon jacket usually tears at the nick, perhaps a bit raggedly, and the sleeve slides off. For short exposed ends, such as a few millimeters of electrode contact, a better trick is to stretch the Teflon sleeve out over the end of the cut wire

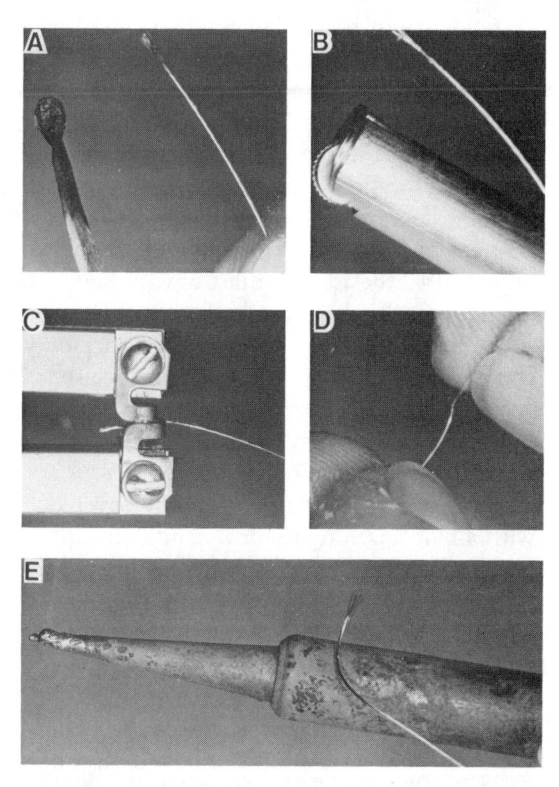

9.1. Various methods for stripping insulation from fine wires without nicking the conductor. *A,* A paper match will burn off many polymers, although this may leave charred remains and the poorly controlled temperature may melt some metals such as gold. *B,* A butane lighter with a small, adjustable flame provides a cleaner combustion. *C,* Best temperature control comes from a thermal wire stripper with a resistance-heated filament applied directly to the wire, melting and/or burning off the insulation. *D-E,* Teflon can often be stretched out over the end of a fine lead, so that the loose sleeve can be cut off and then shrunk back with slight heating. This provides a clean demarcation of the end of the insulating sleeve without leaving a fringe of damaged or charred material on the lead.

(squeezing it between the fingers) and cut off the desired length (fig. 9.1*d*). If the remaining stretched portion is then warmed (carefully)—for example, against the side of a soldering iron—it will shrink back, leaving a completely clean and untouched length of exposed strands (fig. 9.1*e*).

Teflon is not really flammable. Consequently, one should never try to burn off the insulation, as the wire will almost certainly be badly oxidized if it is heated to the temperature at which the charred remains can be removed. It should be noted that the looseness of the insulation makes reliable and stable exposure of a very short piece of wire impossible—that is, the free tip must always be longer than the diameter of the wire. This insulation will deform if autoclaved; hence only chemical sterilization will maintain precise geometrical surface area.

c. Platinum Alloys Coated with Insulation. Alloys of the platinum family covered with any insulation are ideal candidates for burning, because the very high melting point of noble metals protects them from serious degradation up to the ash temperatures of almost all polymers. Because one is usually dealing with fine wires and perhaps with a closely fitting, adherent insulation, mechanical removal of the insulation can be tedious and damaging. The trick is to apply the heat as locally as possible so that the insulation is cleanly removed where desired and intact everywhere else, without an extended intermediate region of degraded material. A hot filament, such as is employed in some thermal wire strippers, can be very useful (fig. 9.1*c*). The experimentalist can construct it easily by attaching a Variac to a short piece of heavy-gauge platinum or nichrome wire. Very high temperatures can also be achieved in so-called Microtorches, which are like very small-tipped welder's torches with tiny compressed gas tanks (see appendix 2; fig. 9.1*b*). However, the edge of a Bunsen burner flame or pocket lighter is often adequate, particularly if the adjacent wire, to be left insulated, is well protected during exposure (e.g., if the wire is held close to the tip in heavy forceps, which will serve as a heat sink).

d. Copper or Gold Wire with Flammable Insulation. The combination of copper or gold wire and any flammable insulation poses a very difficult problem, because these malleable materials have very low melting temperatures. Furthermore, their ductility allows surface nicks to grow into deeper clefts should the wire be bent repeatedly, which generally limits the use of these materials in long-term, implantable devices; however, their extreme flexibility and easy availability may make them attractive for short- and medium-term procedures. Fortunately, they are often coated with readily flammable insulators such as lacquers and varnishes (also making them unsuitable for long-term implantation). A low-temperature flame such as a paper match can often be used to burn off the insulating jacket, if not completely, then at least until the remaining ash is easily flaked or abraded off (fig. 9.1*a*).

2. Soldering

a. General. Probably no other single step causes as much long-term grief in an electrophysiology laboratory as soldering. With practice, the incidence of bad solder joints gets just low enough to induce a sense of complacency; unfortunately, it never drops to zero. With implantable devices, the problem is compounded because (1) the wires involved may be very small, (2) materials such as stainless steel require special fluxes and techniques, (3) the environment accelerates surface oxidation in "cold" solder joints, and (4) implantation removes the joint from testing or service.

Whereas rules and instructions for good soldering techniques abound, in practice they gradually tend to become abandoned unless the practitioner understands enough about eutectic solders to respect the rationale.

The sequence of events is pretty much the same for all metals, fluxes, and solders. First, the chemical flux agent (activated resin or acid solution) is used to strip surface oxides from the metals to be joined, thus exposing the bare metal. Then the metal surfaces are heated to a temperature that will melt (but not degrade) the mixture of metals composing the solder; this mixture has a much lower melting temperature than do the wires. The molten solder "wets" the hot wires so that the solder atoms assume available lattice sites at the surface of the wires (fig. 9.2). In a good joint the solder should flow smoothly out along the surfaces of the metals being soldered, much as water spreads on a hydrophilic surface rather than beading up, as do water droplets on a hydrophobic polyethylene bag. Ideally, the well-applied solder solidifies abruptly when the heat is removed. It then forms a single, homogeneous mass that is both mechanically rigid and electrically continuous with embedded wires.

At each step there is a potential for failure. If the flux is inappropriate or kept from reaching the surface of the metal by surface contaminants, the solder will not wet the surface. A similar problem occurs if the metals to be soldered are not themselves heated past the melting point of the solder. In either case the molten solder may roll onto or around the metal surface, but will not actually bond into the crystal lattice of the metal. Often, an intervening layer of flux solidifies between the solder and the base metal, giving the illusion of a good mechanical joint but forming an unreliable electrical connection. This is the classic cold solder joint, which is apparent to the discerning eye by the sharp convex meniscus at the edge of the solder caused by its surface tension (see fig. 9.2). A cold solder joint may actually have electrical continuity when initially tested, but it is likely to be unstable,

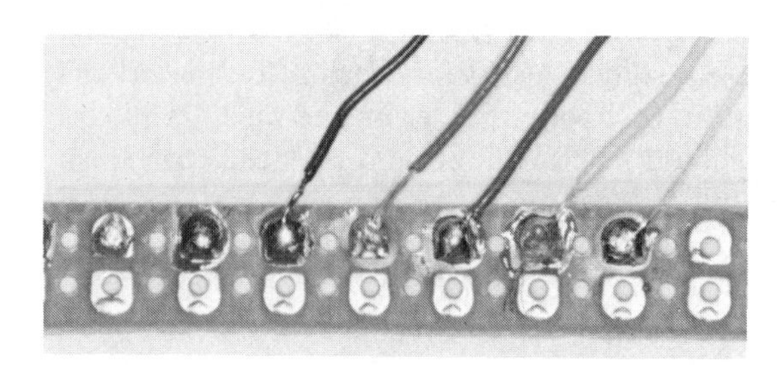

9.2. Various qualities of soldering. The first three pads from left are pretinned and ready to accept tinned leads (*upper row*). The first pad at left has too much solder, threatening to roll over onto the untinned pad below it. The second pad was tinned with stale solder that had been too long on the tip of the iron. It has crystallized, appears dull and lumpy, and has not flowed over the surface of the pad (note crevices and concavities). The third pad is just right. A sufficient amount of bright, smooth solder flowed to the corners of the pad but did not overflow. The first attached lead shows a well-tinned stranded wire inserted into a properly tinned pad. Note that the solder from the pad has flowed smoothly onto the wire strands, coating and combining them into a single structure. However, the insulation has drawn back a bit too far, threatening a short if this lead is bent toward the next one. The second wire was not pretinned, but simply thrust untimely into the molten solder on the pad. Note the puckering of the solder at the front of the pad, suggesting that the only thing holding this wire in place is a skin of fluxing resin; this will make a dandy noise generator. The third wire is soldered just right, with pretinned leads not extending too far out from the insulation. The fourth wire was not pretinned, and the operator tried to fix the problem by adding more solder during the formation of the joint. Not only did this not work, but it left a dome of excessive, partially crystallized solder. Furthermore, the excessive heat melted back the insulation and the operator then inadvertently slid the wire too far across the pad, leaving stray ends that threaten a short circuit. The fifth wire is successfully soldered, stranded stainless steel, showing tinning of the individual strands as they leave the solder pad, producing a smooth, bright joint. Kester 815 acid flux was used before the wire was separately pretinned.

and the interface may deteriorate in a bizarre way with gradually progressing surface oxidation of the metals, producing electrical noise, rectification, or thermal sensitivity.

One common mistake, especially when one is working with small parts, is to be stingy with solder. Just before each soldering operation, the molten solder on the tip of the iron should be cleaned off and replaced with fresh solder from the spool. Old solder is easily wiped from the tip onto a damp sponge, which is an essential tool during soldering. It is best to apply an excess of fresh solder to the tip and then to knock off most of the molten metal by tapping the iron against the counter top (paying some attention to where the hot solder will land). Both of these tricks are designed to combat two processes that are constantly changing the chemical properties of the solder on the hot tip. First, the activated resin gradually evaporates and is oxidized by the heat, leaving a sticky black residue that is not only useless as a flux but likely to obstruct the proper flow of solder on the tip and the joint. Second, the chemical composition of eutectic alloy tends to change locally as a result of melt separation of the components, oxidation, and even dissolution of the base metal of the soldering tip. Signs that one is no longer dealing with a eutectic alloy are that the viscosity of the solder starts to increase, developing little tails as the iron is pulled away, and that the soldered joint forms a rough, lumpy, or dull surface as it solidifies.

b. Stainless Steel. Soldering techniques for stainless steel must start with special preparation of the surface. First strip the wire, remembering that fingerprints and residues such as charred insulation will interfere with fluxing of the surface. Carefully paint the exposed strands with a commercial solution of stainless steel flux. Remember, this contains much hydrochloric acid that will attack electronic parts, tools, and fingers. A moistened cotton applicator makes a good brush and minimizes drips. Alternatively, use a filled eyedropper and briefly slide the wire into its end to wet the surface. Allow at least 10 to 15 seconds for it to work. Apply a fresh layer of a standard 60/40 tin/lead solder to the scrupulously cleaned tip of the soldering iron and allow about 2 to 3 seconds for the excess resin to burn off. Plunge the fluxed wire into the molten ball of solder hanging off the end of the tip. You should hear a sizzle as the excess flux evaporates. Remove the stainless steel wire and examine the cooled surface to see if the solder has flowed well. On stranded wire, the strands should be drawn together by the surface tension of the solder. If there is only a spot here and there where solder appears to be clinging, trim away at least several millimeters beyond the previous attempt and start again with a fresh end of wire. It is useless to attempt to reflux the stainless steel wire because a layer of solder resin will remain from the past attempt, and this will cover the unfluxed surfaces.

c. Platinum Alloys. Contrary to popular misconception, platinum, iridium, and their alloys actually accept solder quite well and do not require special fluxes. However, their surface oxides may require a fairly hot soldering iron, at least during the initial tinning process. As with any solder joint, each part to be joined should be tinned independently before being brought together with the heat and fresh solder to form the joint. Sometimes it is difficult to see whether a thin platinum wire has been tinned, as the layer will be thin and the colors almost identical. The ease with which solder flows to its surface in the joint will usually verify this. If the wire has been subjected to high temperatures (e.g., heat straightening), the oxide may be somewhat thicker and require mechanical scraping with a scalpel blade to provide entry to the bare metal by the solder flux.

d. Gold Wire. Soldering gold requires a very delicate touch and low temperatures, as the gold readily alloys with the solder and simply melts away, even at tip temperatures below its normally low melting point. Low-temperature silver solders (such as 62/36/2 tin/lead/silver) are helpful, and a temperature-controlled soldering iron is indispensable. Pretinning is unnecessary on the gold wire but essential on the mating part because the trick is to make the joint as quickly as possible in order to minimize dissolution of the gold into the molten solder. Fortunately, the complete absence of surface oxide on gold makes cold joints between the solder and the gold almost impossible. However, mechanical support of the finished joint is critical, given the initial low tensile strength of the material and the vagaries of the alloying and melting near the joint.

3. Handling Silicone Rubber

Most of the electrodes described below may be fabricated from insulated wire and various forms of silicone rubber. Whereas the virtues of silicone rubber as a biomaterial are well known (chapter 7), it does have certain limitations that must be kept in mind. These are that it is mechanically friable, that it tends to absorb contaminants onto its surfaces, and that it is virtually transparent to water vapor and many gases. All of these limitations make it unsuitable as a primary, thin-film insulation covering on wires. Silicone rubber can be reinforced with fillers and meshes where their bulk can be tolerated, giving composite materials with the best of both mechanical and biocompatibility features. Its biggest assets to the designer are the variety of forms in which it is available, its claylike ability to be molded, and the capacity of its liquids and pastes to be vulcanized at room temperature.

Silicone rubber is impervious to most chemicals, acids, alkalies, and solvents except those in the benzene/toluene family. Although the latter do not actually attack the chemical bonds, they are readily absorbed by the polymer and actually can dissolve monomer and oligomer. This can be used to advantage to thin uncured RTV materials, such as Medical Adhesive A (Silastic silicone rubber, Dow Corning Corp.). Solvents will thin this normally very viscous, thixotropic gel to the consistency of paint; the solvent then can be driven from the matrix by curing in a hot-air oven (60° C), care being taken with the very flammable fumes. A similar trick can be used temporarily to expand the diameter of silicone rubber tubing so that it will slip over a joint and then shrink snugly into place, perhaps squeezing out excess liquid rubber that has been applied around the joint. However, both of these tricks risk the incorporation of toxic residues. Only reagent grade solvents should be used, and the electrode should then be subjected to a prolonged baking (several days for thick parts).

4. Splicing Implantable Wires

The following splicing technique works quite well for devices that must function for weeks, although the reliance on solder joints and dissimilar metals is probably unsatisfactory for very long-term implants (lasting more than a few weeks). One advantage is that it can actually be performed in the operating room under essentially aseptic conditions (fig. 9.3). First, a short piece of loosely fitting silicone rubber (or polyethylene) tubing is slipped over the wire that has more free play, until it is well past the end. A tubing length about four to five times the length of the anticipated solder joint works well. The end of each wire is then exposed carefully about 2 to 3 mm (first

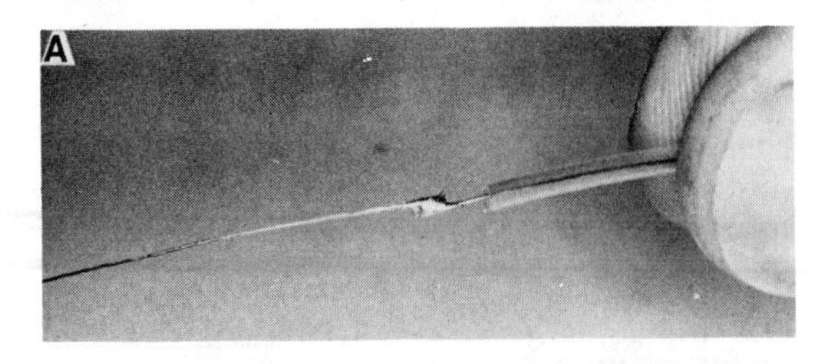

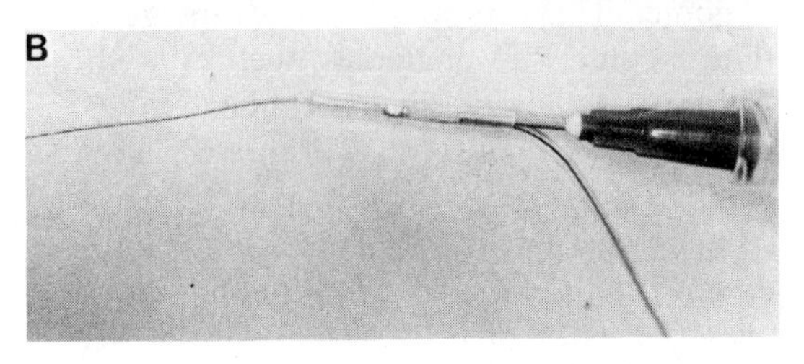

9.3. Splicing technique for two fine wires that will permit them to last in situ. *A,* A short piece of silicone rubber tubing was slipped over one of the leads (grasped at *right*), after which the two wires were carefully stripped and pretinned for about 2 mm. They were soldered while lying in parallel to prevent the ends of loose strands from protruding from the solder. *B,* The silicone rubber tubing was next slipped back over the solder joint and is being filled with silicone rubber RTV, completely encapsulating the joint and a few millimeters of insulated lead on either side.

slipping the tubing over the insulated wire keeps it from getting hung on splayed ends of stranded wire). Any special flux, as for stainless steel, is painted onto the wire and allowed to take effect; then each end is carefully tinned with fresh solder from the hot iron. The two ends are set side by side as shown in figure 9.3 and reheated to allow the solder to flow. The iron should be relatively hot and freshly but sparingly tinned, so that it leaves as small and smooth a joint as possible when it is pulled away; solder tails are particularly bad in this case.

The silicone rubber tubing is then slid down the wire and centered over the joint (fig. 9.3*b*). A syringe filled with Medical Adhesive A is then used to fill the tubing via a blunt needle. The elastomer must completely surround the wires and their joint. For additional security, a small suture can be tied snugly around each end of the tubing and then crimped down onto the lead. The entire procedure can be carried out aseptically. The acid flux is sterile, the solder is sterilized by the heat of the iron, and Medical Adhesive A may be autoclaved in the dispensing tube, as can the wires and tubing. The small amount of liquid silicone rubber will polymerize well, even if the splice is immediately enclosed in the body.

5. Inserting Wires via Needles

Hypodermic needles are generally designed to insert fluids. Their tips are completely beveled to slide most easily and perhaps painlessly through skin and tissues. They are usually not intended to be inserted through hard tissues except through previously drilled holes. More recently, there has been a second category of needles designed mainly to deliver tiny doses of chemicals to various containers; these may have quite different tip configurations. Neither type of needle is ideal for the insertion of the EMG electrodes, and at least three problems need to be considered.

The first is the length of the tip. A needle that is to be used for "painless" insertion of doses of penicillin into the gluteus maximus muscle often has a long bevel so that the "tip" may be three times the diameter. At the very least, this generates a situation in which the needle track extends substantially beyond the site in which the electrode will hopefully stabilize. Fortunately, many types of needles are now available that have a relatively short bevel but are still sharp. The one selected should combine the least angle with sharpness; it should cut, rather than tear, the tissue being penetrated.

The second problem is continuity of lumen.

Percutaneous insertion requires that two twisted wires be inserted via the lumen of a single needle. Obviously, the lumen should be just large enough to accept the wires; the larger the needle (the smaller the gauge number) the more damage is likely to be incurred during insertion. Some fabrication processes leave a bead at the bottom of the tubular portion where this is crimped into the connector, or the thermoplastic of the connector may partially occlude the lumen. This does not provide a problem for liquid delivery systems; it can be critical when one is attempting to thread two fine wires. The experimentalist can test for a bead by passing a trochanter wire (often provided for cleaning better grade needles) down the bore. This may also be used to ream away plastic irregularities and ensure that the wires pass the bore easily without unusual twisting and kinking. Such mechanical deformation may produce abrasion of the insulation, thus weakening the material and providing sites for tissue adhesion and breakage.

The third potential problem concerns the internal bevel. If the needles are cut or ground from a piece of tubular steel, one portion of the internal edge is likely to be just as sharp as the tip. Furthermore, this edge is placed just where the tips of your carefully manufactured electrodes bend into a hook. Any stress during insertion will cause this edge to perforate the insulation or, worse yet, guillotine the protuding tines. Obviously, different insulative coatings have different penetration resistance, and a special supportive coating at the bend may provide further protection (see the discussion of bipolar hook electrodes in chapter 10). Furthermore, some manufacturers now provide chamfered needles in which the internal edge has been dulled during manufacture. If these are not available, one can make a chamfering tool by grinding the tip off of a steel hypodermic needle of equal or smaller diameter; this tool lets one carefully chamfer away the internal edge to a dull or rounded surface (fig. 9.4). This procedure is best done under a dissecting microscope, so that one can check that the process has indeed rounded the edges rather than leaving them jagged. The chamfered needle should then be passed across the surface of an oilstone a few times to ensure that no metal projects beyond the plane of the bevel; any projecting edges will increase the risk of tearing tissues during electrode insertion.

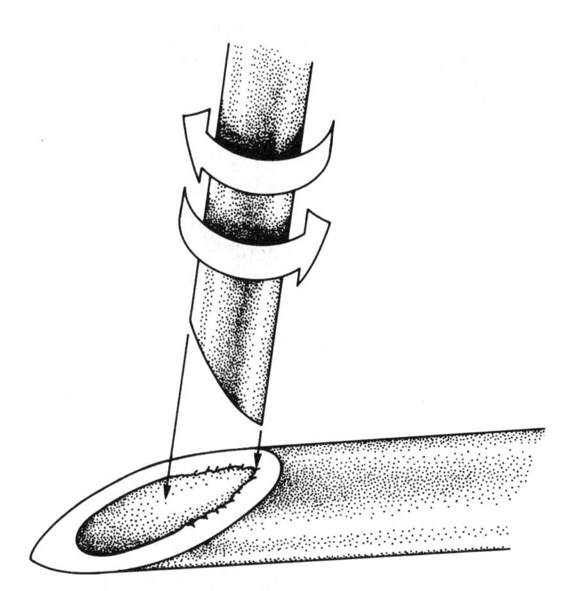

9.4. Chamfering the inside edge of the hypodermic needle. This edge contacts the hook electrode and pulls it along during insertion. In normally manufactured needles the edge is as sharp as the tip, although some manufacturers make special needles in which the inside edge is rounded. A simple chamfering tool for dulling the inside edge may be made from a short-bevel hypodermic needle with an outside diameter that is equal to or smaller than the inside diameter of the insertion needle. It is useful to make the chamfering needle of high-quality steel and repeatedly to polish it on an emery stone; this maintains its sharp edges and abrading capacity. Also the surface of the chamfered needle should be passed across the stone to remove any deformed metal.

Investigators must, in any case, decide whether and when to use the old-fashioned permanent needles or the more easily available throwaway kind. If needles are to be reused, they must be cleaned and checked each time. The cleaning trochanters sold with permanent needles make excellent bore conditioners. There should be a last wash with distilled water, perhaps followed by a surgically appropriate metal conditioner. After washing, the fluid remaining in the barrel should be blown out or removed centrifugally.

Cleaning rather than mere autoclaving is critical, as tissue fluids and other organic materials enter the lumen during insertion and will accumulate there. They tend to cake on during autoclaving and will later interfere as electrode wires are threaded through the lumen. More important, they increase the risk that the foreign protein on the implanted wires will incur tissue reactions as a result of challenges to the immune system. Quite apart from this, needles should always be discarded once their edges show any indication of mechanical deformation.

10 Design and Construction of Electrodes

A. PRINCIPLES OF DESIGN

1. Simplicity

The biological materials forming the bodies of living animals generate variously distributed electrical potentials. Consequently, an almost infinite variety of electrode designs could detect biological signals, and many diverse types have been reported in the literature. In this chapter we first discuss the major techniques for electrode insertion and then organize electrode configurations into six basic types, based on their antenna properties (see chapter 6). For each type we give an example of one specific configuration with which we have had some experience. However, the tissues of animals differ in their consistency and architecture, and animals differ strikingly in their responses to operation and the likelihood of infection. No single configuration will suit all applications. Therefore, our examples are intended to serve mostly as object lessons to introduce general principles and practices of design and fabrication; these can be applied to particular situations that the reader may encounter.

The overall rules of the game can best be stated by a set of three simple guide lines. First, start with something simple. Second, keep it simple. Third, in case of difficulty, try something simpler. Although the environment of the device implanted for the long term is complex, we have found successful implants to be almost always elegant in their simplicity. This is not to say that finding the elegantly simple solution is easy, but rather that one should always be suspicious of solutions that appear to become more complicated as they cope with problems. They are probably faulty in principle. Often, as a device evolves through design iterations, it accumulates features, materials, and construction methods that are ineffectual, counterproductive, or obsolete and that are getting in the way of simplifications that could be made. Often, the reasons for these complicating aspects of a device either have been long forgotten or stemmed from an entirely different application.

One of the most pernicious forms of the unnecessary complication is the attempt to avoid a problem that has never really occurred. For example, the experimentalist may have had a bad experience with stress-fatigued electrode leads; as a result future designs are constantly overspecified, utilizing exotic alloys, bulky coiled leads, and surgical strain relievers. This results in implantable devices that are tedious to construct, traumatic to implant, and apt to interfere with the normal function of the limb or muscle. Although it may be intellectually stimulating to anticipate all the problems that might occur, it is neither necessary nor advisable to try to solve them all a priori. Usually the feature added to solve the nonexistent problem fails itself or gets in the way of a simple solution to a real problem that subsequently does occur.

Frequently, a trial or pilot experiment done in the simplest possible way represents a good test of approaches. Also, whenever a problem occurs with a device, it is important to rethink the entire application, not just the part that is causing the problem. This is likely to reveal unnecessary encumbrances and alternate strategies that cope simply and effectively with the real issue at hand.

2. Size

The next most useful recommendation is to keep the device small. The basic scale of the body is cellular. At the cellular level, the body copes well with many different mechanical consistencies and chemical surfaces. When a homogeneous mass of a particular material exceeds a certain size, it becomes mechanically incompatible with the flexible and constantly remodeling living tissues. Many properties of implants, such as their stiffness, mass, and surface area, are related not linearly but exponentially to lineal dimensions. Large, stiff, heavy hunks of even the most inert materials will cause scar adhesions, compress blood vessels, and provide protected niches for bacteria. If they contain any leachable chemicals, their concentration in the surrounding tissue will be much higher for much longer. When physical strength is needed, it is important to determine precisely where and how much and to select materials that match the desired properties rather than just using more of an inappropriate material.

3. Materials

The experimentalist should use as few different kinds of materials as possible. The weakest points of implanted devices are joints and seams, particularly between dissimilar materials. Dissimilar metals cause battery potentials and corrosion. Transitions between materials of dissimilar mechanical properties become the foci of mechanical stresses. Biomaterials generally have poor adhesion to each other, because their very inertness depends on the lack of surface-reactive molecules required for chemical adhesion. Finally, the use of different materials provides just so many more opportunities by which defective materials, contaminants, and errors of formulation may jeopardize a design.

4. Information

You should understand thoroughly the properties of your materials. Manufacturers are only too happy to provide detailed data sheets regarding the chemical, electrical, thermal, optical, and mechanical properties of their products. Read them! If they are unclear, telephone the manufacturer. Your basic cleaning solvent may seriously degrade your fancy new wire insulation. The catalyst in one polymer may prevent the next layer from curing properly. A brittle outer coating over a relatively soft internal layer may be an invitation to cracks. The attractive color of the encapsulant may come from a toxic particulate filler. The epoxy that cured so nicely last month may have passed its expiration date. All of these problems are subject to the universal caveat "When all else fails, read the directions."

Related to this is the caveat "Never fix a working motorcycle." Never change the fabrication protocol of a device unless there is some very good reason for the change. Then check that the compatibility of the new material or step is carefully considered relative to the remaining materials and steps.

B. INSERTION TECHNIQUE

1. Cutaneous Attachment

In the clinical EMG laboratory, skin surface electrodes are obviously attractive as they represent a noninvasive technique for obtaining gross estimate of muscle activity in large, superficial muscle groups. Human subjects are particularly amenable to this technique, because of their largeness, their generally hairless skin, and their tendency to cooperate with the investigator by performing tasks using particular muscles. These factors do not hold for animal studies, for which there is little call to use such electrodes.

2. Percutaneous Insertion

a. Theory Percutaneous insertion of electrodes is often a quick-and-dirty approach to EMG; this masks its utility in special circumstances. However, percutaneous insertion always requires particular care regarding execution and verification. One instance in which the technique is very useful (and may be essential) is in the study of superficial and even deeper muscles in lower vertebrates. In many such species the skin will lack the extreme flexibility and elasticity seen in the mammalian integument; it may be armored so that operation poses special problems. However, the percutaneous placement of electrodes into deeper muscles poses substantial problems, mainly because the layers of tissue through which the leads must pass may shift relative to each other. The potential difficulties should not be underestimated.

In some ways, percutaneous insertion represents a carry-over from clinical laboratory technique, in which recording from concentric needle microelectrodes is often referred to as "single unit" EMG. Instead of leaving a rigid shaft embedded in the muscle and its overlying tissues, animal experimentalists tend to use hypodermic needles to introduce fine wires into the belly of a palpable muscle and then attach the protruding leads to temporary connectors. The likelihood of success obviously depends on the kind of animal and the nature of its integument, the bulk and position of the target muscle, the number and location of potential wrong sites, and the skill of the operator. Certainly, the percutaneous approach must always follow (and be followed by) a thorough dissection of the whole region. Preferably this should occur in a fresh specimen, so that the changing thickness and orientation of the target muscle may be examined while the skeletal elements are moved and the overlying skin is stretched. It is especially useful to attempt insertion of electrodes during such preliminary dissection; this will well document the elasticity of the tissue and the tendency of aponeuroses to deflect muscle tissues away from the path of the needle. Needles can also be marked for depth and the angle of needle placement determined.

Several critical problems must be kept in mind. The first is that both electrical contacts of the bipolar antenna generally have to be correctly located in the muscle, appropriately separated from each other by a defined spacing and lying on a line more or less parallel to the muscle fibers. This requirement makes it inappropriate to use separate wires inserted individually, as their orientation and spacing would differ among the insertions of a series of experiments; indeed the two poles would tend to migrate independently, changing the sensitivity and selectivity of the electrode pair. Instead, the two leads should be joined mechanically (but not electrically!), so that orientation and spacing of the contacts will at least be fixed with respect to the insertion axis, if not to that of the muscle fibers. Generally, this means that the electrode will have to be inserted tangentially rather than perpendicularly; tangential insertion helps fixation but may hurt positioning accuracy.

Second, there are mechanical problems. Percutaneous electrodes are usually anchored by their recurved tips. The bared contact area tends to be placed on the portion of the wires that protrudes from the tip of the insertion guide (hypodermic needle); this portion is bent backward along the shaft to form the barb. The needle is advanced to the presumed target and then slid back along the leads, leaving the barbs in place.

Several things can go wrong. The wire or its insulation may be nicked during insertion (see chapter 9, section C5). The wire must be stiff enough to form an effective barb, but too much stiffness will cause pain or undue damage if the wire is twisted as the muscle moves during the experiment. If the method is used for deep muscles, the electrode leads may have to cross one or more shear planes between sliding skin and fascial layers; should the wires bind at these places, the wire and particularly the barb will be stressed and flexed each time the animal moves. There may be very poor control over the anchoring of the tips; the barb must be long enough to hold the contacts securely, but this length may introduce considerable uncertainty about the final position of the contact area with respect to the insertion track.

There are solutions to each of these potential difficulties, and we detail these in the section on bipolar hook electrodes. In any case, the signals from percutaneously inserted electrodes should be checked 24 and 48 or more hours after implantation. It is even more important that experiments using percutaneous electrodes be followed by a careful dissection of the site at which the electrode tips rested at the conclusion of the experiment. (Be sure that they did not shift between the time of the recording session and the postmortem examination.) In short, the seeming advantages of the technique are bought at the cost of special care. Shortcuts are likely to lead to uncertain results.

b. Practice. Even if the electrodes are to be inserted percutaneously it is useful first to nick the skin with the point of a scalpel, as this reduces the major stress on the tines during insertion. Even animals with soft mucous skin, such as some frogs, may have a very tough dermis. Its perforation requires very sharp needles, and there is the risk that the needle will slip too far once it has been driven through the skin. Use of a fixed collar will limit this risk.

Preliminary dissection should have indicated the best site of needle insertion so that the electrode reaches the desired muscle with minimal risk of erroneous placement and so that the lead wire will be left along a path that does not induce significant shear as other muscles contract. Once the needle has reached the desired site it is useful to apply gentle pressure with a finger to the outside of the skin, just over the site at which the electrode tip is to rest; then withdraw the needle gently, perhaps rotating it back and forth through an arc of 60° about its long axis. If the topography permits it, you may loop the wire beneath the skin by extending the (partially withdrawn) needle a second time in the loose connective tissue; this will reduce the possibility of tension on the electrode tips implanted in the muscle. Critical for all manipulation is a combination of a smooth needle shaft and unkinked wires. Tension during insertion almost guarantees inappropriate placement.

3. Surgical Implantation

Although surgical implantation is obviously the preferred method from the viewpoint of electrical and mechanical reliability, this technique is also the most complex, time-consuming, and potentially damaging one. Furthermore, it is the most invasive, requiring skill of the experimentalist and a significant recovery time for the animal before any data can be obtained. If surgical implantation is poorly done, there may be residual effects produced by scar tissue or partial denervation. These may cause the animal pain and raise questions about the normalcy of its movement. They may well vitiate any technical advantages.

One simple use of surgical implantation is for properly guiding and fixing electrodes not dissimilar from those used for percutaneous insertion. Under direct visual control, a pair of flexible leads with exposed and oriented tips can be pushed or pulled into a specific portion of a muscle belly and their leads sutured directly to its fascia near the exit point. Assessment of the depth, orientation, and thickness of the muscle, as well as of adjacent muscles that may generate cross-talk, is easiest at the time of surgical implantation. Several electrodes can be implanted via the same incision; this approach may prove useful in the tests for cross-talk described later in this chapter. Electrodes may still migrate, but the extent of such migration is limited at the time that the electrodes are implanted into an exposed muscle and can be accurately assessed when the site is reexposed at postmortem examination.

Surgical implantation is also required for the installation of barrier (patch) electrodes, which prove to be highly effective tools for coping with difficult situations in which cross-talk is likely. Whereas the amount of surgical dissection and the size of the incision necessary to position and fix patch electrodes are often substantial, patch electrodes may be tolerated much better than intramuscular ones, especially whenever the muscles to be tested are small. If implantations are properly planned, the mechanical stresses and potential tissue damage are confined to easily separated fascial planes that are mechanically resilient and tend to be poorly innervated; generally, neither muscle fibers nor nerve terminals need be torn or stressed. Such patch electrodes also facilitate the postmortem localization of the recording contacts without damaging the muscles or limiting future histological or physiological assay.

The surgical techniques that make electrode implantation efficient and well tolerated differ profoundly from those employed for short-term physiological experiments or anatomical investigations. Anyone untrained in clinical surgery should consult chapter 19 and also watch or obtain consultation from the veterinarians responsible for the animal facility (who are too often overlooked as resource persons).

C. SELECTED DESIGNS (fig. 10.1)

1. Skin Surface Electrodes

Skin surface electrodes are commonly used for noninvasive clinical recording but rarely in animals. First a few words must be said about their unique attributes. Although all other electrodes have their impedance and antenna properties essentially defined by the conductive interface they present, the properties of skin surface electrodes are defined primarily by the condition of the skin through which they must detect potentials arising in deeper structures. The electrical conductivity of skin varies greatly, from nearly negligible barriers presented by mucous membranes through the modest values of moist glabrous skin to the essentially nonconductive layers of heavily keratinized skin. A high-impedance barrier between a dipole antenna and a dipole source affects the signal in ways similar to those caused by a large physical separation in a more homogeneous medium. This means lower signal amplitudes, lower frequency bandwidths, and poorer spatial selectivity. To these can be added higher thermal noise and potential motion artifact problems from unstable contact impedances. Even so, careful skin preparation to minimize these barrier effects and still more careful consideration of the anatomical sources of cross-talk and their significance in the experimental design may

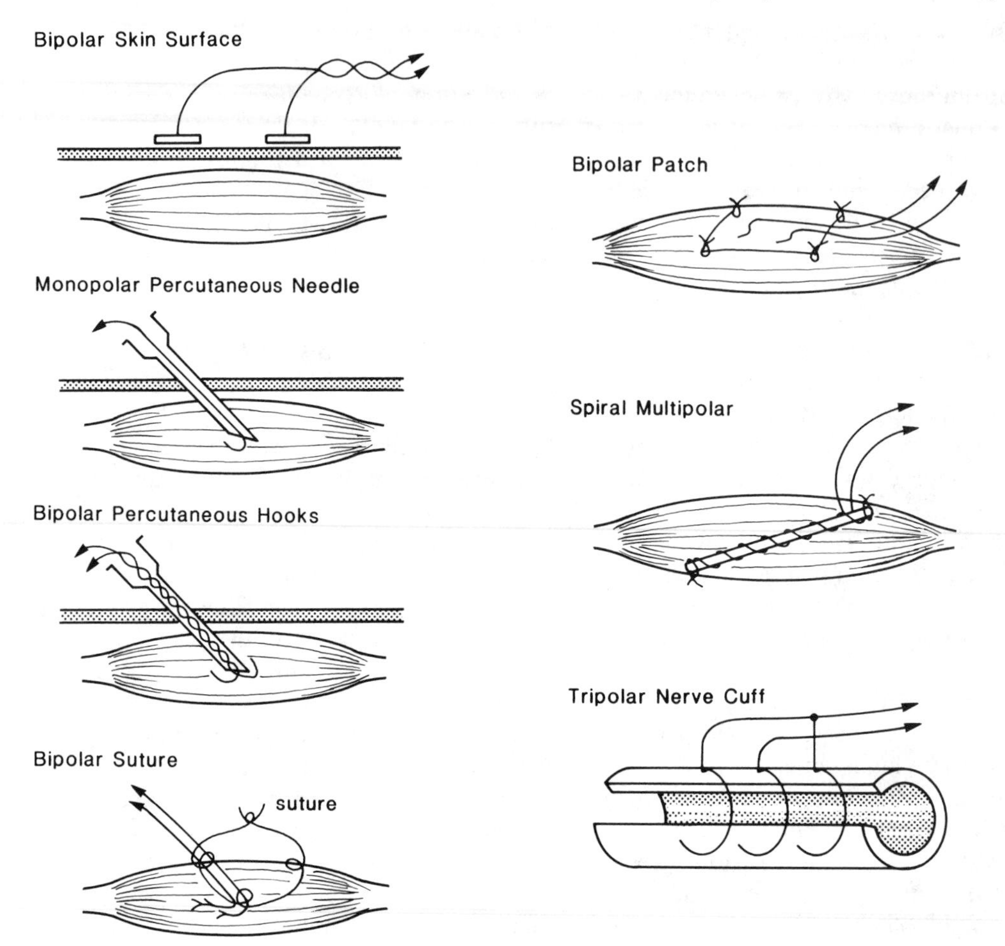

10.1. Electrode designs. These mechanical configurations are frequently used in EMG recording.

permit use of this kind of completely noninvasive procedure.

The physical composition of the contacts and even most details about their shape have little bearing on performance. The electrical properties are similar to those of a patch electrode, so the dipole should be oriented parallel to the muscle fiber, perhaps with a larger interelectrode spacing to allow for the spreading effect of electrical distance on the source dipoles. As always, the contact areas should be as large as practical to minimize contact impedance but not so large as to risk shorting. The shorting problem can be subtle because these contacts are often placed over small blobs of conductive gel to reduce skin resistance, and any continuity of these blobs constitutes a short regardless of whether the electrodes themselves are in contact.

Contacts can be flat or contoured sheets of corrosion-resistant metal (stainless steel, nickel, and silver surfaces are common). Flexible foils are unreliable because they can work-harden and crack, resulting in electrical noise. Silicone rubber impregnated with silver is a superior material if flexing is required; such rubber electrodes with electrical leads attached are commercially available for this purpose. As noted above, the electrode contacts on skin are commonly attached after scrupulous cleaning intended to remove nonconductive oils, often followed by abrasion to remove keratin. Of course, hairy skin must be shaved closely. Almost any fluid or gel will accomplish the required lowering and stabilization of skin resistance. Saline, alcohol pads, and various commercial gels have all been used successfully. The duration for which the electrode must remain attached remains a major consideration in the selection of fluid, since the liquid may dry because of evaporation or may irritate the skin. Some commercially available skin electrodes have a built-in recess for the gel and disposable adhesive strips.

Mechanical fixation is the main problem in clinical studies and can be an overwhelming one with an uncooperative animal. Elastic and Velcro straps work well on distal extremities. Again, care must be taken that the contacts do not slide around, smearing conductive gel into a short between them. Adhesive tape works well on relatively dry skin, and adhesion can be augmented by the standard clinical trick of painting the skin with a tincture of benzoin and allowing this to dry before attaching the tape. There are also commercially available flexible electrodes made of various conductive composite polymers.

2. Single Contact Electrodes

Single contact electrodes have been used and misused extensively. Probably their only justifiable use is when the investigator knows there will be only one active entity and wants to know only the time of activation, rather than relative or absolute amplitude.

Generally these electrodes are made of fairly stiff, springy, single-strand wire with a mechanically tough insulating coating. The wire is passed through the lumen of a small hypodermic needle and the desired contact area exposed in the distal end of the wire as it protrudes through the tip of the needle. The contact area is then bent backward in a hairpin bend at the end of the insulation (this places considerable premium on not nicking the wire when it is stripped). The wire is drawn back into the needle until the bent contact portion hangs out over the lip of the bevel. It is advisable to file or chamfer a dull spot at the base of the hypodermic's bevel where the wire will rest to avoid nicking it during insertion resulting in a weak spot, subject to stress fatigue (cf. fig. 9.4).

For insertion, the needle is driven into the target; the protruding contact acts like a barb and catches in place as the needle is withdrawn over the remainder of the lead. The efficacy of this anchor obviously depends on the stiffness of the wire, the angle of the bend, and the connective tissue matrix of the muscle. An excessively stiff wire will transmit torques and torsions on the lead to the tip, causing considerable motion and damaging the muscle fibers. The hook on a flexible wire may straighten if the lead wire is pulled. With time and constant motion, migration is likely to occur, with the electrode withdrawing or driving deeper as the connective tissue remodels around it in response to even small applied forces. If two such contacts are independently inserted, there are more possibilities for shorts or uncontrolled bipolar spacings.

3. Bipolar Hook Electrodes

The bipolar hook type of electrode is the basic electrode design for percutaneous insertion. It is particularly useful for stereotactic insertion into animals that may have poor tolerance for operation, such as amphibians and reptiles. The stiffness of the wire has an important bearing on the ability to hold the twist. Two versions are commonly used (fig. 10.2): the simple double hook and the offset twist hook. Although the second has different antenna properties, it is much less likely to cause shorts.

The simple double hook produces contact-bearing arms that tend to spread out perpendicular to the path of insertion (fig. 10.2). Care must be taken that the contacts will come to rest with their bipolar axis parallel to rather than across the muscle fibers and that the contacts cannot short against each other. The offset twist electrode produces two contacts with a bipolar axis that lies along the path of insertion. Thus, it is best used when the surgical exposure permits a tangential approach to the muscle belly. If properly made, it eliminates the possibility that the contacts will short and provides a considerably more predictable intercontact distance.

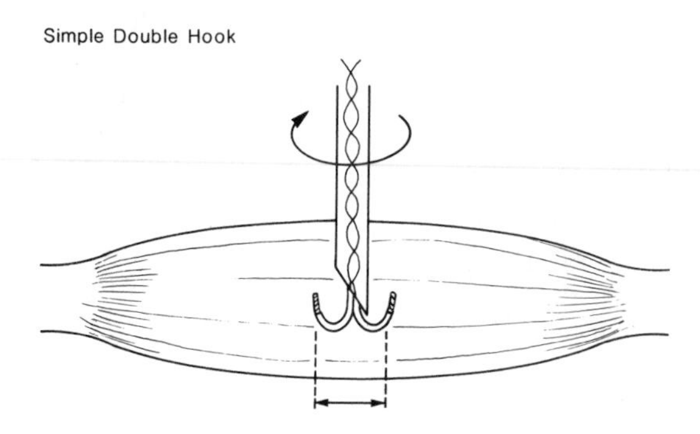

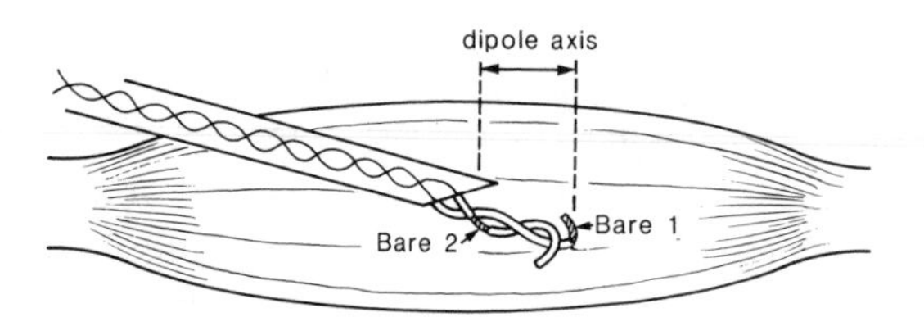

10.2. Two types of bipolar hook electrodes suitable for percutaneous insertion. Note that the simple double hook (*top*) tends to create a bipolar axis that is perpendicular to the insertion angle and that may line up parallel to the muscle fiber dipoles (as shown) or perpendicular to them if the insertion tool is turned 90 ° on the axis shown by the *arrow* (undesirable). The offset twist hook (*bottom*) tends to produce a bipolar axis that parallels the angle of insertion, and it is thus most suitable for muscles that may be approached tangentially (or that are very pinnate).

a. Simple Double Hook. All phases of construction are best handled under a dissecting microscope. Cut a section of insulated wire slightly more than twice as long as the electrode pair to be inserted. Fold it in half and place the loop over the bracket of a rotatable wire twister, such as the chuck of a hand drill (fig. 10.3a). Hold the two wires between the thumb and forefinger of one hand at a place about five cm from the twister, maintain gentle tension, and gradually spin the twister until a variable section of wires has been spun together. Maintain tension by pulling gently, or the turns will be uneven. After an appropriate length has been twisted, move your finger into the loop and push away from the twister in order to set the twisted wires and keep them from unraveling. Lift the loop off the twister and thread the free ends into an insertion needle of appropriate length and preparation (see the discussion of insertion via needles in chapter 9, section C5). Pull the wires through until only the loop projects, trim off the loop to leave two ends of desired length, bend back the two free tines to form a pair of hooks, and strip the tines of insulation for a length established by the antenna properties desired (chapter 6). Stripping of the wire should be performed under a dissecting microscope.

The details of the technique will be affected by the nature of the wire and the length to

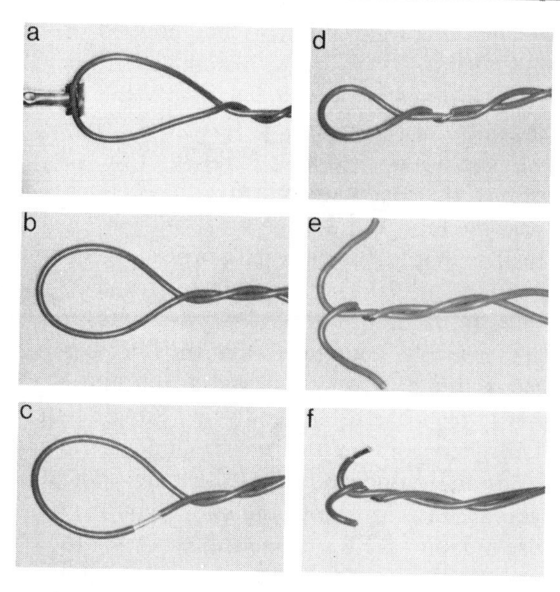

10.3. Model of the manufacture of an offset twist electrode. In this case heavy wire is used to clarify the principles. *a,* A loop of the wire is placed over a hook placed in a rotatable chuck, and the wire is twisted. *b,* The wire is removed from the hook. *c,* A portion (1 mm) of the insulation is stripped off one side. *d,* The loop is replaced on the hook and twisted until the bared area is incorporated in the twisted portion, with twisting continued for at least an equal length. *e,* The loop is removed from the hook, the tip is split, and the tines are bent backward. *f,* A (1 mm) length of one of the tines is bared, thus generating the second contact surface.

which it is to be twisted. After the electrode has been formed, it may be useful to pull it out of the needle and press it flat; if the tines lie in a single plane and face in opposite directions, they are less likely to contact each other after insertion (producing the shorts that are the bane of this approach). A drop of surgical adhesive at the bend further stabilizes the electrode and protects its insulation from being cut during insertion.

b. Offset Twist Hook. Again, cut a section of insulated wire, fold it in half, place it over the bracket of a rotatable wire twister, and gradually spin the twister until a variable section of wires has been spun together (fig. 10.3). Maintain tension by pulling gently. After an appropriate length has been twisted, move a finger into the loop and push to set the twisted wires and keep them from unraveling.

Then lift the loop of wire off the hook and place it flat on the microscope stage; remove the required length of insulation (as measured on a millimeter scale) from the wire on one side of the loop close to the twisted portion. Thereafter, lift the loop back onto the hook (maintaining the orientation!) and twist it in the same direction until at least half of the bared length is included in the twisted portion. Again lift the loop off the hook and bend back both of its ends to form the double hook just at the end of the twisted portion. Remove a length of insulation from the second one equal to that on the first side, starting just beyond the bared area (so that the two bared portions cannot contact each other). Then trim the two tines to the length desired, the cut presumably passing at the end of the bared portion of the second wire.

The electrodes may again be stabilized by application of a drop of surgical adhesive; however, it is important to keep this from spreading over the bared areas. The electrodes may thereafter be stored in individual envelopes or threaded into needles.

4. Bipolar Suture Electrodes

Bipolar suture electrodes are suitable for routine sampling of EMG signals from medium to large fusiform muscles located superficially or deeply. However, they must be implanted by surgical exposure.

Cut two lengths (at least six inches longer than the length between electrode and connector) of stranded stainless steel wire with Teflon or other mechanically strong insulating jacket (fig. 10.4). Strip three to five mm of insulation from each end of both, taking care not to nick the strands (see above). Select a

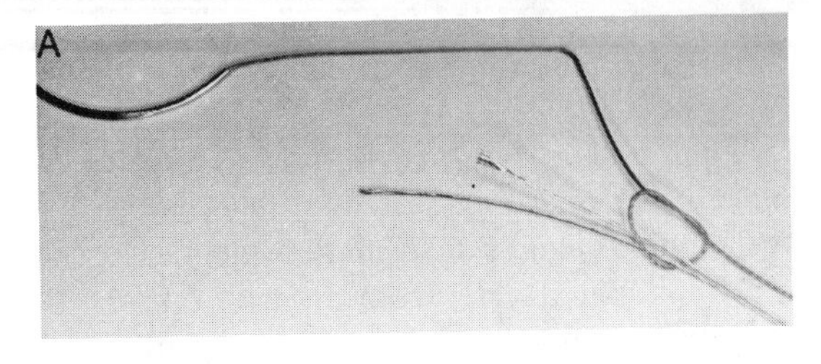

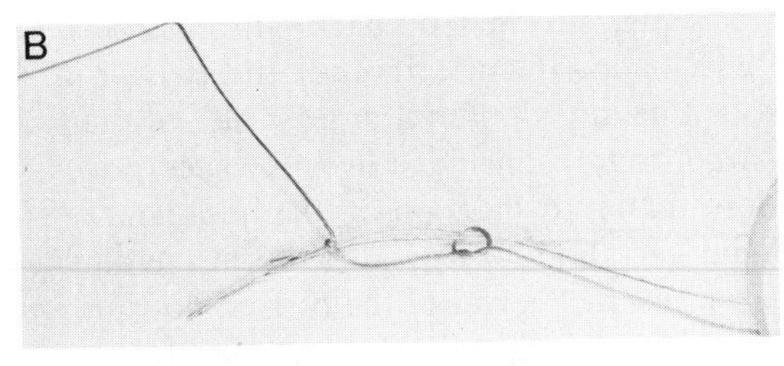

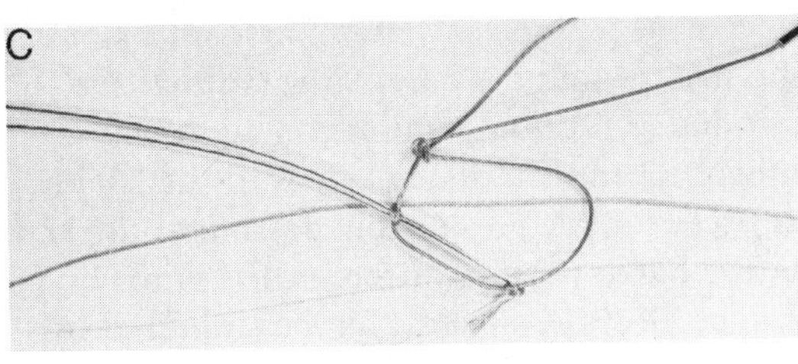

10.4. Steps in the manufacture of a bipolar suture electrode, using a nonabsorbable braided suture material with a swaged-on needle plus two lengths of stranded stainless steel wire with Teflon jacket. Each lead has had 3 mm exposed at the tip by means of the stretch-cut-shrink-back technique (described in fig. 9.1 *d-e*) *A,* The two leads are offset about 5 mm and tied together just behind the exposed regions. *B,* After a square knot is completed, another simple loop knot is placed in the free end of the suture (the one without the needle) about 5 mm back along the wire leads. *C,* The two wire ends have been bent backward to an acute angle just at the distal square knot. The swage needle would be used to create the curved track through the muscle as shown here, dragging the electrode leads so that both knots would be buried in the muscle. The two ends of the suture are tied together loosely on the outside (it is useful to incorporate a tuck of fascia into the knot to prevent migration of the electrode).

strong, nonabsorbable synthetic braided suture (2-0 or 3-0 for most mammals) with a swaged-on needle whose length is about the thickness of the muscle. Position the two wires in a loop tied in the middle of the suture so that one wire ends five to seven mm past the loop and the other protrudes enough further that the deinsulated portions do not overlap (fig. 10.4). Tie a tight square knot there in the suture to bind the two wires together firmly, taking care that the knot lies flat and tight. On the tail end of the suture (away from the needle end), tie a simple loop knot incorporating the two trailing electrode leads about one cm behind the main knot. If you are insecure about the knots, you may use a dab of RTV silicone rubber to reinforce each and prevent the wires from slipping, but a good knot is better. Then bend the two protruding wires back over the knot, crimping in your fingers. The electrode can then be autoclaved. (Watch for deformation of the Teflon after autoclaving.)

To implant the electrode, expose the midportion of the muscle and pass the needle obliquely through the center of the thickest part. A sharp tug will pop the leading knot (with the bent-over contacts and trailing leads) through the fascial covering. Take care not to pull so far that it pops out the opposite side, in which case it is best to pull the whole thing through and try again. Once the contacts are reasonably centered in the mass of the muscle, make a small loop knot of the suture to the fascia at the exit point, taking care to grab only fascia and not muscle fibers. Then loosely tie the trailing suture (coming out of the entrance hole with the leads) to this anchored end so that there will be no tension on the suture even when the muscle reaches its maximal diameter upon full contraction. Pass the leads subcutaneously to the connector, leaving adequate slack in the vicinity of the muscle.

5. Simple Patch Electrodes

Simple patch electrodes are suitable for selective sampling of EMG signals from thin, flat muscles and small muscles lying against large sources that may generate cross-talk (fig. 10.5).

Cut a square of silicone rubber sheet with Dacron mesh reinforcement (fig. 10.6*a*), such as the flexible artificial dura manufactured by Dow Corning Corp. (0.007 inch total thickness). Each edge of the square should be three times the desired interelectrode spacing. Prepare two lengths of insulated, stranded, stainless steel wire at least six inches longer than the maximal path length between the recording site and the connector fixation site. Insert the insulated wire into the back of the shank of a small hypodermic needle that will be used to punch each wire through the silicone sheet as shown in figure 10.6. Note that the two pairs of punch holes describe a square at the center of the sheet, the sides of which equal the interelectrode spacing and the dis-

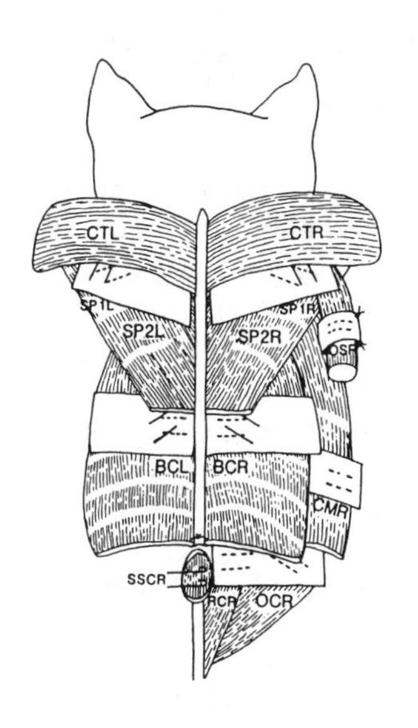

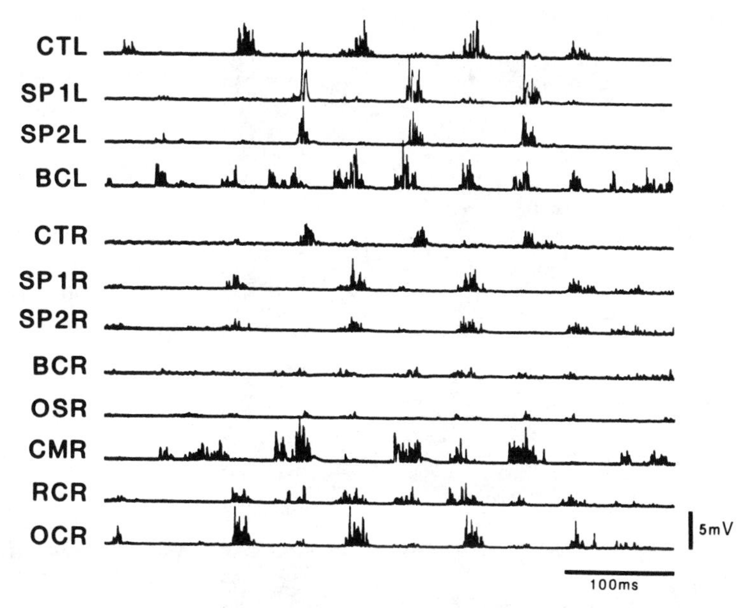

10.5. Use of multiple patch electrodes to record from all of the muscles in several adjacent layers of the cat neck. The EMGs were obtained during a rapid series of vigorous head shakes. The anatomical sketch (*left*) shows the approximate placement of seven separate patches (some with two pairs of contacts) plus one pair of deep hook electrodes (activity not shown). *Dashed lines* indicate downward-facing contacts; *solid lines* indicate upward-facing ones (leads have been omitted). The records illustrated were obtained on the fourth postoperative day; the animal showed no reluctance to engage in vigorous head movements, indicating that the devices were well tolerated. All surgical procedures involved dissection along natural fascial planes, the continuity of which was restored at closure. The records were all obtained at the same gain and are presented here with full-wave rectification (with use of special circuitry to eliminate any biasing from baseline). Note the silence of the records between bursts, although some adjacent muscles had large out-of-phase activity. *CT,* clavotrapezius (*L* and *R,* left and right); *SP,* splenius (first and second series compartments recorded separately from patches on opposite sides); *CM,* complexus; *RC,* rectus capitis major; *OC,* obliquus capitis; *SSC,* semispinalis cervicis. (Courtesy F. J. Richmond and G. E. Loeb, work in progress.)

tance from each hole to the nearest edges of the sheet. At the distal ends of the wires, strip a section of insulation equal to the interelectrode spacing such that a sleeve of insulation remains over the distal few millimeters of unstripped wire. Bend each lead into a right angle as shown in the figure, then trim so that each end has a sleeve about three mm long past the exposed area. Draw the wires back through the holes in the silicone sheet until they lie in the position shown, with the two exposed portions on the same side, opposite to the position of the leads. Apply a small layer of RTV silicone (Medical Adhesive A or equivalent) to the proximal and distal wire exit points on the back side of the patch, taking care to cover any exposed wire, particularly the distal cut end. If some stress is expected, a small suture may be used to relieve strain on the lead wires just as they exit the plane of the silicone sheet. The completed electrode can be autoclaved.

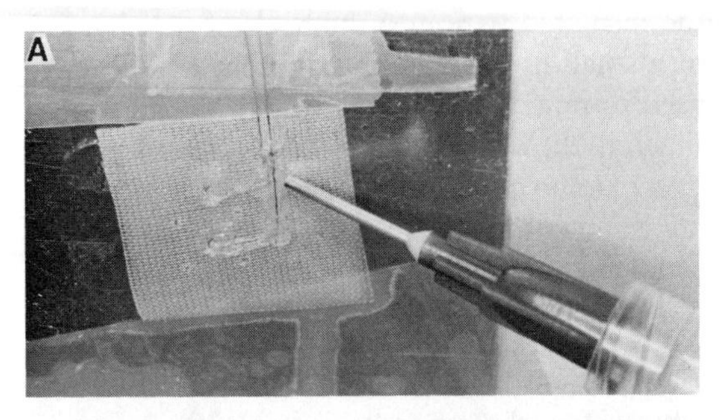

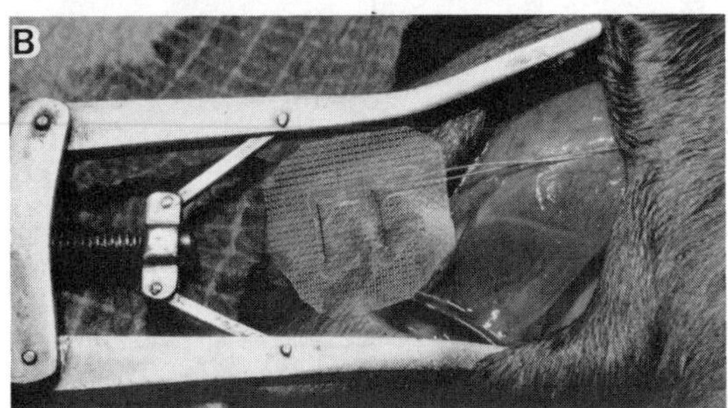

10.6. *A*, Stage during manufacture of a patch electrode. View from the back shows silicone rubber RTV being applied to the entrance and exit points of the leads to provide strain relief and insulate any exposed wire. *B*, Recording surface side of a patch electrode, which will be flipped over and sutured at the corners onto the surgically exposed muscle. *C*, Multicontact patch showing five evenly spaced bipolar recording configurations used to compare recruitment across different parts of a broad, sheetlike muscle.

To implant the electrode, expose the midportion of the muscle and blunt dissect to create a cleavage plane between the muscle fascia and whatever structure represents a source of cross-talk (fig. 10.6*b*).

Then slide the patch electrode into this space with the contacts lying against the target muscle fascia. Place snug stay sutures through each corner of the sheet and the underlying fascia, angling the sutures diagonally away from the patch. If quarters are tight, it may be useful to preplace two deep corner sutures in the fascia, pass them through the deep corners of the patch while outside the surgical field, and then draw the patch into position by tightening up a loop knot on each suture. Be careful to place each suture a couple of millimeters inboard from the edge of the patch, away from the frayed edges of the Dacron mesh reinforcement. For a thin muscle lying between two other muscles, it may be advisable to place a "blank" patch of silicone rubber over the muscle fascia on the side opposite the electrode; this will shield the recording.

Using similar fabrication techniques, it is

possible to make single patches with complex arrangements of multiple leads with various orientations of contacts on one or both sides of the patch (see fig. 10.5 and 10.6). These must be tailored to the specific anatomy and experimental requirements of each preparation.

6. Spiral Multipolar Electrodes

The spiral multipolar type of electrode is suitable for distributed sampling within fairly large muscles.

Cut a length of small-diameter, easily stretched silicone rubber tubing that is equal to the length of the proposed path of the electrode through the muscle (fig. 10.7). Slip the tubing over a wire or string that can be stretched taut, suspended between two posts. Select a stranded, insulated wire that is malleable enough to wind and set into the spiral shape needed without springing back straight. Stranded platinum with 10% iridium is expensive but handles very nicely and allows the insulation windows to be burned in rather than cut into the two lengths of wire. Remove the desired insulation patches, leaving insulated intervals about three times as long as the patches. The intervals between contact points in each lead should be regular and twice the desired bipolar spacing, allowing for the pitch of the spiral.

Using a nonabsorbable, synthetic, stranded suture with a small swaged-on needle, tie the two distal ends of the windowed leads to one end of the spiral as shown in figure 10.7, positioning them so that the windows in each lead are adjacent to insulated segments of the opposite lead. Wind the two leads into a regularly pitched spiral around the tubing, making sure that the leads do not cross and that the deinsulated portions cannot touch each other.

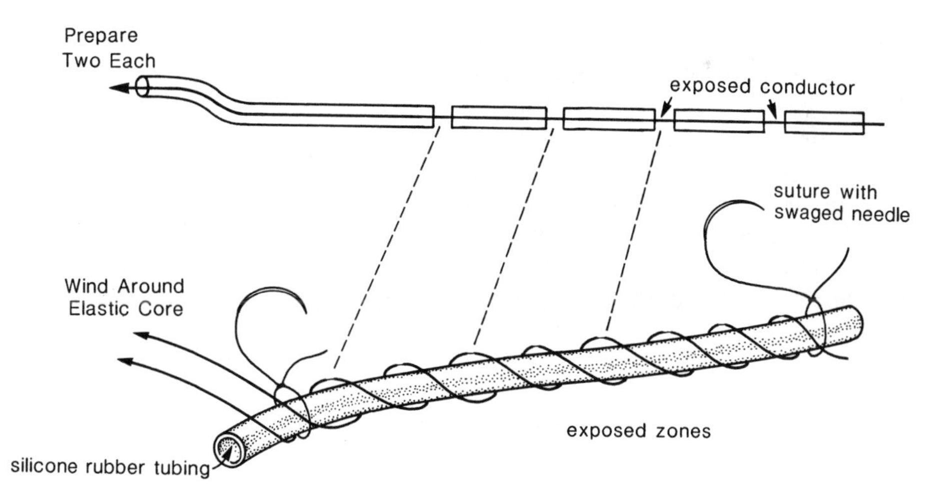

10.7. Construction of spiral multipolar electrode. Two insulated, stranded electrode wires (e.g., Teflon-jacketed, multistranded stainless steel) are prepared by cutting holes in the insulation at regular intervals of a few millimeters. The two strands are helically wrapped loosely around a short piece of very elastic silicone rubber tubing and tied in place at each end by means of a length of suture with a swaged-on needle. The exposed zones are staggered so that there is a regular, alternating pattern of contact to each lead. The electrode is inserted into the muscle at an oblique angle that crosses all of the muscle fibers. It is dragged into place by tension on the distal suture and then tied at both ends to the fascia so that all contacts are embedded.

Tie down the spiral at the opposite end of the silicone tubing using a similar braided suture with swaged-on needle. Apply dabs of RTV silicone to anchor the coils of the spiral at frequent intervals, being sure not to cover any contact areas. The cured assembly is then slipped off the taut wire, and additional silicone rubber can be applied in and around the tubing ends to relieve strain at the suture points, which will be used to anchor the spiral to the fascia of the muscle at either end. If the wire used for the leadout will differ from that used for the spiral, the splices should be made a few centimeters from the spiral itself, because they are likely to be relatively bulky and because the end of the spiral will be a stress point (see chapter 9 for splicing techniques). The completed assembly may be autoclaved.

Insertion of the electrode generally requires an introducing needle to establish a straight path of the desired length and oblique orientation through the muscle. This is used to drag the distal sutures of the spiral electrode into the muscle. After it is positioned so that the ends of the spiral lie just beneath the fascia at either end, the sutures are used to tie each end firmly in place to the fascia, while care is taken not to incorporate any muscle tissue in the suture loops. When correctly placed, the spiral should become slack at the shortest physiological length of the muscle and should not restrict the anticipated physiological elongation.

11 Connectors and Cabling

A. OVERVIEW

Now that an optimized set of electrode contacts has been carefully and securely attached to the muscle, the tiny voltages they pick up must be conveyed to the amplifier for processing and eventual inspection. Generally, the signal must pass some distance within the body from the muscle to a mechanically stable and convenient exit point. If the experimental preparation is a long-term one, there will generally be a connector affixed near this exit point, which allows the animal to be easily connected to and disconnected from the amplifier and its source of power.

The options outside the animal include simple flexible cables, portable amplifiers, and telemetry equipment.

1. Simple Flexible Cable

A simple flexible cable may convey the unamplified signal some short distance to the amplifier. The length of such cable is limited by the tendency of any cable to "lose" small signals and to pick up extraneous noise via stray capacitance (fig. 11.1).

Generally, the animal must be at least somewhat confined if not restrained, such as on a treadmill or in a cage. As we discuss in the next chapter, the tendency to pick up electrical interference is related to the product of the length of cable (in this case the cable carrying unamplified signals outside of the animal) and the electrode impedance. It is generally advisable to reduce electrode impedance to negligible values, thus reducing the level of interference.

2. Portable Amplifiers

If it is impossible to reduce electrode impedance to negligible values, a small, simple amplifier circuit at the exit point on the animal can greatly improve signal-to-noise ratio by eliminating the cable carrying an unamplified signal from the animal. Theoretically, this would permit much longer cables carrying amplified signals and freer movement of the animal, although in reality the approach is usually limited by problems of supporting a lengthy cable and avoiding excessive weight, tangling, and chewing by the animal.

3. Telemetry Equipment

Telemetry equipment implicitly includes amplifiers with stored power in or on the animal (obviously nothing would be gained if the power supply had to be cabled to the animal). There has been great progress in lowpower transmitters and light but powerful batteries. However, this approach requires a careful and sober calculation of the trade-offs among broadcast range, bandwidth, dynamic range, and battery life. If several channels are to be

11.1. Noise pickup by electrostatic radiation from a noise source. The stray capacitance (C_{stray}) between the noise source (E_{noise}) and the preparation forms one element of a voltage divider as shown (*bottom*). The other element is the parallel combination of the impedances of electrode source (Z_{source}) and amplifier input (Z_{in}). Unless the amplifier input impedance Z_{in} is kept much higher than the electrode source impedance, the recorded signal ($E_{recorded}$) will be attenuated. Therefore, the best strategy for reducing noise pickup consists of minimizing stray capacitance (use short, shielded, well-routed cables) and minimizing electrode impedance (use largest contact surface area possible).

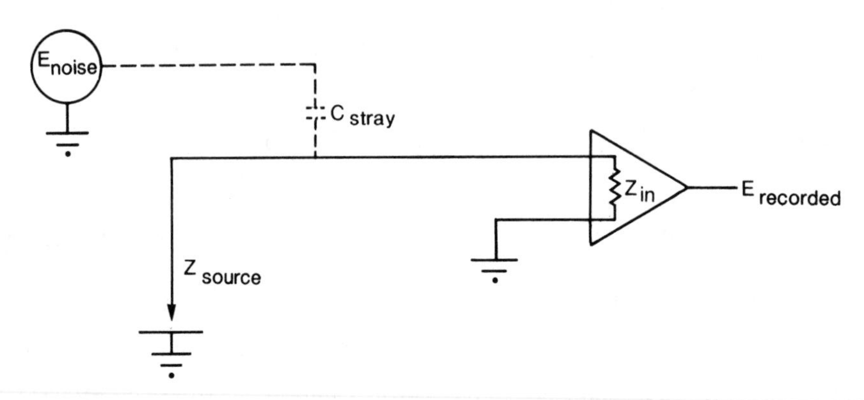

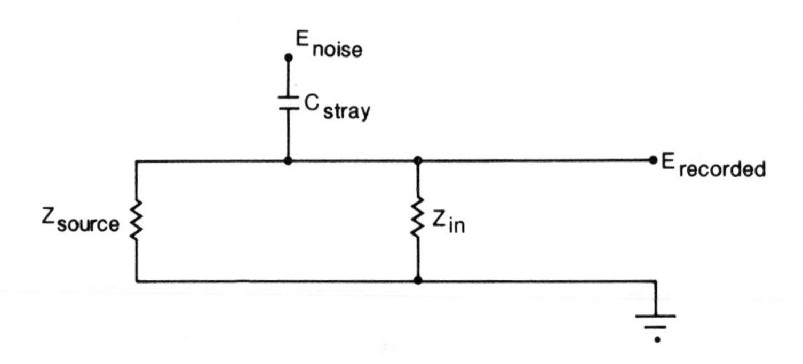

multiplexed, even more sophisticated judgments must be made regarding cross-talk (channel isolation), which interacts complexly with the other parameters mentioned. Finally, the design of radio frequency receiver antennas that can function adequately for any possible position and orientation of the transmitter antenna, which is located on the animal, is an art in itself (see chapter 12).

B. INTERNAL CABLES

1. Concept

Whenever possible, the electrode contact should be a simple exposed extension of the insulated lead wire that will pass to the outside. This minimizes junctions and seams, which are always mechanically and electrically vulnerable both for conductors and insulators.

However, it may well be that the material forming the contacts is mechanically unsuited to the rigors of subcutaneous existence in a constantly moving limb or that, as in the use of stranded platinum for spiral multipolar electrodes, the wire is simply too expensive to be used for extensive connections. In this case a splice must be introduced just outside the muscle, between the local lead from the contacts and the strong, stranded lead to the outside. Of course, splices may have to be inserted in any system to make repairs, correct for inadequate lead length, or handle staged procedures in which mounting of a connector is delayed until the implanted electrode has stabilized (see chapters 9 and 19).

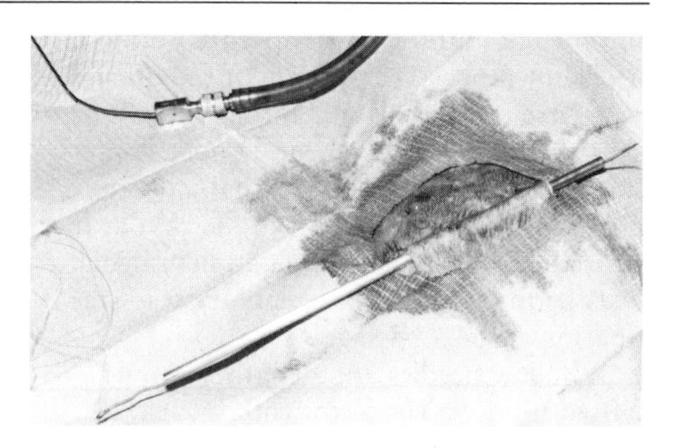

11.2. Subcutaneous passage of leads by means of a hollow tubular probe with blunt-nosed cone (removed after the probe has been passed, not shown here). A malleable probe with a needle eye at one end is used to pass the fine wire leads of the patch electrode (*far right*) through the tube; then the tube will be withdrawn, leaving the leads in place.

2. Subcutaneous Passage Techniques

Often the best exit point of the electrode wires will be some distance from the recording site—for instance, the latter may be out in the limbs and the former on some stable midline structure. Even for percutaneously inserted electrodes it may be useful to widen slightly the entry and pass the wire subcutaneously, closing the aperture with suture or surgical adhesive. Tunneling with the usual surgical instruments, such as blunt scissors and long clamps, is difficult and traumatic. Furthermore, the subcutaneous tunnel has a way of developing blind pockets as soon as you try to push a flexible wire through it.

Several methods for tunnel generation are available. Long spinal needles are one possible device; choose a model with a reasonably blunt tip and insert it carefully (see below). For some animals you can pass sections of polyethylene tubing, which may be stiffened by insertion of a metal wire. An alternative for animals with relatively loose skin, such as cats and monkeys, is the passage of a malleable probe of very fine caliber (fig. 11.2). Such probes may be passed for remarkably long distances, even around complex curves. A needle eye in one end will allow you to pass it long distances and around corners subcutaneously and then drag the electrical leads behind it to the desired point of emergence.

One of the simplest ways to make a supply of such probes with various lengths and stiffnesses is to use copper buss-bar wire, which has just the right malleability and comes in a handy range of diameters. One end of the probe is bent into a crude needle eye, which can be soldered and sanded into a smooth profile, and the other end is sanded into a blunt taper. The whole surface is best polished shiny on a polishing wheel. It can be nickel plated if you are squeamish about having tarnished copper and solder in contact with tissue for even a brief period.

With practice, one soon learns routes that encounter the fewest connective tissue snags. In limbs these snags seem to occur most commonly when passing between ventral and dorsal compartments, such as crossing from shin to calf. It is important to start and stay in the correct subcutaneous plane, which is one reason for not putting too sharp a tip on the needle or probe. The position of the tip and its course can be palpated through the intact skin and probes can be bent by outside pressure to get around corners. Usually the tip is poked through a small nick in the skin at the desired exit point or through an intermediate incision that serves as a temporary way station for the next passage.

If hollow needles are used, they should be passed from the connector end toward the electrode site; probes must pass the opposite way. For needles, the lead wire from the electrodes is then threaded into the barrel and pushed through until the ends emerge. Withdrawal of the needle then leaves the wire in place, although the electrode end of the wire should always be secured so that it will not be pulled out of position if the wire snags during withdrawal. For probes, a few inches of each electrode lead is passed through the eye of the needle like a sewing thread. Then the probe is pulled through from the tip, dragging the leads along. Obviously, this requires that the leads be several inches longer than finally needed and that they be strong and coated with an abrasion-resistant dielectric. The leads should all pass through the eye of the needle from the same direction, and this should be noted by some visible notch on the eye so that the free ends can be pulled out at the destination and the needle disengaged with a minimum of tangling or traction on the electrode ends. Each of the leads is then carefully adjusted for the necessary slack length. Strain to fascia adjacent to the passage of the electrode wires is relieved to prevent traction on the leads from being transmitted to the electrodes. If delicate or irregular devices such as length gauges must be passed, use a thin-walled tube (or a spinal needle) with a removable blunt nose cone to create the tunnel, then use the needle to drag the device into the tube or thread it through and slip the tube out over it, leaving the device in place (fig. 11.2).

One serious but fortunately rare complication of this procedure is the possibility of lacerating a significant vessel somewhere along the blind passage of the probe. Less likely, but more critical, may be the rupture of a nerve; even if the innervation of the muscle being studied is unaffected, rupture of a nerve may hurt the animal and otherwise affect its behavior. Both of these complications pose questions in anatomy and should be considered during the preliminary dissection, which must characterize not only the muscles being investigated, but also the passage and anchoring of all devices to be implanted.

3. Marking Leads

When multiple leads are to be passed, things can get confusing quickly. Some method of identification is crucial. Occasionally one wishes to leave electrodes in place subcutaneously but expects to make the connections later. It is generally much easier to pass the leads back from the implant site than to pass the device in from the connector site, so you want a coding technique that works with loose ends and takes up minimal space. Furthermore, it is much more efficient to have an arsenal of individual devices, any of which can be changed or replaced in the operating room, than to have a precoded array of all devices needed.

Attempts to paint color codes (e.g., with nail polish) can be very frustrating, especially if the wire has Teflon insulation and if you encounter a rough passage during insertion by probe; color codes may work if you are passing the leads through spinal needles. A very simple method for probe-inserted wires utilizes a code of various combinations of knots and loops in the ends of the leads as well as notes regarding the device and its location, orientation, and code. Most devices have two or more leads, so use a double knot to tie them together, with single knots in the free ends to identify the particular device. For a device with a polarity (e.g., stimulating electrode), be sure to make the single knot identifying one of the leads on the device side of a double knot; otherwise, you will have to untie the double knot to figure it out, and it is much simpler just to cut the knots off as you solder each lead to the connector. Naturally, you should trim the knotted sections before soldering the electrode wires to the connectors, because the knotting may damage the insulation. Also you may use simple, sticky, colored dots (available in art and drafting supply houses) for identification while you are assembling sets of electrode wires for soldering.

The small, colored-glass Hishi beads represent a simple but very elegant way of marking EMG wires. These can be autoclaved and strung on the wires and also left in situ without foreign body reaction or harm to the insulation. The beads may indeed be anchored by a loop of the wire or a drop of surgical adhesive. They are particularly useful for identifying leads, either to permit later repair of damage or to confirm device position at postmortem examination.

To minimize loose ends, the two contacts of a bipolar suture electrode can be made from a single loop of wire, with the two leads remaining joined until they are cut apart when the final connection is made. The knots do catch a bit as they are pulled through, but this is generally a problem only when spinal needles are used.

C. EXTERNAL CONNECTORS

1. Principles

Generally, the electrode leads should exit at a point at which the skin shows least motion relative to the deeper layers. For mammals, this tends to be on the dorsal midline, where superficially located skeletal elements, such as the skull and spine, provide connective tissue or bony anchoring for both the skin and the necessary hardware. However, there are many types of connectors and they may be anchored at many sites. Obviously, the nature of the integument affects this anchoring, which is why some of the following accounts have been subdivided by species.

For a while, there was considerable enthusiasm for vitreous carbon buttons, which were billed as permanently installable Biosnaps for percutaneous access. The idea was that the carbon ceramic surface would be colonized by normal body connective tissue and the epidermis would grow up to and adhere to the connector, much like the gums around a tooth. Although carbon buttons remain useful in certain situations, their application introduces two problems. First, epithelium does not stop growing until it meets other epithelium. Thus, it tends to invaginate downward along the percutaneous plug and around the subcutaneous leads. This produces sinus tracks into which keratinized skin and secretions shed and are trapped, thus forming the nidus for infection. Second, if a cable is attached to the connector, its weight will occlude capillary flow on the stressed side of the rather large exit wound required for a multipin Biosnap. Consideration of hoop stresses explains why the living edges of the skin are so quick to necrose (fig. 11.3).

If the connector exit can be made very small, if the connection time for recording is relatively brief, and if there is relatively little stretching of the skin in this region under normal use, then this type of connector can be cosmetically attractive and require minimal care for many months. However, it is almost impossible to repair a carbon button connector; if it does break down, or if the electrical

11.3. Whenever skin is distended elliptically, the greatest tension is generated along the ends of the long axis, whether near the ends of an incision or over objects that push up beneath the skin. These are the zones that would tend to rip and, by implication, the zones in which the circulation is likely to become disturbed by implanted devices and at which necrosis may occur.

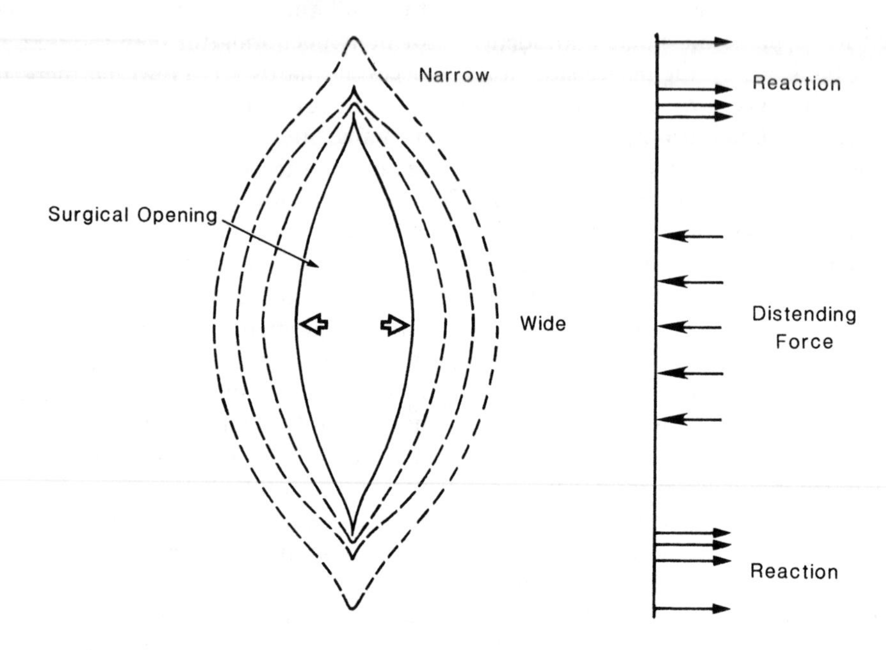

connections at the buried underside of the connector need repair, it inevitably leaves a large, ugly wound.

Some animals have particularly sensitive areas; these may be quite unsuitable for either the exit of electrodes or the attachment of or covering with external devices. Thus, some mammals strongly object to anything the experimentalist fixes to their horns. Furthermore, many horned species are adept at ripping loose such devices (they annually use this motor skill in scraping off the drying velvet); hence, it is best to avoid a potential contest. Various mammals (in which the males engage in sexual combat) may show substantial sex-associated differences in the thickness of the skin. Some conversation with a specialist, such as a zoo keeper, and with others cognizant of the behavior of the potential experimental subject may save continuous trouble.

2. Body Suits and Harnesses

Depending on the species of animal, various harnesses ranging from simple collars to complete Dacron suits with Velcro or zipper fasteners have been successfully used. Sewing supply shops can provide a variety of useful raw materials, such as elastic Soutache braid and soft leather shoelaces. Harnesses are particularly suitable when the electrode implantation procedure is itself minimally invasive and temporary, such as with percutaneously inserted needles and bipolar hook electrodes (fig. 11.4). Harnesses can also be used to support bulky or heavy batteries and transmitters as well as antennas for telemetry. The biggest disadvantage is their limited acceptance by some animals. Considerable training may be required to accustom the animal to the suit, and even then the movement under study may not be performed normally. Cats are particularly likely to reject suits, even well-fitting ones; after an initial frenzy, they often respond by sitting passively and being unwilling to move. The design and fitting of such devices may involve both the animal and the experimentalist with considerable experience.

3. Exoskeletal Fixation

Some animals seem custom-made for the convenience of the experimentalist; they sport spines, shells, horns, carapaces, or other rigid parts that accept connections such as cement, screws, and wires. Unless the study is designed to analyze the motor behavior of a particular species or "preparation," there may be great benefit to the study of the phenomenon in a different, mechanically and behaviorally

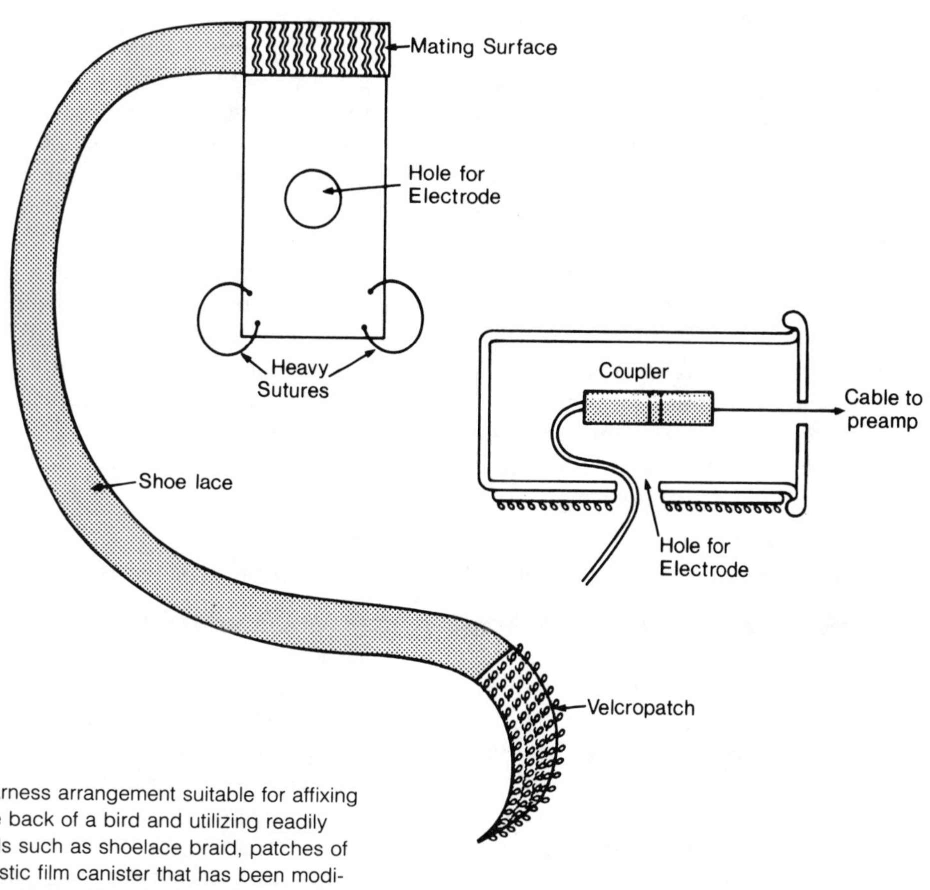

11.4. Simple harness arrangement suitable for affixing connectors to the back of a bird and utilizing readily available materials such as shoelace braid, patches of Velcro, and a plastic film canister that has been modified to hold the connector. The shoelace slips over one wing, through the suture loops, and over the other wing and the terminal Velcro patch; it then contacts the matching one on the saddle. (Courtesy A. S. and S. L. Gaunt.)

more suitable form. Often the "standard" species was initially selected for its availability rather than its suitability. The biological sciences are rife with examples of breakthroughs made possible by a felicitous selection of a kind of organism, rather than the development of particularly complex or innovative devices.

One kind of animal that has been insufficiently utilized for such analyses is turtles. The natural history literature is replete with documentation that these animals have a remarkable tolerance for local lesions of the shell. Its surface readily accepts bone screws, allowing a simple fixation of the connector. If the species being studied is aquatic, the electrode leads are best passed to a hole in the carapace and led from there to the connector, with the exposed zone being potted in silicone rubber (Silastic). Fortunately, experiments on aquatic organisms do not require that the specimen itself be grounded; grounding of the water is sufficient. It should be noted that electrical shunts (short circuits) will occur if any of the wetted cables or connectors have holes or cracks. Electrode connectors can also be glued to the surface of the shell; however, this may pose different problems (see below).

4. Fixation to Avian and Mammalian Skin

The integument of most animals is deceptively tough. Whereas its tensile strength and durability seem ample, this condition is achieved by a quick metabolism and turnover. Consequently, the surface is unstable; keratinized layers are constantly or intermittently shed, and sweat and mucous secretions ooze out of multiple pores. The high metabolic rate is supported by a matrix of fine blood capillaries in the subcutis, which are easily compressed by outside pressure and by stresses on the skin. The extremely dense innervation of the skin by small-caliber afferent and efferent nerve fibers is designed to detect metabolic needs, shunt blood flow where possible, and alert the animal to correct pressure points when necessary. Anything that stresses one place of the skin for a prolonged period is both painful and likely to lead to ischemic damage and necrotic breakdown. If skin fixation is used, it should be assumed to be temporary and care should be taken to distribute tension widely and loosely, as with multiple loose sutures.

Birds and mammals have a mixed curse and blessing known as feathers and hair. Except for a very few species, these outgrowths keep the experimentalist from gluing saddles to the skin for any significant period. However, if the behavior being studied (breathing, feeding) permits this, it is possible to use these structures to hold light (featherweight!) connectors. Connectors and keratinized structures may then be encapsulated in a cone of Silastic, which provides an excellent bond. Such arrangements tend to be useful only for relatively brief periods. Obviously, grooming behavior will have to be considered; also, the subject is likely to become frustrated and may prematurely shed feathers and hairs as a result of tension.

5. Fixation to Mucous and Keratinous Skin

Fishes and amphibians generally have mucous skin containing the openings of multiple glands, although these may discharge only intermittently. Very few adhesives will make much of a bond with such a surface; the glandular secretions wedge off any attachments. However, a few species (e.g. some toads) have local areas of dry integument, although these may contain the openings of poison glands. Such dry patches will accept various glues.

The skin of many frogs passes fairly closely over the sculptured top of the skull with minimal intervening tissues. For such forms we have had luck in molding harnesses out of various dental casting materials, sometimes fixing these with simple bone screws. Obviously, the harness must not occlude regions of nostrils, eyes, and eardrums, or they will induce wiping movements and avoidance behavior. However, animals often ignore well-designed caps.

The keratinous skin of reptiles seems beautifully designed for the attachment of electrodes. Certain contact adhesives form an excellent bond and are well tolerated by the animals. For some applications, patches of Velcro may be fixed directly to the skin, with the matching patches glued to the strain-relieving tether and the connectors. Lizards tend to ignore fine wires that have passed through needle holes in their skin. Such wires may then be glued or taped to the surface and passed to the site at which they may be soldered to a small connector. Both cyanoacrylates and neoprene adhesives provide excellent bonds to keratin (but apply the adhesive as small drops; do not paint the animal with it!).

There are two major problems with the attachment of objects to reptilian skin. The first is that such skin shows differential mobility; thus, the interscalar regions will flex and expand, so that one should allow the connectors to float among several of these, rather than attempting to fix a rigid structure to flexible skin via firm glue points. For that matter, it is useful to attach the harness to places at which there is little relative motion and, more important, to places that reptiles will not "scratch," either accidentally or intentionally. The second problem is that reptiles shed their skin as units, with or without adequate notice. Indeed, manipulation, leading to local integumentary damage, may increase the frequency of the shedding cycle. Glues that tend to induce this process are profoundly to be avoided.

6. Fascial Fixation

Fascial layers, such as those overlying the back muscles, are even more deceptive than skin in their apparent strength and invulnerability. Although their metabolic rate is slower, it is not zero, and the collagen fibers are not permanent. Whenever sutures are placed under continuous tension in fascia (as when they are anchoring an external connector), the connective tissue gradually remodels itself over weeks, thus reducing the tension and loosening the sutures. Whereas this does not produce the soupy mess of devascularized skin, the relaxation can result in a loose device that then does stress the skin. Also, this assumes that the sutures have been placed through pure fascia; if muscle fibers have been incorporated, pressure necrosis can be rapid and is likely to be followed by infection.

7. Skeletal Fixation

a. Skull Pedestals. Transferring the stress to skeletal segments is the best strategy for long-term implants. Bone has the highest tensile strength and lowest metabolic turnover. Whereas bone will remodel under stress and bone screws have been known to work loose, well-designed bone fixation will generally outlast even the lengthiest experiments. One special application of this method has already been mentioned in the discussion of exoskeletal fixation.

The techniques for constructing skull pedestals are fairly well developed and simple (fig. 11.5). First, the skull site is prepared over a fairly wide area by careful scraping of the site down to solid, smooth bone, with removal of periosteal layers and cauterization of perforating vessels if necessary. Bone wax should be avoided as it may interfere with the adhesion of methacrylate. Then one or a few small bone screws are placed into the skull to

11.5. Typical construction of skull pedestal connector. Features include bone screws (short enough to avoid penetrating the skull and with heads not turned completely down to permit acrylic flow underneath), connector (flanged at both ends with right-angle solder pots and recessed pins), and solder joints to leads protected with silicone rubber. Skin edges are drawn under the connector flange and over the cured acrylic to minimize exposed edges (using a purse-string suture that is here omitted).

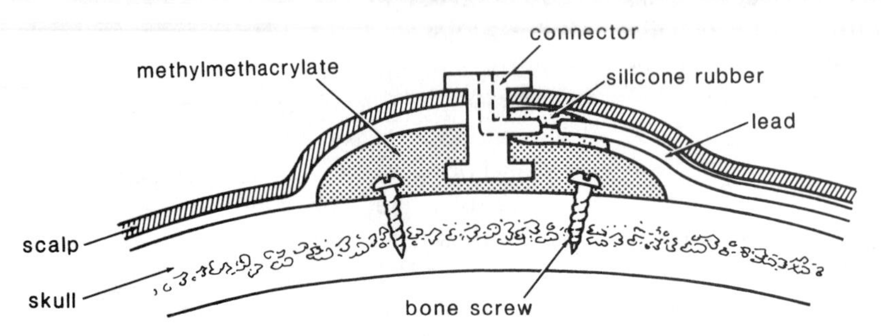

act as reinforcing rods between the cranial bones and the methylmethacrylate polymer, which will be polymerized in place to build up the connector site.

It is difficult to find surgical screws that are small enough; self-tapping, stainless steel, industrial grade sheet metal screws work well, although it may be helpful to grind flat their pointed tips. Be sure that they are really stainless steel and not electroplated with "corrosion-resistant" and highly toxic metals, such as cadmium. Drill a hole through the cranium at each site using a dental burr slightly smaller than the tap drill size for the screw (minimal diameter between the threads). Take care to support the drill bit so that it does not vibrate horizontally. Also, use a sharp drill bit so that very little pressure needs to be applied and consider using a collar stop on the bit to reduce the risk of punching through the meninges or even into the brain. A screw-holding screwdriver is very handy for turning in the screw. Be careful not to advance the pointed tip past the inner table of the cranium. The screw head should sit 2 to 3 mm above the outside surface of the skull so that the acrylic will run under it and be mechanically anchored. If possible, place two or three screws near the perimeters of the acrylic, where the stresses between the rigid bone and rigid acrylic will be the greatest.

The entire region in which acrylic will contact bone and screws must be carefully cleaned and dried to promote adhesion. The acrylic is best applied in layers, starting with a thin first layer of material of low viscosity prepared by use of additional solvent. Ideally, the acrylic should not contact skin or other soft tissue; excess acrylic should be wiped away before it sets. The curing reaction is very exothermic; a large volume can actually burn adjacent tissue. Generally, the connector is positioned in soft polymer and mechanically anchored into the matrix as it sets, with multiple layers painted on as needed. If a large connector is needed, it may be better to anchor mounting brackets to the acrylic so that the exit wound can be made smaller than the connector (see fig. 11.5). The scalp generally settles into a fairly stable perimeter around the connector, but as with any percutaneous exit, the smaller the opening, the fewer the potential complications.

Many researchers make the electrical connections to the back of the connector pins at this point and bury them in the acrylic as well; this will present a smooth, tough external profile. However, if shorted, or open circuits develop because of infiltration of fluids or poorly made joints the connections are then inaccessible for repair. One compromise is to coat the connections and wires with Medical Adhesive A and to cure this in place. This coating better resists the infiltration of free water and can be cut away with a scalpel if necessary. Another compromise is to leave the joints bare and dry, enclosing them with a detachable cover made from aluminum sheet for protection. With this approach it may be necessary to clean and dry carefully the connector for the first few days to prevent accumulation of serous drainage.

b. Vertebral Saddles or "Backpacks." The experimentalist can anchor connectors to the vertebral column in the same rigid manner, by building up acrylic pedestals that bridge multiple vertebrae. However, the scarcity of good, dry bone anchor points and the much larger relative motion of skin to bone makes this very difficult. We have had great success with a nonrigid anchoring system by which an external connector "saddle" is tied to small holes drilled in the dorsal spinous processes via short percutaneous suture loops. It works well with the large processes of the lumbar vertebrae, which are easily exposed via short midline incisions followed by periosteal elevation of the muscle attachments (fig. 11.6).

Use a dental burr to drill a small hole near the center of each exposed process (generally one near each end of the saddle is sufficient). Very large-gauge synthetic sutures are available with large, curved, swaged-on needles, such as #5 Ethibond (braided polyester, Ethicon Ltd.). The suture is passed in a curved path through the bone hole, from entrance and exit punctures in the skin about 2 cm lateral to each side of the midline incision. The incisions are completely closed after bone chips are removed by irrigation. These suture ends are passed through holes in the saddle and tied together so that the saddle is held securely but not tightly against the skin surface.

Usually, the leads are first passed up through one or more stab wounds located under the saddle, so that the percutaneous leads and the sutures have separate holes and are not incorporated into the midline incisions. The leads can be attached to the connector after operation without sterile precautions and either potted with Silastic or covered with a detachable metal shield for protection. As the exit points are so small, there will be a minimum of secretions and the danger of infection will be greatly reduced.

Even bulky and relatively heavy saddles are very well tolerated by the animal, as they can shift slightly as posture changes and as the stress is all referred to insensitive bone. The large diameter of the sutures minimizes the chance that the suture will erode through the bony spine. The only long-term attention needed may be the removal of fur or other debris, which can collect under the saddle. It is advisable to close the midline incisions either with absorbable or very nonreactive sutures or with use of a running subcuticular stitch that can be removed from the periphery of the saddle; conventional, simple silk sutures can irritate as they will not be accessible for removal. If the saddle is to be used for more than six months, the suture holes through the spinous processes should be lined by metal grommets to prevent spicules of growing bone from abrading the sutures.

D. SELECTION OF CONNECTORS

1. Problem

Probably no single component is the source of more trouble in electronics than are connectors, whether in biomedical applications or

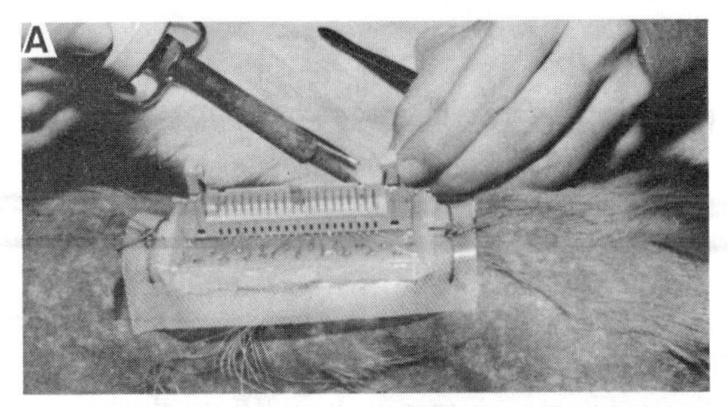

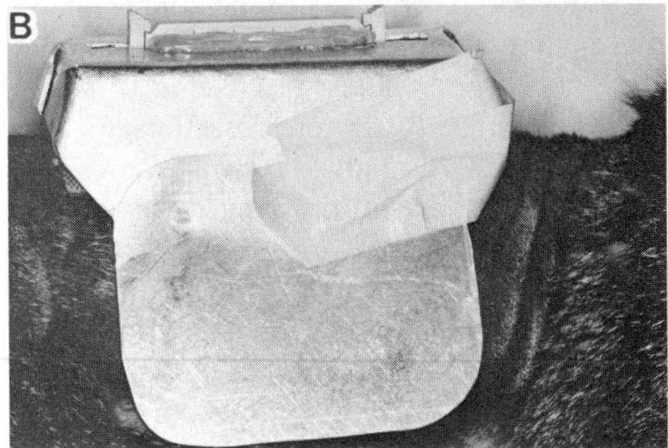

11.6. Saddle connector system for bringing large numbers of leads out through small incisions on the back. *A,* Saddle in place showing ribbon cable connector header mounted on small, custom-printed circuit, in turn attached to a sheet of Dacron-reinforced silicone rubber by silicone RTV adhesive. The sheet is held on the back of the animals by means of heavy-gauge sutures at either end that pass through skin and fascia to holes drilled in dorsal spinous processes. Fine leads from electrodes pass subcutaneously to exit puncture wounds under the saddle and are then soldered to pads on the printed circuit. *B,* Aluminum cover designed to protect connector between recording sessions. *C,* Ribbon cable attached to saddle in recording configuration. *D,* Ribbon cable attached through intervening printed circuit board with 12-channel FET hybrid amplifier.

purely inanimate systems. The very notion of a connection between two electrical conductors that is mechanically and electrically secure at some times and easily detached at others should sound ominous. Combine this with a situation in which the part to be connected is constantly moving and attached to creatures that shed fur or feathers, perspire, leak mucus, and urinate, then consider that the electrical level of the signals is in the microvolt and microampere range, and you should develop a healthy respect for the problem before learning about it the hard way.

2. Rules of Thumb

1. *Make sure the connector system is expandable.* Whereas your initial experimental design may be modest, requiring perhaps two leads plus ground, it is part of the natural history of experimentation to keep pushing success to the point of impracticality. Fortunately, connector technology is usually built around families of similar and compatible connectors with gradually increasing numbers of pins.

2. *Mechanical and electrical functions should be separate.* The mating pins should not be

the sole or even a significant part of the mechanical attachment between the two (mating) connector assemblies. If there is relative motion between these pins, there will be electrical noise and abrasive wear of the contact surfaces, which are usually formed of corrosion-resistant but soft metals, such as gold. Ideally, the molded support around the pins should provide for lateral stabilization to prevent rocking of the pins or the imposition of lateral or twisting stresses.

3. *Beware of locking mechanisms.* If the animal panics, the mechanism holding the two connectors together should release before the connector fixation to the animal fails. At the very least, the experimentalist should be able to release the connection quickly, so that the animal will not have to be unduly restrained. It is better to remove any but the simplest locking clips (unless these are designed to protect the electrical connections).

4. *Pins and assemblies must be cleanable.* If there are inaccessible recesses, sockets, or threads, they will inevitably plug up with debris. This is particularly likely to occur on the connector half that is chronically mounted on the animal, as this cannot be removed for cleaning or replacement. The decision as to whether the male or the female connector is to be mounted on the animal must be based on the part that is most rugged and easiest to maintain. It is best to design a cap or noncable connector that may be placed onto the connector that is firmly mounted to the animal.

5. *Insertion forces should be minimized.* Pins that fit their sockets tightly generally form more secure connections with less mechanical noise. On the other hand, a connector mechanism that handles well with three or four contacts may become impossibly stiff in a 30 pin configuration. Plan carefully for how you can grasp the connector on the animal both to draw it together with its mate and to separate them. If it takes a lot of force to detach a connector and it gives way suddenly, you will probably have a very unhappy animal on your hands.

Often it is worthwhile to equip the connector housing with grasping points, perhaps for use with a pair of needle nose pliers to make sure that the forces are applied between the connector mates and not across the anchoring points to the animal. At other times, one may design a special tool to hold the connector placed on the animal so that force may be exerted against it, rather than against the animal. Draw-down screws solve this problem only at the expense of neglecting rule of thumb 3.

6. *The attachment of leads to pins must be reliable and accessible.* Generally, the leads are attached to the connector during or after the operation to facilitate their subcutaneous passage. Making reliable solder joints at an awkward angle on a nonrigid mounting while trying to maintain aseptic technique can be frustrating. Although closely spaced, hexagonally packed pins make for a very compact connector, they can be very difficult to solder, particularly with fine-gauge wire. This is an application in which a resistance soldering tool proves ideal (one lead can then be mounted by a special plug on each minipin in turn; the other end of the pretinned minipin can then be touched by the second lead, only long enough to melt the solder and slide the flux-coated lead into place). Connectors with only two rows of pins are helpful, because you can always work from the outside to avoid crossing previous joints. Of course, there are other techniques for terminating wires besides solder, and many workers will be tempted to experiment with mechanical

crimping, conductive epoxies, and spring clips. Suffice it to say that nothing compares with a well-executed solder joint for reliability and repairability.

7. *Try to find a connector family that includes mass-terminated ribbon cable.* Ribbon cable is unmatched for flexibility, low levels of electrical noise, and ease of termination, so it represents the natural solution for the cables from the animal to the amplifiers and perhaps elsewhere in the system. The computer industry has spawned a well-developed connector technology providing fast, reliable attachment of lightweight, multipin connectors to ribbon cables. It then becomes feasible and relatively painless to have a variety of cables of varying length and for special functions such as impedance testing. Also, these cables and their connectors are subject to considerable wear and tear, not to mention chewing and clawing, so it is nice to be able to replace them at the first sign of trouble.

8. *Strain relief must be adequate.* No solder joint or mechanical crimp can be expected to last long if it is subjected to constant flexing and strain (fig. 11.7). Typically, the stressed lead will soon break just behind the joint. If small joints and splices must be protected locally, a short piece of heat-shrinkable tubing, just tight enough to encapsulate the joint snugly, can be very effective. (Note that small, hot-air blowers are specially made to shrink the tubing in place.) The connector must incorporate some clamping device that grips all of the insulated wires some distance back from this joint so that the stresses pass from cable to connector body, not from the cable to the pins. This may make the connector unacceptably large or heavy. An alternative is to strain relieve the cable directly to the animal, separately from the connection itself. A spring, wire, or rubber band from the cable (proximal to the connector) can be attached via a harness, bone screw, or other device to a site on the animal that is suitable for mechanical attachment but that might not have been convenient for electrical access (e.g., horn, fin ray, carapace, exoskeleton, but remember to check on the animal's behavior).

9. *Utilize modern connector technology.* Connector technology is a highly evolved art and science. Commercially available connectors are composed of complex assemblies of lightweight flanges, interfaces, locking devices, strain reliefs, and gaskets. Each of these aspects is likely to have been derived through careful design, testing, and experience. If you attempt to modify such an assembly (e.g., by removing a locking mechanism), you must approach it thoughtfully, because modification may have unforeseen consequences for reliability. Manufacturing your own connectors de novo or from mismatched parts will usually take much effort without achieving either the reliability or the miniaturization of a commercial device. In this case your time probably will be spent better by obtaining product catalogs and making thorough inquiries of sales representatives from some of the manufacturers listed in appendix 2.

E. CABLING FROM THE ANIMAL

1. General

The factors that determine the acceptability of the external cables that connect an animal to recording equipment are largely mechanical; however, some birds may display psychological factors, as they tend to treat such cables as snakes. Obviously, the cable must be light and flexible enough not to interfere with the behavior of the animal but durable enough to take constant flexing and sudden stresses. Much is heard about "cable noise," which is

11.7. Strain relief techniques depend on production of a smooth, gradual transition between the mechanical properties of highly flexible wires to rigid connectors, thereby minimizing the concentration of stresses at the junction.

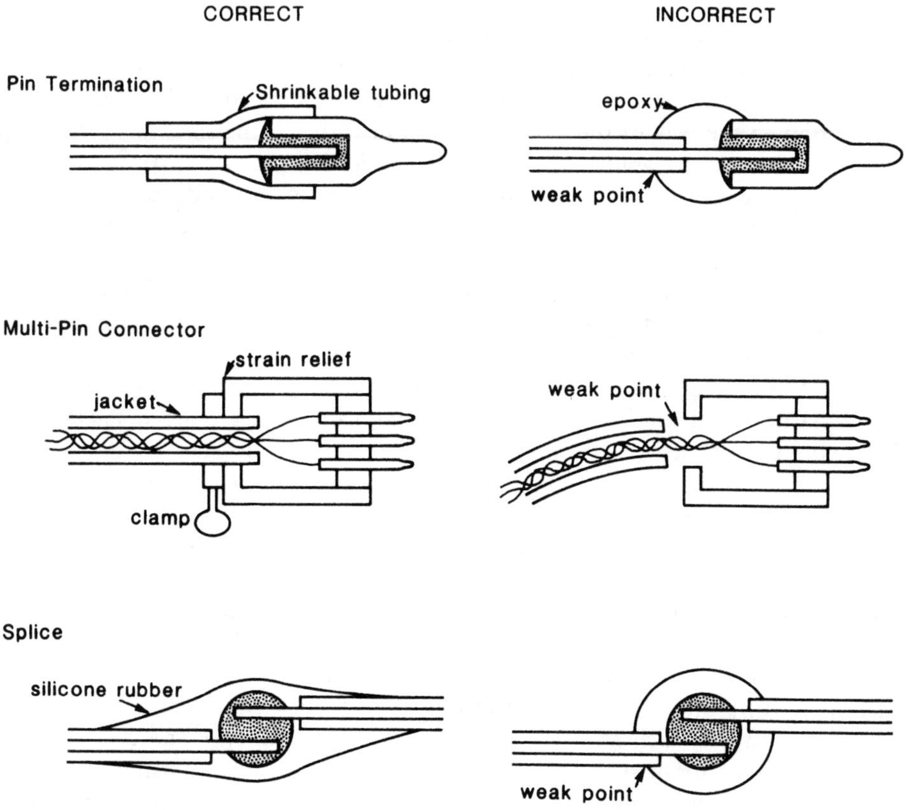

actually the mechanical motion of individual leads with respect to each other and ground. Any given lead has a certain electrical capacitance with respect to all conductors in its vicinity. The capacitance is proportional to the surface areas facing each other and is inversely proportional to the distance between them. If leads shift position with respect to each other, their capacitance will change, causing a fluctuating signal in the leads. Although the capacitive conductance is actually quite low, it may not be negligible in a high-impedance (low-conductance) electrode circuit. Most EMG electrodes are not in this high impedance range, but this potential source of noise should not be ignored, particularly if the cable passes near power lines or other high-voltage sources such as surfaces that retain static electricity (e.g., Plexiglas enclosures). (This provides another reason for separating the various cable arrangements from each other so that power cables to amplifiers, oscilloscopes, and tape recorders do not cross the leads from animal to preamplifier.)

2. Wire Types

We have already discussed the virtues of ribbon cable, particularly when large numbers of leads are present. Its biggest disadvantage is that it tends to hang limply in the way of the animal and to get easily twisted if the animal circles. Obviously, something with the mechanical properties of a coiled telephone cord would be ideal. However, such cable is apparently unavailable for more than a couple of conductors. One alternative is flat conductor ribbon cable, which consists of printed copper conductors in a relatively springy plastic film sandwich. The thermoplastic jacket can be twisted into a spiral and set into a self-telescoping spring after being heated briefly in boiling water. Unfortunately, the termination methods and connectors for this type of cable are neither as reliable nor as easy to use as are those for standard round wire ribbon cables.

Round cables made up of bundled round conductors are often quite stiff, although they may incorporate desirable features, such as braided outer shields. The flexibility of such cables must be evaluated by tests not only for bending but also for torsional properties (the ability to absorb a twist at one end without transmitting it to the other). Such torsional capacity depends on the orientation and packing of the individual conductors rather than on their total caliber. "Low noise" cable is often made deliberately stiff in rotation to reduce interconductor motion.

One common mistake is to misuse coaxial cable, which is often seen as a panacea for preventing noise and cross-talk in electronic equipment. This may be true for single-ended signals, such as those from monopolar electrodes and the outputs of electronic devices, such as amplifiers. It is definitely wrong for differential inputs and outputs, such as those from bipolar recording electrodes and isolated stimulator outputs. The whole point of differential amplification is the cancellation of common mode noise by subtraction of similar noise signals picked up by the two electrodes and their leads. If one lead is the center conductor and the other is the shield of a coaxial cable, the shield will pick up extrinsic noise while shielding the inner conductor from it; thus, the effect of noise is maximized and common mode rejection is circumvented. The best way to ensure that both leads pick up the same noise is to make sure that they go to the same places—that is, to use twisted pairs of similar conductors. If it is necessary (and it rarely is), you can add a grounded shield over the outside of the twisted pair.

The same arguments in reverse demand the use of twisted, paired cable instead of shielded coaxial cable for the output of isolated stimulators, which should not be shielded at all. Two principles are involved in minimizing stimulus artifact, and both are defeated by the coaxial cable. The first is preventing spatial electric and magnetic fields from being generated by the stimulus voltages and currents in the leads. This is best done by using twisted, paired conductors so that the field generated by the flow of current in one direction in one lead is exactly cancelled by the opposite flow in the other lead. The second principle is to maintain isolation from ground reference circuits. This is defeated by the high stray capacitance of coaxial shielding, even if it is used over twisted pairs.

3. Commutators

The problem caused by twisting of the cables can be quite severe, particularly if the animal is unrestrained and must be left unattended for any length of time. There are a number of commutators available commercially, derived largely from the aerospace industry. In fact, they are often distributed as odd-lot items by various scientific distributors who get them as overruns from military production. Some are quite compact, allowing up to 30 slip rings in about 1 cubic inch; some even include a commutated fluid path. In evaluating such a device, you should look particularly at breakaway torque, which is the force required to get the rotor turning. This may be much higher than the running torque, in which case the release of cable torsion will be sudden and erratic. In addition to irritating the animal, this may introduce electrical noise. There is no sure way to anticipate whether the commutator noise will be significant in your application, so a trial demonstration is advisable if possible. At the very least, the contacts should be gold to gold.

An electrically and mechanically superior, but commercially unavailable, low-friction commutator employs platinum pins that spin in circular mercury-filled channels (fig. 11.8). In designing such commutators one should reduce both the bearing friction and the inertial mass of the rotating part. Note that for a given torque, required to turn the assembly, one can minimize the force on the animal (and its attachments) by increasing the length of the lever arm. However, the cable will tend to twist should the animal circle at radii much smaller than the lever arm. The method represents a compromise, permitting one to obtain long-term records, for instance from medium-

11.8. Rotating commutator suitable for transmitting the signals of four electrodes from a freely moving animal to the recording apparatus. Each channel is filled with mercury, and the spinning arm carries platinum pins that penetrate the pools. The mechanical connection to the animal proceeds via a wire attaching to the hook at the end of the arm. The wire is always shorter than the electrode connections, so that these will not be unduly stressed.

sized aquatic organisms. Unfortunately it works poorly for small creatures that make tight circles.

Of course, the simplest solution is a paradigm that discourages or prevents circling, such as a narrow treadmill (or a large activity wheel). Another version forces the animal to travel in a narrow annulus with a radius equal to or greater than that of the lever arm of the commutator. When twists do develop in a cable, a design that permits an easy and rapid disconnection, untwisting, and reconnection of the connectors is a big help, although this can only be used on animals that do not respond adversely to the presence of the investigator.

4. Faraday Cages

If the cable or the animal itself picks up AC power line hum or any other noise from equipment that must be in the vicinity, it may be necessary to use a Faraday cage. In simplest terms, this is an enclosure of conductive material that is independently connected to ground. As light and ventilation of the contents may be a problem, one tends to fabricate it of screening or slotted metal. The overall efficacy of any Faraday cage is a function of the overall conductance. Any significant resistance within one of the panels, among panels and doors, or among any of the units and the ground cable will permit the currents to generate substantial voltages within the system. Hence, it is obviously necessary to shield all six sides of the cage and electrically to interconnect the sections of the Faraday cage so that all are grounded. The ground connection must be of very good integrity, both at the cage and at the ground cable. Note that a screened cage does not produce an absolute shielding of the contents. For instance, metal-framed aquariums often serve as secondary antennas. If this generates a problem, ground the aquarium frame or, better yet, use an all-glass aquarium. If nails are used to connect the screening of the Faraday cage to a wooden frame or to connect sections of the wooden frame to each other, be sure that these are grounded in turn, so that they do not act as antennas that breach the integrity of the shielded space.

In the practical case, one often finds that many signals are very directional, so that one may ultimately keep all or part of one side open. This permits one to film the animals without showing a superimposed screen on the image. Although it is critical that appliances using line currents (such as heating pads, illuminators, and stimulators) be kept outside of the cage, it is often useful, indeed appropriate, to place battery-powered preamplifiers into the cage.

The design thus far mentioned serves as a good shield against line noise and other low-frequency electrostatic fields. It does not protect against high-frequency (radio and television) signals. For these, special shielding materials (mu metal) must be used; happily, this kind of shielding is generally unnecessary.

F. CABLING IN THE LABORATORY

1. Theory

An EMG facility generally starts out with one or two channels of EMG data passing from a connector on the animal to a preamplifier and on to a chart recorder or tape recorder. Perhaps an oscilloscope is tied in to keep an eye on the signals during recording. Next a few more channels are added, which means more amplifiers, more recorder channels, and perhaps a multipole switch so that the oscilloscope trace can be scanned through the various channels. Then it is decided that the EMG signal needs to be processed by analog integration or digitization before the experimental results can be published, so new boxes, interfaces, and cables are added.

Electronic stimulation or behavioral training devices (and their timing circuits) greatly expand the scope of the experiments possible. Of course, one needs a video or cinematographical record of the complex experiments, which means synchronization signals, voice annotation tracks, speakers, and lighting controls. Now there are cables carrying signals ranging from DC logic levels to audio frequency EMG signals, to radio frequency video information, all of which are likely to interfere with each other by electromagnetic radiation and many of which can damage valuable equipment if inadvertently interconnected.

It is no wonder that many investigators live in fear that some enterprising tyro will rearrange a few cables and connections and thereby cause chaos. At the same time, few laboratories can afford the luxury of duplicating expensive equipment such as computers, cameras, and tape recorders just so that each investigator can leave his or her setup unaltered over a long course of possibly infrequent experiments.

There are two general approaches to having several projects coexisting in the same facility. The first is to train the personnel such that they all understand completely how everything should be used and the state in which it should be left, and then to rely on the fear of God (or the departmental equivalent) to keep them following the rules, no matter how hectic the experiments become. The second approach is to plan carefully and construct the laboratory in such a way that practically anything anyone might want to try can be accommodated in a foolproof way.

Most short-term experimental physiology laboratories rely primarily on the "people" approach. It saves time and money, at least initially (when these commodities are scarce), and it leads to a laboratory with that traditional look of really exotic work in progress. It has been our experience that this approach quickly breaks down under the methodological diversity of kinesiological experiments and the unpredictable situations that arise when experiments deal with unanesthetized, freely moving, behaving and misbehaving animals (and investigators).

2. Systems Design

Although the second approach, that of foolproof design, does not eliminate the need for some discipline among the experimentalists, the design is much more forgiving of human nature, animal nature, and Mother Nature. Successful systems design requires primarily careful thought, thorough consultation among the users, and an ongoing commitment to keep the system current with needs, all of which you will have to do for yourself. In the following section we try to provide some techniques, guidelines, and options for your consideration.

a. Standardization of Connectors and Wire. Establish one or two types of connectors and wires for all interconnections among equipment. Equipment manufacturers seem to delight in decorating front panels with several unique, highly specialized connectors that do not mate with anything else in the laboratory. In a typical laboratory, you can find BNC, UHF, banana, minibanana, pin-type, microphone, telephone, phono-plug, microcoaxial, twinaxial, Cannon, Hex, Winchester, Blue Ribbon, Scotch-flex, and some other connectors no manufacturer will even own up to. If you start collecting cables with one connector on one end and a second on the other so you can interconnect each piece of equipment, you will soon look like *Prometheus Bound.* If you have to add a tee to branch to two pieces of equipment, you may end up in a padded cell.

Instead, decide on your favorites: we favor banana plugs for short individual leads where shielding (in or out) is not necessary, and BNCs for long-distance routing of signals that are sensitive to or that can radiate noise (fig. 11.9). Buy or make adapters for every single other type of connector in the laboratory and leave them on the equipment permanently. If you cannot find a simple adapter, make one by wiring two individual connectors together and then wrapping some epoxy clay around the whole assembly to give a rigid single piece. Then invest in a large selection of different lengths of well-made patch cords, tees, and unions using the connector type(s) you have chosen, and find a neat, convenient place to hang them.

It is also useful to standardize wire type, particularly the coaxial wire used with BNC connectors. The various sleeves and crimp parts of these connectors are formed in specific sizes, which work well with the proper coaxial cable and miserably with the wrong sizes. It is expensive and confusing to keep a stock of BNC connectors for a variety of wire sizes. However, if one has standardized all connections for a single connector type and cable size, it is well worthwhile to invest in one outrageously priced crimping tool for making perfect connections each time. The smallest size RG-174 is relatively inexpensive, easy to handle, and suitable for audio frequency signals. However, although it does have a relatively low capacitance per foot, it is not an impedance-matched transmission line. This may cause high-frequency losses or instability in the output stages of low-power devices (especially video signals) whenever very long lengths (more than 100 feet) have to be used. Whereas such lengths do not normally arise in the laboratory, the extensive use of patch panels (see section G, this chapter) can lead to long lengths of parallel paths.

The only exception to this—and it is an important one—concerns basically incompatible signals. Power lines should always have their own connector system, and unduly low and high voltages must be rigorously separated. High-frequency signals should also be kept separate, as they may induce unhappy effects if accidentally patched into an EMG circuit. Remember that the spectral content of digital pulse signals with sharp edges will include very high frequencies, even if the pulse rates are low. Very high frequencies in video equipment require special low-loss coaxial cable, which is bulky and too expensive for general use. Such circuits should indeed use a distinct connector type and, furthermore, should be labeled clearly.

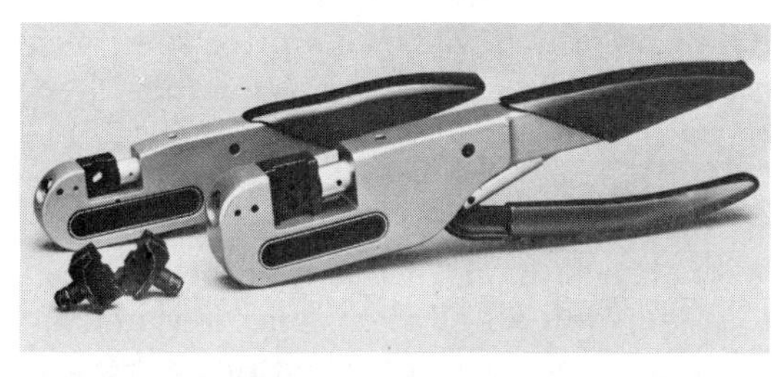

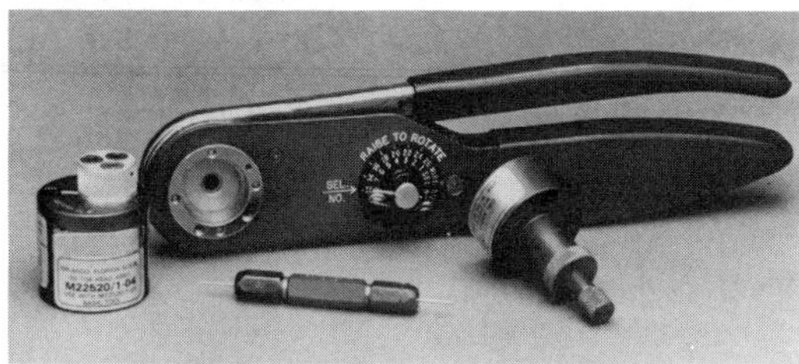

11.9. Crimping tools designed to fasten the BNC connectors directly to the wire by the application of mechanical pressure. Various devices may demand different kinds of trimming of the several layers of insulation and more than a single application of the crimping tool; however, the costs of tool and connector will differ significantly. (Courtesy of Daniels Manufacturing Co., Orlando, Fla.)

b. Labeling of Cables. Label all standard cables completely and permanently. Many connections between pieces of equipment are quasi-permanent, in that they are needed in almost all experiments. This does not, however, prevent them from being disconnected, if only inadvertently or in the process of tracking down some technical problem. Any such cable should have a clear, permanently affixed, unsmudgeable label on each end stating where it comes from and what it normally goes into. Do not put this vital information on a torn scrap of masking tape or paper stick-on label when any electronic supply house can supply you with perfectly lovely, blank, write-on plastic labels for a few cents. This kind of label should be applied to any cable that is routed such that both ends are not immediately obvious. This includes cables from rear panel connectors and lines between equipment racks, which lines may normally be left in place even when unconnected at either end.

11.10. Groupings of RG-174 coaxial cable carrying various kinds of signals, organized into hierarchical bundles by pop-type cable ties that facilitate tracing of connections and addition or deletion of cables as needs change.

c. **Routing of Cables.** Route cables coherently. If several cables form some kind of group (e.g., all the inputs to a multichannel tape recorder), use cable ties to bind them into a neat bundle at frequent intervals to support the bundle at convenient anchor points (fig. 11.10). The reusable pop-type ties are very handy, because they accommodate the addition or removal of a line from the bundle. If you ever have to remove or replace a set of cables (e.g., to accommodate an increased physical distance between pieces of equipment), this will prevent the formation of a Gordian knot of intertwined cables from every piece of equipment in the laboratory.

d. **AC Power System.** Consider an AC power distribution system. AC power line radiation and transmitted "glitches" are always a serious potential problem, so you want a situation that, once it is working, is reasonably stable. Before routing any signal lines, collect all the power cords from your equipment in some out-of-the-way corner of a metal rack, coil up the excess length of each, and put an identification label near the plug. Then plug them into an AC distribution box with a three-prong, grounded socket for each, and run the single line from the box into a wall outlet on a circuit that has not caused problems. In large installations, it can be very convenient to have a separate line and circuit breaker from a central box to each rack (or group of simultaneously needed equipment). This allows all equipment to be turned on or off from the central box, so that the many individual power switches can always be left on.

e. **Map of Cable System.** Keep an up-to-date map of any cable systems in use. This takes real discipline, at least on someone's part. Yet it will easily return the investment in time when you introduce a new investigator to your laboratory or consider additions or modifications to the system. It will simplify cable labeling because you can always use the map to explain shorthand codes. It will greatly facilitate the tracking down of interference problems and equipment and connector failures. It will alert you to situations, such as excessive loading on output circuits caused by cable length or distribution to too many parallel loads. If you hang the map in a prominent place, it will be available to those who need it and you will be motivated to make a decent job of it.

f. Suspension of Cables. Suspend your cables when they are not in use. Several manufacturers make inexpensive cable racks consisting of a wall-mountable bracket with slots just wide enough to pass the cable, suspending it by its connector. This keeps the shielding from being kinked as you cram it into a drawer and avoids bending the cable next to the connector, at precisely that spot at which it is most sensitive to failure. Even if such racks (which can also be constructed out of strips of scrap metal or plywood) are not labeled, a glance will tell whether you are reaching for a 12-inch or a 6-foot cable.

You should also suspend all cabling off the floor. Assuming that cable racks will protect the cable connector junction, there are two remaining failure modes for the heavy shielded cables used to pass signals among electronic gadgetry—namely, sharp bends and local compression. If connecting cables are allowed to rest on the floor, you will step on them. The resulting failure mode is nasty, as the cable does not suddenly become inoperative but only noisy, so that it may take much time to discover the difficulty. Sometimes the noise only appears when the cable is bent into a particular configuration and it is certain to wait its debut for a critical experiment.

One other reason for keeping cables off the floor is important in a general laboratory setting. Even if the experiment does not involve obvious sources of liquids (stills, aquariums, temperature baths), there is the plumbing system. Entering a flooded EMG laboratory can easily become a shocking experience.

G. PATCH PANELS

1. Concepts

The selection, installation, and use of patch panels is an art in itself, highly developed in the aerospace and telecommunications industries and sporadically misapplied by physiologists who sense that it could help them but do not know how. There are two basic kinds of patch panel—programmable and telephone distribution types—and the wrong choice or a poorly designed implementation can wipe out any potential utility.

For our discussion here, we define a patch panel as any collection of connection points that brings together at one location the input and output lines of several physically separate pieces of equipment. This can be a fully engineered commercial installation or a homemade panel with a few similar jacks that are cabled to several signal sources and their potential destinations. The patch cords complete the circuits as desired in some flexible way that avoids having to deal with long cables and nonstandard connectors going to the equipment itself. A glance at the patch panel (if it is well laid out and labeled) immediately shows the configuration of the equipment without the need to trace lines. The patch cords and jacks on the patch panel can utilize a single connector scheme that is economical, rugged, and convenient and that provides good signal isolation, shielding, and termination loads if necessary (e.g., video lines, see chapter 15).

2. Panel Types

There are two general schemes of patch panel wiring: matrix and point-to-point. In point-to-point panels each piece of equipment generally has a group of connection points allocated to it, with each input and output having its own single, labeled line. Connections are made by having the patch cord bridge between a signal source point and a destination point. As it is frequently desirable to be able to fan-out a given signal to several parallel destinations, signal sources can be arranged to occupy several jacks wired together. Another scheme is to have every input and output point occupy two jacks wired together, permitting a signal to be "daisy chained" from one device to the next. Yet another scheme is to have several groups of jacks wired together but with no assigned input or output. They can then be used to fan out any signal that is patched to them.

In matrix panels the source and destination lines, respectively, make up the rows and columns of an array of shorting points. When a plug is inserted into a single panel location, it bridges a row line to a column line, making a connection between a source and a destination. This system obviously allows for the most versatile fanned-out design, minimizes patch cable clutter, and reduces the chance of two outputs being patched together, a constant danger in point-to-point panels. The main disadvantage is that it requires N^2 interconnection holes whereas a daisy chain, point-to-point system uses only 2N holes (where N is the number of connectable lines). Matrix type panels work well when there are few connections requiring intensive parallel interconnection. They are cumbersome and can be confusing to read when used with diverse pieces of equipment operating in serial manner on a signal (e.g., amplifier to filter to tape to monitor to integrator to computer).

Both the matrix and the point-to-point types are available as programmable and telephone distribution types. The telephone distribution type looks just like its description (fig. 11.11). The panel consists of labeled jacks that are arranged in some convenient but essentially arbitrary scheme. Each jack is labeled, with notation of the signal present on it. The operator then inserts a single plug, making a row/column connection (matrix type) or a patch cord between two jacks that require connection (point-to-point type). Each connection is made individually, and parts or the whole of the ensemble can be removed by another user wishing to change the configuration. The jacks and patch cords are selected for their ability to be frequently reconnected with a minimum of operator effort and without becoming worn or noisy. The more practiced the operator becomes with the layout of the system, the faster and more reliable this reconfiguration becomes, but there is always the danger that some connection will be omit-

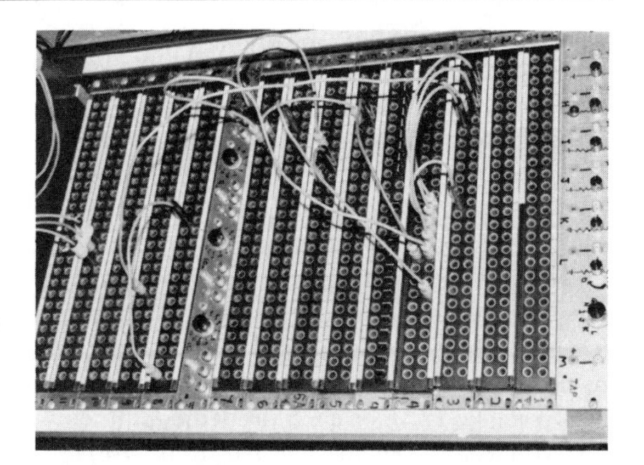

11.11. Telephone jack patch panel provides a convenient and expandable way to facilitate on-the-fly reconfiguration of laboratory equipment for recording and processing multichannel signals. Signal connection points for inputs and outputs to and from all equipment converge here, so that all combinations can be made with one set of coaxial, high-reliability patch cords. Special features include attenuator pots and external connection points with BNC and banana jacks (*above top row*), double throw switches between some telephone jacks (*fifth row*), and rotary scanning switches for rows of jacks (*between ninth and tenth rows*).

ted or incorrect. Thus, the system functions very well for highly versatile, on-the-fly reconfiguration, but it presents dangers if it is relied on for unrecoverable situations (e.g., processing data prior to recording) or for interconnecting pieces of equipment that can damage each other if connected incorrectly.

The programmable panel consists of a carrier board into which all the necessary shorting pins (matrix type) or patch cords (point-to-point type) are inserted before the system is used (fig. 11.12). For each needed equipment interconnection scheme, a separate board is arranged with the necessary pattern. At the time of use the board is inserted into the panel where it simultaneously makes all of the desired interconnections. This scheme has essentially converse advantages and disadvantages to the telephone type. If carefully prewired, it eliminates errors and omissions that might arise during the heat of an experiment. It is also very awkward to rearrange on the fly. The connection points are usually small and closely spaced with little room for labels and less for fingers. The connectors are more delicate and may have relatively high insertion force to resist being dislocated when the board is inserted into the panel.

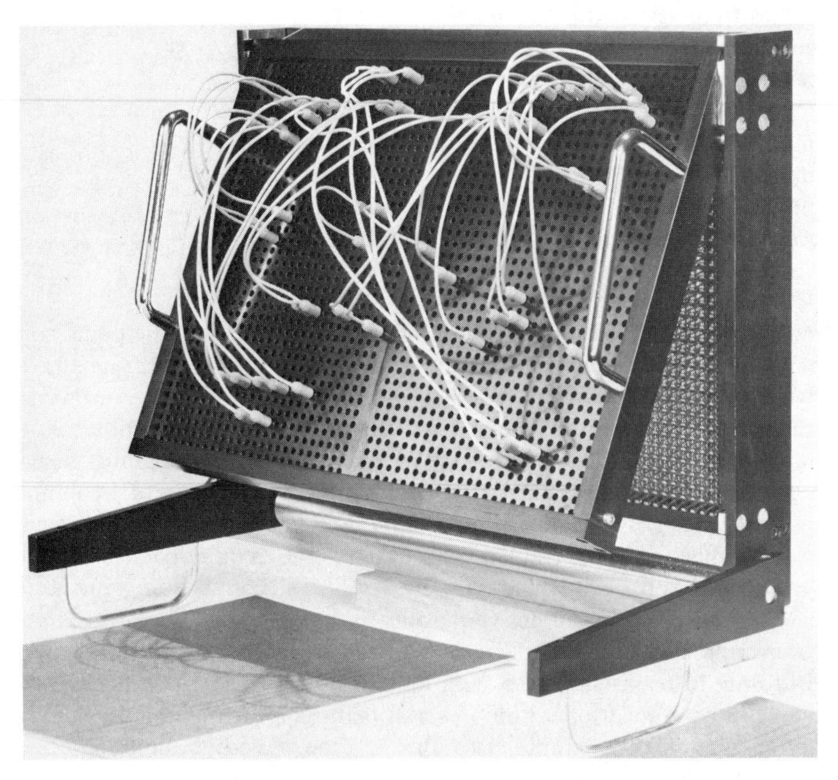

11.12. Programmable patch panel showing patch cords arranged for a complex interconnection scheme among various pieces of laboratory equipment. The panel containing the patch cords can be removed from the locking frame, as shown. Consequently, several different panels, each containing the configuration for a particular experiment, can be readily exchanged. Other types of patch cords, including shield pair and coaxial cords, can be accommodated by this and similar designs. (Courtesy MacPanel Co., High Point, N. C.)

3. Rules of Thumb

Obviously, the decision to install either (or perhaps both) type of patch systems in a laboratory depends greatly on the number of pieces of equipment, on their physical separation, on the number of investigators using the laboratory, and on the nature and diversity of their experiments. However, once the decision has been made, a number of factors concerning implementation must be considered if the investment is to pay off.

1. *The system must be complete.* All of the sources and destinations that might ever need to be interconnected must be present on the system. Otherwise, it quickly degenerates into an unreliable muddle, with cords snaking off the panel to mystery locations and cables between the panel and the standard equipment disconnected to accommodate ad hoc connections. Of course, there will always be unforeseen circumstances: a new or temporary piece of equipment may be required (e.g., for testing or trouble-shooting). Thus, it is always a good idea to leave a few unassigned locations on the patch panel with unconnected rear jacks or loose cables (carefully labeled) that can be used for temporary additions to the patching system.

2. *Avoid incompatible signals.* Most of the EMG laboratory signals are DC to audio frequencies. After their initial amplification, these signals should all have similar amplitude ranges, usually based on the dynamic range of the A/D (computer analog-to-digital converter) or the tape recorder. Both digital signals and video signals contain high-frequency components that readily generate cross-talk by stray capacitance. As there is little need to route these specialized signals to the same devices as the analog data, they are best left out of the patching system or confined to a system of their own. The same consideration goes for specialized equipment that might generate dangerously high voltage levels or that needs to be isolated from the chassis grounds of other equipment (e.g., stimulators).

3. *Leave room for expansion.* Select a patch panel system that either is expandable with additional modules (telephone distribution type) or has at least three times the number of connection points that you think you need.

4. *Facilitate expansion and rearrangement as needed.* The jack or matrix lines on the panel can be directly wired to the cables going to the terminals on the equipment, but this makes rearrangement a very difficult operation. Some patch panels include a rear panel connector jack for each patching jack, which is very convenient but tends to be more expensive and requires a mating connector on each cable to the equipment. You may find a happy compromise by using fixed lines for equipment unlikely to change, reconnectible lines for miscellaneous equipment, and lots of unused patching locations in between to accommodate changes and additions.

5. *Use floating shield connectors and cables.* A patch panel can be an ideal way to control the problems of ground loops and cross-talk or an almost infinite source of them. All of the jacks and patch cords should accommodate coaxial cable and should connect both the core conductors and the shields when two pieces of equipment are patched together. However, the patch panel itself must not be a ground point for any of the cable shields or patch cords or metal connector housings in contact with shields. This means using either floating ground jacks (plastic housings) or nonconductive chassis.

6. *Plan layout carefully to minimize patch distances.* Frequently used general purpose

equipment should be near the center of the panel face, where it can be more or less equidistant from the various sources of its inputs arrayed around it (e.g., oscilloscopes). Arrays (such as tape recorder outputs) that are frequently structured in a fixed order to inputs (such as A/D lines or chart recorder inputs) should be arranged in parallel rows for which the connections can be verified at a glance.

7. *Some connections should be independent of the patch panel.* For example, if the time code equipment always interacts with the same tape recorder channel (and it should), then the connections from the output of the time code generator to the tape recorder input and those from the reproduction output to the code reader should be explicit and permanent outside the panel. However, both signals (input and output) should appear on the patch panel by tees from the permanent cables so that they can be monitored or perhaps patched to other equipment as well. In fact, when possible there should be a "skeleton configuration" that provides a certain minimum level of laboratory function that is accessible, but not changeable, through the patch panel.

Going through the experimental protocols to establish such invariances will help the experimentalist to standardize procedures. It will minimize the chore of documentation and the chances of error during recording sessions as well as shorten the apprenticeship period for new users. However, these invariances must really become universals, never to be changed under pain of death, because they are the critical connections that, by their omission from the patch panel, are not readily subject to inspection before each new experiment.

8. *Provide ample interpanel channels.* You may want to have one patch panel for the signal conditioning and recording channel selection (probably a programmable one) and one for general purpose data processing to be done off-line (probably a telephone distribution type). Make sure you have a set of lines that allows signals to be passed back and forth between equipment accessed by one patch panel and the other.

9. *Document and label religiously.* The labeling of jack positions on the patch panel must be legible, unambiguous, and unquestionably accurate. It should be easy to distinguish inputs from outputs, especially to avoid connecting two outputs together. There should be a paper map of the panel layout on which more detailed labels can be affixed and that facilitates the planning of changes. Cables to and from the patch panel must be labeled at both ends with their origin and destination (i.e., patch panel coordinates and equipment termination). Whenever you make a change or addition, be sure immediately to change the front panel labels, the map, and the cable labeling.

10. *Select or modify equipment to be suitable.* Particularly with telephone-type panels, the configurations that may actually arise during use are quite unpredictable. Equipment should tolerate electrical abuse, such as grounding of outputs and unusually large amplitude inputs. Output drivers should be of low impedance, preferably under 100 Ω, to handle what may be very large, parallel loads and many yards of highly capacitive coaxial cable. Most modern equipment is well designed for these attributes, but it pays to read the specifications and be aware of the potential sources of problems. The design of patching systems for high-frequency video and digital equipment requires special attention to transmission cable impedances and line terminations, which is best provided by a qualified engineer.

11. Combine discipline with adaptability. Once novices realize how convenient the system is, they will use it properly. However, the convenience is predicated on their being able to accomplish absolutely everything they ever need through the patching system. If you find that an investigator is bypassing the patch panel to make direct connections between pieces of equipment (and thereby jeopardizing their patch panel cabling), then you, as designer, are obligated to demonstrate how the desired configuration can be achieved within the system. If such a configuration cannot be established, then you had better modify the system immediately to accommodate this need.

12 Basic Electronic Equipment

A. LEVEL OF UNDERSTANDING

Once upon a time, everyone involved in any way with electricity relied on the same set of fundamental laboratory tools to amplify, transmit, record, activate, and power their signals. A physiologist could borrow an oscilloscope from the television repairman and an amplifier from the physics laboratory and use them without questions or problems. In fact, the equipment would probably be identical to that eventually bought for the laboratory. All of this made sense, because electromagnetic phenomena, from DC to blue light, are based on the same fundamental physical processes, whether they occur in atoms, animals, or television sets.

Currently, every scientific subspecialty has its own little industry of equipment manufacturers, busily repackaging, rediscovering, and specializing equipment for particular applications. This has utility, because it makes for efficiency in setting up and performing experiments that are vastly more complicated than those possible in ancient times (circa 1930 to 1970). However, as investigators have become more removed from the electronic fundamentals, we may witness a new generation of experimentalists who do not know what is happening to their data, as the signals go through these much more complex black boxes. Whereas no one should now have to build oscilloscopes, everyone should know in general how one works and what limitations it might have. There is simply too much time and money at stake to justify selecting or operating equipment by rote or by chance.

Some basic knowledge of electronic equipment is also important to its care and repair. For example, you should know that it is useful to "burn in" new transistor-based equipment by leaving it turned on for a few days or even weeks, because most potential faults show up in the first hours of use and because its subsequent functional life will depend much more on avoiding dirt in the switches and knobs than on the intrinsic life of its electronic components. Conversely, tube-operated equipment (yes, you can still find some) fails from use-dependent aging of its tubes. Also you should know that the heads on high-performance instrumentation tape recorders rely on a tiny, nonconductive gap between the poles of an electromagnet in contact with the tape, which can be destroyed by the abrasive action of particles of oxide from the tape unless the heads are frequently cleaned and the system is kept closed and free of dust. Whereas there is no need to learn how to repair a defective oscilloscope or tape recorder, you should be adept enough at the use of common test equipment (such as a volt-ohmmeter) so that you can decide whether and where a problem actually resides in a complex system. More than a few high-priced service calls are made to repair equipment that was plugged into a dead outlet or connected to a defective input or output cable.

If you are going to go poking around the inside of electronic equipment, you should keep a few things in mind regarding self-preservation. Most transistor-operated modern instruments such as amplifiers and filters generate only low voltages, so most of the potential damage from careless practices is to the equipment rather than to the investigator. However, any piece of equipment that plugs into the wall will have 115 VAC line voltage present in its power supply, which is more than enough to kill you. Whenever possible, confine your probing to equipment that has been unplugged, preferably long enough for any power supply filter capacitors to have discharged (a few minutes). Certain items actually operate with large DC voltages, including anything using a vacuum tube (including oscilloscopes). If you are in doubt as to whether a circuit is still "hot," invest in a simple line tester, but always be sure that you keep parts of your body out of any potential circuit. Television displays are notorious and rightly so; you cannot be very useful working inside one, so best stay out. One often overlooked hazard is the electrical stimulator. Even battery-operated stimulators can sometimes put out many millamperes at over 100 V—enough to give you a stimulating experience.

B. AMPLIFIERS, VOLTAGE, AND POWER GAIN

1. Principles

The word "amplifier" conjures up images of a fairly expensive and bulky box that takes low-voltage signals and turns them into higher voltage signals. However, this is only one manifestation of the general process of amplification, by which signals with low power are boosted to levels that allow them to perform some useful work. There are two basic reasons for amplifying a signal. First, amplification will make the information-carrying signal large enough so that it is not swamped by electrical interference during transmission over a distance. Second, storage or display media such as chart recorder pens, loudspeakers, and oscilloscopes all consume considerable power in the transduction of the electrical signal into mechanical or light energy; if the initial signal lacks this power, it is likely to be transduced inadequately.

The transmission problem is actually the one that concerns the experimentalist the most, because modern display devices all have built-in amplifiers for developing whatever combination of voltage and current is required to drive them. In transmission the key concepts are signal power (not voltage) and signal-to-noise ratio. Signal power is the product of current and voltage. More commonly, we know the signal voltage and the electrode impedance from which it was obtained; these allow us, via Ohm's law, to write the power as the voltage squared divided by the impedance. Immediately, we see the rationale for the emphasis on well-designed electrodes that span the spatial voltage gradients of the target optimally and that have as low a contact impedance as possible. The tendency of a transmitted signal to become degraded by significant percentages of noise is related to the ratio between the power in the desired signal and the power in the noise-radiating sources, which we assume for the moment to be beyond the control of the investigator. Thus, the amplification of the signal prior to its transmission can perform a useful service by simply changing the effective source impedance without actually changing the voltage level. This is important, because such so-called voltage-following-amplifiers, which affect current but not voltage, are very stable and extremely easy to build in miniature configurations.

In considering the signal-to-noise level of a

signal, we must differentiate among noises that are present within the signal source (biological noise), noises that are introduced in the antenna or electrode (thermal noise), noises that are added by the amplifier itself (amplifier noise), and the extraneous electrical interferences we discussed above. The biological noise is related to the orientation of the electrodes, whereas the thermal noise is related to their contact impedance. The extraneous interference represents a threat only to the extent that its amplitude is a significant proportion of the power of the transmitted signal, hence the rationale for preamplification.

The intrinsic amplifier noise turns out to be potentially important (fig. 12.1). It is added into the signal before it is amplified, and it is often large compared to the thermal noise in low-impedance electrodes. A first stage of preamplification with a modest voltage gain will boost the level of the output signal to the point where the amplifier noise of the next stage represents an insignificant percentage of the input signal to that stage. In recent years amplifier noise has been reduced drastically even for the inexpensive miniature devices that are convenient for biological work, so the importance of this consideration will depend on the actual signal levels anticipated. One must differentiate between current noise and voltage noise. Whenever one is recording from high-impedance source electrodes, even small amounts of current noise can generate large noise voltages because these currents must flow through the recording electrodes. Because of their low current noise, junction field effect transistors (JFETs) are preferred in the input stages of high-impedance amplifiers, even though they have relatively high voltage-noise specifications. Conversely, bipolar transistors make better input stages for low-impedance electrodes because their voltage-noise is lower and because they are less prone to damage if large voltages are inadvertently applied to the input (e.g., from stimulation circuits). The third major transistor technology—namely, metal oxide semiconductor field effect transistors (MOSFETs)—has little place in biological amplifiers, because they have been optimized for applications requiring high frequency rather than low noise.

Finally, the levels of extraneous electrical interference that are actually experienced by your signal-transmitting cables can be considerably reduced by cable shields, Faraday cages, and even elimination of certain offending pieces of equipment and AC power supplies. These approaches are covered in chapter 13. In the present chapter we emphasize the priority and efficacy of sound electronic design over ad hoc and cumbersome protective measures.

2. Bandpass and Filters

Every amplifier, in fact every component and wire, is effectively a filter, as no electronic component transmits all the frequencies from DC to blue light with equal facility. The electrodes themselves have much lower impedances for higher frequencies than for lower ones. However, a frequency-dependent impedance is not, in itself, a filter. The simplest filter element is a frequency-dependent voltage divider—that is, two components in series that divide an imposed signal into a voltage drop across each of the components. If one of the components has an impedance that is frequency dependent in a manner different from that of the other, then the percentage of the signal across that component will depend on the frequency spectrum of the signal. The output of the filter (the voltage drop across one of the components) is thus some frequency-distorted representation of the input (fig. 12.2).

There are always actual or implied compo-

12.1. Relative effects of noise at various amplifier stages. Noise (here shown as 2 $mV_{p\text{-}p}$ of line hum) can be injected directly into the signal at the source by a stray capacitance; it then significantly adds to and degrades the recorded signal. A similar stray capacitance from the noise source into the output of the first stage and input of the second stage has much less effect because the first stage output impedance is much lower than the electrode source impedance. However, because there is no voltage gain in the first stage, amplifier noise in either stage is equally deleterious to the recorded signal. Intrinsic amplifier noise in the third stage has one tenth of the effect on the recorded signal because of the second-stage gain. Thus, it is useful to have both impedance conversion (first stage) and a small gain (second stage) combined into a single first stage as close to the preparation as possible, to minimize the problems both from intrinsic and from radiated noise.

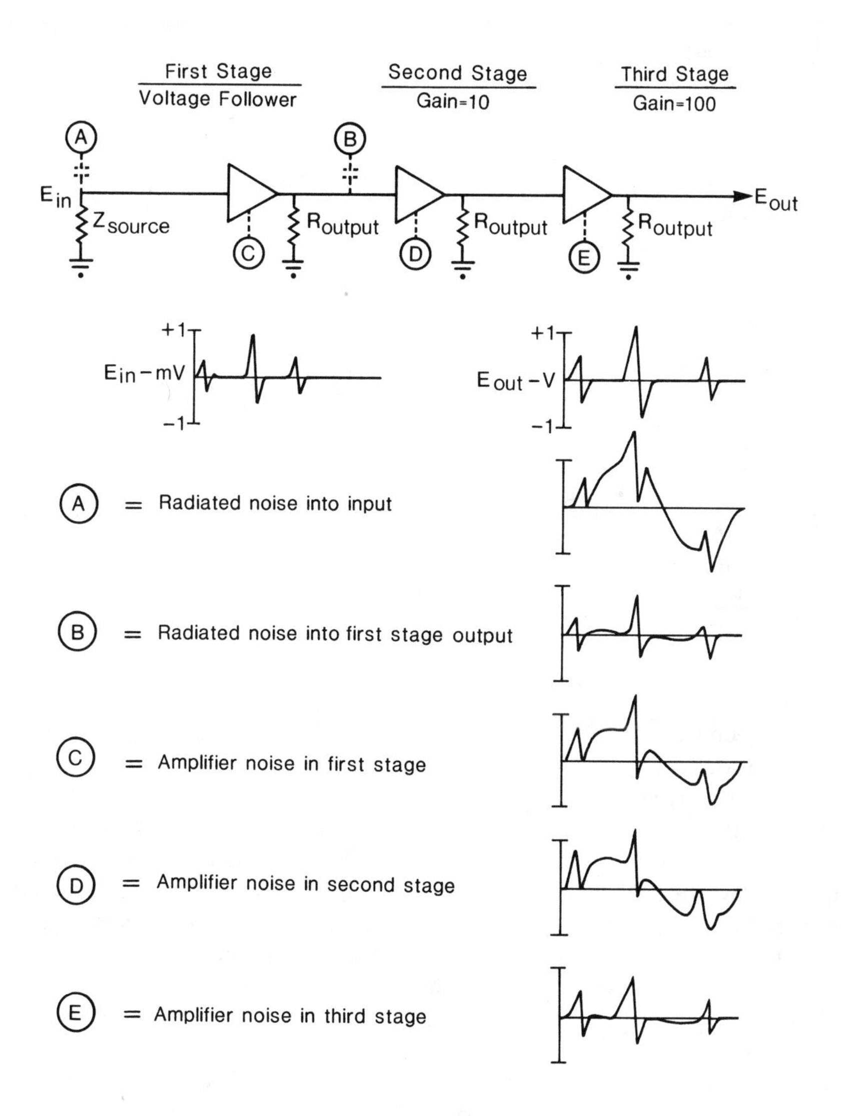

12.2. Filters. The simplest possible filter circuits are made up of a capacitor and a resistor, either as explicit components or as effective terms disguised as stray capacitance or amplifier input impedance. Depending on their arrangement, these components selectively attenuate either low frequencies or high frequencies (sinusoidal test signals, *right*) and distort the shape of more complex waveforms that are actually made up of multiple frequencies (square-wave test signal).

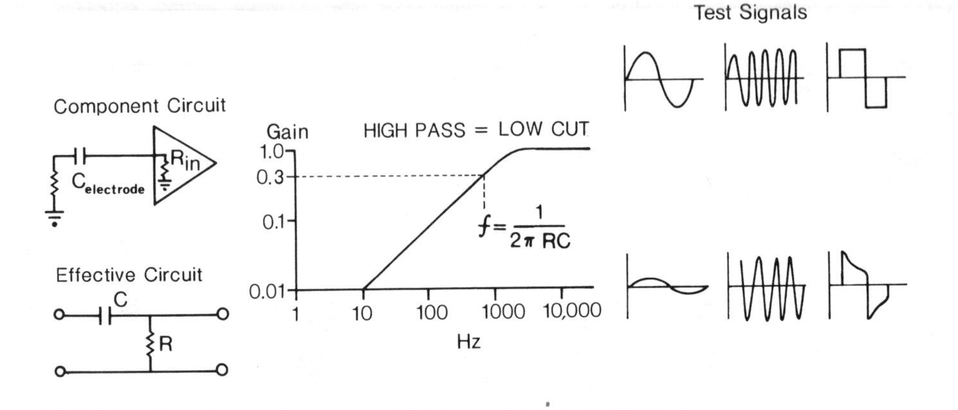

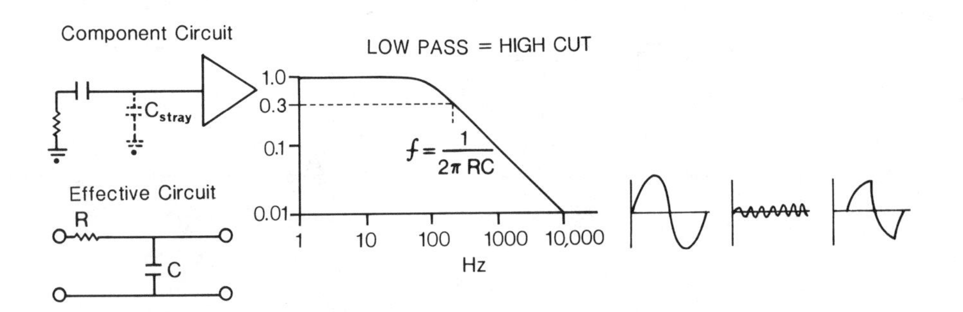

nents around to complete the required series circuit of the voltage divider. The electrode contacts are in series with the amplifier itself, which acts more like a resistor. Because the contacts are somewhat capacitive, the low frequencies will tend to have their largest voltage gradient across the contacts rather than the amplifier. This will attenuate the low frequencies. On the other hand, the cables from the electrode to the amplifier may have considerable capacitance to ground, particularly if they are closely shielded. This capacitance will provide a low-impedance shunt for the higher frequencies picked up by the electrode, and the amplifier will also see high frequencies that are attenuated.

These implicit filters can be quite troublesome if they go unrecognized. Explicit filters with real rather than stray components are based on the same principles and can be very useful in improving the signal-to-noise ratio. This is because the various sources of noise operate throughout the frequency spectrum, while the signal of interest is likely to occupy a fairly narrow band. Selective filtering allows one to get rid of most of the noises that lie outside this band without significantly distorting the desired signal.

There are other reasons for filtering. The implicit filters in various devices do not simply attenuate certain frequencies, they also delay them, causing phase distortions. Amplifiers often produce well-defined, stable gains by the use of feedback in which a percentage of the output signal is led back to the inputs. If this feedback is delayed, it causes the amplifier to overshoot or undershoot the desired gain. That, in turn, causes further distortions of the output signal that is being fed back to the input. Eventually, the amplifier breaks into oscillation. As oscillation is likely to occur in any system that shows a combination of high gain and delayed feedback, it is useful to lower the gain of amplifiers for those frequencies at which phase distortion can occur—that is, high-cutoff filtering.

The implicit filters play an important role in the design of differential amplifiers and make it desirable to locate such amplifiers near the signal source. The goal of differential amplification is to extract signal differences between two recording points and to exclude signals in common, hence common mode rejection (fig. 12.3).

If the two contacts are electrically similar and located in proximity, they will pick up virtually identical amounts of signals generated by remote sources, because the field gradients from these sources are so small in the vicinity of the electrodes. Theoretically, this makes it possible to extract small, local signal sources that are orders of magnitude smaller than large, remote ones. However, this extraction depends on being able to convey the common mode signals to the differential amplifier in a completely symmetrical manner. Even a tiny difference in the distortion of the leads will seriously degrade performance, because it is the difference between two large voltages that matters, and the difference

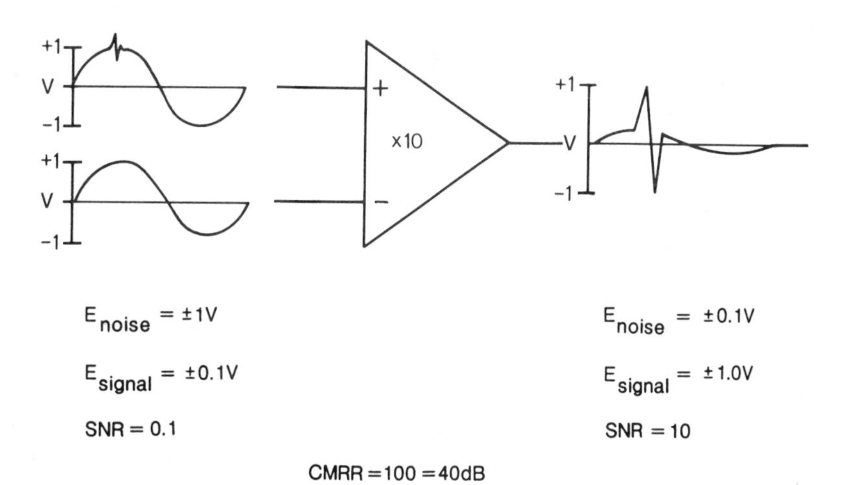

12.3. Common mode rejection by a differential amplifier. A small input signal (brief impulse [E_{signal}] on positive input waveform) riding on top of a much larger signal in common to both inputs (E_{noise}) is selectively amplified, whereas the noise largely cancels out through the electronic subtraction of the differential amplifier. The ratio of the two signal-to-noise ratios (*SNR* output to input) provides a figure of merit for the amplifier called the "common mode rejection ratio" (*CMRR*), usually given in decibels of power [20 log(SNR_{out}/SNR_{in})].

should be zero. Thus, stray capacitance that might not noticeably affect overall signal gain can seriously degrade common mode rejection, particularly for higher frequency signals. Such stray capacitances are also the couplers for radiated, electromagnetic interference into the system; there is no point in precisely subtracting the noise levels present on each lead unless they are identical. All of this provides the rationale for locating the differential amplifier stage as close as possible to the experimental subject and carefully designing leads and connectors to minimize stray capacitance and, particularly, asymmetries.

3. DC versus AC

Raw EMG signals are AC signals occupying a bandwidth of a few tens of cycles per second to perhaps 3000 cycles per second. Therefore, it may come as a surprise if you find a large DC voltage coming out of a preamplifier. Before you simply push the "AC-coupled" button on the oscilloscope to eliminate this voltage, it might be useful to understand where it came from and what it might mean.

First, you should understand that pushing the AC-coupled button eliminates the large DC voltage by interposing a capacitor between the input jack and the rest of the vertical deflection circuitry. At any stage, a series capacitor can be used to pass the AC component and remove a DC offset, so there may be similar offsets occurring in other pieces of equipment that do not appear, simply because they have been removed by a coupling capacitor at some stage. As the AC frequency of interest gets lower and lower, the size of the coupling capacitor (relative to the input impedance of the next stage) must get larger and larger to avoid attenuating the low frequencies and distorting the overall signal. Conversely, signals that are mixtures of AC and DC components (e.g., those from strain gauges and rectified EMG signals) will always be distorted if they are AC coupled, often producing bizarre scallops and undershoots of the baseline as the coupling capacitor charges and then discharges out of phase with the applied signal.

A DC potential can arise between the recording electrodes themselves, particularly if they are made of different materials. Dissimilar metals in contact with a conductive electrolyte solution constitute a battery. Although such a battery is inadequate for powering a flashlight, it can overpower a sensitive amplifier, or worse, cause rapid electrolytic degradation of the electrode surfaces and surrounding tissues. For these reasons coupling capacitors are often built into the input of AC amplifiers, although they tend to store charge and thereby prolong the effects of noise transients, such as stimulus artifacts.

Active electronic components, such as transistors, tend to work with either positive or negative biases, but not with both. There are many electronic techniques for using balanced pairs to hide this fundamental asymmetry, but it is almost impossible to remove completely residual DC offsets from their output. This is a particular problem for older equipment, which may have been built with less stable components or with components that have drifted far from their original properties. For this reason, DC-offset-adjust potentiometers are often placed on the panel or into an easily accessible position within sensitive equipment; consult the instruction manual for the proper procedures and sequence of adjustment. Also, be aware that DC offsets are particularly dependent on temperature. They may go away after a few minutes of warm-up; never try to adjust them until the equipment has been on for 15 to 30 minutes.

A DC offset visible on an oscilloscope trace may easily have arisen within the amplifier circuitry of the oscilloscope itself, as well as in the piece of equipment attached to its input. Grounding the input or AC coupling should not make such an offset disappear. Often, such internal offsets make it impossible to use an oscilloscope channel at the higher sensitivity ranges, because it will be inpossible to return the beam to the screen with the vertical positioning knob, itself a form of offset adjustment. Again, consult the instruction manual for procedures for correcting such a problem. FM tape recorders also tend to generate DC offsets that drift with time, because of instability in the center frequency for both the recording and reproducing circuitry. If large, these offsets can cause asymmetrical clipping of the input AC signal, such that large excursions of one polarity are unattenuated but the opposite polarity is cut off above small excursions.

DC offset can also be a useful tool, and some pieces of equipment, such as EMG integrators, specifically include front panel controls to introduce such offsets. If the output of such a device can only have one polarity (e.g., a positive full-wave rectifier), then you can obtain twice the dynamic range by starting out near the negative limit of the power supply instead of at zero volts. This trick is often used to improve the resolution of noise-limited processes such as FM tape recording and digitization (A/D conversion).

4. Preamplifier Circuits

Figure 12.4 shows a typical, single-ended, low-gain preamplifier. This device is based on an operational amplifier chip configured for a gain of 10. Sometimes such "op amps" are configured as voltage followers with a unity gain—that is, the output signal has approximately the same amplitude as the input. However, the output impedance will be on the order of a kilohm whereas the input impedance will be many megohms. This impedance ratio means that the voltage follower can be used without significant signal loss, even though the contact impedance of the electrodes ranges up to megohms. The transmitted output signal will be very much less prone to noise pickup by capacitive radiation from nearby noise sources, as the low output impedance acts as a voltage divider in series with this stray capacitance. Of course, the stray capacitive impedance may also be quite low for very high frequencies, but such frequencies can be filtered out easily in the next stage because they are higher than the band of the desired biological signals. The device is so small and simple that it can be installed in the connector at the head of the transmission cable.

Figure 12.5 shows a typical differential amplifier for use with low- to medium-impedance electrodes. It is based on a hybrid, instrumentation-amplifier chip (Analog Devices AD521K) with input capacitive coupling and impedance limiting (10 MΩ to ground on each input) and a gain of 10. This is followed by a set of operational amplifier circuits that perform additional gain and high- and low-pass, first-order filtering at a variety of rotary-switch-selectable frequencies. The overall gain can be continuously varied in two switched ranges from 10 to 1000. There is an output drive (LH0002CN) to permit the output signal to drive long lengths of coaxial cable or many parallel loads.

Figure 12.6 shows a transformer-coupling stage that is very useful for dealing with very small signals from low-impedance electrodes. It was originally devised for work with nerve cuffs, in which signals are typically in the microvolt range and amplifier noise tends to be much greater than the thermal noise of the signal source. Under such conditions it is useful to boost the signal amplitude before it encounters an active amplification stage. The voltage level of any AC signal may be pas-

12.4. Simple, inexpensive, and small head-stage amplifier circuit that is suitable for high-impedance, single-ended recording. As configured here, the operational amplifier (*OP AMP*) provides an input impedance of 1000 MΩ and a voltage gain of 10. The output impedance is only a few kilohms, so the amplified signal can be transmitted over long distances without picking up noise. Pin *8* is attached to the case of the tiny, transistor-can enclosure (TO-9 package) and can be used as shown to provide a driven guard signal to any shield around the lead to the input (this shield then carries a signal equal to the source signal to prevent shunting to ground, so it must not be connected to ground at either end). Any symmetrical source of DC power from about ±5 to ±18 V would be suitable at pins *7* and *4*. The power supply ground or reference must be connected to the animal via an indifferent electrode to complete the circuit. The 100 pF capacitor on the input blocks any electrode polarization potentials from saturating the amplifier, and the 10^9 Ω resistor behind it provides a path for the tiny input bias current required by the amplifier. The 100 pF capacitor in the feedback circuit prevents high-frequency oscillation; larger values can be used to provide some high-frequency cutoff filtering, but this is better done in the subsequent amplification stages. (Courtesy M. J. Bak.)

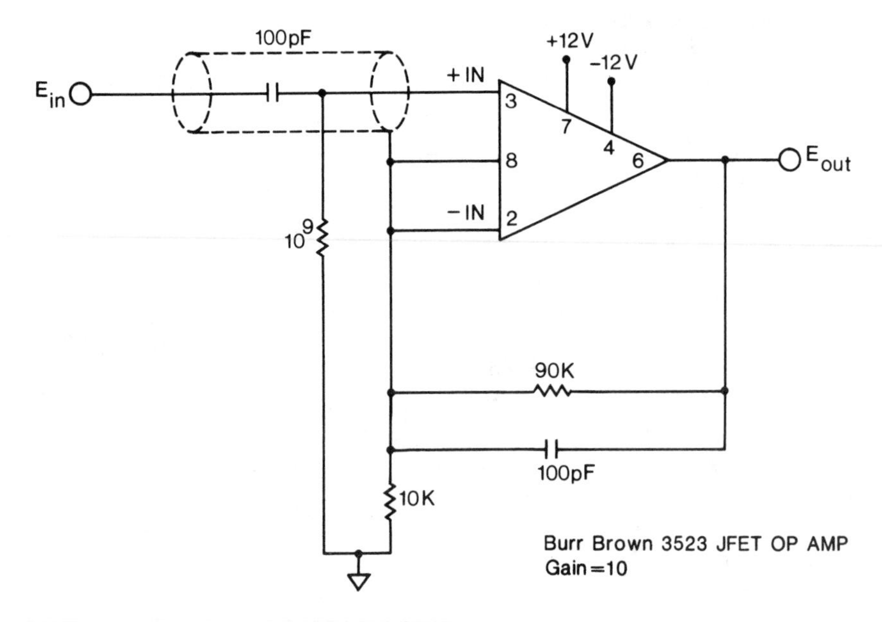

sively amplified with a transformer. However, there is no such thing as a free lunch. The transformer takes the current-driving capability inherent in a low-impedance signal source and converts it into a higher voltage with proportionately less current capability. Thus, assuming that the operation incurs no losses, the total power (E^2/Z) available from the signal remains constant.

To see how this works, assume that the electrodes have a tip impedance of 1 kΩ and a signal size of 10 μV. Suppose that all sources of noise present on the signal total only 1 μV. It looks like we have a very nice 10:1 signal-to-noise ratio, but the amplifier noise is likely to be about 10 μV, and that noise is added to the input signal before amplification. However, the amplifier has an input impedance of greater than 1000 kΩ, far more than needed. If we connect the electrodes to a voltage-stepup transformer with a 10:1 turns ratio, the output side (which will go to the amplifier inputs) will have a 100 μV signal with 10 μV noise. The noise of the amplifier will now be no greater than the noise inherent in the input signal. The transformer has the property of multiplying the effective source impedance by the square of the turns ratio, or 100. Still, the 1 kΩ electrodes now represent only a 100 kΩ source, which is still far enough below the input impedance of the amplifier to cause only negligible loading.

There are two hidden specifications in the selection of the coupling transformer. First, the design of the windings will cause the transformer to have a certain real input and output impedance of its own, which is essentially parallel to that of the loads to which it is connected. In fact, such transformers are usually specified by this impedance ratio (e.g., 10 kΩ: 1000 kΩ) rather than the turns ratio. It is important to select an input impedance that is equal to or greater than the electrode impedance to ensure good coupling of the electrode signal into the transformer. Second, these impedances and the implicit voltage gain are valid only for a limited bandpass, dependent on the precise configuration of the windings and core. Fortunately, such transformers are most widely used in audio equipment, and their bandpasses tend to bracket the EMG band quite well. In fact, they may constitute useful bandpass filters in and of themselves.

All things considered, it will be necessary to calibrate the actual voltage gain rather than depending on theoretical calculations of turns

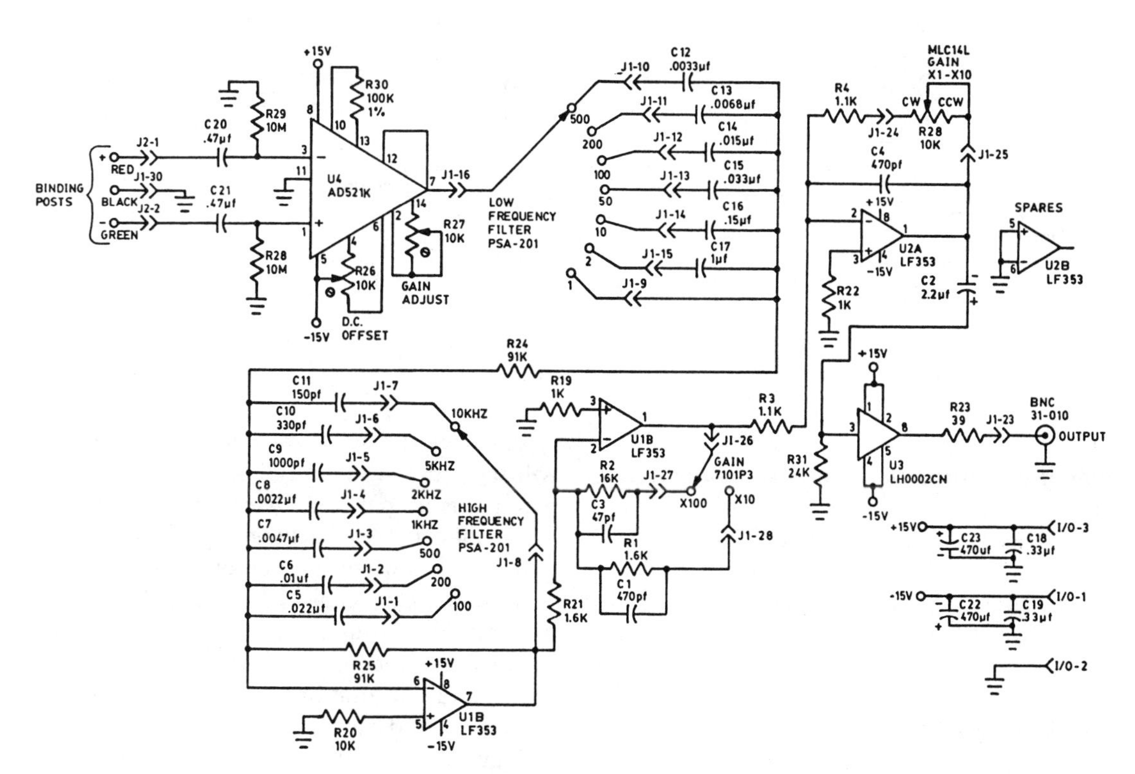

12.5. Circuit diagram for a commercially available differential amplifier with 10 MΩ input impedance, continuously variable gain from 10 to 1000 in two stages, and several independently settable filter frequencies for both low- and high-frequency attenuation (6 dB per octave rolloff). (Model MDA-2, courtesy Bak Electronics, Inc., Rockville, Md.)

12.6. Circuit diagram for a commercially available differential amplifier suitable for use with low-impedance EMG and nerve cuff electrodes. The total gain of 100 derives from a passive transformer coupling stage (noise-free gain of up to 10 as described in text) and an active, differential amplifier (gain of 10, adjustable at trimpot) with high CMRR (Analog Devices AD521K instrumentation amplifier). Diodes on inputs of power supply prevent blow up of amplifier from incorrect connection. Capacitors *C1* and *C2* provide capacitive coupling of the input if the transformer stage is omitted. (Models ADT1 and AD1, courtesy MicroProbe, Inc., Clarksburg, Md.)

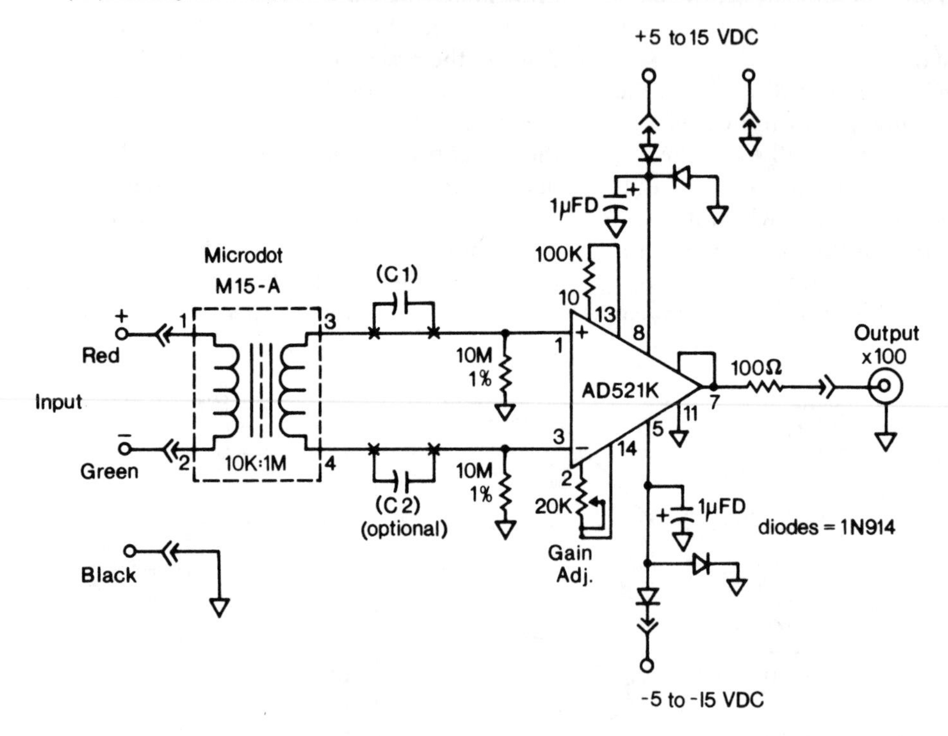

ratios. Note that this can be done correctly only if the source impedance of the calibrating voltage is approximately equal to the electrode impedance anticipated and the test waveform is a sinusoid or other narrow-band signal within the range of the EMG signal frequencies.

C. TELEMETRY

1. General

Anyone who has spent hours soldering fine wires into a percutaneous connector and cable system only to have them eaten by the subject will sense the attraction of telemetry. Unfortunately it is no panacea. Telemetry is essential for studies such as tracking migratory caribou, and it is useful for certain simple studies on bioelectrical signals. Its successful use depends on a thorough understanding of its limitations and trade-offs as well as careful design and selection of both the telemetry equipment and the rest of the experimental configuration. Furthermore, once constructed, such a link to the preparation is far more difficult to modify than simple wires, should you want to add an extra channel, a higher gain, or a longer range. All of the following factors tend to interact competitively in the design of such equipment.

2. Size and Weight

The use of telemetry to transmit a signal from the animal means having the animal physically carry a great deal of circuitry normally kept in an equipment rack. In addition to the electrodes themselves and any leads or connectors to the external fixation point, the electronics include the amplifier, a radio transmitter, its antenna, and a battery power supply. With modern hybrid circuitry, these can be made amazingly small, particularly if only a single channel with a limited range is all that is needed (fig. 12.7). However, even the smallest

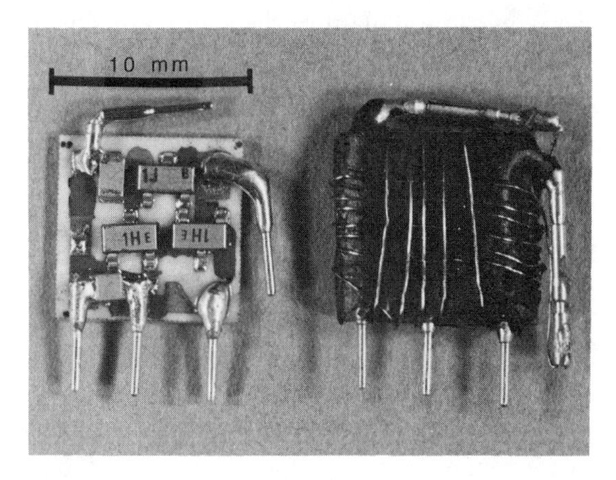

12.7. Hybrid microcircuit transmitter for biotelemetry of EMG data. The unencapsulated view on *left* shows surface-mounted component packages soldered to ceramic chip with silk-screened conductor and resistor patterns. The device on the *right* has had the transmitter antenna coil wound in place over the encapsulated chip, thereby defining its transmission frequency in the FM radio broadcast band. The three pins at the bottom provide the differential input from the electrodes and power from a miniature battery. The assembled unit is dipped in beeswax and overcoated with silicone rubber to waterproof the connections. (Courtesy MicroProbe, Inc., Clarksburg, Md.)

commercial devices constitute an appreciable load, for instance for a good-sized laboratory rat.

3. Range and Orientation

The descriptive literature may read "up to 75 foot range," but this should often be read as "down to 5 feet" if the animal turns the wrong way. It is virtually impossible to make an omnidirectional transmitting antenna, and the limited sizes and geometrical shapes suitable for mounting on an animal's back are usually far from ideal. Even with a highly distributed receiving antenna, there will always be some null points at which the signal strength is much lower than typical. With FM transmission, the signal fidelity tends to be all-or-none; the signal is very clean until the carrier drops abruptly below a critical point and reception breaks up very badly. Some modern telemetry equipment includes dual receiver systems in which an electronic switcher picks the antenna with the best signal. However, it is always better to have more power than you need, particularly if you need clean, continuous data. On the other hand, in a well-designed transmitter the transmitted radio frequency power will be the limiting factor in the battery size/life trade-offs; also, the Federal Communications Commission has some rules about this sort of thing.

4. Channel Number

We have frequently touted the virtues of recording simultaneously from several EMG channels (and/or force and length transducers), particularly if cross-talk between muscle potentials is suspected. This poses a serious problem in a telemetered system because of the wide bandwidth of EMG data. The usual solution to multichannel telemetry is multiplexing, in which the output of each amplifier channel is sampled in turn at a very high rate and the serial information is then transmitted to a receiver that reconstructs (demultiplexes) the several separate channels of output. However, the rate of multiplexing must be several times the product of the number of channels times the bandwidth of each.

Handling these very high frequencies and avoiding cross-talk between channels (particularly if adjacent channels have signals of very different amplitude) is almost impossible with the limited space and power available. The more practical solution is to have a separate FM telemetry link for each channel, tuning each transmitter to a different frequency and having a separate receiver for each. This is obviously cumbersome for more than two to three channels, but if you need more channels than this, you should probably give up on telemetry anyway. The transmitters can share a single battery, but be sure to separate the transmitting and receiving antennas from each other in space to avoid stray inductance and capacitance effects, which can seriously degrade tuning. The output frequencies of most microtransmitters are set by slight physical adjustments of the inductance of their antenna coils, making them less than rock stable. Some new FM radio tuners (conventional audio equipment) incorporate an automatic frequency tracking circuit; this handles the drift problem nicely, unless the transmitter drifts into a local radio broadcast frequency. However, many digital display tuners synthesize their tuning frequencies to match broadcast frequency standards; fine-tuning them to follow a drifting nonstandard frequency may be difficult or even impossible.

5. Bandwidth

We have discussed above the problem of bandwidth in multiplexed systems. It is also important to check on bandwidth when purchasing single commercial receivers and transmitters. Many microtransmitters are made for lower frequency applications, such as clinical electrocardiograph and electroencephalograph monitoring. These applications may have wider commercial markets than EMG; consequently, they may require modification in the amplifier or modulation stages to be suitable for much higher frequency EMG signals. Many FM radio receivers, designed for audio equipment, have capacitively coupled amplifier stages that strip DC and low-frequency components from the received signal; check to see how far through the circuitry such signals would survive before trying to use them for strain gauge or other DC signals.

6. Sensitivity and Dynamic Range

With a conventional rack-mounted preamplifier, you can usually turn the gain up or down if the signal picked up by the electrodes happens to be a little too small or too large. Adding a gain adjustment on the input to a telemetry system may involve more weight than the whole remainder of the circuit, so gain is usually fixed. You can always turn up the volume on the receiver, but this will not correct noise and distortion, which are problems fundamental to a poorly modulated carrier. Undermodulation means noise, the same kind of static you get when trying to tune in a distant AM station. Overmodulation means distortion; at best large spikes are clipped, but the whole signal may break up from saturated demodulation stages. Before specifying the gain desired for the telemetry link of a preamplifier, you must have a very good idea of the amplitude (referred to input) of the EMG signal to be recorded under all possible circumstances.

7. Noise and Cross-talk

Even a well-modulated transmission signal is subject to both white noise and specific interference from adjacent telemetry channels, broadcast frequencies, and general radio frequency (RF) radiation from electrical equipment, such as computers, switching power supplies, ignition systems, motor brushes, and strobe lights. You can reduce much of this noise by increasing the power detected by the receiver antenna, either with a more powerful transmitter or a more directionally selective receiving antenna. However, if more than one telemetry link is in operation, these solutions tend to aggravate the cross-talk problem by overwhelming the frequency selectivity of the tuner (much as happens when you try to use a radio near the broadcast antenna of a commercial station). Interchannel cross-talk is particularly likely to occur if one transmitter is being excessively modulated by an out-of-range input signal; this will generate undesirable sidebands of radio frequency transmission. If you do see correlated signals in two channels, you will have to determine whether they arise from cross-talk of the potentials recorded in the preparation (particularly if the amplifiers are single ended, as is often the case) or from cross-talk introduced in the telemetry process.

8. Battery Life

With space and weight at a premium, most microtransmitters do not come equipped with on-off switches. Usually they are activated by connection of a disposable battery of the smallest size possible, consistent with the duration of a single recording session. As battery voltage goes down, radio frequency signal power and amplifier gain decline, and, to a lesser extent, transmission frequency may drift. The experimentalist must keep all of this in mind when designing the protocol and in the heat of battle. A larger battery with the same voltage usually means a longer life, but with the proliferation of new electrolytic systems (alkaline, lithium, etc.), this should not be taken for granted. Also, mass and size are not necessarily related in a predictable way. The basic specification of battery life is the product of current and time (usually expressed as milliampere-hours). However, you must be sure that the battery is capable of supplying the current actually needed by the transmitter (some long-life batteries have high internal resistances and are intended only for very low-power circuits). Also, you should get some idea about how the voltage drops as the life limit is approached (some batteries decline gracefully and others die abruptly or may even produce potentially damaging reverse voltages if overworked). The various electrolytic single cells (and their combinations into batteries) produce somewhat different working voltages. This may be used to advantage, because the simplest way to boost transmitted power in many systems is to use a somewhat higher supply voltage. However, remember that power transmission and consumption increase with the square of supply voltage. Excess voltage can damage components directly by exceeding their rated bias voltages or their power dissipation. Also, microtransmitters lack space for protection circuits; a few microseconds of the wrong polarity may produce a very tiny but expensive piece of junk.

D. TAPE RECORDERS

1. General

Of all the pieces of equipment likely to be purchased, a tape recorder probably represents the most expensive, the most useful, and the most critical single item. Obviously, it is difficult, if not impossible, to process data represented by chart recorder traces. It is impossible to process multichannel signals on-line during an experiment without seriously compromising the ability to reconstruct what actually happened after things settle down. It is inefficient to digitize multiple channels of wideband EMG signals into megabytes of computer storage; however, it is easy to save hours of data on a single analog tape reel. So the major decisions are how many channels and whether to use direct recording or FM (frequency modulation).

A second decision, which is more a matter of maintenance than of electronics or experimental design, concerns the housing of the recorder. Shall it be mounted in a wheeled instrumentation rack or in a smaller set of transportable cabinets? Unless service calls are of little concern, we would recommend the latter option. Not only will you be able to move the recorder among laboratories, but you will be able to load a malfunctioning unit into the trunk of your car for delivery to a repair center for inspection. However, remember that such sensitive pieces of equipment demand a firm base during operation, so invest in a solid laboratory cart that is larger than the recorder in all dimensions.

Decisions about acquisition of recorders obviously require a modicum of familiarity with their principles of operation, specification trade-offs, and failure modes.

2. Principles of Operation

All tape recorders work by using a tiny electromagnet to impress a variable magnetic field into a film of magnetic metal oxide that is carried over the electromagnet at a constant velocity, leaving a "track" of data. Direct recording channels add a high frequency bias signal to the AC input signal to compensate for nonlinear magnetic properties of the tape. This form of recording is efficient in terms of the bandwidth of recorded data that can be obtained for a given tape speed, but the amplitude that is reproduced from the tape may be influenced by factors other than the strength of the recorded signal (e.g., defects in the tape or misalignment between the track and the heads); these cause amplitude fluctuations, noise, and even dropouts. However, modern high-quality recording tape is remarkably good, particularly if it is used sparingly and not replayed too often. The major drawback of direct recording is the inability to record low-frequency or DC signals, such as those produced by transducers.

FM tape recorders differ from AM recorders only in that they incorporate an additional stage of encoding and decoding of the recorded information. In fact, you can convert most instrumentation tape recorders from one mode to another (or even a mixture on different channels) by substituting modular channel amplifiers. The frequency modulation device simply encodes the incoming voltage fluctuations as frequency fluctuations around the center frequency of an oscillator (fig. 12.8). The center frequency of this carrier oscillator must be 6 to 10 times the highest frequency of the input signal. The modulated carrier is recorded as a fixed amplitude signal in the same manner as the AM signal would be recorded normally, but its high frequency requires faster tape speeds and usually much more expensive recording and reproducing heads. If only low-frequency signals must be recorded in the FM mode alongside one or more direct channels for EMG, you may want to consider inexpensive stand-alone modulation and demodulation equipment to convert a direct channel on a multichannel tape recorder. Because of the encoding scheme, it is relatively simple to combine several carriers with sufficiently dissimilar center frequencies, thereby multiplexing more than one signal per channel. This approach has been used by some manufacturers to build multiplexed FM encoders that take advantage of the wide bandwidth of inexpensive video cassette recorders. You should consult the manufacturers of such equipment for details regarding the selection of carrier frequencies, modula-

12.8. Transformation of a low-frequency input signal into a modulated, high-frequency carrier suitable for transmission or storage by magnetic waves (e.g. for telemetry or tape recording).

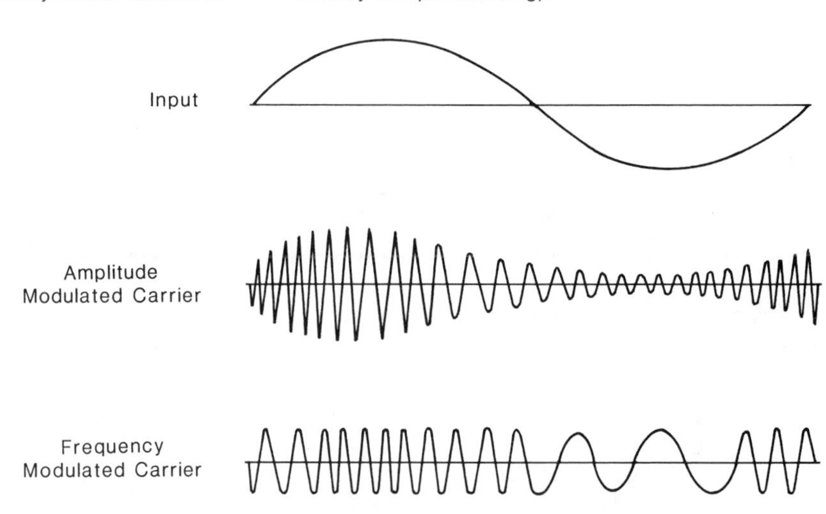

tion bandwidths, and other important matters.

3. Specifications

a. Needs. Tape recorders come with a confusing array of specifications and options, making for difficult and expensive decisions. In the following sections we note some of the more important ones and mention some points you should keep in mind, but your best hope is a thorough understanding of exactly what you want and expect from the machine as opposed to which features are irrelevant to your application.

b. Number of Channels. Tape recorders seem to come in two sizes: too big and too small. A machine that has more channels than you need will be an unnecessary expense, extra work, and a constant invitation to make a simple experiment into a complicated mess. On the other hand, a machine with too few channels will leave you constantly deciding which of the several important channels of data should be ignored. When you are making this decision, be sure to account for channels required for synchronization with video or cinegraphic equipment and to consider other important timing and annotation information such as stimulus synchronization pulses, time codes, treadmill speed, and voice (see chapter 21). If you are investing in an instrumentation recorder, check on the possibility and expense of adding other channels later; much of the electronics cost per channel is usually in modules rather than mainframe. Avoid using so-called edge tracks for critical timing or data; these channels usually have low-quality, marginal performance, suitable for voice annotation at best.

c. Tape Handling. The general format of your machine can make a big difference in cost and convenience over years of use. Half-inch tape is less expensive and easier to handle and store than 1-inch tape, but quality suffers whenever more than seven channels are squeezed into the narrow format. Large reels must be changed less often than small reels, but make sure that the speed of the fast forward and reverse modes is adequate to avoid annoying delays in recording or analyzing data. Coaxial reels and compressed or self-threading tape paths conserve space, but often with a steep cost in reliability and maintenance. Multiple recording and playback speeds may seem desirable, but they tend to be expensive and potentially confusing with FM channels, they cannot be used for voice tracks, and they may be used much less often than originally anticipated. You may need only a single standard recording speed with adequate bandwidth (10kHz max.) plus one-fourth or one-eighth playback speed for transfer to chart recorder. Instrumentation tape itself is expensive, but this is probably not a good item on which to pinch pennies, as heads worn prematurely from poor-quality tape are even more expensive (see below).

d. Resolution. Resolution is usually specified as signal-to-noise ratio in decibels, often with a cross-talk specification as well. (Cross-talk here differs from the use of that term in the biological preparation and refers to spilling of signals among the channels of the recorder.) These numbers interact complexly with different types of heads, tape widths, tape speed, bandwidth, and recording format. By using so-called wideband heads, it may appear that you can use one half or one fourth of the tape speed with only a small decrease in performance. However, there is a very large difference between noise levels (e.g., of −35 dB and of −47dB); this will become apparent as soon as you try turning up the gain on a recorded low-amplitude signal.

It is customary and very convenient to set up a multichannel tape recorder with a fixed 1/1 ratio of input/output sensitivities on all channels, perhaps ±1 V nominally. The EMG signals from many muscles tend to have a very large dynamic range, from first recruitment of a few units to maximal effort involving much larger motor units. Obviously, you do not want to risk saturating an input channel with excessive gain, and it is inviting disaster to change amplifier gains or recorder sensitivities in the middle of a recording session. Therefore, you will often find yourself analyzing signals that occupy only one fourth to one tenth of the input dynamic range. Suddenly, 1% noise (−40 dB relative to full dynamic range) looks like 10% noise (−20 dB relative to peak recorded signal), a completely unacceptable situation.

4. Preventive Maintenance

Preventive maintenance boils down to one vital and often ignored piece of advice: *clean the heads*. Recording tape has the structure of very fine sand paper, consisting of a hard, abrasive metal oxide coating on a thin flexible substrate. Grains tend to work loose and clog up the tiny gap of the recording and reproducing heads. The fine deposit of dust often seen around the capstan, pinch rollers, heads, and tape paths rapidly accelerates abrasive wear of all moving parts and the surface of the heads. Replacing heads is expensive (several thousand dollars on a multichannel system), requires many hours of service technician time, and necessitates completely recalibrating channels, perhaps even losing critical amplitude and interchannel timing compatibility with previously recorded tape. You can reduce the frequency of this disaster by up to an order of magnitude by simply swabbing the heads and other metallic tape path parts (never the rubber pinch rollers) with 90% isopropyl alcohol before each daily use (recording or analyzing). Whenever long reels are to be rewound, you may be able to reroute the tape path to bypass the heads.

Also, use only the best tape, store it carefully according to manufacturer's guidelines, and discard it at the first sign of excessive wear or oxide shedding, signal dropouts, or stretching (indicated by concave cupping across the tape width on the wound reels or scalloping at the edges). Whenever the machine is first installed and whenever it is serviced, get your money's worth from the technician by asking lots of questions. These highly experienced people can be real allies, but they tend to divide their clientele into two groups: those who really understand and care about keeping their equipment in first-class condition and those who pester them constantly with crises that could have been avoided.

Also remember that the equipment should be calibrated regularly; the frequency of calibration should depend on the criticality of accuracy to the particular experiments and the history of the particular machine. FM channels in particular have a tendency for both gain and DC offset (center frequency) to drift a few percent over weeks to months. For many biological applications this is not critical, but you should have some simple procedure for at least a cursory check of the general performance of each tape channel periodically, as this will often give warning of impending disasters. A simple procedure is to record a high-frequency AC signal with zero DC bias on each channel, using a frequency near the high end of the bandwidth and an amplitude near the maximum rating for each channel. You can use this to generate a standard calibration sequence at the beginning of each recorded segment. You can even use it to verify the gain of preamplifiers by arranging to put a low-amplitude version of this test signal into the input of such amplifiers and using their gain to boost it to the desired amplitude. When you are looking at the playback, check for overall amplitude, DC offset (make sure the scope is DC coupled), asymmetry between positive and negative phases, and clipping of either or both peaks. If you have any doubts about a channel, as a result of either this calibration or examination or monitoring of recorded data, leave a note on it and plan to run a complete calibration as soon as possible. This will usually involve a few specialized pieces of service equipment, perhaps a special module from the manufacturer, plus careful adherence to the procedures outlined in the user's manual.

Most instrumentation tape recorders (but only some consumer audio recorders) allow monitoring of the recorded signal through the reproducing heads simultaneously with the original recording. There will be a slight time lag as a result of the tape path, but otherwise it should look just like the input. Ideally, you should check the input against the output on each channel at the beginning of each recording session. If you only plan to look at one channel at a time, be sure it is the reproducing signal; the recording signal only shows you what is going in, not whether anything is being recorded on tape.

Finally, be sure you understand what erasing capabilities may or may not be built into your machine for writing over old tapes. Some machines have no erasing heads and are designed to overwhelm any previously recorded FM traces with new carriers, which may be a problem if you have a direct data channel or voice track. Others erase poorly, making a bulk tape eraser a modest and cost-effective investment to protect the quality of your data.

E. STIMULATORS

1. Principles

Probably no aspect of electrophysiology is more confusing to the novice than electrical stimulation. Despite the fact that the biophysical aspects of the process are actually quite well understood, a large body of folklore and empiricism continues to resist theoretical explication and underlies the design and use of much instrumentation. You may wish to consider the following important points when selecting and using such equipment (fig. 12.9).

2. Active Ingredients

Most of the confusion, such as about constant current versus constant voltage, biphasic versus monophasic, bipolar versus monopolar, arises from failure to appreciate exactly which aspect of an applied electrical stimulus actually causes the target to generate action potentials. Let us trace this backward from the critical event, which is, of course, the depolarization of a local area of the cell membrane from its resting potential of about 80 mV (inside negative) to its threshold of perhaps 50 mV (inside negative). If we had an electrode inside the nerve or muscle fiber, we could simply pass positive electrical current into it (with respect to a reference on the outside), thereby directly depolarizing the membrane immediately in the vicinity of the positive electrode (anode).

However, with both stimulus poles on the outside of the cell, it is not immediately clear how we can accomplish anything in terms of the transmembrane potential. The trick lies in the nonequilibrated potentials that arise when brief electrical pulses of current pass longitudinally through extracellular fluids and the axons or fibers parallel to this current. Because of the resistive and capacitive nature of the cylinder of cell membranes (and myelin, if present), the current density within a fiber differs from that on the outside. Current flow through a resistive medium generates potential gradients. If different current densities are flowing inside and out of the cell, the resulting gradients of longitudinal potential will be different, giving rise to changes in transmembrane potential. Any current that enters the fiber must exit someplace (to complete the circuit), so there will be regions of both hyperpolarization and depolarization along the fiber.

Thus, the critical aspect of a stimulus is the amplitude of the *difference* in the potential gradients induced by the applied *current*. We cannot change the manner in which current transients partition themselves among extracellular and intracellular spaces; this is determined by the physical dimensions and electrical properties of cell membranes and fluid spaces. These factors account for the general rule that fibers of large diameter are easier to stimulate (have a lower threshold for a particular stimulus waveform) than fibers of small diameter, and myelinated fibers are more easily excited than unmyelinated ones. What we can influence are the time course of the current transients and their spatial distribution in the vicinity of the target structures. The former is regulated by the constant-current stimulator (and its pulse-shaping circuitry); the latter is determined by the physical position of the stimulating electrodes.

3. Harmful Ingredients

The passage of electrical current in biological tissues involves a conversion from the motion of electrons in metal conductors to the flow of ions in an aqueous electrolyte. The mechanisms for passage of direct current (or any waveform having a net flow over time in one direction) all involve the class of electrolytic reactions that are irreversible processes. That is, they involve transformations in the valence of atoms coupled with a migration of those atoms away from the electrode surface (usually by diffusion), which leads to a permanent change in the chemical composition of the system. The most commonly occurring reac-

12.9. Stimulus current waveforms. The temporal pattern of stimulation current actually delivered to the preparation depends on the waveform produced by the current generator (assumed to be either monophasic or biphasic square-waves), coupling capacitance used to effect charge balance and prevent net DC current passage, and the limitations of the equipment actually used in the face of particular patterns of stimulation (phase duration, *d*; interstimulus interval, *t*; stimulus period, *p*).

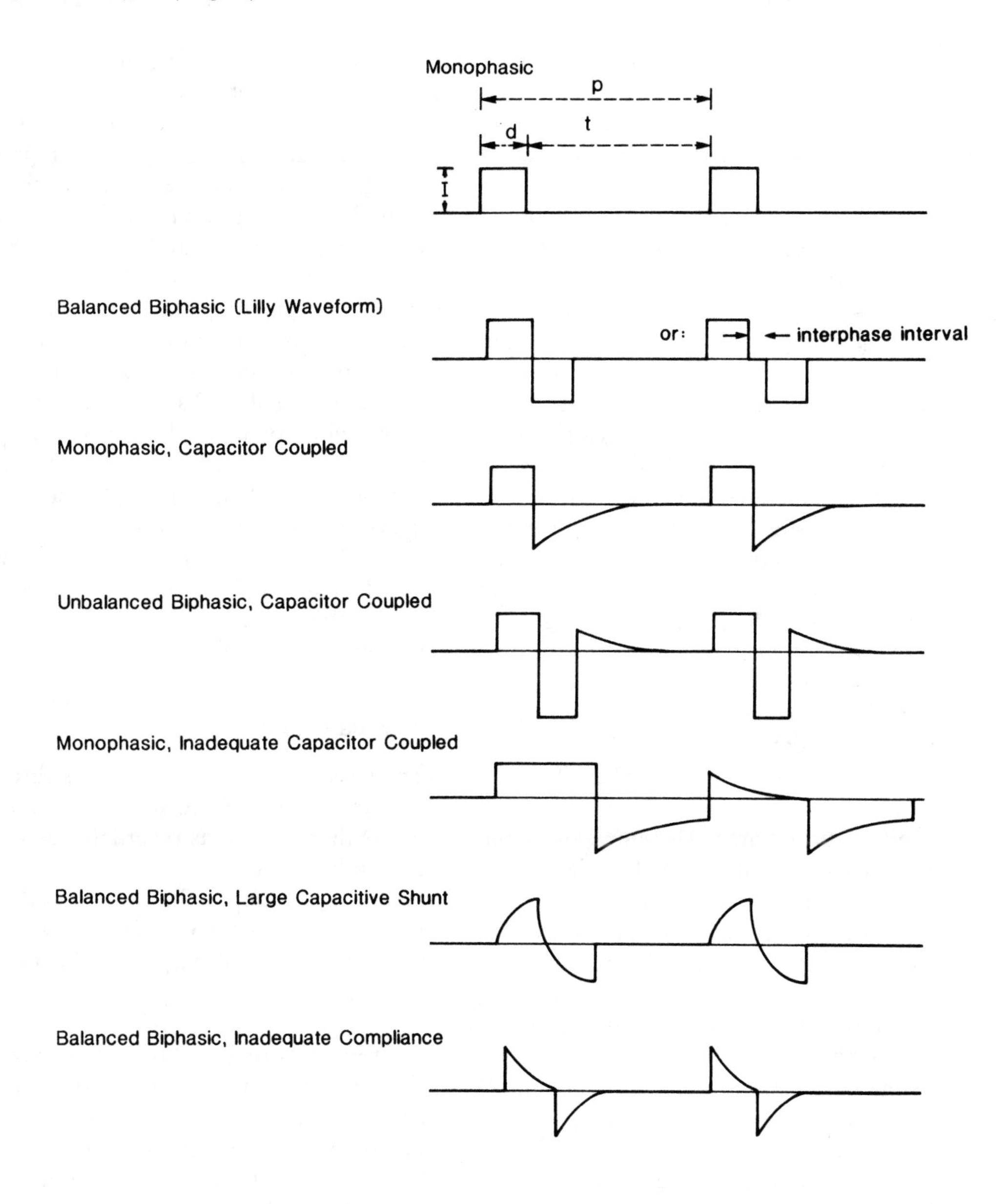

tions involve the electrolysis of water (leaving residual hydroxyl or hydronium ions that alter local pH) and the corrosion of metal (leaving heavy metal ions in a solution, which is usually toxic). In addition to damaging tissue, such reactions may alter the working properties of the electrode by changing its surface area or the exposure of the surface (e.g., accumulating residues of denatured organic molecules). For biomaterials considerations, see chapter 7 and table 7.2.

One can escape these problems by noting that the active ingredient in electrical stimulation is the brief, depolarizing transient of current flow, not the continuous or net DC current. Once a portion of membrane (of sufficient extent) reaches threshold (for a very brief period), the activation cannot be reversed by current flow in the opposite direction. This means that a biphasic stimulus, passing two equal amounts of electrical charge in opposite directions, is effective. If the amount of charge (stimulus current times pulse duration) is low enough and the surface area of the stimulating electrode is large enough, the current flow can occur via any of several reversible reactions that do not permanently change the composition of the electrodes or adjacent tissues. Thus, the critical measure of the potentially harmful effect of an electrical stimulus is the charge per phase per unit of surface area. Therefore, the stimulating electrode should be made as large as possible (consistent with inducing a sufficiently localized current gradient and avoiding physical damage of the tissues), and the stimulus charge should be as low as possible. The current flow must be balanced so as to reverse the electrochemical reactions between successive stimulation pulses (see the discussion of capacitive coupling below, section 5).

There is a well-known hyperbolic relationship between duration of the stimulus pulse and threshold current. As the pulse width increases, the current decreases monotonically, to an asymptote called the rheobase. However, the product of current times pulse duration, the charge, goes through a minimum. For most nerve and muscle fibers, this minimum is at 100 μsec or less.

It is also worth noting that the susceptibility of various metals to electrolytic corrosion with the passage of any level of current is not well correlated with our general impressions about their corrosion resistance in the passive state. Iridium appears to be the least corrodible material known, and its surface can be activated with a bound hydroxide that permits very high charge densities without the occurrence of irreversible processes; however, it is expensive, difficult to work with, and not generally available. Platinum and its various alloys with noble metals such as iridium and tungsten are generally acceptable for all but the most critical chronic applications in humans. Stainless steel can be reasonably safe and stable at modest charge densities with well-balanced pulsing. Gold corrodes relatively rapidly; silver and copper are completely unacceptable.

4. Side Effects

The passage of electrical currents through a preparation is likely to set up potential gradients that are just as recordable as those caused by the active processes of bioelectrical tissue—for example, the EMG signal. This is the so-called stimulus artifact, arch-nemesis of many experimentalists. It must be attacked on three fronts.

a. Stimulus Strength. Stimulus strength should be minimized. It may take much more current to activate a nerve trunk with stimulus source and sink (cathode and anode) on a line perpendicular to the fiber direction rather than parallel, because the resultant perpendicular voltage gradients do nothing to cause longitudinal current flow to depolarize the fibers.

The trade-off of stimulus current versus pulse duration is somewhat more complex, because you may be able to make the artifact smaller by using a longer, albeit less than optimal, pulse duration. However, both the size and the duration of the stimulus artifact are often related to the charge per phase; this has complex electrical reasons having to do with the charging of various capacitive surfaces. Furthermore, it is often the duration rather than the amplitude of the artifact that causes problems, as it obscures responses with short latencies. Therefore, it is generally advisable to use brief stimulus pulses. The capacitive discharge technique described below for balancing the output of a monophasic stimulator tends to put long tails on the stimulus artifact. You may need to invest in a true biphasic stimulator that produces the so-called Lilly waveform (equal and opposite square wave phases).

In addition to the parallel orientation of the stimulation electrodes, the spacing between anode and cathode should be of the same order as that used to record from the target element—that is, they should have approximately the wavelength of the action potentials that occur in the target, perhaps somewhat larger values if the target is distant.

b. Orientation of Electrodes. The potential gradient along the recording antenna dipole should be minimized. If possible, the orientation of the bipolar recording electrodes should be perpendicular to the axis of the bipolar stimulating electrodes; the reason for this is that the stimulating electrodes should not be perpendicular to the target structures. Bipolar recording electrodes connected to a differential amplifier detect current gradients in much the same way as nerve fibers; if there is no gradient, then there is no artifact or response. When nerve hooks or cuffs are used to stimulate muscles, the naturally perpendicular approach of most nerves to the muscle fibers can be exploited to produce optimal stimulation conditions and optimal recording conditions with minimal stimulus artifact. Often, small changes in the orientation of stimulus or recording electrodes can greatly improve stimulus artifact pickup without compromising the efficacy of either the stimulation or recording processes.

c. Stimulation Currents. The stimulation currents should be kept out of the amplifier. You would not knowingly connect one pole of the stimulus-generating box to an input of your amplifier. However, that is precisely what you are doing if you use an unisolated stimulator, which outputs a current or voltage with respect to the same ground used by your recording equipment. The stimulus current generated by such a stimulator has no more reason to travel to ground through the stimulation electrodes than through the recording electrodes leading to the indifferent or common point in the amplifier.

Two techniques are available to generate transient currents between a source and sink having no electrical connection to ground or ground-referenced power supplies. In transformer-coupled systems, the stimulus is inductively coupled into an isolated winding leading to the two stimulus electrode contacts. Whereas such output stages will not be perfectly isolated (because of interwinding stray capacitance), they may be adequate for isolation and will lack a net DC component no matter what the input. However, it is difficult if not impossible to regulate current output, the critical parameter; control of output voltage is undesirable as most of the voltage is dissipated across the stimulus electrode/electroyte interface, which tends to have an unstable and nonlinear impedance.

Most modern stimulus isolators use photoisolation, in which a trigger or waveform-controlling signal is passed to the output stage by the light flux transferred between a photoemitter (usually a light-emitting diode, LED) and a photosensitive element (usually a phototransistor). The light-detecting side of the system must have its own power supply plus circuitry for generating and regulating the current output and balancing the waveform to prevent net DC. To avoid defeating the isolation, this supply is almost always a self-contained battery, often one that has a large voltage to provide the necessary compliance to drive large currents through electrodes with significant impedances. It is essential to avoid breaking isolation; never connect any part of the output circuitry or its power supply to any other equipment in the laboratory, via recording electrodes, monitor circuits, or indifferent electrodes. To check the performance of a constant-current stimulator, disconnect it from the preparation and use a conventional oscilloscope to measure the voltage generated across a precision resistor to ground while generating a particular current. Then disconnect the resistor and the monitoring scope and reconnect the output to the bipolar stimulation electrodes only, with no ground path.

Remember that stray capacitance constitutes part of an electrical circuit as do real physical wires, particularly for the fast transients used in stimulation circuits. Therefore, keep output leads to minimum length and do not run them in shielded, ground-referenced cables. You can effectively minimize radiated artifact by using twisted pair leads, but still keep these short. If you are stimulating through electrodes of relatively high impedance, remember that the precisely regulated stimulus current delivered at the output terminals of the stimulator can pass equally well through interlead capacitance as through the preparation, but the effect will be very different.

5. Capacitive Coupling

Capacitive coupling—the wonder drug—is a simple technique that can be used to save your preparation from disaster and make relatively simple equipment work quite adequately. A capacitor presents an infinite impedance to direct current and, if adequately sized, a negligible impedance to fast transients. This protects the preparation from any DC offset that may be coming from the stimulator. Furthermore, any charge that flows in one direction must eventually flow back in an equal and opposite effect to discharge the capacitor. Thus, even a monophasic pulse generator can be made to generate a biphasic pulse, in which the current flow of the first phase is actively produced by the generator while the second phase follows the passive discharge properties of the capacitor and the rest of the circuit.

Unfortunately, that is where things can go awry. If the monophasic generator is a constant-current source, it probably has an effective output impedance that is very high. In fact, one simple if inefficient way to make a voltage source into a current source is to hook it in series with a resistance that is much larger than the electrode impedance, so that it dominates the current flow. The discharge current from the capacitor has to flow through the series circuit made up of the capacitor itself, the impedance of the stimulating electrode, and this potentially infinite source impedance. If very little discharge current can flow between pulses, the capacitor will simply keep accumulating charge until the voltage across it reaches the output compliance voltage of the stimulator, which will prevent further stimulus output current. The circuit needs a bleed resistor that will discharge the capacitor at a reasonable rate. However, this bleed resistor lies in parallel with the stimulating electrodes, so that the stimulus output current can also flow through the resistor, thus degrading output accuracy. Some compromise is required in the selection of the bleed resistor and the coupling capacitor. Their selection will depend on the stimulating electrode impedances and the ratio between the duration of the stimulus pulse and the interpulse interval available for discharge. Figure 12.10 provides some simple rules of thumb for reaching this compromise.

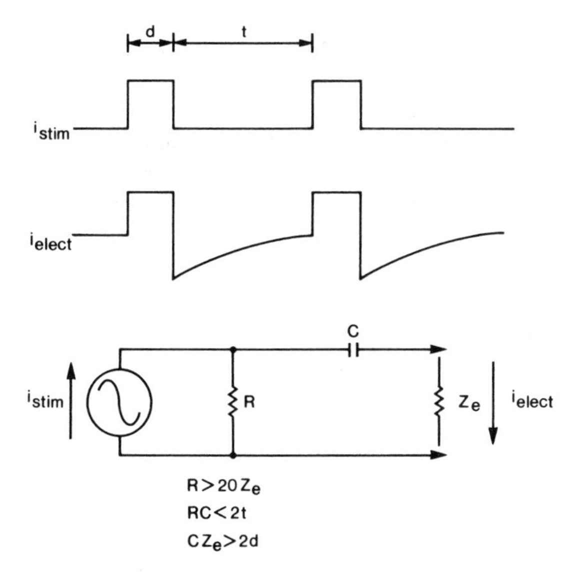

12.10. Capacitor coupling to charge balance a monophasic current stimulator. If pulse width is *d* and interpulse interval is *t*, then the appropriate values for coupling capacitor (*C*) and bleed resistor (*R*) can be calculated from the electrode impedance (*Ze*) so that the current flow through the electrode will have a net value of zero. In order to prevent loss of significant stimulating current output through this bleed resistor, its magnitude must be much greater than the electrode impedance. The coupling capacitor must be large enough to store the charge for a single stimulus pulse without approaching the compliance voltage limit of the stimulator but small enough so that it will have time to discharge through the bleed resistor and electrode impedance between pulses.

In summary, the best approach to stimulation is usually to use brief, constant-current pulses from an electrically isolated, charge-balanced stimulation circuit.

F. POWER SUPPLIES

The electrical power supplied by the local utility company comes in a form that is efficiently transmitted over long distances. Unfortunately, this power comes at voltages and frequencies that are unsafe, unsuitable for most instrumentation, and prone to generating severe interference. The art and science of turning this raw energy into useful, stable DC energy is complex and highly evolved, almost to the point where it can be taken for granted. Unfortunately, the aspects that can be taken for granted decrease as the application becomes more critical.

Most laboratory equipment comes with its own internal DC supplies, but there are some important choices to be made. If you buy modular equipment or build your own, do not overtax the power supply. The rated output is the point at which things start to melt; the regulation and immunity to transients on the incoming power line or to transients generated in the equipment itself will drop drastically at less than half the rated output. If a piece of equipment has a special ground jack on the back, it should probably be connected to a common ground point such as a metal water pipe. However, make sure the ground goes to ground, particularly in these days of glass and plastic drain pipes. Do not, as a rule, disconnect the ground plug of conventional three-prong AC plugs and power distribution boxes (although on rare occasions such as noise on the house ground, this lead may have to be replaced with a true external ground with the advice and consent of the building engineer).

For critical applications or for field use with uncertain power supplies, batteries can provide a clean source of DC power, the reliability of which tends to be directly correlated with the experimentalist's obsessiveness with maintenance. A pair of conventional lead-acid storage batteries provide an exceptionally noise-free bipolar DC power source, although the working voltage will tend to vary between 11 and 14 V depending on the state of charging. Most modern equipment can tolerate this range fairly well, with little loss of accuracy, although imbalance between the two batteries providing a ±V DC supply may cause DC biases or even produce damage if the imbalance is too great. Inevitably, someone leaves the equipment on over a weekend, resulting in two problems: delay while waiting for a recharge and possibly irreversible damage of the battery as a result of deep discharge. For these reasons you may wish to keep both a fast, "hot shot" type charger as well as a trickle charger that avoids "cooking" the battery dry and also can be used routinely between working sessions. You should also invest in the special "deep discharge" batteries made mostly for recreational and marine use, as such batteries tend to be more resistant to this sort of abuse. It is not without reason that manufacturers of automotive batteries automatically void their warrantee for any battery that has been used in anything other than a conventional automobile.

The other two types of batteries often found in specialized pieces of equipment (such as head stages of low-noise preamplifiers and stimulators) are rechargeable and disposable batteries. Disposable battery technology is improving steadily, with more and more milliampere-hours in a given package and less and less tendency to go stale sitting on the shelf. However, like the light bulbs of slide projectors, you can expect them to die specifically when you are out of replacements. Keep plenty of spares and prolong their shelf life by storing them in waterproof containers in a refrigerator. Rechargeable batteries are becoming more expensive and may constitute the single most expensive part of a piece of equipment. Most of them are so-called nicad (nickel-cadium) batteries, which have the perverse property of losing longevity the more compulsively they are recharged (the opposite of the lead-acid storage battery). Nicad batteries should only be recharged when they are clearly discharged, and then should be fully recharged just before use. If regularly kept at full charge, their ability to hold charge declines rapidly, although you may restore this by briefly overcharging them to burn off the internal shorts to which they are prone. Always remember that recharging equipment may cause certain types of batteries to explode. Hence, keep rechargeable and disposable batteries in separate containers and consider setting up the recharging operation in a protected area.

13 Noise and Artifact

A. CONCEPTS

In our theoretical discussion of the electrode as an antenna we have concentrated on optimizing its ability to record selectively from a subpopulation of the multitude of bioelectrical signal generators in the body. However, electrically active cells are not the only sources of electrical fields in the body and much less outside of it. In chapter 12 we indicated how any electrical fields in the vicinity of the electrodes and their lead cables can become electrically coupled into them and generate the signals we call noise. We also mentioned the intrinsic noise of passive electrical components, such as resistors and capacitors, and of active devices, such as amplifiers; these noises are a consequence of the random motion of electrons caused by thermal vibration. Noise is inevitably present in any electrical signal and affects the useful information that can be derived therefrom. This chapter concerns minimizing the degradation by noise of the desired bioelectrical signal.

B. SIGNAL-TO-NOISE RATIO

There are no criteria for what constitutes an acceptable signal amplitude or an acceptable noise level. The information content of a signal is determined by the simple ratio of the amplitude of the desired signal to the amplitude of the added noise. Studied conversely, this means that we can tolerate an amount of noise that is proportional to the amount of signal picked up by the electrodes. Thus, optimization of the electrodes, according to the design principles so far elucidated, will reduce the chance that you will encounter noise that compromises your ability to interpret your records.

Of course, some signal sources are inherently of low amplitude and some environments are unavoidably polluted by noise sources of large amplitude. We have already seen that improving the selectivity of biological electrodes frequently entails decreasing their dipole separation and the surface area of their contacts. Placing the electrode poles more closely improves selectivity but reduces signal amplitude, as the closely spaced poles span less and less of the potential gradient available in the tissue. At the same time, the thermal noise increases because it is proportional to the square root of the electrode impedance, which rises with reduction of contact area. Furthermore, high source impedance makes the electrode-cable-amplifier circuit more sensitive to stray capacitive coupling of extraneous noise sources. (See chapter 20 for special considerations regarding monopolar electrodes and semimicroelectrodes.)

At what level will the signal-to-noise ratio begin to interfere with analysis of the results? The answer to this question depends on both the nature of the interfering noise and the type of analysis to be performed. Continuous wideband "white noise," such as that produced by thermal motion, may form a smooth baseline; a desired signal of only twice the noise amplitude will then stand out clearly. However, periodic noise or pulselike events with a pattern similar to that of the desired signal may prove confusing even when the signal being recorded tends to be 10 times greater than the amplitude of the noise. Noise that is time locked to events under study may seriously degrade results even when it is invisible on the raw records; it will interfere particularly with signal analysis by averaging or other statistical methods that are also time locked to the event. Whenever the noise sources interact randomly with each other and with the recorded signal, the probability that the record will include an event large enough to be confused with a "real" signal increases with time (i.e. with the duration of the record being examined). The study of rare anomalies of behavior may require detailed statistical measurements of the nature of the noise and consideration of the probability that noise alone could have given rise to the observed anomalies.

The quantification of noise and signal-to-noise ratio is a science in itself, outside the scope of this book. The interested reader is directed to the references. The experimentalist primarily needs methods for recognizing the types of noise likely to be encountered during EMG work and for minimizing their amplitude and effect.

C. RECOGNIZING NOISE

Everyone knows what noise is: the part of the record that does not agree with what was expected. Unfortunately, this definition also characterizes the most interesting results of research. The first step in identifying whether a wiggle in a trace represents a significant biological event or noise is to determine its absolute amplitude referred to input. If the amplitude is much greater than that produced by similar biological sources recorded with similar electrodes, then one must be suspicious. If the amplitude of an entire record is much smaller than expected, there is a significant chance that much or all of it results from noise sources.

The waveform of the signal can also be very useful. However, it may be misleading if the investigator is unaware of the range of distortions that amplifiers, filters, and display devices are likely to cause. The naturally occurring EMG signals tend to be fairly smoothly graded collections of "spike" events that have similar positive and negative contours. The frequency band occupied is fairly narrow, although the actual frequencies may vary considerably depending on both source and electrode configurations. Signals with unphysiologically fast components (or "spike-and-wave" asymmetries) or slow, regular undulations are likely to derive from noise sources or motion artifact (see below). Unfortunately, narrow-band filtering will turn any of these characteristic noise signatures into a signal that appears very physiological. However, even when visible signatures are ambiguous, the ear can be a much better discriminator of temporal patterns, hence the utility of audio monitors in the physiology laboratory.

Whereas filtering is an excellent way to minimize noise and enhance the biological component of a noisy signal, it can also distort noise signals to the point that they look entirely physiological. The "raw," unfiltered signal coming from a preamplifier should be inspected regularly with a high-fidelity display, such as an oscilloscope with a fast sweep speed (around 1 msec per centimeter). Inspection will reveal whether unphysiologically fast, large noise spikes have been clipped and smoothed into convincing EMG impostors and whether the observed regular discharge signatures of motoneurons are really the scalloped square waves typical of bad fluorescent light fixtures. Remember that almost all noise sources contain a large range of frequencies in their spectra, which means that any filter will distort the shape as well as attenuate the amplitude. The tighter the bandpass of your filtering system and the faster the rolloff of the filters, the more they will tend to homogenize disparate inputs into indistinguishable outputs.

D. STABILIZING THE ELECTRICAL ENVIRONMENT

In some laboratories it is more difficult to deal with noise because of its capricious nature. The same preparation may produce excellent signal-to-noise ratios one day, only to be unusable the next. Intermittency is a distinguishing feature of certain categories of noise sources, but it must be distinguished from inconsistency of the preparation.

We have already discussed in detail the importance of balanced electrode impedances, which can only be achieved by well-controlled and standardized fabrication procedures. However, one primary factor that often goes uncontrolled is the physical arrangement of electrical devices and connecting and power cables in the laboratory. With several channels of amplifiers, recording devices, displays, and test equipment, it is quite easy for the laboratory soon to become a maze of tangled cables, some with low-level signals, some with high-voltage line power, and some temporarily unconnected and ungrounded. The exact configuration of this maze is, of course, completely unreproducible from experiment to experiment, giving rise to a variability that is incorrectly ascribed to fate. The fault lies not with our stars but with ourselves.

At the very least, some simple method for physically separating the leads that carry signals of different types, should be utilized. Inexpensive plastic cable ties, conduits, and other easily modified systems are readily available (see fig. 11.10). More complex or frequently reconfigured systems will benefit greatly from carefully designed patch panels (see chapter 11), which limit the possible range of variability.

One variable that often goes uncontrolled is the experimentalist. Any ungrounded electrical conductor can act as an antenna for electrical noise, both receiving and reradiating signals. A grounded electrical conductor can act as a shield if interposed between a noise source and the electrodes and input cables. As the investigator moves about, perhaps occasionally touching grounded surfaces (such as metal tables, the chassis of an instrument, and the frame of the Faraday cage), his or her contribution to the equivalent electrical circuit is constantly changing. Similar, even larger changes occur as the animal moves from one place to another or makes variable contact with grounded surfaces. It is usually best for the investigator to be well grounded if it is necessary to be physically close to the preparation during recording. Conversely, it is usually best to keep moving animals from contact with any electrically conductive surface and to rely only on a single, implanted ground connection to provide a stable, low-impedance ground point. However, such conditions can pose a safety hazard to both animal and experimentalist if high-voltage lines, such as AC power, are carelessly maintained.

Finally, it is important that all other electrically conductive items (unused cables, small electrical devices, stray surgical instruments) be either removed or physically and electrically stabilized. Electrical leads that must swing freely should be coaxially shielded and their shields grounded and jacketed by a rugged nonconductive coating to prevent intermittent ground loops as they slide and scrape over conductive surfaces. Even then they should be allowed to hang freely and should not be touching during recording sessions.

E. IDENTIFYING NOISE

1. Principles

Once one has a clear notion of the amplitude and shape of the unfiltered noise signature, one is a long way to pinpointing the source. The following categorization of noise sources includes descriptions of the sorts of waveforms that may be encountered (fig. 13.1). This is followed by a "decision tree" series of steps that should facilitate an orderly narrowing of the possibilities (fig. 13.2).

2. Bioelectrical Noise

Although any biological source of potentials in the body of the experimental animal is actually a source of cross-talk rather than noise, some sources are so unlike the EMG sources that they are usefully considered as noise sources. In this category are electrocardiograph and respiratory waves as well as neural evoked potentials. The first two are most obviously identified by their characteristic temporal pattern and correlation with easily observable processes.

The slow potential fluctuations accompanying the expansion and contraction of the chest wall are actually forms of motion artifact; potentials are generated as the spatial relationships between electrodes and charged structures change. Slow-motion artifacts are quite unlike EMG signatures and can usually be filtered to negligible size. However, if the method of analysis includes rectification and integration, even small residual, low-frequency signals may generate large outputs because of their large area-under-the-curve. If the study is one of cyclic events such as activity of the respiratory muscles, both EMG signals and movement will be time locked to the respiratory cycle; thus the biasing effect, even of small artifacts, may be quite significant.

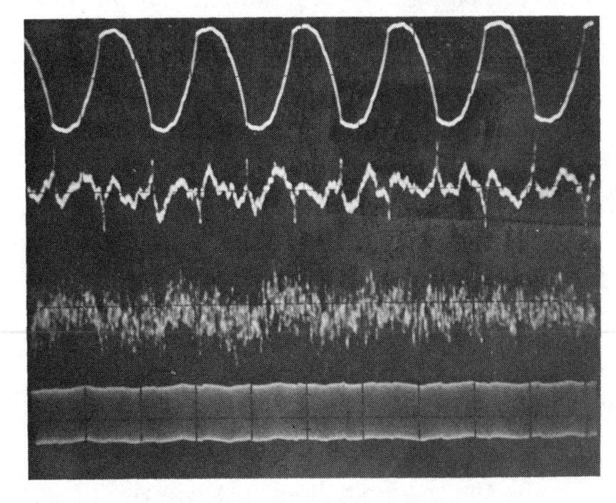

10 ms/div

13.1. Oscilloscope tracings of four common types of noise, shown at 10 msec per division with the sweep trace time locked to the power line frequency. *Top trace,* Simple power line hum (note 16.7 msec cycle of sinusoidal signal, corresponding to 60 Hz). *Second trace,* Also line-locked noise, but the narrow spikes occurring twice each cycle come from a nonlinear device producing transient surges of current flow whenever critical power line voltages are reached (e.g., fluorescent lights or a silicon-controlled-rectifier electrical motor). *Third trace,* Wideband or white noise, such as that generated by any source impedance or in the amplifier itself (a random combination of all frequencies that can be passed by the bandwidth of the system). *Bottom trace,* High-frequency oscillation noise. If the trace were expanded greatly, the thick line would appear as very high–frequency oscillations of the type shown in the first trace. The frequency might be indicative of some particular source (e.g., AC carrier used by a length gauge bridge circuit or horizontal flyback frequency in a video screen), or it might simply represent the intrinsic oscillation frequency of this or an adjacent amplifier channel not properly connected to an input source or ground.

13.2. Noise decision tree showing an efficient set of questions to ask about interfering noise and to converge quickly on the likely source of the problem.

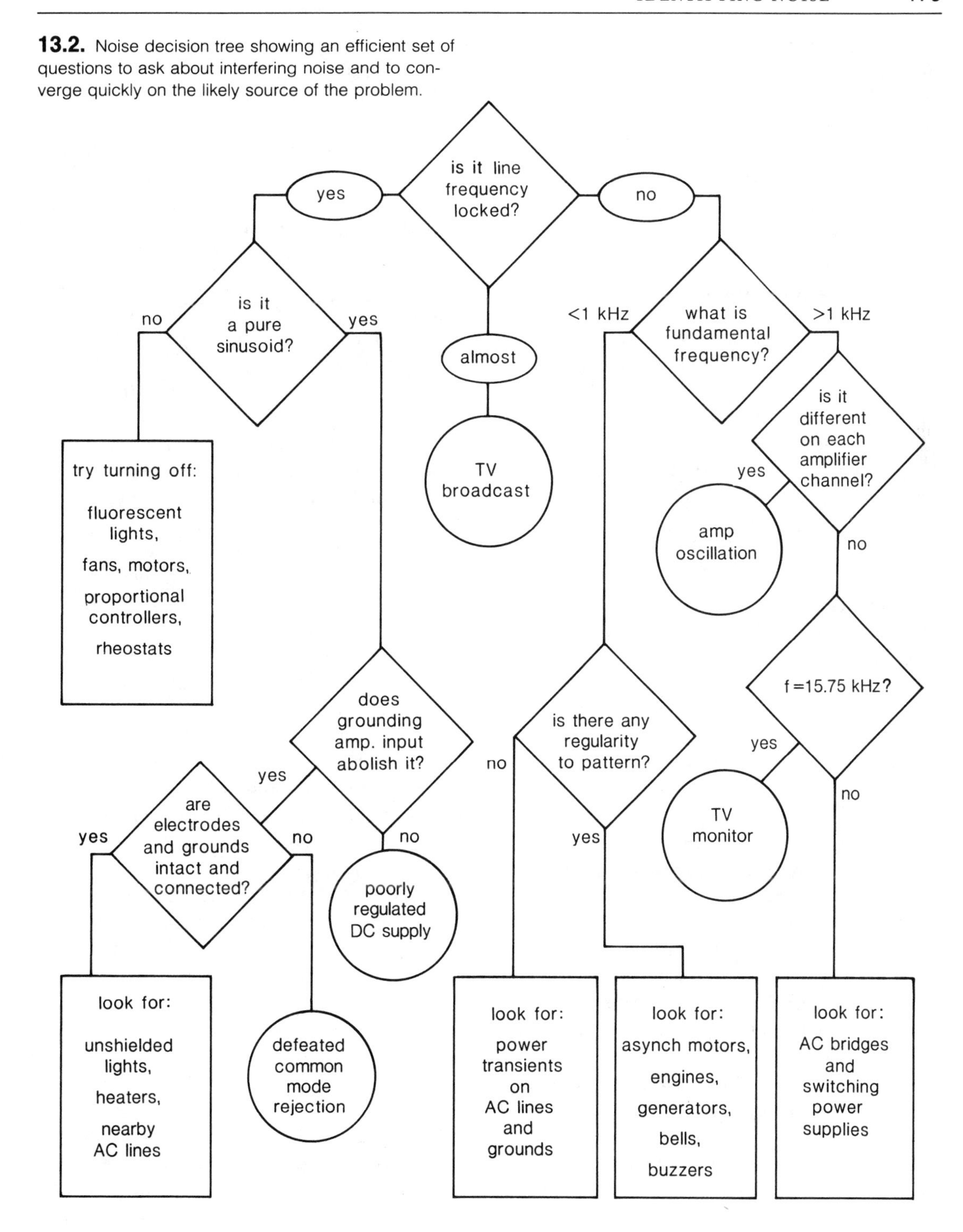

Neural potentials are usually much smaller in amplitude than EMGs. However, experimental paradigms involving synchronous activation of large populations by external stimuli can give rise to significant waves in organized neural structures, such as tracts and cortices. The experimentalist should keep this in mind when working with cranial and facial muscles.

3. Biomechanical Noise

Motion artifact can be a particularly difficult source of randomly occurring, large-amplitude, low-frequency waves. A record that looks perfectly clean when initially inspected and while amplifiers and recording devices are adjusted may suddenly erupt into unusable chaos when the animal begins the motor behavior being studied.

The mechanism behind this pernicious timing is really quite clear. Remember that all bioelectrically active cells are charged to a resting potential of some 60 to 100 mV. This DC potential is some two orders of magnitude larger than the AC potentials typically recorded by extracellular macroelectrodes such as EMG probes. Now recall that the surface of the EMG electrode acts like one plate of a capacitor, the other plate of which is the volume-conductive extracellular fluid. If the electrode surface is pressed up against one of these highly charged cells, a slight motion between the cell and the electrode translates into a large capacitive change in the coupling between the DC resting potential of the cells and the electrode. This translates into a transient motion of the electrons in the electrode surface, which constitutes a current through the amplifier.

In another scenario the electrode surface may have sharp edges that occasionally puncture surrounding cells, suddenly releasing the stored resting charge into the extracellular fluid. Again the fluctuating potential, sensed by the surface of the electrode, is amplified and becomes part of the record. Obviously, both of the above elements are most likely to occur during periods of vigorous movement of the muscle or the electrode. They may cease to be a problem for chronically implanted electrodes if one simply waits a few days for wound healing to stabilize the mechanical milieu.

It is usually possible to provide macroelectrodes with a large enough surface area so that the fluctuations caused by puncture or motion with respect to a single cell do not represent significant aberrations in their stable mean environment. However, electrodes located near large structures carrying a coherent surface charge, such as visceral organs with air or liquids, can pick up large potential transients as a result of relative motion. Whereas these potentials are usually of lower frequency than the EMG signal, they may be difficult to filter to the point at which they do not bias analysis based on rectification and integration. Analysis may be further complicated by the above-noted coupling between artifact and mechanical action.

As the contact surface of the electrode becomes smaller and impedance becomes higher, the frequency spectrum and amplitude of the random motion artifact from individual cells start to rise. Whenever very small electrodes are used for single-unit recording, even microphonic motion of the electrode, induced by sympathetic vibration to acoustic noise sources in the laboratory, may become a problem. It is important not to confuse motion artifact with connector noise, which may look and act similarly.

4. Intrinsic Electrical Noise

Intrinsic electrical noise includes all the sources of noise that are part of the physics of electrons and electronic components and, consequently, an integral part of any recording process. Most of them are described in discussion of the impedance of electrodes (chapter 14) and of the design and construction of preamplifiers and amplification systems (chapter 12). Generally, these phenomena produce random, very wideband noise that lacks apparent pattern or predominant shape. Of course, this means that when narrow-band filtering is used, signatures quite like real EMG potentials will occasionally appear. The probability of their occurrence will relate mostly to the signal-to-noise ratio. There are also some noise-generating processes in amplification devices; such noises have a frequency distribution that is not flat, but biased toward low frequencies or random bursts of larger amplitude "shot noise." It is rare that these phenomena have a significant amplitude in modern physiological equipment, but they can plague old tube-type amplifiers and dirty and corroded potentiometer wipers, as well as switch contacts in equipment of any vintage.

Most such intrinsic noise effects can be evaluated by the simple expedient of examining the output generated in the normal recording configuration at a time when no signal is expected. The simplest way to achieve such conditions is to place the electrode in a beaker of saline (properly grounded, of course).

One nonrandom source of noise that does appear to occur intrinsically in amplification equipment is line frequency ripple, but this is actually a failing of the power supply that converts AC power into DC levels. As amplifiers can always be run by ripple-free batteries, this is dealt with under external noise sources below. Another nonrandom noise occasionally seen in wideband amplifiers is high-frequency oscillation. Although oscillation usually causes an amplifier to produce a completely saturated and unusable output, occasionally there may be a persistent, small-amplitude oscillation with a frequency near the high cutoff frequency of the amplifier. This oscillation is typified by its high frequency and sinusoidal waveform, but it must be distinguished from extrinsic sources of high-frequency oscillation such as video displays and switching power supplies (see below).

5. Extrinsic Electrical Noise

By far the most common and most troublesome sources of noise are those that have nothing to do with the electrodes or equipment actually necessary to obtain the EMG signals. In fact, some of these sources (e.g., stimulators, stroboscopes, displays, storage devices, or lighting, heating, and anesthesia support systems) may be part of the experiment. Others are completely unrelated and may even be physically remote from the preparation. However, even if its source is physically remote, electrical interference from these devices may be transmitted close to the preparation by power cords, control lines, and fluid-filled tubes, such as circulating pump or perfusion systems. In most cases their effects may be minimized by careful design of optimal, low-impedance electrodes, balanced and shielded cables, differential preamplifiers, and properly set filters. However, each laboratory is likely to have sources of noise that are its unique plague and that may require extraordinary measures. Eventually these become part of the folklore of the laboratory, but when they first occur, their identification requires all the powers of observation and deduction of Sherlock Holmes. We here collect all those of which we have heard, organized by the distinctive features that facilitated their eventual identification.

a. Line Frequency-locked Sinusoidal Ripple. We have already mentioned the sinusoidal ripple that can occur in the output of an amplifier if the DC power supply does not completely remove all traces of its sinusoidal AC input from the mains. This is less of a problem with modern, regulated power supplies, but it can still be seen if small-amplitude signals require very high gain. It is more likely to become a problem if a single power supply must serve several devices, such as in modular systems, because the ripple starts to rise rapidly as the rated current output of the power supply is approached. Simply adding a device unrelated to the recording in question may precipitate a problem.

Most external sources of sinusoidal ripple at the power line are likely to be electrically separate from, but physically close to, the electrodes, lead cables, and preamplifiers. An AC power line has 115 or even 240 V of oscillation at the line frequency, which is well within the low-frequency boundary of the signal to be recorded. Even a small amount of stray capacitance between the conductors in the power cable and those in the recording circuit may couple in significant potentials. An AC-operated device radiates both electrical and magnetic fields. The electrical fields are related to the level of the actual operating voltage. The magnetic fields are related to the amount of current flowing through the device. Heating and incandescent lighting devices consume very large amounts of current and therefore radiate large magnetic fields at the power line frequencies. Typical offenders are microscope and operating lights, flood lights for photography, and heating pads and baths. Thermostatically operated devices will obviously generate fluctuating levels of interference as they switch on and off. Shielding is usually required unless adequate physical distance can be achieved.

b. Line Frequency-locked Complex Waveforms. Many devices drawing power from the AC lines do not draw current equally from all phases of the sinusoidal line frequency. Devices such as fluorescent light fixtures begin to conduct abruptly on the peaks of each cycle. Essentially continuous devices, such as heaters and incandescent lights, may be provided with power regulators that, in a similar way, admit power on an adjustable percentage of the complete waveform. Silicon-controlled rectifiers (SCRs) turn on quite abruptly, generating fast spikes of noise with each cycle or half-cycle. Still other devices rectify the line voltage, turning the smoothly sinusoidal profile into a sharply scalloped wave.

All of these transformations of the sinusoidal fundamental line frequency introduce higher frequency harmonics that almost certainly encompass the bandwidth of the EMG signals. When processed through filters, the characteristic signature with the prominent, line frequency fundamental may be so distorted that the noise looks quite similar to EMG signatures. If the amplitude is being regularly modulated by some feedback controller, unbeknownst to the investigator, the whole effect seems quite biological. Equipment with infrequent cycles (e.g., frost-free freezers) can be particularly exasperating.

A particularly pernicious form of this problem arises with motorized equipment that provides the actual load against which muscles are working. This includes treadmills, wind tunnels, torque motors, and the like, which may generate interference levels that are tightly coupled to the expected biological signal amplitudes. Treadmill motors with SCR speed controllers are notorious offenders in this category.

c. Almost Line Frequency-locked Noise. The nominal power line frequency maintained by the power company is quite accurate over the long run, enabling clocks with synchronous AC motors to keep accurate time without adjustment. However, the continuous changes in load on the power generators cannot be instantly compensated to prevent short-term fluctuations that are considerable (up to ±3% of nominal line frequency). Systems that depend on more accurate, short-term timing frequently employ internal time bases that are precisely set to the same nominal frequency (or a multiple thereof) but drift much less. As the power line frequency drifts around the nominal, these frequency sources are continuously drifting in phase with respect to the line. If such a source is generating interference, this interference will be seen to drift continuously through a standing wave display on an oscilloscope that has been synchronized with the power line frequency (typical "line trigger" mode). Conversely, if the scope is triggered by the frequency source, the line-locked ripple will drift through the display.

The most common source of such interference is television transmission. The basic vertical scan frequency for each field of view is equal to the line frequency but not synchronized by it. Because the transmission frequency of the broadcast carrier is many megahertz, it comes as a surprise when television signals show up in amplifier systems with audio frequency bandwidths. Of course, what is showing up is the demodulated signal of the picture envelope, which contains continuously changing frequency components from the vertical field rate to the horizontal resolution rate. The demodulation can occur in any component of the system that picks up the radiated electromagnetic interference of the carrier asymmetrically, thereby partially rectifying it. Typical offenders are the metal/electrolyte junctions of an electrode, poorly made solder joints, or amplifiers provided with diode protection circuits on their inputs. One of the things that make this type of interference particularly difficult is the unpredictable nature of the effective antenna. The high-frequency carrier has a short wavelength that penetrates shields and chassis that normally screen out audio frequency interference. The amplitude of the noise is not solely related to the amount of carrier coupled into the effective antenna, but also to the degree of the nonlinearity that demodulates it. Thus, it is not uncommon for significant levels of such interference to arise in amplifier stages after the input connections and preamplifier, even though the instrumentation has been carefully and properly designed and built.

Fortunately, this problem occurs only close to high-power transmission antennas, but that is scant consolation if your laboratory is located near such. If the pickup is in the preparation itself, you can try the old trick of installing a pair of carbon resistors (value equal to about 1% of the amplifier input impedance) in series with the input pins en route to the first stage in the chassis. This type of resistor has a high inductive impedance at television carrier frequencies, and so acts as a choke without noticeably affecting amplifier noise or gain. If the pickup is further along in subsequent amplifiers or their cables, it may be useful to install more than the usual degree of shielding, placing fine mesh screening over ventilation openings and using braided coaxial cable on low-impedance interconnections, which normally do not need shielding. Remember that the mesh size for effective shielding must be much smaller than the wavelength of the interference, which starts to approach the physical dimensions of the preparation in the UHF bands.

One empirical solution to which you may be driven is to arrange all amplifiers into a physical configuration in which the various couples simply cancel each other. Television waves are highly directional, as anyone who has spent time adjusting a rabbit-ear antenna knows. Just as you can optimize pickup for your television set, so can you virtually cancel pickup in your preparation. Of course, if the offending cables must be connected to a freely moving animal, this approach may not be useful; however, simply orienting various pieces of equipment and cables can be surprisingly effective. Finally, there is the radical solution of deferring experiments until the station signs off in the evening.

d. Steady, High-frequency Noise. If your recording is contaminated by high-frequency noise sources, the usual appearance at the usual oscilloscope sweep rates is the "fuzzy trace," in which the desired signals are seen riding on a trace that appears unfocused. As the sweep speed is increased, the fuzziness resolves itself into a high-frequency oscillation with a basically sinusoidal shape. This may be difficult to appreciate until the sweep is triggered by the oscillation itself. (Use the internal trigger AC–low-frequency reject setting, if available.)

The precise frequency of this oscillation will be a big help in pinpointing the source. We have already mentioned the possibility of an intrinsic high-frequency oscillation of the amplifier. This can be ruled out if the frequency is unchanged when the signal is passed through another amplification channel. Even if the second channel also oscillates, it is unlikely to do so at precisely the same frequency whenever all other channels are turned off.

The most common radiators of high–audio frequency noise are switching power supplies and video monitors. Both produce noise because they incorporate the same electronic design trick. It involves one efficient way to build low-ripple, DC power supplies, particularly for high voltage levels. The output of a sloppy high-frequency oscillator is rectified and integrated into DC power. The running frequency of such power oscillators tends to be in the 10 to 25 kHz band, and the radiated electromagnetic interference may be enormous. The high DC voltages needed to accelerate the electron beam of a television monitor are obtained by rectification and integration of the high-energy fields in the horizontal sweep oscillator and deflection coils, which run at 15.75 kHz (63 μsec period between cycles). Whereas this frequency is considerably higher than the usual band limits of EMG amplifiers, it is important to remember that the rolloff of frequency response above the cutoff frequency is not infinite. The usual 6 dB per octave filter with a 3 kHz high-cutoff frequency will only remove about 85% of 15 kHz interference. If the interference starts out several times larger than the signal, it will still dominate the filtered record.

Further filtering is possible, but it is always better to clean up a record at the source rather than to try to recover after the fact. These frequencies of electromagnetic interference are readily shielded, either at the preparation or at the source. Because it is always cumbersome to shield the preparation, particularly if it involves a moving animal being photographed, it is preferable to shield the source. Sometimes all that is required is the connection of a ground lead to an already present chassis ground point. The proliferation of plastic rather than metal enclosures has aggravated the radiated interference problem, so it may be necessary to add metal screening around the offending device. Sometimes you can accomplish this discreetly by painting the inside of a plastic chassis with conductive paint and attaching a ground clip. If the radiation is coming out through the front of a video screen, it may be possible to turn or move the device. Both of these simple approaches are often surprisingly effective.

Often you can identify the source of such noise simply by turning off power supplies around the room (and perhaps the adjacent rooms) until the noise goes away. Because these pieces of equipment are usually not integral to the recording session, this procedure represents an obvious strategy for dealing with them at critical recording moments.

One last caveat: make sure that you are not actually injecting the offending oscillation into the preparation via the AC bridge carriers used in some length gauge and footfall detection schemes or via impedance testers.

e. Variable-frequency Pulses. Frequently, one encounters noise pulses that, while irregular in amplitude or frequency, are clearly not random. They often come and go in regular waves of gradually changing frequency and amplitude. These noises are most usually caused by motors and engines that run asynchronously, changing their speeds in response to loads. Ignition noise from internal combustion engines, either in vehicles or auxiliary generators, can be a serious problem to trace because the interference is radiated over long distances and from inconsistently present sources. Most vehicles effectively shield their own radiation, but open power-generating equipment such as lawnmowers, motorcycles, compressors, and service vehicles may not. A quick look out the window when intermittent noise occurs can be revealing. If the source is outside, simple shielding, such as with aluminum foil or metal screen over nearby windows, is often effective, therefore you may not need to resort to cages around the preparation. Brush noise and pulse noise from pulse-width-modulated controllers of electrical motors can look similar. We once encountered such a problem with the bell-activating circuit of a telephone connector box.

Identification of these sources requires careful observation of the pattern, thorough inspection of equipment in all nearby areas, and deductive reasoning. Anything with a motor in it, from a centrifuge to a refrigerator to an automatic specimen changer in a scintillation counter, must be considered a potential source. Shielding at the source is often effective, and simply turning off the source at the critical moments is always a course of last resort. The more distant the source, the more effective is good differential amplification with balanced inputs.

f. Variation on Power and Ground Lines. Thus far, we have assumed that the source of electromagnetic interference was radiating through the air into the preparation by stray capacitive coupling. This condition at least reduces the significance of noise sources as the distance from the preparation increases; it generally permits effective Faraday cage shielding at either source or recording site. However, you may actually be importing electromagnetic interference from distant sources on the AC power and ground lines that run to all your equipment (or on any unconnected, hence ungrounded, cable). When a heavy piece of equipment turns on or off, the sudden change in current flow introduces spike transients that radiate through the entire circuit. These waves generally do not pass back through the step-down transformers at the main feed, but the physical location of all the outlets from one feeder circuit may be very widely distributed, particularly in older, repeatedly renovated buildings. DC power supplies are designed to remove ripple efficiently from the AC power line frequency, but their response to high-frequency input transients may be much less imperturbable.

Some pieces of equipment not only change their loads abruptly but also may dump large amounts of current into ground lines. No electrical conductor has zero resistance (except at superconducting temperatures); therefore, the product of these large currents and nonzero resistances may produce significant potentials, at least equivalent to the microvolt to millivolt range of bioelectrical signals.

These kinds of interference tend to be the

least regular in wave form and temporal pattern and least responsive to the usual shielding and grounding maneuvers. Once you suspect them, you can easily confirm them by monitoring the power and ground lines themselves, although this may require special equipment when searching for infrequent, brief transients. Dealing with them is less easy.

If the offending signal arrives through the ground lines, it may be necessary to remove all equipment from the power supply ground (the third wire of the typical three-pronged safety plug), and replace the service ground with a private ground consisting of an actual low-impedance connection with the earth via a buried pipe. This should be done only in consultation with the plant engineers, because it is expensive and potentially hazardous if not done properly. Sometimes you can get away with the simpler expedient of uncoupling the DC power supply commons from the chassis and service grounds. Because most power supplies now employ transformers with separate primary and secondary windings, the DC output is technically floating with respect to AC ground. At the rear of many pieces of scientific equipment there will be a switch or jumper that connects the common to the ground. Of course, there are many routes for the connection to be remade, via the shields of output lines to other equipment or in the preparation itself—for instance, via fixation pins. Unless a few simple variations produce an immediate effect, a systematic study of the connections and chassis ground and shield conventions of all related equipment may be called for.

If the transients are buried in the AC power, the simplest expedient is battery operation of all critical, low-level signal-processing stages. Once the raw signals have been amplified to the usual storage and display levels, most equipment is adequately regulated to be immune to power line transients. The actual power draw of modern integrated circuit preamplifiers is so low that even a large multichannel array of them can be run for hours by a modestly sized battery. Most such amplifiers will work well with a symmetrical supply ranging from ± 10 to ± 15 V DC; this is easily provided with a pair of automobile batteries and a trickle recharging system to be used between experiments (see chapter 12). Such batteries have very low source impedance and high source capacitance. This makes them even more effective than regulated DC power supplies at preventing cross-talk among multiple amplifiers operating from the same supply. Isolation transformers can be purchased to clean up AC power lines, but they are expensive and tend to be relatively ineffective on the high-frequency transients that cause the most problems in an EMG laboratory.

Always remember that transients in power and ground lines can be radiated through the air, as can those generated in local equipment. Shielded raceways and power cords may still be needed in the laboratory to cope with this part of the problem, even if battery power and separate ground circuits are used in the preamplifiers themselves.

F. DECISION TREE FOR NOISE IDENTIFICATION

Figures 13.1 and 13.2 provide a simple strategy for efficiently categorizing and eventually identifying noise sources. The primary decision regards the relationship of noise to the AC line frequency; this is quickly established by examination of the preamplifier output on an oscilloscope triggered by the line frequency with a sweep of about 10 msec per centimeter. Line frequency–locked noise is usefully separated into waveforms that are pure sinusoids and those that also contain pulses or harmonics of higher frequency. This separation is best made with an unfiltered signal, as the actual unadulterated shape of the noise waveform provides useful clues to the experienced eye. It is always worth investigating the integrity of electrical connections both to grounding points and to electrodes. This is best done with an AC impedance meter that will be sensitive to any asymmetry of differential electrode impedance; the asymmetry may be degrading the common mode rejection of the amplifier.

In the absence of such obvious failings, the best procedure is to begin turning off unnecessary equipment until the noise is reduced. Note that several devices may be contributing to the radiated field, and the contributions of the lesser sources will not be evident until the largest sources have been dealt with. Because of vectorial phase addition, some sources may actually be canceling each other. You may have to go through the laboratory several times in different sequences until the significant culprits have been rank-ordered.

Noises that are unrelated to line frequency may be divided into high- and low-frequency ranges. Low-frequency noises are either regular processes arising from devices such as motors and engines with their own cycle times or irregular processes arising from the occasional operation of one or more intermittent loads. The latter usually appear on AC power lines and grounds and may act like antennas collecting such glitches from equipment all over the building. High-frequency noises almost always have a single fixed frequency, which is generated by the normal operation of some piece of equipment with an integral fixed oscillator. Some culprits actually use carrier frequencies as part of biological measurement such as impedance bridges (have you left your electrode impedance tester turned on?). Others generate them as part of their power supply operation; these include television monitors, time-code generators, computers, and other (usually digital) equipment. Again the strategy is to turn things off until the noise goes away, usually quite abruptly.

This decision tree (fig. 13.2) is by no means all inclusive, but it does cover the major categories of noise and suggests the sort of orderly strategy that is needed to minimize wasted time.

G. ELIMINATING NOISE AND ARTIFACT

Obviously, the identification of the source will dictate the range of measures that will be effective and feasible. However, it does seem useful here to summarize the types of measures available in a general order of desirability.

1. Optimal Electrode Design

You are probably getting tired of this by now, but the most common source of poor signal-to-noise ratio is poor electrode design with unnecessarily low-signal amplitude, compounded by unnecessarily high-impedance antennas for noise and motion artifact.

2. Differential Amplification

The first stage of the preamplifier is the crucial one. Make sure that you have optimal common mode rejection and linearity by using a simple, symmetrical head-stage design. Do not clutter it up with filtering, input limiting, or excessive gain that will cause nonlinearities and asymmetries. These things belong in later, preferably physically separate stages.

3. Grounding

Keep a simple and clear path for the ground circuit between amplifiers and preparations.

Avoid multiple grounds from multiple devices connected to the animal, because these lead to the infamous "ground loop," in which potentials in one part of the circuit cause current flow and potentials in the other parts. Make sure the ground contact has a low impedance and the ground leads are large-gauge wires so that their E = IR potentials from currents flowing in the ground are minimal. Be certain that all the equipment in the monitoring circuit is commonly grounded. Many modern oscilloscopes have floating shields on their inputs, and you must explicitly ground them to the preamplification system via their ground jacks rather than relying on the coaxial cable connection.

4. Shielding

Try to shield out noise at the source, rather than building giant Faraday cages that will make your life miserable. Try shielding the connectors and cables from preparation to preamplifier. If you have to have a cage, do not defeat its function by bringing AC power lines and AC-operated equipment inside it. Remember that an unused electrode lead may serve as a radiating antenna inside your cage.

5. Unnecessary Equipment

AC line hum is often a cumulative process in which no single device is at fault. Rather, there is a complex summation of the electromagnetic radiation from all the AC-operated equipment near the preparation, much of which is not actually in use when the recording must be made. Do not forget to turn off or ground the input of unused amplifiers, which are prone to breaking into high-frequency, large-amplitude oscillations that radiate through the air as well as into power supply lines.

6. DC power

Many resistive devices, such as heaters and incandescent lights, work just as well with DC as with AC power. If you cannot turn off or move these devices away from the preparation during the recording, try operating them from automobile batteries or well-filtered DC supplies.

7. Battery Power in the Amplifiers

If there is variation on the AC power lines or hum or cross-talk between amplification channels resulting from inadequate regulation of the power supply, use batteries to power at least the preamplification stages.

8. Filtering

If you cannot clean up the signal at the source, you can always go to ever more elegant (and expensive) filtering techniques. Simple bandwidth filtering with RC filters (6 dB per octave rolloff) that have cutoffs outside the frequency spectrum of the signal to be recorded are always advisable for limiting broad-band noise but are rarely effective for removing narrow-band noises from specific sources. Special high-rolloff filters, such as notch filters and "brick wall" filters, inevitably distort the signals in question and often work surprisingly poorly on the noise, simply hiding it better in the desired signal rather than eliminating it.

14 Electrical Testing, Calibration, and Failure Analysis

A. OVERVIEW

The reader of the methods section of a scientific paper is asked to assume that the devices and equipment therein described were, in fact, operating according to specification when the experiments were conducted. In the days of simple, direct reproduction of relatively unprocessed signals, such as traces on smoke drums, both the types of malfunction and their nonpresence were relatively easy to ascertain. However, equipment has become increasingly complex and published papers rely more often on highly processed graphical extracts from the original signals. This puts the responsibility for the largely unrewarding task of testing and maintaining equipment squarely on the shoulders of the experimentalist, who often thinks he or she has better things to do.

In fact, this aspect of research is often overlooked, not because it is difficult or even particularly time consuming, but because the investigator has not set up or does not know about efficient and simple tests of the data collection system. In this chapter we describe such test procedures, based on use of inexpensive and generally available test equipment, some of which can even be homemade. The purpose is not to make accurate factory calibrations or diagnostic tests, but rather to facilitate the kind of go–no go assessment that must become part of the data collection routine.

The most important distinction that must be made is between problems arising in signal-processing equipment and problems at the signal source, usually implanted electrodes or transducers. Whenever a signal does not look quite right, it is important to make this distinction immediately and unequivocally, because the methods of failure analysis and potential for recovery are likely to be quite different. The single most useful tool is the AC impedance tester, as almost all of the potential problems of implanted signal sources are reflected in changes in the impedance measured between leads of the device or between a given lead and a reference electrode in the preparation. As we shall see, this and other tests are readily incorporated into routine experimental procedures.

B. SIGNAL SOURCE PROBLEMS

1. Bench Testing

EMG electrodes are, for the most part, extremely simple electronic devices, often consisting of little more than a bit of insulated wire with an exposed end. However, as we have seen, a great deal of experimental validity will depend ultimately on the exact extent and location of the exposed contacts that form the recording antenna.

No electrode fabrication bench should be without the equipment for performing a bubble test, particularly as the test requires only a battery and a bowl of saline solution (fig. 14.1). Simply dip the completed device into the saline solution, using a good light and if necessary a dissecting microscope, to facilitate direct visualization of the contacts and lead wires. Provide a large, surface area reference electrode of some inert metal (e.g., platinum or gold; the shell of a stainless steel bowl will do as well). Connect the reference to the positive (+ or anode) side of a source of 3 to 12 V DC. A 9 V transistor radio battery works fine. With the electrode submerged and in view, touch the lead of one contact to the negative (− or cathode) pole of the battery. Immediately, you should see a stream of gas bubbles coming from the exposed electrode surface. If the surface is very small, one bubble may be trapped on the surface and will slow or stop further electrolysis until it floats free. To clear the surface of bubbles and to repeat the test, briefly withdraw the electrode from the saline.

The test is an extremely sensitive indicator of poor contact exposure (uneven or absent regions of bubbling), holes in the leadout insulation (extra streams of bubbles along leads or coming from the wrong surface of patch

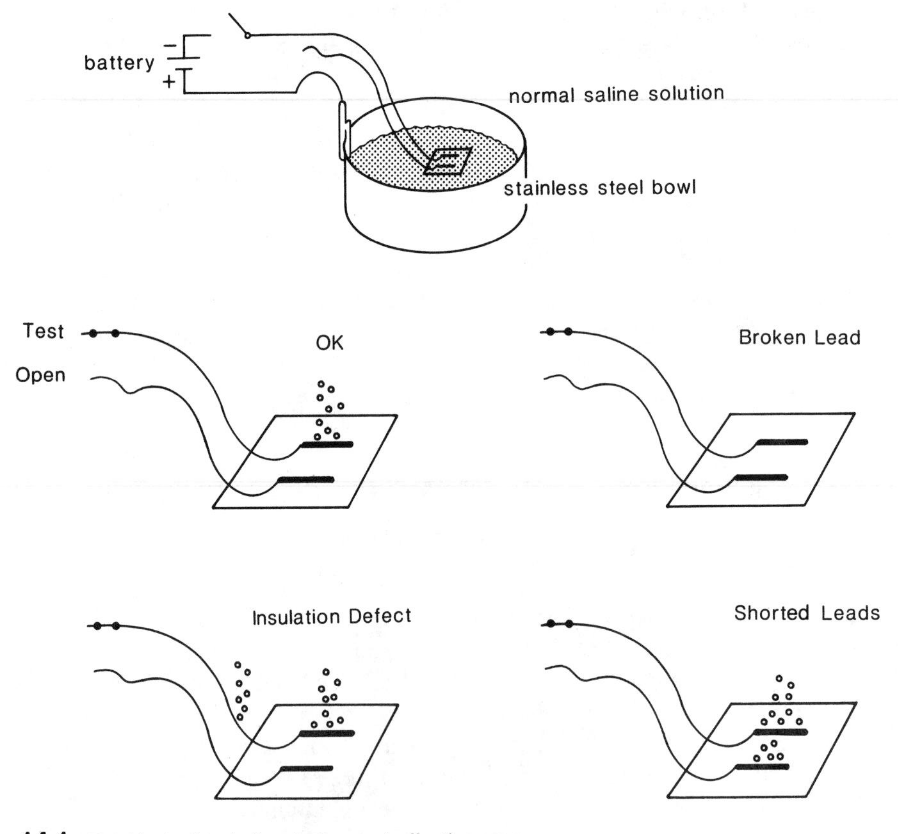

14.1. Bubble testing is the single most effective way to reveal defects in an electrode ranging from broken leads, to shorted leads, to insulation defects.

electrodes), and shorts between contacts (bubbles forming on two separate electrode contacts when only one has been connected to the battery). Furthermore, the scrubbing action of the cathodal electrolysis reactions will often clean and stabilize contact surfaces otherwise contaminated with residual oils and oxides.

Four minor caveats should be noted. First, never connect an electrode contact to the anode of the battery, as this can lead to rapid electrolytic corrosion of many metals, including copper, stainless steel, silver, and gold. Large surface area stainless steel and gold reference electrodes are generally acceptable as anodes, only because the current density is very much lower. Furthermore, the bubble test is intended to be used only for brief bursts and should never be left connected for long periods. (Microelectrodes also pose special problems, see chapter 20)

Second, the electrolysis of saline at a cathode causes release of hydrogen gas, leaving behind hydroxyl groups in high concentration. The highly alkaline solution in the immediate vicinity of the contacts may degrade some polymers, which is another reason to make the current test as brief as possible. If the electrode incorporates any materials that are highly pH sensitive, consider using a carbonate-buffered test solution. (Teflon, silicone rubber, and polyimide insulators are all quite pH stable).

Third, never try this test in vivo, for obvious reasons.

Fourth, never use this test on saline-filled length gauges; the electrolysis bubbles then will be trapped inside the tubing.

Do not confuse bubble testing or AC impedance testing with the common laboratory DC volt-ohm-meter, which provides no useful information about either insulation leaks or contact impedance. This is because there is no simple relationship between the property of the electrode and the DC potentials imposed by various volt-ohm-meters (VOMs).

2. In Vivo Testing

EMG electrodes, once implanted in the target structure, can be characterized functionally by AC impedance testing. An impedance tester is a specialized form of ohm-meter that uses an alternating current of a particular frequency in place of the DC test current generated by laboratory volt-ohm-meters. As we have just noted above, DC current results in electrolysis of water, with generation of gas bubbles and severe pH changes; both are deleterious to living tissue. Furthermore, the sudden onset and offset of DC current flow will activate excitable structures such as nerves and muscles.

Commercially available AC impedance testers range from special biomedical devices with display meters for a few hundred dollars to general purpose impedance bridges for several thousand dollars. The important features are the ability to generate the appropriate test frequency at a current level low enough to avoid stimulating tissue. The appropriate test frequency is one within the bandwidth of the anticipated signals, usually 100 to 3000 Hz.

Neurophysiological impedance testers usually employ a 1 kHz test frequency, although some older 60 Hz models are around. The latter can be expected to read considerably higher impedance values from a given electrode, and they are somewhat more sensitive to line noise interference, particularly when used with low-impedance electrodes that result in lower voltages from the constant test current employed. Test current should be well below 100 μA, particularly if the tester will be used on nerve cuff electrodes in vivo. Capacitor coupling is also important to prevent the tester from inadvertently administering any net DC current with the AC signal should it be imbalanced or malfunctioning.

An AC impedance tester consists of an oscillator, a constant-current source, an amplifier, and a display of the voltage across the test leads (fig. 14.2). You can turn any suitable AC voltage source (signal generator, oscilloscope calibrator, even a Variac or filament transformer) into a constant-current source by hooking it in series with a resistance much greater than any test load anticipated. If the maximum electrode impedance anticipated is 100 kΩ, a 1 MΩ resistor in series with a 10 V source signal makes a dandy 1 μA test signal (see fig. 14.2). Always add a series capacitor (1 μF will do nicely) to block any DC current. The same amplifiers that you are using to record the EMG signals can detect the resultant signal across the electrodes; the expected voltage will be equal to the product of the test current times the electrode impedance. A 5 kΩ electrode contact being driven by a 1 μA test current will generate a 5 mV test signal. With a gain of 1000, the resulting 5 V signal can be displayed on an oscilloscope or used to drive an AC volt-meter.

By using a pure, sinusoidal test waveform and triggering an oscilloscope display at a particular point on that waveform, it is actually possible to characterize a test impedance by its phase angle as well as its magnitude. The angle provides detailed information about the electrochemical conditions on the surface of the electrode and possible shunt effects through insulation layers or electrostatic shields. However, this information is rarely important for low-impedance EMG electrodes. In fact, only an approximate value of the impedance amplitude (±50%) is needed in most cases, as electrodes are likely to vary over this range as a result of slight differences in fabrication, cleanliness, and remodeling of tissue encapsulation. Nevertheless, it is advisable to keep a couple of calibration resistors handy. You can then make a quick check of the ability of your particular impedance-testing configuration to signal accurately the impedance ranges anticipated. It is not uncommon, particularly

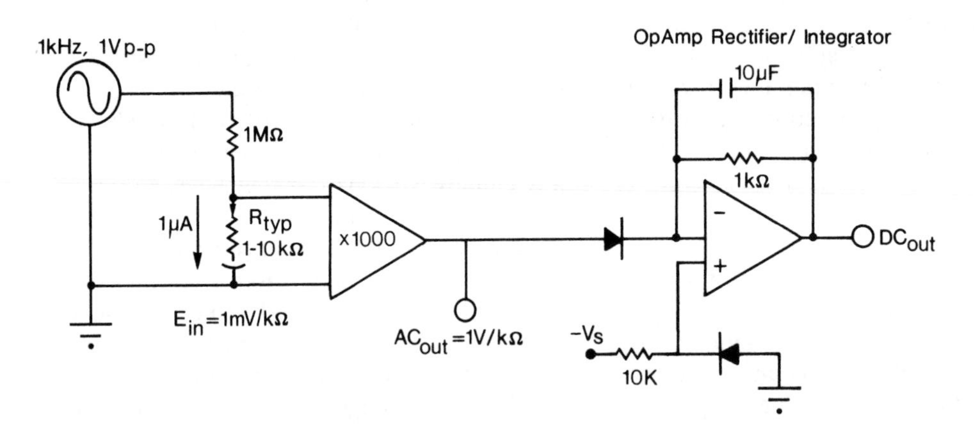

14.2. One simple way to test electrode impedance requires only a sine-wave generator and the same amplifier used for recording. If only low-frequency displays such as a chart recorder or volt-meter are available, the circuit shown on the *right* will convert the 1 kHz test signal to a DC level that is proportional to electrode impedance (the same circuit is an effective zero-offset rectifying integrator for EMG, although the time constant given by the 1 kΩ resistor paralleled by the 10 μF capacitor (10 msec) may need to be adjusted depending on the application.

for homemade equipment, for the effects of stray capacitance or noise sources to limit severely the linear range of the tester; open circuits then do not really read very high and shorts do not really read zero.

A daily record of the impedance of each electrode contact with respect to the reference electrode and between pairs of differential contacts can be an invaluable aid for recognizing problems whenever they arise, correcting them during an experiment, and preventing them in future preparations (see fig. 8.1 for an example). If the equipment is left set up (and it should be), it will take less than 1 minute per electrode to perform and record the test in a notebook. Such a record can be interpreted to look for the following types of problems.

a. Breaks in Lead Wires. Typically, a break in the lead from a low-impedance electrode, such as an EMG probe, will cause a sudden increase in its measured impedance to the reference point, because the test point is no longer connected to a large, surface area recording contact. Breaks in well-protected junction regions, such as solder joints, may even read as open circuits, particularly if the surrounding insulation has not been breached. Conversely, a break in a high-impedance probe, such as a microelectrode, will often lead to an impedance decrease because of an associated break in the insulation. Obviously, some potential for missed breaks exists at intermediate values, but fortunately such coincidences are rare. Be particularly wary of intermittent connections, such as at defective solder joints or loose connectors, that may read normal values unless and until some particular mechanical stress occurs, such as limb motion or traction on a lead cable. Remember that the mechanical protection of joints will reduce but not eliminate such problems.

b. Shorts between Contacts. If two separate electrode contacts have become shorted together, the impedance measured between them obviously will fall to almost zero. However, the impedance of either contact with respect to the reference point will still read about half of what it was before, assuming that the two contacts have similar individual impedances. Obviously, detecting this problem requires some care. Given the inherent, but generally gradual, variability of electrode contact properties over time, the occurrence of sudden changes should raise suspicion, even if the changes are relatively small. An exhaustive test of all possible pairs may be necessary to find the short, although suspicion will usually focus on wires close to each other in a connector assembly or coursing together in tight spaces.

c. Insulation Breakdown. Breakdown of insulation is a difficult problem, as small defects in insulation do not cause significant decreases in contact impedance unless they lead to exposed surface areas of the same order of magnitude as the actual contact. Of course, the tendency to pick up spurious signals is similarly related to the ratio of the surface areas, but this is scant consolation if the break happens to sit right over a large source of crosstalk. If this problem is suspected, it may be necessary to defer definitive diagnosis until the devices may be retrieved post mortem, at which time a bubble test can be employed. You can locate an insulation defect along an electrode or lead by monitoring the impedance as you lower the lead into a grounded beaker of saline or pass it through a grounded wire loop suspending a soap film in air.

d. Adverse Tissue Reaction. It is a common fiction among electrophysiologists that electrode impedance is a highly sensitive indicator of the nature and condition of the tissue surrounding an electrode contact. Unfortunately, this is generally not the case, and even the grossest pathologic state may be overlooked by those who religiously follow the ups and downs of AC impedance values over time. Certainly such curves cannot serve in lieu of a postmortem visual examination of the implant site. The problem is that most of the impedance measured by an AC impedance tester (particularly at the EMG frequencies) represents processes occurring in the first few angstroms of the gap between the electrons in the metal and the ions in solution. Changes in the resistivity of the tissues intervening between the contact and the reference have relatively little effect. Even the densest, nonbony scar tissue has less than twice the specific resistivity of healthy muscle, and the worst pool of pus is only about twice as conductive as muscle. Most of the fluctuations of impedance from the test bench values in saline, observed in vivo over time, represent changes in surface oxides and absorption of protein and other contaminants that can come and go capriciously.

C. PROBLEMS WITH RECORDING EQUIPMENT

1. General Awareness

There are a number of general properties of the anticipated signal that you can keep in mind during experiments to raise the subliminal awareness that things may not be quite right. Much of the art of electrophysiology comes from this sixth sense of experienced expectation. An audio monitor is particularly effective at conveying maximal information without distracting your eyes and hands from more immediate tasks.

Be aware of the acceptable amplitude and character of the noise levels that should be present when the amplifiers are turned on with inputs shorted and with source impedances typical, although no signal is present. If the noise seems a bit high or unusually nonrandom whenever the preparation is connected to the amplifier, try shorting to ground alternately the positive and negative inputs of differential amplifiers to look for an open lead (the one that reduces the noise the most when shorted). Also check the integrity of the ground connection and make sure that unused amplifier channels are turned off or shorted at their inputs. Similarly, if the noise seems unusually low, look for shorted inputs (perhaps you left a dummy source impedance or shorting jumper in place).

During recording sessions, listen for the distorted, high-frequency sounds of clipping at saturation levels, easily appreciated on an oscilloscope trace as the level (positive and/or negative) beyond which no waveform ever goes.

If two differentially recorded signals have unexpectedly tight correlations in their microstructure (as viewed on a dual trace oscilloscope at fast sweep speeds), this may indicate cross-talk between electrodes. However, it may also be the result of mislabeled leads or miswired cables connecting a single amplifier to one contact from each of two bipolar electrodes.

2. Continuity Testing

It is well known in the electronics industry that no matter how complex the equipment and how elegant the components, most failures occur in connection points. This is the property that allows so many television sets to be "fixed" with a well-directed blow to the chassis. If the signal does not look right, first check the on-off switch and then examine the integrity of the connections to the preparation.

Whenever multiple devices are implanted in

animals and whenever specialized cables and connectors are employed to facilitate free movement of the animal, the number of potential failure points in the connection scheme can rise dramatically. Some method for quickly testing their integrity—both continuity and isolation from adjacent circuits—is essential. The simple laboratory multimeter (VOM, VTVM, DVM, etc.) is usually more than adequate. Inexpensive models (under $100), now available, include an audio signal (usually called a diode tester) that permits you to evaluate continuity without having to look up from the probes, which is a very handy feature. Note, however, that all these devices use DC signals, and so they cannot be applied safely to the circuits when the animal is connected. This means either probing as far down the cable system as feasible (and risking damage to delicate connector pins) or devising a simple test load to simulate the presence of real electrodes in the animal. The latter strategy is very useful for a number of tests, and later we present it in greater detail (fig. 14.3).

Once the animal has been connected to the apparatus, it is useful to perform the in vivo AC impedance tests noted above via all the usual cables up to the amplifier input. In fact, with high-impedance input amplifiers and low, constant-current testers, the amplifiers can be left connected and turned on without degrading the test (monitoring their output may be

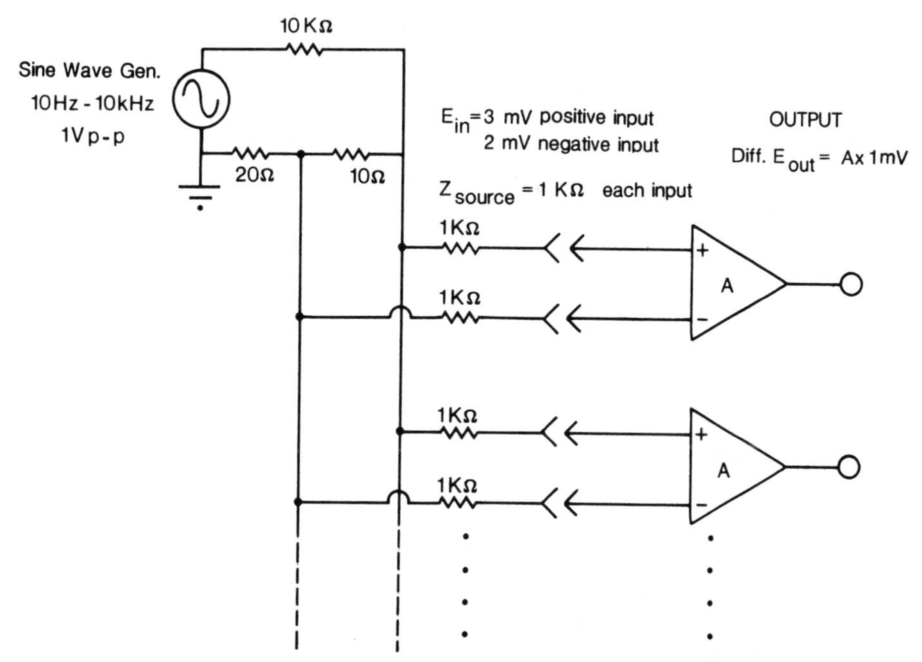

14.3. Dummy load and calibration circuit. A single sine-wave generator can provide a set of test signals that facilitate simultaneous confirmation of function and calibration of a battery of amplifiers and connector cables. If properly configured, each amplifier should produce an output equal to the product of its nominal gain times 1 mV. If there is a short or open condition in the negative input of an amplifier, the output will be three times too large; a short or open condition in a positive input produces an output that is two times too large. The circuit also simulates realistic source impedances, permitting an assessment of noise and artifact pickup from radiated and intrinsic sources.

all that is needed for the test, see above). Note that this cannot be done with low-impedance amplifiers such as transformer-coupled devices. Obviously, this will not differentiate between breaks in the cables and breaks in the animal, nor is this a good way to test for shorts in connectors and cables, particularly if they involve unrelated channels. All cables and connectors should be exhaustively checked when first made and at regular intervals with a DC ohm-meter for correctness of pin assignments, continuity of leads, and possible shorts to ground and between adjacent pins and conductors. Discovering an error in a multilead cable after a series of experiments will cause untold confusion and probably require the entire data set to be discarded.

3. Gain and Bandwidth Calibration

The single value of gain usually quoted for (or beautifully engraved on the front panel of) an amplifier is, in fact, an approximation of a highly complex set of properties of the amplifier, its control settings, and the connected cables and signal source properties. Such a complex system is best characterized and tested in as complete and realistic as possible a simulation of its normal configuration. This is best done by connection of the electrodes to a dummy source impedance of the approximate value and generation of a set of test signals at that source to cover the range of amplitudes and frequencies of interest (fig. 14.3). Similar procedures of applying a known test signal and examining the output of various stages can and should be used regularly to calibrate all signal-processing equipment. This includes tape recorders, chart recorders, oscilloscopes, and computer digitization equipment.

Start by looking carefully at the amplifier output with no signal input using a display with sufficiently high sensitivity and bandwidth, such as an oscilloscope. There should be a low-amplitude, random pattern of wiggles representing the thermal noise of the amplifier and the source impedance. For source impedances under 100 kΩ (typical EMG electrodes), the amplifier noise will usually dominate, and there will be little significant reduction of this noise by shorting of the amplifier inputs to ground. However, if there is a nonrandom component, such as sinusoidal line noise or spiky interference, this noise is often highly related to the source impedance and its physical orientation with respect to the source of radiated electromagnetic interference. Consult chapter 13 for advice on coping with this problem. Note that if the noise is much lower than the signals anticipated, it may be safely ignored, although pickup of radiated noise can often become drastically augmented (or reduced) whenever the actual preparation is connected in place of the dummy source impedance.

Next, pick a signal frequency in the midrange of the EMG bandwidth (e.g., 1 kHz as used in many impedance testers), and apply variable-amplitude voltages at these frequencies across the dummy source. See the discussion of dummy source for voltage-dividing networks (below) to convert the output of a waveform generator to physiological levels. Examine the amplified signal to be sure that the ratio between the amplitude of the input signal and that of the amplifier output accurately reflects the predicted gain of the amplifier. Be aware of nonlinearities at either end, such as low-signal dropout (a problem of certain classes of old-fashioned push-pull amplifiers) and high-signal nonlinearity and saturation of output (whenever the product of the gain times the input signal approaches the voltages of the amplifier power supply).

At some midrange amplitude expected for the signal to be recorded, shift the frequency of the test signal first toward the upper boundary and then toward the lower boundary of the desired bandwidth. Check to be sure that the output stays constant in amplitude for a constant-amplitude input until close to these boundaries and that the output amplitude then begins to decline at a gradually increasing rate as the boundary frequencies are passed. Check carefully both for attenuation of the signal within the supposed bandpass region (which will degrade the recorded signal) and for failure of the signal to attenuate outside the bandwidth (which will result in unnecessary noise). Devices such as notch filters and transformer couples may cause local dips and wiggles in the overall gain profile at certain frequencies. Be sure that these variants are acceptable and recognized as potential sources of distortion of signals including these frequency components. One particularly instructive exercise for the electronics tyro is to apply some simple nonsinusoidal waveform (e.g., square-wave or sawtooth) at these same fundamental frequencies and observe the waveform transformation at the output as a result of the filter properties of the amplifier.

D. FAILURE ANALYSIS OF IMPLANTED DEVICES

If all else fails, at least be sure that you learn from your mistakes. Chronic electrophysiological experiments are entirely too expensive of time and other resources simply to hope that a problem will go away next time. It takes discipline to refrain from disposing of a failed preparation, electrodes and all, but that is the time to be most careful in testing and analyzing it.

If the electrode recording areas and the connection points to the animal are some distance apart, it is best to expose both carefully with minimal disruption of the leads between them. This facilitates direct probing for continuity with an ohm-meter to make certain that the contacts are correctly and securely connected to the appropriate connector pin. Even the lowest impedance pathway through fluid will not read as a zero resistance on a DC ohm-meter. Having established continuity, it is often easiest to cut the leads near the connection point and evaluate the connector and the electrodes separately. With the leads cut and the severed stumps dry, the impedance between any two connector pins should be infinite. If it is not nearly so, look for fluid seepage into and around joints in the connector assembly. Obviously, this test is virtually worthless if the connector is left to dry out overnight before study.

The electrode site should be examined for gross signs of trauma to or dislocation of either the device or the surrounding tissue. Be sure to document carefully the orientation of the contacts with respect to nerves and muscle fibers. Suspected points of leakage in the insulation can be confirmed by bubble testing in a bowl of saline.

E. DUMMY SOURCE

The amount of time you will be able to use profitably recording normal activity from a cooperative subject is the difference between the patience of the subject and the time wasted setting up items such as connector cables, amplifiers, gain and filter settings, patch panels, tape channels, and cameras. In a multichannel recording system this difference can easily become a negative number unless most of the system can be configured, checked, and calibrated without the subject's being present. This calls for a dummy source that will simulate the electrode impedances and anticipated signal levels (fig. 14.3).

This dummy source should simulate both

the physical and electrical properties of the preparation as closely as possible, so that cables connect it to the equipment in the same manner and location. Figure 14.3 shows a circuit diagram of a simple resistor network that can be installed on a connector of the type used on the preparation. With no input signal, the passive device itself simulates differential source impedances on each channel, providing a good check of noise levels and common mode rejection of radiated noise pickup.

The inputs can then be attached to any waveform generator to verify the gain and bandwidth of the amplifiers and to calibrate recording equipment. The voltage-dividing resistor network provides two different pseudo-floating millivolt-level signals to test the two inputs to each differential amplification channel. The correct throughput signal will be observed only if both the inverting and noninverting connections to the amplifier are intact and the amplifier is working with the proper common mode rejection and gain. Any number of amplification channels can be hooked to the test signal in parallel; the simultaneous and identical calibration signals received by all channels should result in simultaneous and identical outputs of the appropriate magnitude.

15 Visual Correlation Techniques

A. PROBLEMS AND DECISIONS

1. Correlation

The actions of muscles exert forces within animals and tend to produce displacements that we see as movements. Yet neither the force nor the displacement of a muscle will be simply correlated with the magnitude of the EMG signal. Absence of linearity requires us to test and establish the correlation for each combination of muscle and movement, rather than permitting us directly to extrapolate force from the primary electrical signal.

Even the simplest movements will be complex and difficult to appreciate, because animals are multijointed creatures and normally move more than a single pair of elements by contracting more than a single set of linking muscles. Furthermore, many animals tend to move rapidly, and the action of muscles proceeds in milliseconds, much faster than the response time of the human retina. Hence, the EMG signals and their effects must be recorded for review in a step-by-step fashion. Recording also permits the events to be reviewed at a variable rate, so that the observer can perceive the dynamic interrelationships.

The decisions regarding which movements and effects to select for correlation and how to record them may well be as or more complex than the decisions regarding how to record EMG signals and which muscles to study. There are many possible techniques, depending very much on the results required. Consequently, in this chapter we present basic principles and note the currently simplest or most promising technologies. Other approaches are available and multiple variants may be found in technical reports and industrial journals (see the references).

The closer mechanically the movement being recorded is to simultaneously active muscles, the greater the probability that the two aspects are associated. However, it is often possible to estimate the displacements and forces occurring within an animal from movements of its surface. Such observations of the external surface of an animal do not require an invasive approach; thus, the animal is less likely to be disturbed by the recording process, and its behavior is more likely to be normal. The resulting record may provide a critical baseline that permits us to check whether anesthesia and invasive approaches, such as those required for EMG, have introduced behavioral artifacts. Such possible artifacts are easiest to discern from a global view of the system, such as a cinematographical record.

All such global optical techniques are suitable for correlation with EMG tracings from an unrestrained animal. They provide a record of the integrated action of multiple muscles, rather than telling us about mechanical effects of the activities of any individual muscle. Film analysis of the summed effects of muscle activity may provide the most meaningful initial approach to the biology of the animal.

2. Recording and Behavior

The behavioral tendencies of animals are likely to have an important influence on the success of an experiment. Consequently, you should learn how your animal is likely to behave. Its general physiological and psychological state can often be assayed by external observation. Discerning owners are able to recognize the meaning of slight behavioral alterations in a favorite pet, and any good farmer can tell whether his cow or horse is contented or excited. It may be useful to collaborate with or employ a specialist who is trained in interaction with the particular species being studied.

An observational phase should be part of any experiment. As the animal performs, you should note how it acts. If the aim is coordinated EMG, you may later switch the various EMG signals sequentially to a speaker and use the sound to coordinate the visual records. Whenever using such preliminary approaches, you must ensure that the audio system correctly reproduces the bandwidth and dynamic range of the signal and that the possible time delays (such as that required for the tape to pass from the record to the playback head, chapter 17) do not affect synchronization.

It is often possible to condition semidomesticated animals to the circumstances involved in well-conceived experiments. However, conditioning may be more difficult when you are dealing with wild animals from populations that have not been selected for tolerance of humans in their vicinity. Animals that respond to disturbance must be treated carefully. A pet cat will watch the hand of the owner and may arch its back against it, welcoming the promise of contact. A diving animal may respond to the same cue by reducing its heart rate in bradycardia, a fright or withdrawal reaction, whereas other species may become excited and show this by tachycardia. The same pet that welcomes contact may respond negatively or by attack to loud noises, such as thunder, suggesting that it experiences an irrational (to us) fear.

Some animals do not respond to disturbance by an obvious fright reaction, but only by freezing or becoming immobile. For instance, some snakes show minimal visible response to loud sounds; however, their autonomic reactions change markedly in response to disturbance. In other species it may be appropriate to estimate the degree of disturbance by assaying heart or ventilatory rate. In any case, training and conditioning may increase the number of phenomena that can be effectively studied, and information about normal responses may permit us to recognize and avoid disturbances that are likely to induce modification of behavior (and potentially of results).

3. Concepts in Recording

Animals are three-dimensional and often subtly asymmetrical, as are their movements. Motion may be recorded continuously or intermittently; in the latter case, the recording frequency (frame rate in filming) must be matched to that of the events to be analyzed. Sometimes it is possible directly to monitor the force output of a particular muscle by placing a strain gauge in series with it, or to monitor the displacement by placing an indicator in parallel with the muscle. However, in most cases one tends to measure more distant events that may result from the actions of not only one, but many muscles.

Muscles induce stresses, deformations (strains), and pressures, as well as accelerations, velocities, and displacements of parts, more or less in that order. We often find it easiest initially to record displacements and use these to establish velocities and accelerations by mathematical or electronic differentiation with respect to time. In some cases we deduce causative pressures, strains, and forces by the rules of statics (steady state) and dynamics (changing state). Static and dynamic analysis by extension introduces the concepts of momentum, work, and energy. In the next chapter we provide some methods for measuring the dynamic aspects directly, whereas in the present one we emphasize image records suitable for kinematic analysis (fig. 15.1).

Kinematics is the study of movement independent of the dynamic effects that induce it. It deals with the displacement, velocity, acceleration, and jerk of an object and its parts. Kinematics does not by itself provide information regarding the forces required to induce

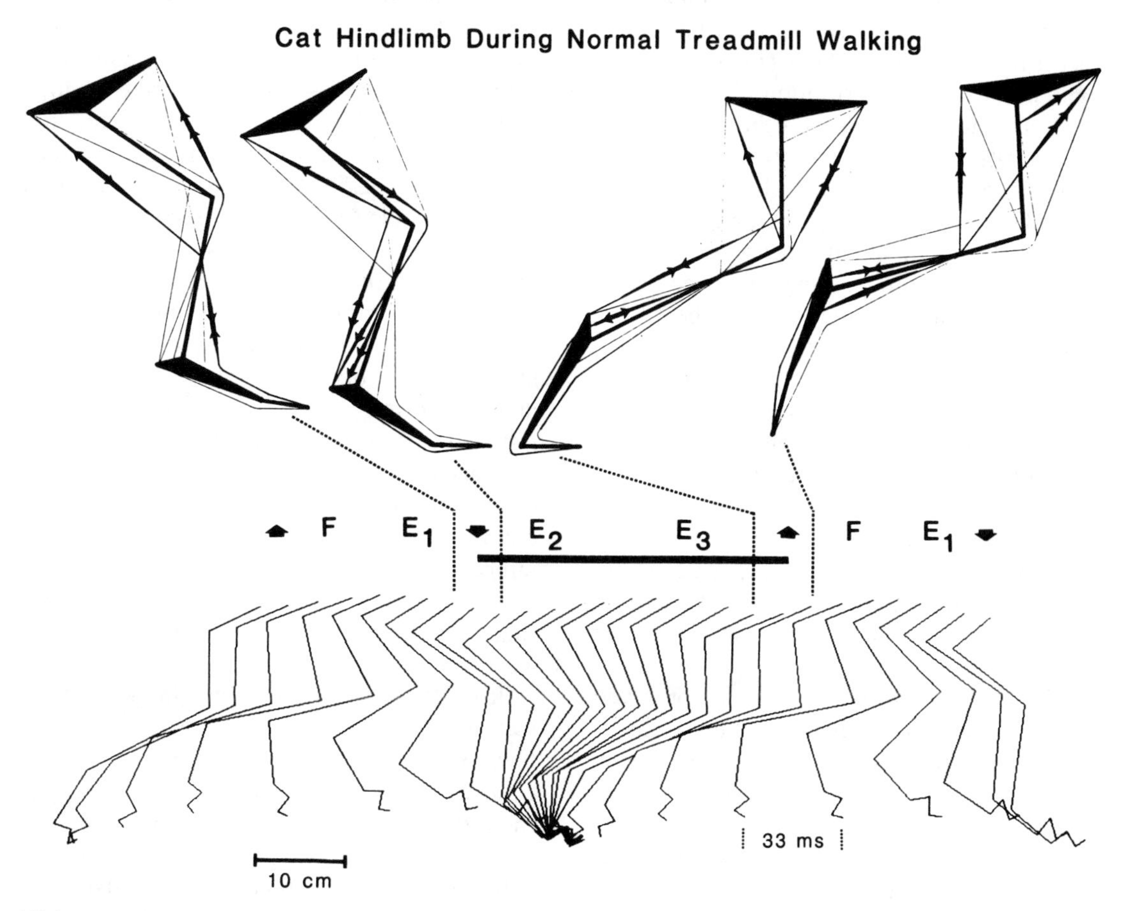

15.1. *Bottom,* Sequence of computer-generated stick figures taken from a hand-marked sequence of video image stills of a cat walking on a treadmill. Only every second image of the 60-field-per-second recording is shown. *Top,* Classic Phillipson step cycle phases. Four figures have been selected and redrawn with the electromyographically active muscle groups shown as *thickened lines* across their anatomical pathways. *Arrows* indicate whether the motion applied to the muscle is lengthening (pointing away from origin) or shortening (pointing toward origin), by the instantaneous direction of motion at each joint that the muscle crosses. (From Loeb, 1986.)

the observed displacements and accelerations, although kinematic data may occasionally permit their estimation from incidental observations, such as the tensing or swelling of the mass of a muscle or the deformation of zones against which the animal presses.

Movements of a single rigid body in three-dimensional space may be characterized by resolution of them into six components or six degrees of freedom. Three of the components describe translational (i.e., straight line) movement; thus, the linear displacement of a body can be described by resolution along three axes at right angles to each other. A three-dimensional object can also rotate about each of the three axes. Whereas solid geometry provides us with conventions for mapping the space in which these movements occur and kinematics defines the parameters of the movements, descriptive geometry treats the appearance of the object as if projected (photographed) onto plane surfaces at specified angles to each other. It also establishes conventions that permit reconstruction of the position of an organism and any of its elements from a limited number of views, provided that these are appropriately selected. Introductory textbooks on topics such as kinematics, statics, dynamics, and solid and descriptive geometry deserve review and belong in our reference sets.

The accuracy of spatial and temporal resolution required will generally establish the cost of analysis in terms of equipment, materials, and time that must be invested. Before deciding among analytical approaches and recording devices, you may find it useful to reconsider the biological question to be answered. An important item may well be the need or utility of knowing the absolute position of the elements being moved. This may be essential if you are dealing with processes of locomotion but less important, for instance, in the study of mastication, in which most of the interest is in the relative positions of skull and mandibles. Absolute measurements pose problems with defining reference frames and achieving adequate resolution over wide dynamic ranges. Relative measurements may require more continuous attention to the field of view relative to the moving object.

Factors that must be considered include not only the information that is to be recorded but also the way recording is to be approached. Generally it is useful to align one axis with that of gravity in recording of displacements or forces. In other words, try to film the animal with the camera mounted horizontally (0°) or vertically (90°) rather than at angles of 11° or 54° from the horizontal. This may not influence the cost of recording but reduces that of analysis. You may further reduce the cost of analysis by recording only the restricted set of movements of interest to the particular study, rather than recording the many sets that occur. For instance, the opening and closing patterns of the mouth may be signaled by the output of externally applied mercury displacement gauges (e.g., in reptiles) or of accelerometers fixed to the chin (in mammals). This limits the information obtained to the mechanical axes of the devices, but it is likely to facilitate analysis by permitting comparisons among the multiple bites of a sequence.

It is often useful to combine images. For instance, a global overview of a galloping cat or horse may not permit detailed recognition of changes in the joint angles as the foot strikes the ground. A close-up camera may be difficult to keep in focus and provides only one detail. However, combination of the overview with a close-up image provides both kinds of information. One may make similar use of instrumentation that furnishes only relative values of displacement or acceleration by calibrating these intermittently against the output of more fine-grained (and costly) approaches, such as digitized film analyses.

Methods that seem extraneous may be useful for developing decisions about scale of activity. For instance, trackways will provide details about stride length and the placement of the feet under relatively or absolutely unrestrained conditions. One can even map such tracks by applying dye to the animal's feet and allowing the animal to track the dye on paper surfaces (however, the texture of the surface may itself affect the behavior).

Restricted records are obviously much more cost effective. For instance, they allow one to synchronize the EMG signal with particular aspects of cyclic events; consequently, they facilitate later, simple, possibly automated, analysis of large numbers of cycles. It is even better if one is able to record a computer-readable analog signal permitting automatic synchronization for comparison.

B. VISUAL RECORDS

1. Principles of Recording

The principles of a movie camera and, for that matter, of a video camera are those applying to any other camera with the complication that the recording medium is presented with multiple images in sequence. The resolution of the lens and the grain of the detecting surface determine the sharpness of the image, which is called the resolution; this may be expressed as the minimal spacing of a grid of equally spaced lines that can be detected. The relative proportions of object and image should remain equal across the field.

A few basic rules define the simplest conditions for movement analyses. The position of three points placed at known locations (noncolinear) on any rigid body will define the position and orientation of this body. If any animal or part thereof moves significantly in all three dimensions, the position of each point will have to be recorded in all three dimensions. The number of parts of an animal that move relative to each other determines the complexity of recording. The parts of an animal that are rigidly fixed to each other can be treated as a single unit. In contrast, if an element deforms during movement, the motions of its parts must be recorded separately.

In general, the number of degrees of freedom observed in movement determines the kinds and number of images that must be generated for analysis. The number of film images required to establish the position of aspects in three dimensions must be established from the principles of descriptive geometry. In simplest terms, the position of objects restricted to movement in a plane may be determined from a single image, whereas the consideration of objects traveling in three dimensions requires a second image. Obviously, only visible points can be analyzed and this requires additional views when one is dealing with opaque and asymmetrical (or asymmetrically moving) structures. Whenever motion of one element is to be established relative to a second one, both must appear in the image. In section D4, below, we deal with some simple rules for combining images and generating data about positions in three dimensions.

Corollaries of such simple rules indicate that if motion is limited—for instance, because a symmetrically moving animal is constrained along a straight track—two views (or sometimes one) may be sufficient. (However, parts of an animal may move at angles to the component directions of movement, and this may complicate the selection of a preferred direction from which to film.)

The possibility of parallax must be considered, particularly when one is making fine measurements of objects that move across the image; the closer one goes to wide-angle filming, the greater the convergence angle of the light beams from the edges of the image. The effect of parallax obviously varies from the center of the image to the periphery. Objects that are closer to the camera may shade out those that are further away.

2. Principles of Photography

a. Light Gathering. Cameras utilize the light reflected from an object to produce an image. Objects obviously reflect light in many directions. The lens collects part of this light and projects it on the light-sensitive surface placed at the focal plane. The incident light energy here sensitizes a film emulsion or activates the phosphors of a vidicon tube. Consequently, the greater the light-gathering capacity of the lens, the greater the energy available for image formation; hence, less time is required to obtain sufficient energy for sensitization.

b. Aperture. The capacity of a lens to gather light is indicated by its greatest possible aperture, or minimum f value. However, the light gathered by a particular lens may be decreased by changing its aperture. Sets, such as the one numbered 64, 32, 16, 11, 8, 5.6, 4.5, 3.2, 2.8, 2.1, 1.8, 1.4 or some subset thereof, will be marked on the lens housing; the apertures of each set normally represent fixed steps for each of which the light incident upon the film doubles (fig. 15.2).

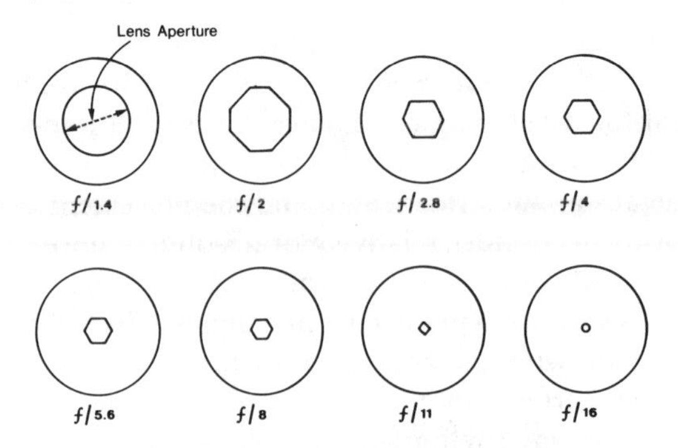

15.2. Relationship between lens opening (referred to as f-stop or aperture setting) and the amount of light reaching the focal plane (that at which the film should lie to obtain sharp images). The set of f-stops is normally inversely correlated to exposure time. Thus, each increase in stop value reduces the incident light by one half; hence, the shutter must remain open twice as long.

c. Reflection. The direction toward which light is reflected by an object forms an important aspect of the available level of energy. Simple experiments with a mirror, which provides almost perfect reflection with relatively little lateral scattering, shows that the closer the image axis of the camera to the direction of incident light, the greater the portion of reflected light that will be available to sensitize the film; this is one of the reasons why still photographers are urged to take pictures with the sun at their back. (Direct light incident on the lens will also increase the light/dark contrast, disturbing light meter readings and causing glare and refraction at the light/dark transition line.) The experiments with the mirror also show that the amount of light reflected per unit of real area is less for surfaces facing sideways than for those at right angles to the light path. Resolution of detail in surfaces facing obliquely may require additional illumination or lateral reflectors, which give these zones equal or more light per real area.

d. Exposure. Over a limited range of light, the sensitization reflects only the absolute amount of incident energy; consequently, a single stop increase in aperture should permit a halving of the exposure time. However, this reciprocity between aperture and exposure time may not apply for very short, intense pulses of light; similar reciprocity failure may also occur for particular emulsions and wavelengths.

e. Resolution. The amount of information that may be withdrawn from an image depends on its sharpness or resolution, which reflects the scatter of light and is calibrated as the closest distance over which adjacent lines may be separated. The resolution of a lens ideally should be constant across the field. The lens must also provide constant magnification so that the representation of an object on the film plane will occupy an area of equal size and shape, regardless of where it lies. (This rule is obviously broken in the design of so-called fish-eye lenses, which permit photographs over an extremely wide field). The sharpness of the image will also depend on grain size of the light-absorbing material. Hence, the larger the film format and the slower the film, the better the resolution. Movements of the image on the film plane during exposure obviously affect sharpness. Hence, mount the camera on a heavy tripod and reduce vibrations of the light beam on the image plane.

f. Focus. The resolution or sharpness of an image is affected by the ability of the lens simultaneously to focus light of different colors and wavelengths onto overlapping positions on the image plane. In modern compound lenses this is generally optimized by construction of the individual elements of different materials. However, focus represents a compromise and will differ for light of different wavelengths. Lenses tend to be calibrated for a mean value established for white light; the calibration value must be modified if the image is to be formed by monochromatic light, perhaps in the infrared or ultraviolet. Resolution also reflects variation in the refraction at glass/air and glass/glass surfaces, which tends to be dealt with by coating of the surfaces of the compound lenses.

g. Depth of Field. An image will be relatively sharp not only whenever the lens is focused at a single plane but for some distance away from this in both directions. This is the so-called depth of field, which increases as the aperture closes (f stop increases). Depth of field varies with the kind of lens used, increasing from the telephoto to the wide-angle range (fig. 15.3). As depth of field is least with the lens wide open and at maximum magnification (telephoto setting), zoom lenses should be focused at this condition; then, they should be stopped down and the field area increased to the desired setting. Screw-on lenses, extension tubes, and bellows may affect the zone at which a lens is in focus. Hence, through-the-lens focusing and examination of the image with a secondary magnifier are desirable options.

3. Guide to Filming

Cine cameras are designed to record the movement of objects. They are also required to achieve sharp images. This generates an intrinsic technical conflict that must be faced in selection of equipment and materials. One must also be cognizant of the aesthetic conflict between the artistic aims of the professional and amateur cinematographer and the need of the experimentalist, whose primary hope is that the records will be informative and facile to analyze.

a. Panning. Cine cameras can be panned or rotated during filming, thus indicating the context of an object and its surroundings. Use this feature *only* if the panning feature can be regularly coupled to the path of the animal, as during tracking of a running horse. Even then do not change focus during filming; thus, try to have animal and camera move on parallel tracks (or let the animal run in a concentric circle around the camera position). Treadmills and activity wheels are similar options, although they introduce some artifact into locomotor behavior. Analysis will become easiest

15.3. Focus of the camera is affected by the f-stop. Whenever the light is focused on the image plane (*right*), change of aperture does not affect sharpness. However, the wider cone of incident light produced by opening the lens produces a wider region of uncertainty (*circle of confusion*); hence a given error in focusing provides a fuzzier image.

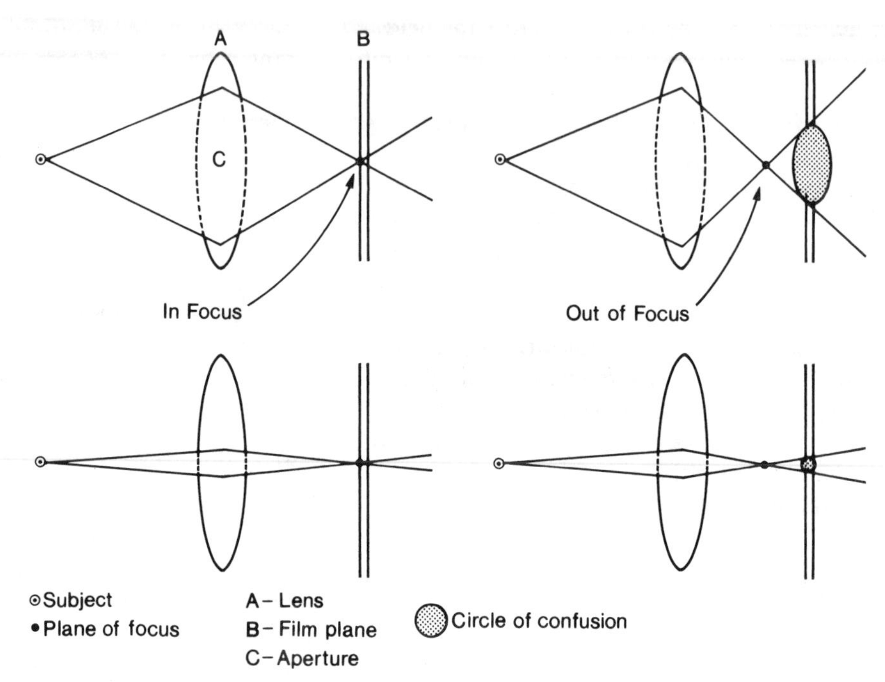

if the camera can be mounted on a solid tripod, with the pan head locked and the focus fixed.

b. Zoom Lenses. Zoom lenses permit change of image magnification. *Never* use this feature while the film is being exposed if the film must later be analyzed. Displacement calculations become extremely complex if the magnification keeps changing, even if the rate of change is regular. None of the putative benefits of the zoom justify the trouble they cause in analysis. Lock down the system!

c. Transport Mechanisms. For slow filming rates, the transport mechanisms required for a cine and a still camera need not differ much; however, even the standard, home-movie rate (24 frames per second) exceeds the capacity of motorized film advance in still cameras. The film must remain stationary and lie flat at time of exposure; also, it must move rapidly (without risk of rupture) between exposures. There is also the matter of register or of the constancy of distance over which the film is transported between frames. You can best estimate this constancy by viewing the width of the unexposed strips between frames. Irregularity may not pose much of a problem during recording but may, as noted below, generate substantial problems during analysis. It is mainly affected by the kind of transport device. In many cine cameras there will be two transport mechanisms, one moving the reels and the second, likely to be a sprocket-driven or pin-feed mechanism, advancing the film frame by frame past the shutter. Certain high-speed cameras never stop the film but use specialized devices such as rotating prisms, to form stationary images on moving film.

d. Shutter Speed. The shutter speed of a cine camera is affected by the filming rate. Although it cannot be slower, it can be much faster, in which case the aperture (e.g., the blades of a rotating shutter) is only open for a fraction of the time during which the film remains stationary. In such a moving slit shutter, portions of the image will then be exposed in sequence, as a slit-like beam sweeps across the image plane. This may pose problems in analysis as different portions of the resulting image may have been exposed at different times (certainly check the direction the slit travels). Recently, it has become possible to couple a stroboscope (high cycling–rate electronic flash tube) to the transport mechanism. The duration of each single flash, rather than the dwell of the film opposite the lens or the shutter speed, then defines the time during which the film is exposed. However, such flash tubes can be significant sources of electrical noise and require special shielding and circuit separation.

e. Filming Rate. Most critical for obtaining sharp images is the movement of the projected image relative to the stationary film, rather than the absolute velocity of the object. Hence, the closer the distance from an object moving at a constant speed to a particular lens or the greater the magnification of the object, the more rapid its apparent travel across the film plane. What should be the filming rate? If it is too slow, the projected image will have moved a significant distance across the emulsion during each exposure, leading to streaking and lack of temporal resolution. If it is too fast, there will be multiple frames with almost identical images, and the increased cost (in terms of film and time required for analysis) will not be matched by an increased benefit. There are equations that match the velocity of an object, plus distance and lens type, to emulsion type and filming rate. However, a relatively simple rule suggests the use of a trial film of relatively rapid filming rate and the preparation of displacement plots from alternate frames. As long as those deriving from odd- and even-numbered images do not differ significantly, it is possible to halve the filming rate without losing information.

Always remember that the cost of rapid filming extends far beyond that of the film. The faster the filming rate, the less the time required to expose a standard (50 or 100 foot) reel and the greater the need to predict the onset of the event being studied. Consequently, for high-speed photography of noncyclic or irregularly occurring events you may have to utilize an intrinsic triggering device that starts the camera once the animal engages in a stereotyped behavior (perhaps an intention movement), just prior to the event of interest. Remember that the time required to get the camera to full speed increases with the filming rate. Selection of the trigger may be complicated; for instance, when prey and capture activities are being studied there is substantial benefit to carrying out the experiments in isolation, so that the animal will not be distracted by movements of persons in the laboratory.

f. Lighting. The use of high-speed cameras does have the disadvantage that exposure of the rapidly moving film requires substantial illumination. Such high light levels and the resultant heat may disturb experimental animals. This is one reason why the relatively "cold" light of an electronic flash may be advantageous. However, some species of animals (certain fishes, frogs, and turtles) react markedly to the light emitted by a flash, and the behavior of others may become entrained in the rhythm of the flashing lights. This suggests some cautions as such lights may influence animal behavior. A simple circuit may drive the flash continuously even though the camera is still. Whenever the trigger is pressed, the shutter activation takes over the synchronization, but the animal is not disturbed by the sudden onset of the light.

g. Site Enhancement. Various simple tricks facilitate later analysis of images made with limited light exposure. Even though the coat or integumentary color of an animal may be dark, one may enhance specific sites by applying light-colored pigments that have increased reflectance. Even better are small spots of selected high-reflectance materials, such as silvered beads or colored reflecting tapes. Then there are tiny light-emitting diodes, which may be attached to the surface of the animal and which emit steady or pulsed light in several colors. If carefully attached to significant sites and matched to the properties of the recording system, such devices may speed up analysis.

h. Third Dimension. Information about the third dimension can be obtained in several ways. As long as the absolute size of an object is known, either its distance from the camera or the angle at which it lies can be determined by simple proportionality, but only if one of these variables is known. Alternatively, at least one additional view is needed to recognize all positions in space. This second image may be obtained by simultaneous filming with two cameras; however, analysis requires later synchronization of the records, which is, of course, easier with the "mixers" available for video systems than for film cameras. Remember that analysis will be easiest if the two views are at right angles to one another.

Perhaps the simplest way of obtaining two views of an animal is to insert a mirror into the field of view. A single (preferably a surface-silvered) mirror, placed above or below an animal being filmed from the side, may, for instance, let one establish the position of the animal on a treadmill and consequently let one determine the position of multiple sites in three dimensions. Theoretically, the image of more than a single mirror may be included in the visual field; however, remember that increase in the number of images will reduce the area available for each, which may be critical if resolution is limited. This also suggests that the overall image should be composed so as to reduce uninformative empty space and to fill most of the image plane with usable information.

The use of mirrors complicates focusing, as the light path from camera to mirror to animal is likely to be longer than the direct one from camera to animal. Hence, the two images are unlikely to be in focus simultaneously. If the light level is adequate (relative to the sensitivity of the film emulsion) so that the lens may remain stopped down and the depth of field adequate, one may focus the camera midway between the two positions and still obtain images sharp enough for analysis. A better approach is to equalize path length by artificially lengthening the "direct" lightpath (perhaps by the use of two mirrors). This has the further advantage that the magnification of the two images will be equivalent.

4. Choice of Medium

First, is is necessary to face questions regarding the relative costs and advantages of various kinds of equipment. There is the obvious choice between the ideal and the affordable.

Table 15.1 Relative Advantages of Film and Videotape for Motion Analysis

Property	Film	Videotape
Resolution	Relatively high, limited by emulsion, distortion of image	Limited by vidicon type, (No. of lines) relatively low, possibility of distortion by electrical circuits
Cost		
Camera	Low to moderate	Moderate to high
Medium (film or tape)	Low to medium (one use only)	Medium to high (reusable, hence equal to film)
Frame-by-frame analytical systems	Moderate, multiple formats	Relatively high, limited by model
Time till review	Days (depending on development facility), (20 min. for Polaroid film)	Minutes (immediately after filming has stopped)
Ability to combine images	Poor—requires complex lenses and mirrors or multiple cameras and special projection	Excellent—images may be combined, cameras taking views may be far apart
Amplification	Easily changed by modification of projection distance	Limited by screen size, electronic amplification complex and expensive

The first decision must obviously be between cinematic and video tape recording (table 15.1). Cinematic recording has the initial advantage of higher spatial resolution. In contrast, video recording has the advantage of its capacity to deal with very low light levels and to combine the images of several cameras on a single screen. Also, it is much easier to achieve instant playback, permitting a rapid check on the results of experiments. However, the Polaroid Company now offers "instant developing" Super-8 format cameras; it has recently marketed a medium-speed (300 frames per second) version that has good resolution and also provides projection-ready Super-8 film in less than 10 minutes. Equipment selection then depends very much on the nature of the application. One suspects that well-equipped laboratories must have at least limited capacity for both cine and video recording.

Standard cine cameras expose film at rates of 16 frames per second for silent and 24 frames per second for sound. Most cameras have special overrides that produce rates of 56 or 64 frames per second. Whatever the filming rate, it is essential that the camera incorporate a shutter contact that signals the instant that the shutter is open and that the image is being formed. This may have been designed for an electronic strobe synchronization or a mechanical device that permits an external indication of shutter state. Such devices should be coupled to an electrical signal generator that may produce a marking for the EMG tape and thus facilitate later synchronization.

Higher filming rates normally require so-called medium- or high-speed cameras, the former restricted to rates below 500 frames per second and the latter extending beyond 10,000 frames per second. Some of the better and more expensive models are driven by microprocessors so that one may dial particular rates.

Once the decision has been made between video tape recording and film, one may approach the question of film format. For most investigators this is reduced to the choice between 16 mm film and Super-8 film. The standard 8 mm film has long been outdated, and 35 mm is too expensive for all but the most complex projects or for those who have nearly unlimited funding. The advantage of 16 mm film is a much greater selection of cameras, lenses, analyzers, films, and emulsions. It is the medium of choice for much news and motion picture photography. Consequently, almost every country has facilities for developing and copying and the advice of professional editors.

Other things being constant, resolution will be limited by reduction of image size. However, the Super-8 format does have some compensating advantages for the experimentalist. Although 8 mm film involves a reduction to half of the linear dimension of 16 mm film, the sprocket holes of the Super-8 format have been redesigned so that it provides an image area some 60% of that of 16 mm film. The public's acceptance of the format has led to mass production of film and cameras of exceptional quality, as well as substantially reduced cost for film, processing, and filming and editing equipment. (However, the recent advent of home video is rapidly changing this market.)

C. CINEMATOGRAPHY

1. Types of Cine Systems

a. Sixteen Millimeter Cameras. There seems to be no maximum but only a minimum in the cost of 16 mm cameras, as the 16 mm film format is the industrial workhorse. Cameras come mounted on tripods or suitable for being held in hand, with spring-driven or electrical motors, a seeming infinity of lenses (fixed focus and zoom, the latter hand powered or remote controlled), vibration-inhibiting saddles, and magazines holding 100, 200, 400, or more feet of film. Lenses, films, editors, and analyzers are equally diverse.

Although 16 and 24 frames per second are the standard filming rates for silent and sound films, many cameras exceed these rates in various ways. Synchronization is critical for films that must later be analyzed. Most medium-speed cameras use sprocket drives (advancing the film a frame at a time), so that the film is in good register and remains immobile when the shutter is open. For high-speed cameras the time required to expose each image may become so short that the stop-start action may induce film rupture. In such systems there may be multiple drives reducing the strain on the film rolls; also the so-called Hi-Cam camera utilizes a rotating prism assembly, the light beam being truncated, while the film advances continuously.

b. Super-8 Cameras. The Super-8 film comes packaged in cassettes that can cue most cameras to the exposure factors required and that can be reloaded rapidly. Standard Super-8 cameras tend to have an electrical motor drive and zoom lenses, whereas others have macro features allowing closeup photography. Two filming speeds are common (16 and 24 mm), and a fast speed (56 or 60 mm) is often included. Genuinely high-speed Super-8 cameras are available at substantial cost. The Polaroid 300 frames per second, instant-development camera rapidly produces a developed Super-8 film record.

Many Super-8 cameras have automated exposure control that generally averages the light over the entire field and then sets the aperture accordingly. In most cases this metering is through the lens, although a few cameras use external light meters—for instance, in association with macro features, which may be affected by objects outside of the field of vision. Very light and very dark areas will affect exposure determination and may lead to underexposure or overexposure of the remaining area. Also check that the speed of the meter and aperture control is adequate to match the rate at which the incident light is likely to change during a typical experiment. The automatic exposure may lag a bit when the light changes.

Important considerations in selecting a Super-8 camera are the general sharpness of the lens at commonly used settings and the existence of a shutter sync for driving stroboscopes. A connector for a remote-control cable is also advantageous, both to eliminate vibration and to activate the camera jointly with other equipment. Do not bother with sound cameras as the extra features reduce reliability without adding benefits for the experimentalist.

Major disadvantages of the Super-8 format are the limited availability of emulsions and the poor copying facilities for Super-8 films. Copies generally yield much poorer color values and lower resolution than do copies from 16 mm films. Furthermore, there is substantial variation in the quality of service provided by vendors, so that it pays to shop about and find a company prepared to pay special attention to one's problems. In any case, it also pays to generate longer or repeated sequences on the relatively inexpensive Super-8 film, so that some original film can be dedicated to projection at meetings.

2. Projection Systems for Frame-by-Frame Analysis

Whenever experiments are designed to combine the use of multiple recording media (chart recorders, tape records, films, cinematography, video recording), all of which are likely to be run intermittently during a long session, it is useful to select one medium as the master and use it to record the times the other equipment is turned on and off. For example, written remarks on a chart or a voice track on tape or film can provide a continuous overview of the entire recording session.

It is often best to have the film copied prior to repeated analysis; this reduces the risk should the transport mechanism fail or other events occur that may scratch or chew up the most significant sequences (which incur highest risk as they are likely to be examined most often). Copying also forces you to consider the analytical equipment. For best resolution the film should run with the emulsion facing the lens; remember that processing laboratories return the film wound emulsion outward. Always ask to have the film copied onto stock with double sprocket holes, but note that your projector will determine whether it is to be reel-mounted or spooled on a plastic core. For splicing, use abrasion plus glue rather than the plastic tape kind. Editing and splicing of films should follow rather than precede copying. As the start and stop of sequences are critical, it is useful to maintain short sections of leader between sequences. The "fade in" approach had best be left to home-movie artists.

As you start analysis you will also encounter the merits of three procedures that should have taken place during filming. The first is the inclusion of a test grid of rectangles of known size. This should always be filmed briefly at the start of each sequence. The second, which may be superimposed upon the first, is the record number matching the film to the tape on which EMGs and other analog signals are stored (this image had best include date and camera settings). The third is a baseline of standard length and orientation, somewhere in each field. The last item is particularly important whenever one uses zoom lenses; it is so tempting to zoom in onto a single site without any indication of the place of adjacent objects. Finally, it is useful to check that the interframe interval is constant; the included baseline permits one to compensate for shifted images which usually suggest that the camera may need repairs.

The images of a film must be projected in order to analyze the sequential movement of particular points. Projection may be forward onto a plain white surface or backward onto ground glass. A 30 × 30 cm sheet of frosted glass mounted over a 45° mirror represents a remarkably simple and inexpensive analytical tool (see fig. 21.4). Specific points may be identified by marking them automatically or by hand on successive frames. It is advantageous to use acetate or other transparent overlays; these protect the screen and facilitate later comparisons among multiple runs, if you have taken the trouble to film and trace a baseline onto your records. Note that the distance between projector and screen will affect the magnification, and you should fix it by bolting down both projector and image surface. In setting up the projector, always have the centered light axis normal to the projection field.

Individual frames of cine films may be projected by any number of devices. One inexpensive version may be produced by simple modifications (masking) of 35 mm slide or by film strip projectors. Some of the now outmoded manual slide projectors have excellent optics and provide more than adequate resolution, even for the small area represented by Super-8 film. Whatever the approach, it is useful to project more than the minimal image area; the edges of the film can then be color-coded by hand or perforated to facilitate recognition of frame numbers. Commercial frame-by-frame projectors are obviously easier, and multiple models now on the market can handle the several formats. Many of them are modeled on standard cine projectors but provide special transport mechanisms and filters or blowers to keep the film in the gate from overheating when it is stopped.

Any frame-by-frame projection system should also incorporate variable speed and reverse projection. Whereas this is not particularly useful for EMG correlation, variable-speed projection may permit recognition of motor phases that are not otherwise obvious, and thus provide preliminary indication of phenomena of possible importance in future correlations.

Rather than deal with specific models, let us review some principles of variable-speed projectors. What is the framing rate? Does the image start to flicker as the projector slows down below 8 frames per second? Does the projector provide adequate light levels when stopped, and is the cooling system adequate to keep the film from buckling or does it require frequent refocusing? Some systems transport the film horizontally and pass the light downward through it by an array of mirrors, thus separating the film from the hot light source. Is there adequate protection for the reeled film (the heated air leaving the bulb of some projectors impacts directly on the takeup reel)? Does the single-frame feature advance the film a frame at a time or does it give an unregistered boost that results in a short but variable advance? Is the mechanism relatively tolerant of splices, neither tearing these nor causing the film to jump sprockets? One high-priced model is notoriously sensitive in the latter regard, so take a spliced film along when selecting such equipment. How good is the register (placement of successive images) in forward projection? Does it change when the film is reversed?

A remote control switch that lets the projector be operated from the screen or a digitizing pad is an extremely advantageous feature. A frame counter is an absolutely essential item. An expensive but useful innovation is to have this counter located on the remote-control switch. It is important to check that the frame counter system really provides a good indication of the absolute position and does not just note the number of times the single-frame button has been pushed. In some systems the counter jumps multiple numbers when a splice runs through; this may markedly affect count accuracy. Particular care is necessary whenever the film is to be run back and forth repeatedly.

Finally, there are systems that project the film onto the back of a frosted sheet of glass and that incorporate a cursor system that may provide analog or digital readout of the points on the surface. This is an expensive but highly desirable option because the string of values can generally be stored automatically—for instance, in a microcomputer—and it is then easy to generate derivative information (see the discussion of cinefluoroscopy, below). Whenever you are using such automated analysis it is critical that you keep checking the projector registry (between the forward and reverse directions). Also remember that most such systems have a fixed projection distance, and this will establish the relation between the size of the object on the film and that on the screen. Will the resulting magnification be sufficient for all envisioned applications? Many kinds of graphics tablets now on the market will accept forward- and reverse-projected images from an existing projector system. This potentially allows variation of the distance from projector to screen and, with this, digitization of segments rather than of an entire image. These segments then may be combined electronically, thus markedly increasing the resolution of the resulting record. (Remember that the beam of the projected image must still intersect the digitizing surface at a right angle.)

D. VIDEO TECHNIQUES

1. Cameras and Image Formation

a. Concepts. There are certain similarities between video and film cameras, particularly regarding the basic optics of lenses, image size, depth of field, and illumination. With a film camera the temporal resolution is simply the frame rate, and the spatial resolution is simply the grain size of the emulsion compared with the frame size. However, the process by which the video camera image is scanned and turned into a series of intensity values has complex implications for the picture quality that can be recorded and reproduced.

The lens system focuses the image on a light-sensitive screen analogous to the film in a camera. In most video cameras this screen is a coating that emits electrons in proportion to the light that hits it (the inverse of a television picture tube). A scanning beam of electrons detects the local densities of these emitted electrons, and the amplified values eventually become the picture information in the video signal. Recently, solid-state sensors made up of arrays of light-detecting photodiodes have become commercially available. This technology has great promise for eliminating some of the sources of distortion that plague the scientific use of video cameras, but currently available devices are generally not as good as the better vidicon vacuum tubes. The discussion here concentrates on black-and-white tubes, with occasional comparison of solid-state and color technology.

The most serious limitation of video systems tends to be temporal resolution. This is somewhat surprising, as the 60 images per second of the North American video format seems considerably higher than the usual home-movie frame rate of 16 or 24 frames per second. However, each film image is a complete and independent picture, with no blurring except for motion occurring during the shutter period. One can virtually eliminate even the latter by delivering all of the light flux needed to expose each image in a brief burst using a strobe light. In a video camera the light-sensitive coating is continuously collecting and bleeding off electrons, with a time constant that depends mainly on the material used in the coating itself. Even the fastest coating used in the expensive plumbicon tubes (called thus because the coating is lead based) has a time constant somewhat longer than the time between successive images. Commonly available silicon-based tubes tend to persist for three to five images, so even a strobe-illuminated scene will appear to be composed of several distinct ghosts, each somewhat dimmer and further back in time than the current image. Without strobing, fast movements by small objects (such as markers on the skin) will disappear entirely because the object image has not been located on a particular part of the screen long enough to have affected significantly the total light flux over the period during which the image accumulated.

With strobing, fast movements may be separated by identification of the leading ghost, but moderate-speed movements will not result in adequate image separation, so the partially overlapping ghosts will not be distinguishable. In theory, solid-state tubes should not have this property; however, the commercially available ones that we have tried have a quite similar lag that is apparently the result of some electronic trickery to improve the light sensitivity by integrating over a longer period. For a number of complex reasons, color cameras always have more persistence and poorer spatial resolution as well as distinctly higher price tags.

Video equipment sales personnel are used to catering to nontechnical consumers and the special needs of broadcasters and designers of surveillance equipment. We have yet to meet a field person who could provide even the relevant specifications, let alone understand the implications of image persistence, scan linearity, and other video quirks important to specialized research applications. The best quality of these salespersons is their willingness to obtain and demonstrate on your premises a variety of pieces of equipment from the several manufacturers that they usually represent. This is crucial. There is no way you can evaluate a video picture without trying out the camera in the exact setting with the exact illumination in which you will be using it. Furthermore, you cannot properly compare cameras and their pictures without looking at the still image as it will be recorded and played back through the rest of your system. A cooperative subject going through the actual motion to be studied is the best bet. If it is too extravagant to arrange this, a very simple and telling test is to hang a meter stick like a pendulum in front of the camera and swing it at various amplitudes. The blurring of the various numbers and markings and the ability to resolve the leading and trailing edges will be most instructive.

The spatial resolution and the temporal resolution have a frequently overlooked interaction in the interlaced, raster scan system used by commercial television equipment. Each *frame* of the video image is made up of 525 scan lines, but only half of those scan lines (every other one) are actually scanned in each of the 60 images per second. Video recording generates only 30 complete frames each second; there are two interlaced image *fields* for each frame. This obviously has great implications for the playback of still images. If the still image consists of a single field, the image will be a rather grainy 213 lines of vertical resolution. If the still image consists of a complete frame, the image will be the overlay of two screen scans taken 16.7 msec apart. If you use a time-code generator with a video image inserter, you will need to know this characteristic of your playback system; otherwise, the least significant digits (frame or field count or milliseconds) may be unreadable or unchanging between images. Also, when you are dealing with fast movements, it is important to remember that the part of the picture at the top of each field was scanned at the beginning of the field scan time, whereas the activity seen in the bottom of a single-field image actually occurred up to 15 msec later. (This is equivalent to the moving slit exposure of some cine cameras; however, there it occurs at high speeds and tends to be less of a problem).

Cameras are rated in terms of horizontal resolution, which means the ability to resolve a vertically oriented grating with a certain number of lines filling the whole field of view. This translates into the rate at which the intensity signal can be modulated as it scans the image at 15,750 horizontal sweeps per second. A minimal value seems to be about 300 lines of resolution, and better quality black-and-white cameras will often resolve 500 lines or more. However, before putting much weight on this factor, frequently stressed in the sales brochure, you should be certain that the rest of your video system can handle the wide bandwidth signals implicit in high-resolution scanning. Frequently, nonprofessional tape recorders are limited to 300 to 350 lines of resolution, and a poorly designed set of connecting cables and terminations (see section D4), this chapter) can drop effective resolution even below this.

Finally, there is a set of special questions to consider prior to purchase. Must the unit be portable? If so, bulk and power consumption must be minimized, as the batteries must be portable as well. Alternatively, is the camera to be set up in the laboratory and rarely moved? In this case there is little reason for paying the cost (in price and performance) of miniaturization. How many cameras may be used simultaneously, and are electronic mixers available for combining their images onto a single screen? Does the mixer allow overlaying of the full image at reduced size (a sophisticated special effects generator), or does it combine portions of the images of two cameras without changing their magnification (a simple screen splitter, see fig. 15.4)?

Next are options regarding lenses. Standard C-mount lenses are commonly available. Special lenses integral to a camera are much less flexible. Least flexible are cameras that have only a standard, built-in lens system that cannot be removed. Whereas resolution is rarely a factor (being limited by the camera and recording medium), field of view and focus range can be quite important.

Less standard video recording components may have substantial advantages for particular applications. Examples are image-intensifying circuits and special vidicon tubes that substantially improve the sensitivity of cameras under low light intensities; some cameras may work under conditions that pose problems for unassisted vision. Other lens and tube combinations are sensitive to wavelengths of light far outside of the visible spectrum (even x-rays for fluoroscopy), so that one may record images in the zone invisible to humans (but not necessarily to particular animals). Note that many ordinary lenses selectively absorb infrared and ultraviolet light, so that special kinds of glass may have to be used if the system is to record at these wavelengths.

b. Operation. The scanning and reconstruction of the image by means of a detecting tube and a display tube with continuous surfaces are the sources of two important types of spatial error virtually unknown to film equipment. In both tubes the vertical and horizontal motions of the scanning electron beam are controlled by independent deflection circuits. These analog circuits are subject to errors of amplitude and linearity, and they are frequently adjustable in both the camera and the monitor. Thus, there is no guarantee that a square object will actually appear square or that it will be the same size or even have the same proportions in all parts of the image. Sophisticated cursor systems incorporating digital x-y position counting can overcome such errors in the picture tube when measurements from still images are being made (see section D2, Recording and playback, this chapter), but nothing can eliminate this problem if it occurred in the camera tube, and hence in the stored image information.

It is always advisable to place a test grid of equal squares of known size in the field of view of the camera at the beginning of taping of an experiment so that the extent of such errors will be known and any correction for aspect ratio or screen position–dependent magnification can be made later. For fixed-camera and subject orientations this will also permit correction for various parallax and

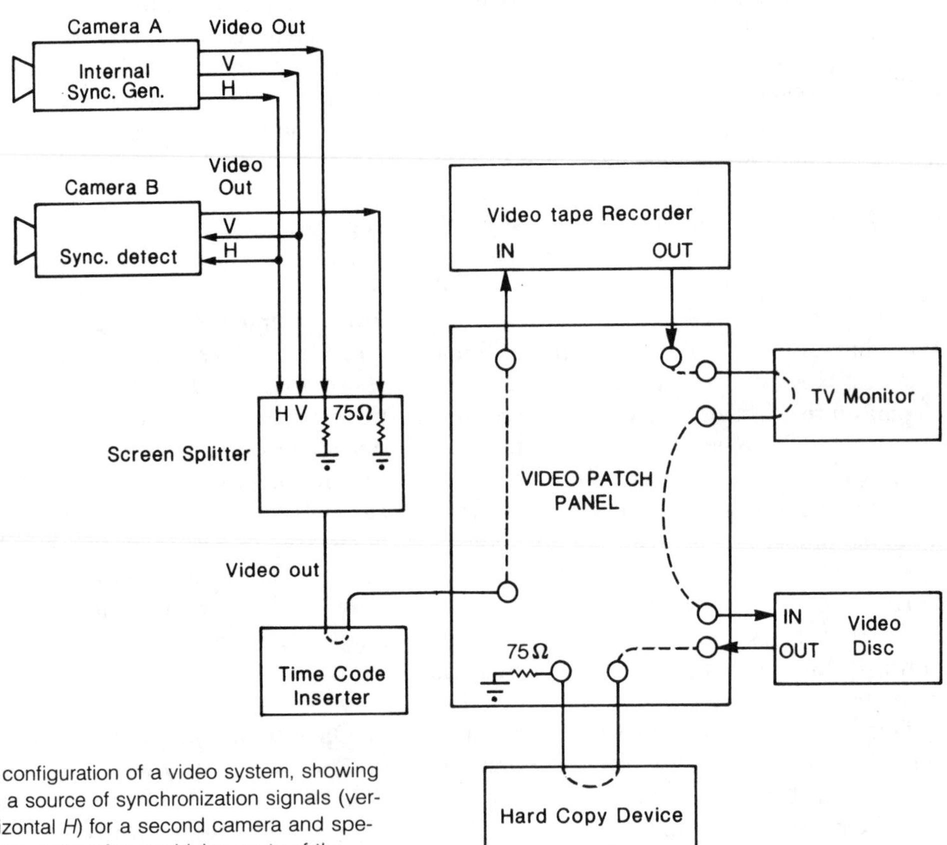

15.4. Typical configuration of a video system, showing one camera as a source of synchronization signals (vertical *V* and horizontal *H*) for a second camera and special effects screen splitter for combining parts of the two images. Note the strictly serial course of the video signals through various processing, storage, and display stages and the 75 Ω termination resistor at the end of the line.

foreshortening errors caused by lens type and camera position. Because of the limited spatial resolution of the video image, you will encounter the tendency to make optimal use of the image space by positioning the camera as closely as possible to your object, which has the side effect of magnifying such errors. It may be better to use a telephoto lens (and extension tubes) and avoid parallax; this also keeps your work space uncluttered by allowing you to position the cameras some distance away, although it increases the chance that something or someone will steal the scene during the recording session.

The illumination of videotaped scenes often requires more attention than that required for filming. The dynamic range of video cameras is quite limited, and frequently they employ automatic circuits to adjust the camera sensitivity to the mean light level of the image. If you deactivate these, you will find that you have to be quite careful to adjust the lighting and the lens iris to keep an optimal picture. Otherwise, the dark grays and blacks will merge, or the whites will wipe out local detail. This is because the entire range of intensity information must be encoded into a strictly limited range of +1 V for the whitest white and −0.3 V for the blackest black. Voltages above the positive limit are clipped off and voltages below the negative limit are used exclusively to transmit the synchronization pulses needed by the horizontal and vertical sweep circuits.

Unless you have a dedicated cathode ray tube monitor of the video composite signal to facilitate its adjustment, you will probably want to work with the automatic gain control of the camera turned on, but you will have to work around its little quirks. First, be sure to open the iris as far as you can—that is, until the brights in the picture start to "bloom" and cause white smudges with poorly defined edges. The automatic gain control will produce a reasonably illuminated picture with the iris much more closed, but the picture will be very grainy or snowy, as you will be forcing the camera to work near its noise level. Next, try to fill the image area with the object to be photographed or with a background of similar overall luminance. Frequently, one is tempted to improve image contrast by using a very dark or very light background. Although this will improve silhouette contrast, it will wipe out internal object detail. Remember that the automatic gain control circuit is hunting for the mean illumination in the picture. It can be forced into an inappropriate sensitivity by any large area of significantly brighter or darker luminance. A similar problem occurs in the display end (particularly with monochrome systems), because the net DC part of the video signal is often thrown away at various stages in its transmission, storage, and display. Thus, the pictures are always forced into the same mean luminance level, even if you really wanted a large dark or light background area.

Whenever multiple cameras with screen splitters (rather than image inserts) are used, these effects become very complex in their interactions. In such a case the whole field of view of each camera will affect its automatic gain control operation, but only the luminance of the part of its image selected for display will contribute to the mean luminance of the composite picture. Thus, changes in background that are actually out of the field of view of the recorded image can change the contrast and absolute illumination of all parts of the recorded image, even those parts coming from the camera with no change in its field of view.

Be careful of highly reflective or self-illuminated markers. When the photon density at a particular point on the camera coating reaches a critical level, the excess electrons generated spill over into adjacent areas, causing the bright blotches known as "blooms." The critical level for this phenomenon is not fixed, but depends on the overall sensitivity setting of the tube, as controlled by the automatic gain control or a manual override. For this reason, black spots on a relatively light background are often better resolved than the converse, assuming the coloration of the subject affords you the choice. Obviously, natural markings can be a great aid or a great hindrance, and you should consider this when selecting subjects such as cats and dogs. (Do note that human hair dyes work well on many animals.) When using illuminated markers or displays, particularly the light-emitting diodes that are so conveniently sized, remember that the spectral sensitivity of video cameras is not nearly so flat as that of our eyes or even conventional film emulsions. For some camera tube types bright red light-emitting diodes will be almost invisible, but for others they will bloom into impressionistic poppies. Filters may then be in order.

Video camera tubes have limited life spans. They should never be left on when unneeded, and the lens should be capped at all times to prevent premature aging, even with the power off. Plumbicon tubes are particularly fragile and should never be tipped face down (loose particles trapped inside can easily punch holes in the free-standing screen).

One useful trick for stroboscopic illumination is to make use of the similarity between the 60 field per second rate of some cameras and the 60 cycle per second power line. Even when the video system uses its own crystal time clock instead of power line frequency–locking circuits, the discrepancy will be less than 3%. Most inexpensive AC synchronous motors operate at exactly one half the line rate, or 1800 rpm. Such a motor can be used to rotate an opaque disc with two equidistant holes in its perimeter that permit the passage of light from a high-intensity spotlight behind the disc. This avoids the expense and acoustic noise of strobe lights and permits you to find an optimal trade-off for hole size relative to disc size, which determines the proportion of the light source that is used and the sharpness of its occurrence in time. If the video synchronization is not exactly time locked, there may be some phase drift across images, perhaps with some flicker, but this is usually not apparent because of the persistence of the camera screen. With low-persistence cameras, such as those using plumbicon tubes, special care must be taken to synchronize the strobe precisely during the interval between field scans in order to avoid breaks in the middle of the image.

2. Recording and Playback

There is a rapidly changing set of options regarding the recording of video signals. Although this has undoubtedly made video technology more accessible to the average investigator, it is worth remembering that almost none of this equipment was designed with the needs of experimental biologists in mind. Thus, it takes real diligence and discipline to figure out which of the many flashy features are actually useful, which are hindrances, and which are unnecessary expenses.

If you are going to use the video record simply to provide a general indication of the nature of the activity being performed, then almost any of the consumer-type video cassette systems will be suitable. When you are dealing with familiar, repetitive activities, particularly when recording analog signals from motion transducers as well, this kind of record provides a simple and invaluable way to keep the experiment "honest" and to verify what appear to be anomalous results. However, if you are going to be making detailed temporal and spatial measurements of body parts from successive video stills, things are much more complicated.

The best medium currently available is the video disc, best known to the televised sports fan as the "instant replay" device. These hold some brief segments, perhaps 10 to 40 seconds, in the form of individual magnetic tracks, each representing one video field (60 fields or tracks per second). Once the segment is recorded, it can be played back at any desired speed in either direction, including single-stepped fields that are completely clear and stable. Most such recorders do not have removable discs, so analysis of the stored sequence must be completed before another sequence is recorded over it. Most currently available systems are designed for professional broadcast use and tend to be quite expensive. Their most common use is to permit careful analysis of a short segment transferred from a long videotape.

Inexpensive discs intended for consumers and sports education were driven off the market by the new generation of tape cassettes featuring "stop acton." To appreciate their capabilities and limitations, one has to be familiar with the storage format used to write the picture information on a continuous magnetic tape. The record heads (usually one for each alternate field) are mounted on a spinning drum that is tilted with respect to the tape path. The tape is wrapped around at least half the circumference of the drum as it moves at a fairly modest speed, causing the successive fields to be written as a closely spaced set of parallel, oblique lines across the full width of the tape. Much of the mechanical complexity and many of the failure modes of video tape recorders come from the Rube Goldberg devices that pluck a loop of tape from the cassette and wind it around the complex path in the tape recorder.

The biggest problem in freezing a single field for display from a videotape is the fact that each field is only a few thousandths of an inch from its neighbors and there are no sprocket holes to align the recorded information to the display. The newer players use some fancy electronics and multiple, switched reproducing heads to generate a stable picture that does not have the "tear line" of interfield synchronization pulses located in the middle of the displayed image on the monitor. The best of these can step fairly reliably from one field to the next, but only in the forward direction. Many simply tick the capstan motor forward by some small, somewhat random increment, causing perhaps one to four fields to be skipped. Others read two adjacent fields simultaneously to give a full spatial resolution frame, with consequent temporal blurring.

Even the best currently available players occasionally stop with a tear line obscuring the middle of the displayed image. Whereas this may not seem catastrophic, the inability to step in reverse makes the analysis of any complete series of adjacent fields almost impossible. Remember that to reverse at all, most tape recorders must drop out of the play mode entirely, rewrap the cassette loop, and go into fast reverse. No matter how good you are at the controls, the field that is displayed when you get back into stop-action play is likely to be dozens if not hundreds of fields away from the field you were analyzing. Before buying any machine, try it out carefully, simulating the type of analysis you intend to pursue. Then look carefully into the quality and availability of service, because the fine overlap between optimal and adequate performance can ony be maintained in an exquisitely tuned machine. Also, remember that heavy use of a machine in the still mode is very stressful to both heads and tape, so use the best quality media available and clean the heads frequently.

As we said before, video technology is in a state of flux. Long-playing discs that can be recorded and stored in archives by the user are in the research and development stages now. High-speed digitization and massive electronic memories are on the verge of making digitized video images a practical medium. Prices will fall and quality will rise. In this sense, selecting a video system is much like selecting a computer: know exactly what you need now, buy only what you need now, and keep reading trade magazines and snooping around trade shows. Items such as cameras and monitors are somewhat more stable and can be considered longer term investments, the quality of which will persist and even become more apparent through successive generations of recording and playback equipment.

One stopgap measure that should not be overlooked is the market for used broadcast equipment. The need to stay competitive in the face of rapidly improving technology has caused much highly desirable equipment to be considered obsolete. Often the older devices with their larger tape formats employed both heavier construction and looser tolerances, making it possible to advance the tape manually or with analog capstan motor controls. A knowledgeable audiovisual engineer, often to be found in a university environment, may be willing to guide you through the somewhat more perilous unknowns of compatibility and condition. The engineer may also be able to advise whether the equipment can be serviced and whether and for how long parts are likely to remain available.

When procuring and maintaining video equipment, you should remember it is not just equipment that is in a state of flux. Distributors, salespersons, and service personnel change affiliations frequently and range from qualified engineers to ex–used car salesmen. As you are not likely to represent anyone's "big account," it is essential to develop a working relationship with one trusted supplier and obtain all of your purchases and service there. It is really worth your time to locate and consult with any other users of specialized video equipment in your area, such as hospitals, educational program designers, physical education departments, and other kinesiological researchers. Once you have made your choice, keep track of the key people with whom you interact and consider switching suppliers if they leave the particular establishment that has handled your business.

3. Measuring Images

Once you have a stable, still image on a video monitor, you are faced with the problem of making measurements. If you are looking for particular temporal events, such as times of foot lift and fall during walking, then you will want a reproduction system that can be easily stepped in forward and reverse (e.g., video disc, see above) and an image that includes an unambiguous time code (e.g., video character display, see below). If you need to make spatial measurements of limb position, the choice of equipment is less clear-cut.

The simplest approach is to place a piece of tracing paper over a television screen. This will only work if the picture tube face is actually exposed and relatively flat. Avoid monitors with a protective or antireflective window, because any distance between the paper and the phosphor will diffuse the image and cause parallax errors. It is a big help if the monitor is mounted with the display surface facing up in a work space, but be very careful about protecting the picture tube when you are not working with it; an imploding picture tube is an awesome safety hazard. Also, be aware that some playback machines may not lock onto each successive field in exactly the same vertical position on the display.

Video projection systems are becoming somewhat less costly, although most produce very large images with rather poor contrast, about the reverse of what you need. This is an area that may improve as video technology finds its way into more specialized applications, such as interactive instruction.

The biggest advantage of the video format is the ability to combine one image with other images, such as from another camera, a time-code character generator, or a cursor. There are two basic types of cursors, both of which use a hand-operated positioner such as a joystick to position a dot or crosshair in the actual image displayed. The simpler type takes the analog voltages from a two-axis controller and converts them into delays from the horizontal and vertical sync signals, adding a voltage pulse into the video signal at that instant to cause a white or black mark or line on the picture. The analog voltages from the joysticks can then be read out by a digital voltmeter and scaled to indicate the position of the cursor in x-y coordinates. The problem is that there is no guarantee that the nonlinearity of the deflection circuits in the monitor will match the nonlinearity of the potentiometers in the controller. It is mandatory that you quantify these nonlinearities and errors of aspect ratio by calibrating the cursor system against a display of a test pattern with known dimensions. Test all parts of the image area that you intend to use, because these errors are often much worse at edges and corners.

The better quality cursors digitize the analog voltages of the controller to generate a digital count of the x-y position and use these digital numbers as both the output of the coordinates and counts of the vertical and horizontal units of time at which the display marks will be generated. This guarantees that the cursor position is linear in terms of the recorded signal, regardless of the nonlinearity of the display. As nonlinearities and errors of aspect ratio in the camera scan are often minimal or can be well controlled, such a cursor may provide accurate enough x-y values without correction factors. However, the burden of proof is still on the user, and a test grid should always be available, at least to verify absolute factors of scale that are completely dependent on camera position. In addition to a numerical display, such cursors usually provide a parallel binary (or binary-coded decimal) output suitable for interfacing with a computer system, which is an enormous advantage if many numbers have to be tabulated. They allow one to store only those aspects of interest and limit the memory required.

If you have a computer with a video format graphics display, you may be able to synchronize its dot-addressable display with the video display from your single-field playback device. This allows the computer to generate the cursor, fill in complete stick figures, and perform any linearization or calibrations on the internal data stream. The question is whether to use the sync from the playback system or from the computer graphics, and the trick is to get the slave system to accept the external sync signals. This will probably require consultation with at least one of the engineering teams at the two factories responsible for your particular mix of equipment, as well as a special effects box to add the two images together without wiping out either one. It is definitely not a job for amateurs. Also, beware of computer graphics generators with margins that prevent pointing to the edges of an otherwise acceptable image.

4. Systems Engineering

Experience with audio frequency circuits such as are used in amplifying and processing EMG signals can make one a bit cocky about electronics in general. Once low-level audio signals have been properly shielded and amplified, little can go wrong as long as you deal with complete circuits and avoid short circuits. With radio frequency circuits in the megahertz band, all your instinctive assumptions are likely to be wrong. Quick-and-dirty connections and equipment casually strung together will rapidly degrade video signals to an unusable blur of ghosts, snow, limited dynamic range, and poor contrast.

The central point to remember is that all video equipment is engineered to operate with 75 Ω source and load impedances and that each box, each connection point, and each cable must be treated as if it constituted a discrete device, each with an input and an output. Certain types of coaxial cable are designed so that their internal inductances, resistances, and capacitances are balanced so as to look alike a 75 Ω load regardless of the length of cable employed. When properly terminated at both ends (both the central conductor and shield with a coaxial radio frequency connector, such as BNC or UHF type), such cables can be used to connect devices without signal degradation. Note, however, that not all coaxial cable has this impedance characteristic, and even some types that do may have high-loss factors due to intrinsic properties of the materials used in them. In particular, the small-gauge, flexible cable commonly available in physiological laboratories and used for its shielding properties is rarely suitable. Be sure to maintain a stock of a suitable cable type, such as RG-59, as well as the proper coaxial connectors and any special tools needed to make those connections reliably. (See fig. 15.4 for a typically configured video system.)

If each connection point has to maintain a constant 75 Ω input and output impedance, it should be obvious that you cannot create a T junction to send a video signal to two separate pieces of equipment. That is why each piece of video equipment always has two signal jacks, so that signals can be passed in a "loop-through" chain without changing the load at any point. Indeed, each piece of equipment with such a loop-through chain looks like an open circuit unless and until its second connector is connected to a cable. How do you terminate the line? Most such equipment will have a switch under the connector that indicates whether it constitutes a "Hi Z" or a "75 Ω" load. Whenever the piece of equipment constitutes the end of the line, the switch associated with its input connection should be set to the 75 Ω position, causing an internal 75 Ω resistor to be placed across the connector terminals. When a piece of equipment is in the middle of such a chain, it should have any such switches set to high impedance (Hi Z) positions to avoid loading the line. If a cable needs to be left unconnected at one end (e.g., in a patch panel), install a termination connector that temporarily puts a 75 Ω resistor between the core conductor and the shield conductor.

Many pieces of professional equipment require separate sync lines as well as picture signal lines. Such equipment usually works with a central, high-quality sync generator that keeps all the equipment locked together. In a research laboratory it is more common to derive sync from a source, such as a camera, during recording sessions or a playback device during signal analysis. Sync signals must still be routed among all of the pieces of equipment requiring them and should be treated just like video signals, with attention to terminations and avoidance of T connections. Pieces of equipment that can both generate and respond to external sync generally need some switch setting to go from one condition to the other, so you will need to keep track of where sync originates and where it goes.

Certain pieces of surveillance equipment, including some low-cost cameras and most

European equipment, work with sync standards that differ from those used by the American broadcast industry. Because a video system represents an ongoing investment in modular pieces of interchangeable and, hopefully, upgradable equipment, such bargains should be avoided. There are also some specialized high-resolution or high–field rate video systems available. Be sure they will do everything you need and that you can afford to be locked into a completely specialized set of high-cost components before you take this plunge.

Video equipment requires frequent inspection, cleaning, and adjustment to perform to the demanding requirements of scientific use and under demanding conditions such as constant replay, still image work, and animal fur in the air. You can perform some maintenance, such as head cleaning, once a service technician shows you how. Some adjustments such as sync level and picture dynamic range require modest equipment such as a video synchronizable oscilloscope, plus familiarity with video signal specifications. Some servicing, such as adjusting tape recorder drives and tape path, must be done by qualified service personnel. Ignoring obvious symptoms, such as unusual noise, diminishing picture quality, and dropouts, can lead to catastrophies, such as damaged head, torn tapes, and unplayable records, that are likely to show up at the least convenient time. Regular preventive maintenance and a good working relationship with a local service department are essential to protect what usually becomes a major investment.

E. CINEFLUOROSCOPY

Cinematography tells us what is happening on the outside of the animal; cinefluoroscopy utilizes x-rays to tell us how the internal elements shift during animal motion. Unfortunately, cinefluoroscopy is much more complex and expensive than cinematography. Also, x-rays are dangerous, and it is likely that a responsible person will have to devote full time to ensure that safety guidelines are maintained and that neither animal nor investigator receives an excessive exposure. As the equipment is too dangerous for offhand use, and even the rooms used have to be modified and shielded to protect people in adjacent spaces, we only present a general introduction to its utilization. There are, of course, multiple regulations controlling the installations and use of such equipment. Some simple guidelines for evaluating plans for prospective installations are as follows. (1) The lower the voltage of the source the less the potential hazard. (2) Rays are directional, so that the source should be shielded electrically or physically to generate radiation only toward the target. (3) Avoidance is better than shielding, so keep human parts out of the direct beam. (4) Remote control may have advantages (if the animal can be motivated to perform). (5) Sources and shields may deteriorate. Check regularly that the design conditions still apply. Insist that everyone wears film badges and that these be checked at regular intervals.

The principles of cinefluoroscopy are similar to those of photography, except that the image is formed by x-rays deriving from a more or less collimated source that pass through the object rather than being reflected by it. Therefore, the basic processes of radiology may be conceptualized as shadow casting. The object itself as well as the differential density of its intervening tissues produce differential shadows, and the detecting surface hence may form an image or overlapping images outlining peripheral and internal elements. The energy reaching any portion of the detecting surface is affected by the voltage of the x-ray source (positively), the atomic number of the intervening materials (negatively), and the density of such material (negatively). Detecting surfaces generally consist either of a film emulsion (which produces a permanent chemical change that must be fixed and made visible by development) or a fluorescent screen (which will glow at a rate reflecting the amount of energy that reaches it). Standard video records and cine films are generally taken off some kind of fluorescent screen; hence, they fit the second category.

The resolution of film is established by its grain size, and fine details about small structures may be perceived; however, this may demand a substantial exposure time per frame. In contrast, the image formed on phosphorus of the fluorescent screen reflects not only the grain size of the phosphor, but also the persistence of the local reaction. The glow continues and even a pulsed x-ray film will suffer some temporal blurring because of this effect.

As in any shadow casting, the sharpness of the image can be increased by reduction of the effective size of the x-ray source. The experimentalist may proceed with this by reducing the effective size of the x-ray target or by increasing the distance between tube and image. The latter process will have the additional advantage that the rays will be more nearly parallel, reducing parallax so that objects of equal size at different levels in the animal will appear more nearly of equal size in the final image. However, the approach increases the requirement for x-ray power and space and the cost of source and shielding.

X-rays incident upon a body may be reflected, pass through parts of it, or be absorbed. Reflection or scattering tends to be a nuisance as the scattered rays tend to fog the image and reduce contrast. The beam scatters on the way from source to film or screen (front scatter) and after passing the emulsion (back scatter). The effect of back scatter may be reduced by placement of the film on a sheet of lead. Front scatter is reduced by reduction of the beam diameter to a minimum. Toward this end, sources are equipped with a cone, and in extreme cases (vibrating) screens may be placed above the film so that they will pass only rays traveling approximately parallel to the line from source to the center of the plate. Both reflection and absorption reduce the fraction of the original energy that reaches any portion of the detecting surface. Their combined magnitude hence establishes the energy available for forming the image.

The screen may be photographed directly—for instance, by means of an integral 16 or 35 mm camera—or the signal may be enhanced electronically and recorded on a video tape circuit. In each case the resolution remains limited beyond the differences imposed by the slight density variants of the tissues being differentiated.

A variety of techniques permit improvement of these procedures. If only a limited number of images need to be taken at a rate of less than 10 per second, one can utilize various kinds of rapid film changers that yield a limited number of fairly sharp images. On the other hand, there are electronic image-enhancing processes that increase the contrast of x-ray films but at a very substantial cost. Much of this cost would seem to reflect the fact that the technology is generally intended for use on humans in hospitals; mass production of low–x-ray dosage, electronic image-enhancing systems (perhaps variants of those used for airport baggage inspection) carries the potential for drastic improvement of image at lowered cost.

More effective are procedures that enhance the contrast of particular tissues or internal locations. First establish the structures that are likely to move and those movements that are likely to be significant to the study. This determines the positions that should be enhanced. The most obvious enhancement techniques utilize implantation of coated pellets of lead (undesirable for any long-term use because of the toxicity should the coating fail) or of other inert metal (gold or platinum is just as radiopaque). Point sources in bone may be generated by dental techniques with use of implantable pins and amalgam fillings. Hollow spaces may be enhanced with barium emulsions. Tendons may be visualized with materials, such as palladium wire, that will have to be anchored to the soft tissues (remember that neither the operation nor the implanted device should have a significant effect on the resulting movement). The principles of kinematics and descriptive geometry again must be kept in mind during site selection, so that the three-dimensional movement of structures may be determined from the projection of enhanced points.

In complex situations one may wish to use two cinefluoroscopy systems to allow determination of the position of internal objects in three-dimensional space. However, each such system is likely to be extremely costly, and a combination of x-ray study and cinematography may offer an acceptable substitute. Whatever the approach, cinefluoroscopy is a technique that requires substantial technical and fiscal commitment. It requires more planning than we can outline in the framework of this book.

F. ON-LINE MOTION ANALYSIS SYSTEMS

1. Cautions

There are several high-technology approaches to automating the measurement of object location in two- and even three-dimensional space. Automated measuring equipment is generally expensive and usually unsuited to animal work because of its bulk or requirement for unobstructed, high-contrast images. However, it is worth keeping up with such technology because automated measuring equipment is becoming more available in human kinesiological research and because it is evolving rapidly and may someday save researchers much painstaking measurement by hand.

2. Reflective Marker Tracking

The simplest system uses bright markers on the points to be tracked and processes the picture so that only those brightest points remain. This usually requires special illumination and a reasonably homogeneous subject and background, often difficult to arrange in a physiology laboratory. With a single camera, the possibility of missing one mark, because it disappears behind a part of the subject or because an electrical cable dangles in front of it, is fairly high. Once a spot is missed, the computer-based analysis usually cannot figure out which one it is and produces meaningless results regarding the whole shape. More sophisticated systems with multiple cameras and more intelligent software are starting to become available, but the extremely high data rate involved in digitizing video images on-line to find their bright spots requires expensive, dedicated processors.

3. Light-Emitting Diode Tracking

The commercial systems of diode tracking (Selspot and its several competitors) are based on a switching system that serially illuminates individual light sources positioned on the subject. This simplifies the tracking problem because the sensor array (usually a solid-state photodiode array) knows for which spot it is looking, and special wavelengths can be used to avoid false reflections from ambient light. The major drawback is the mounting and cabling of the light emitters, which are unsuitable for many small experimental animals.

4. Polarographic Reflector Tracking

This clever idea requires the use of polarized light to illuminate the subject and attachment of polarized reflector strips onto the limb segments rather than at the joints. The orientation of the strip with respect to the reference axis can be monitored by the amount of light reflected, allowing the processor to reconstruct a stick figure by connecting up the free segments. We have not seen this system in laboratory use, but we can imagine problems with attachment sites, skin slippage, and uneven illumination and shadowing. These must obviously be considered in application.

5. Acoustic Detectors

Nonoptical position-detecting systems have been developed that use the transmission time of sound through air to determine the distance between a sound emitter and a sensitive microphone. In practice, an electrical spark gap is used to generate a very brief ultrasonic click that is picked up by one or more microphones for a determination of latency. Spatial resolution can be extremely precise (better than 1 part in 10,000), and various triangulation schemes are available for determining location of objects in three-dimensional space. Whenever multiple, separately triggered sound sources are multiplexed, the orientation of complex shapes can also be followed.

However, there is a limitation on temporal resolution because of the need to allow acoustical echoes from the first source to abate before the next is activated; this usually requires about 5 msec in a laboratory environment. Other problems that must be considered in experimental design are shadowing of the second source by body parts (usually requiring redundant detectors and sophisticated algorithms for picking the best signal) and electrical interference from the spark gap (which can be completely enclosed in an electrically, but not acoustically shielded case).

G. DISPLACEMENT SYNCHRONIZATION

1. Frame Counting

How are the EMG signals to be synchronized with the recorded displacements? Synchronization by counting of the image frames is more useful for film analysis than for video, in which time-code generators and image combination techniques are more facile. However, the principles are similar for both systems. Basically, frame counting consists of starting the camera after the the instrumentation tape (or chart recorder) is running and placing a single shutter synchronization signal on the tape (or paper) each time the film is exposed. By counting the markers one should be able to establish the actual position of a particular frame in a sequence and correlate it with the vagaries of a particular EMG signal, keeping in mind excitation-contraction coupling and other time delays (see the discussion of temporal correlation in chapter 16, section F).

Several cautions apply. Always make sure that the sequence filmed starts after the poorly exposed (overexposed) leader of the film has passed (this section also tends to be scratched in development). Always end the sequence before the film ends. Intersperse experimental sequences with blank or labeled intervals. If two "takes" are run together, the transition may not be immediately obvious. These simple precautions will allow you to confirm counts by counting both forward and backward and thus to check the position of any frame from two directions. In any case, the count can be duplicated as both the film frames and the frame markers provide an independent record.

The increased use of microprocessors facilitates automatic counting circuits. For instance, the EMG digitization (see chapter 18) may be triggered by a signal from the shutter synchronization pulse recorded on the tape. If this method is to be used, check regularly that the waveform of the frame markers is adequate to trigger the counting circuit and that there is no ringing or noise that might generate false counts. Strobe photocell arrays may become misaligned and are notorious collectors of noise, degrading the signal.

The use of microprocessors to analyze the EMG signals also makes it easy to store the results of a parallel analysis of the mechanical events derived from the cine records as a series of columns in a combined data matrix. For instance, each row of such a matrix might represent a numbered temporal interval and the values within it the summed attributes of the EMG signal for multiple muscles during this time frame. Film analysis can then be used to add supplementary information for one or more points to the end of each row.

Analyses of simple cine images easily provide information about displacements along two axes. Information about the filming rate or the time between images permits automated calculations of velocity and acceleration, either for the direct distance (angle) that the object moved between successive frames or as resolved along any pair of coordinate axes. Analytical programs may also be devised to provide information about the position and displacement of objects in three-dimensional space. Examples are the use of two or more images (from an included mirror), perhaps taken at 90° to one another. Also the change in the relative size of an object of known size lets one determine its angle to the plane on which the motion proceeded or its distance from the camera (but not both).

The positional information derived from the first film image is stored in the first line of the data table. The shutter pulse coincident with the first image presumably triggers the EMG counting circuit. The pulse of the second image terminates the first count and starts the second one; thus, the displacement values observed between these two images are the ones that will appear in the second row of the table, and these values, divided by the filming rate, will yield the velocity over the interval. In short, line 1 of the table contains only positional, but neither EMG nor velocity data.

In contrast, values of acceleration (second derivative of position) can only be determined after the second velocity (row 3) has been established and, like it, will be placed in row 3, even though they pertain to the intervals between rows 1 and 3. The frame interval for 24 frames per second is 42 msec, which is more than twice the maximum estimated value for slow excitation-contraction coupling. This makes it likely that the EMG signals seen during a particular interval generated some of the forces responsible for the accelerations then noted. Indeed, the EMG signals during the first half of such a temporal slice are likely responsible for the motor changes of the second half. However, the EMG signals of the second half will only exhibit effects during the following interval. Thus, such analyses exhibit some temporal lag that can be ameliorated but not eliminated by shorter sampling intervals. Also, each differentation step (two are needed to get from position to acceleration) greatly accentuates any noise in image position. Often the position data will need careful digital smoothing techniques to minimize this noise without attenuating or delaying the real data.

It is important that the critical portions of an animal can be recognized and their position in space determined in successive frames; this requirement may be solved by a sequence of supplementary techniques, such as the use of various kinds of mirrors and spot reflectors. Selection of such techniques may save substantial effort during the terminal analysis. Whatever the method of observation, it is essential to engage in a series of trial runs in which the elements to be checked are clearly defined, the possibility of enhancing sites in animals is considered, mirrors and other devices are installed, and the results are analyzed.

2. Time Codes

The most unambiguous way of keeping track of the time of a visual event is to include some kind of time display in the field of view, such as a stopwatch or a digital clock. The problem comes in relating such a record to a particular instant of a continuous analog signal on an instrumental tape recorder. Several specialized digital clocks are available that generate various time codes that can be recorded on a track of multichannel tape, such as IRIG-B code. Some of these codes can be read by eye from analog chart records made with even low-frequency pen galvanometers. Often the time-code generator has a translation mode that will reproduce the time display as the tape is played back. It may also provide a binary code to a computer that can use this information to start and stop digitization at points previously selected by review of the film or videotape.

Time-code generators have a way of becoming central to the recording and data analysis phases of the work, so they should be selected carefully. Generators used with video equipment frequently have a video signal loop-through that will insert a numerical display of the time, resolved down to individual fields or milliseconds. This can be much handier and more reliable than trying to keep the time display panel in view of the camera and worrying about the persistence of the camera blurring the rapidly changing display units. Some time-code generators will translate the analog-code signal, even if the playback of the tape is not at the same speed as the recording; others will not. Some time-code readers are very sensitive to the kinds of amplitude fluctuations and brief dropouts that occur in inexpensive AM tape recorders and edge tracks of instrumentation recorders; others are more tolerant. Careful reading of the brochures from the several manufacturers of such equipment is called for. Their sales personnel are used to working with users whose expertise does not extend to the sometimes arcane world of time calibration, so ask plenty of questions.

16 Analog Correlation Techniques

A. APPROACH

All of the imaging techniques described in the previous chapter provide us with a plethora of data regarding the position of many points on the surface of the animal, most of which are probably extraneous to the study. Many of the analytical procedures listed there serve mainly to winnow the useful from the chaff and to extract data related to velocities and forces of body parts significant to the study from the series of images that likely illustrate more than necessary.

In the present chapter we introduce different approaches intended to generate direct readings of particular displacements, forces, and pressures for correlation with the EMG signals. These will always be restrictive, measuring specific factors in particular axes and under defined conditions. Consequently, the recording method and site must be selected carefully to monitor items of relevance to the action of particular muscle groups and to the biology of the system under study. The tools available are multiple—some adapted from engineering, others from kinesiology and physiology. All must indicate physical responses within an animal without unduly disturbing it and without significantly affecting, much less inhibiting, its behavior. Several kinds of measurements may be used. The first kind establishes positions and dimensions, the second records forces directly, and the third notes the effects of forces in terms of indirect pressures, deformations, and sounds.

Force and stress (force per unit of area) are invisible and are measured by the kinds of deformations and strains (deformation per unit of distance) they induce in materials with known physical properties. Generally, a secondary calibration allows the results to be correlated back to the stresses and forces that presumably induced the deformation.

In correlating EMG activities with forces in moving animals, one must always take the total system into account. Much of the force transmitted across the forelimb bones of a galloping mammal is generated by the hindlimb muscles and may be stored as momentum in the trunk. Similarly, forces at the wrist may be generated by elbow and shoulder muscles. Such considerations become even more important in the analysis of two joint muscles and in systems in which muscles exert forces while being stretched by the action of others. In short, a moving animal does not operate by a series of independent actions, but the bridging muscles of linkages must be treated as elements in dynamic balance.

Perhaps the most critical consideration of direct-reading implanted devices is that they must deal with dynamic, rather than static situations. They must, furthermore, do this in an environment composed of mixtures and patches of liquids, gases, and solids, the architecture of which is extremely complex and temporally changing. The biological materials involved may behave plastically or show elastic responses that need not be linear; for instance, they may vary with the time course of load application. This makes it imperative that we understand the intrinsic properties of the sensors and the way these may be affected by their surroundings.

Measurement of force involves measurement of deformation, either static or dynamic. Energy is imposed upon tissues and work is done in deforming them. In static situations the system tends to proceed to equilibrium, whereas it tends not to achieve this if the situation is dynamic. In theory an entire chain of linked bones, cartilages, and tendons is loaded simultaneously as force is applied. However, even if all elements in the chain are identical, the end of the chain closest to the site of force application tends to deform first; if the chains are composed of links of differing stiffness, viscosity, and inertia, the system will have less predictable properties, and the actual deformation resulting from forces applied at a particular point will, in a dynamic system, be a function of the number and kind of elements in series.

Consequently, the deformation will have a time course, as will the force generated by the muscle fibers. As long as the force is generated faster than the system complies, the maximum values increase. In a dynamic situation the relaxation of the twitch and the action of antagonists are likely to overtake this process before an equilibrium is reached, so that peak values are attenuated (dampened) and the instability is amplified. Attenuation of force peaks during dynamic loads affects not only the time course at which the force rises at multiple sites within the system, but also the correlation of EMG with observed stresses; the peak stresses may be delayed substantially (particularly in complex systems) beyond the time estimated from excitation-contraction coupling by itself.

In testing any dynamic system, one has to determine the time constant of the sensors. This is normally defined as the time required for the output of the gauge to drop to 37% of its initial value, once the load is completely removed. Obviously, this value represents the maximal temporal response of a system at which performance is already seriously degraded.

Fortunately, we rarely see undampened oscillations in biological systems, except for sound-producing mechanisms and special cases, such as insect flight muscles. However, it is likely that many cyclical activities will occur near the natural oscillatory frequencies of the structures involved.

B. POSITION SENSING

1. Stride and Gait Patterns

There are no general electronic methods for determining the physical placement of the feet; visual inspection and measurement from film and videotape records are excruciatingly tedious. Furthermore, one usually wants rather high spatial resolution over extended physical distances of many gait cycles. One simple and often overlooked way to obtain such information is to generate tracks or spoors. Spread a long sheet of paper along the animal's path and dip its feet in ink or press them onto an ink pad (of a different color for each foot if you are uncertain about your qualifications as an Indian scout). The resultant track documents stride length. Alternately, cover large sheets of paper or cardboard with a film of soot or other deformable material such as talcum dust or plasticine. You will soon have an accurate, permanent record of the sites at which the feet were placed. This record is easily cross-checked with cine or video records taken at a distance, to provide the missing timing information. Some deformable materials such as plasticine will also provide an indicator of the force magnitude and force application pattern.

2. Footfall and Contact Timing

The exact timing of contact between a limb and a fixed point on a platform or manipulandum can usually be determined by a force or acceleration sensor. Keeping track of multiple contactors such as limbs on a continuous surface such as a treadmill or walkway is a more difficult problem. One solution is to use the conductive properties of the limb to transmit signals from a conductive coating on the surface in question. The scheme shown in figure 16.1 relies on the fact that a limb in contact with the signal source will pick up some signal through the conductivity of the skin touching it, and this signal will set up a voltage gradient along the axis of the limb on its way to a reference ground in the body core. Body segments not in contact at their distal extremity will show little or no such gradient. Differential amplification from two recording sites oriented longitudinally along the limb axis will pick up a signal whenever that limb touches the ground. If a high-frequency carrier signal forms the source, conventional EMG electrodes will pick it up. The specific signal may be filtered from the wideband signal, so that the EMG electrodes, implanted for the usual purposes, thus provide two outputs, each of which may be processed further.

3. Manipulanda

Many behavioral experiments require the animal to move a manipulandum or lever arm, sometimes against an external load on the device. The position of such a device is usually sensed by potentiometers attached to its points of rotation. High-quality, linear, wirewound potentiometers should be used and precautions should be taken to seal their working parts from dust and liquids. This kind of position sensing can be used in more than one axis by the principles of the draftsman's arm. Once such device is attached to the point to be monitored, either by grasping or physical attachment, the freely moving object can be tracked accurately and continuously with minimal loading.

C. BODY LENGTH AND POSITION TRANSDUCTION

1. Muscle Length and Joint Angle

So-called length gauges operate on the principle that a fixed volume of conductive fluid in a distensible tube will increase its resistance from end to end as the tube stretches and its cross-sectional area decreases. The general form of the relationship is parabolical, but such gauges often operate over near-linear ranges. The gauge itself is made out of thin-walled elastic tubing, such as silicone rubber, with metal electrodes sealed in each end. Two filling fluids often used are mercury and saline.

Mercury gauges work at very low DC resistances (below 10 Ω), requiring relatively heavy-gauge electrical leads but permitting simple DC bridge circuits to be used for the transduction. Because of the toxicity of mercury, such gauges are usually attached to tethering points outside the body.

Saline-filled gauges work at 5 to 50 kΩ and usually have stranded stainless steel leads as electrodes (figs. 16.2 and 16.3). However, they must be excited with AC currents to prevent electrolysis, and they require the use of AC bridge circuits, usually at frequencies above those likely to interfere with physiological signals. Because water diffuses rapidly through silicone rubber, these gauges must be kept submerged at all times (including the time of autoclaving) and are used only as im-

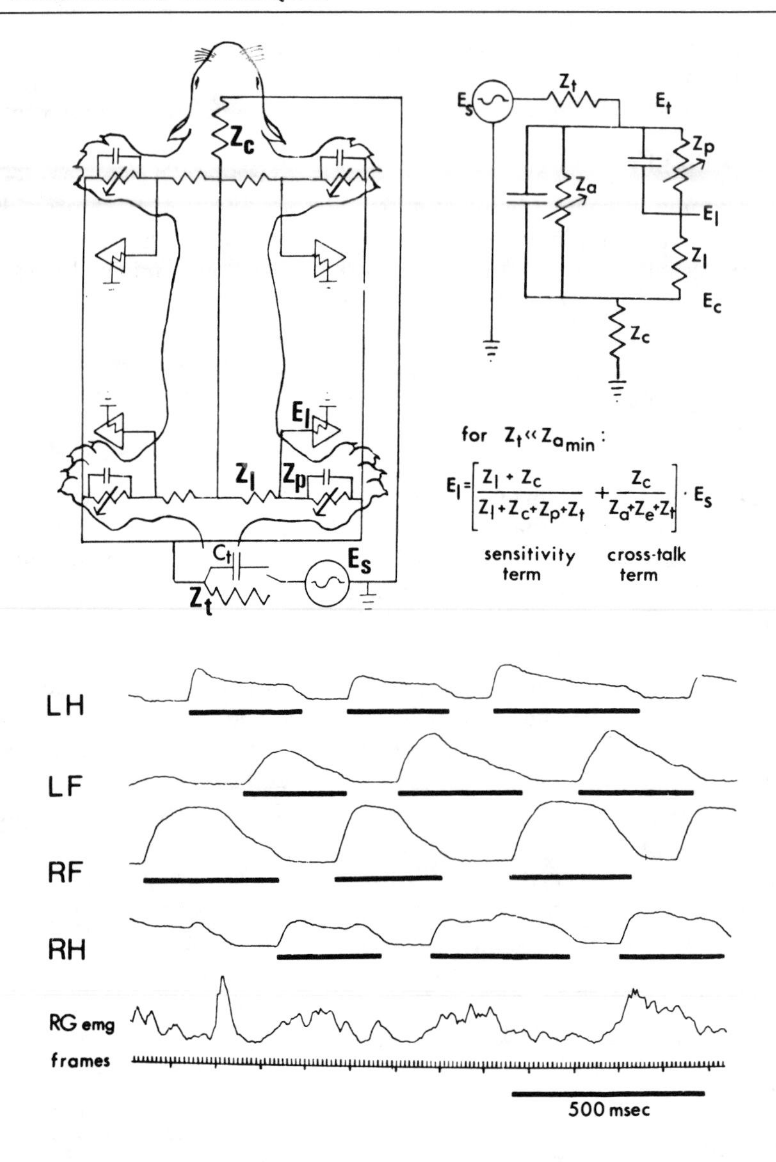

16.1. Footfall indicator. *Top left,* Schematized circuit for currents flowing through a rat from a conductive treadmill surface to a reference electrode in the animal. *Top right,* Detail of the circuit for a single limb only. The sensitivity term of the equation *(center)* specifies the change in amplitude of the carrier signal associated with the treadmill contact of the single limb in which the recording electrode is implanted. The term for cross-talk specifies the change in amplitude associated with position of the other three limbs that may contact the treadmill. *Bottom,* Sample record. Comparison of the footfall patterns as determined by the electronic method (*upper four analog traces*) and by film (*heavy bars under each trace*). Records shown are for left hindlimb (*LH*), left forelimb (*LF*), right forelimb (*RF*), and right hindlimb (*RH*). The fifth trace (*RG_{emg}*) shows the smoothed EMG of the right gastrocnemius. The bottom trace (*frames*) indicates the synchronization pulses for the movie frames (90 per second). (From Chapin, Loeb, and Woodward, 1980.)

16.2. Typical saline-filled length gauge. Note details of its construction (*top*) and typical installation (*bottom*). The Bergen wire rope used in each end is a commercially available, Teflon-insulated, stranded stainless steel composition. Care must be taken to avoid air bubbles and prevent leakage at the ends. The device must be drawn tightly enough at time of operation to prevent its becoming slack at minimal physiological joint angles while avoiding undue torque on the limb segments or excessive tension on the gauge itself over the physiological range of motion.

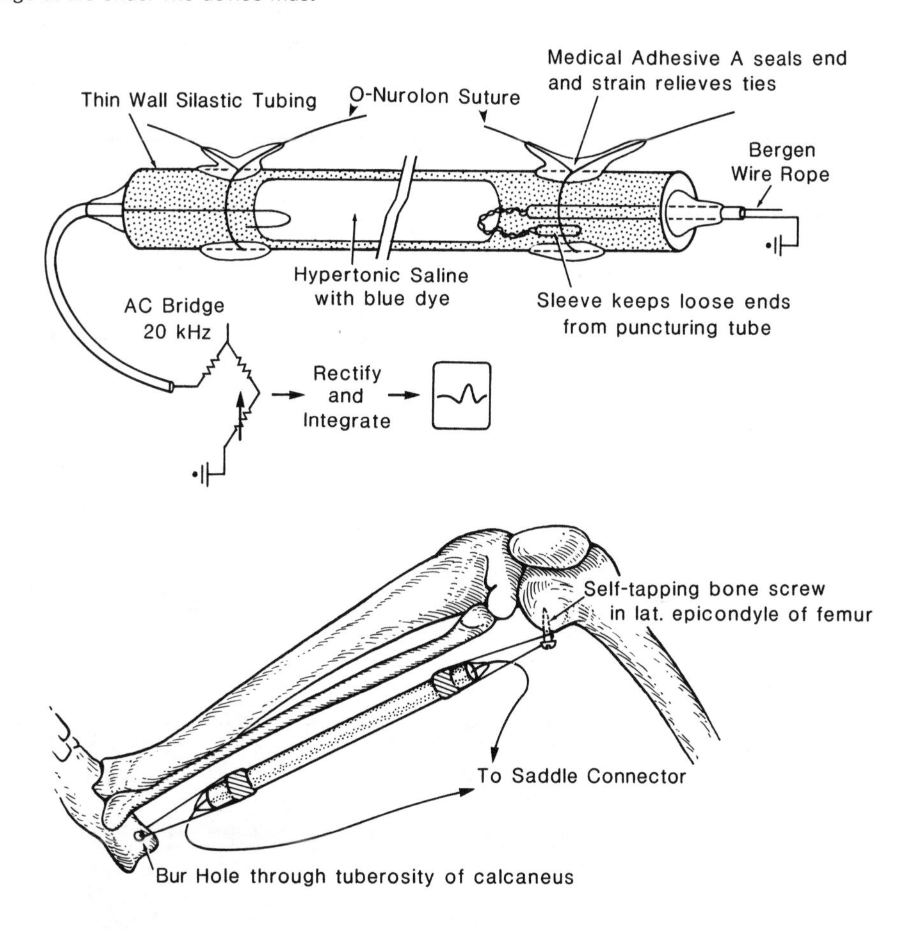

planted, biocompatible transducers. One useful trick, which takes advantage of the selective permeability of silicone rubber to water, is to fill the tubes with hypertonic saline (about two to four times isotonic concentration). The gauge will then pressurize to the point of hydrostatic equilibrium with the osmotic pressure; this reduces the chance of kinking or compression of the gauge as a result of lateral pressure from surrounding tissues. A bit of blue vegetable coloring in the saline solution facilitates visualization of the gauge during operation and at postmortem examination; it also provides the means to check on leakage during prolonged storage. Whenever stranded wires are used, be sure to include part of the stripped wire in the sealing plug of silicone adhesive; otherwise, the internal fluid may leak under the jacket and up the leads by capillary action.

2. Goniometers

A whole variety of physiological devices permits one to measure displacements directly in various ways. Many of them record the positional changes of particular elements; however, the signal they emit may not be rectilinearly correlated with displacement. Others use sonar or similar vibrational pulses. Some variants deserve special attention.

Impedance measurements are useful for correlating the state of distance or inflation between two measurement sites—for instance, the sides of the rib cage of a breathing animal. In this case both distance and mass between the indicators change so that a large, but mixed, record of change is provided.

A different kind of approach imposes a very high frequency signal field across the trunk of an animal. The response of the signal is picked up by the paired EMG electrodes in the same way as previously described for footfall indications. After separation from the EMG signal by differential filtering, it may provide an indicator of movement at specific sites that is effectively correlated with the action of the particular muscle recorded by the EMG electrodes. The signal measures the movement of the pickup site relative to the position of the source. The signal source should be placed relatively firmly; this may make the approach more applicable to recordings of large arthropods, which have a firm exoskeleton, than to recordings of vertebrates.

Ultrasonic piezoelectric emitter and detector materials, such as polyvinylfluoride, are beginning to be used for proximity and position detection in robotics and prosthetics. They can be used in pairs to detect the trans-

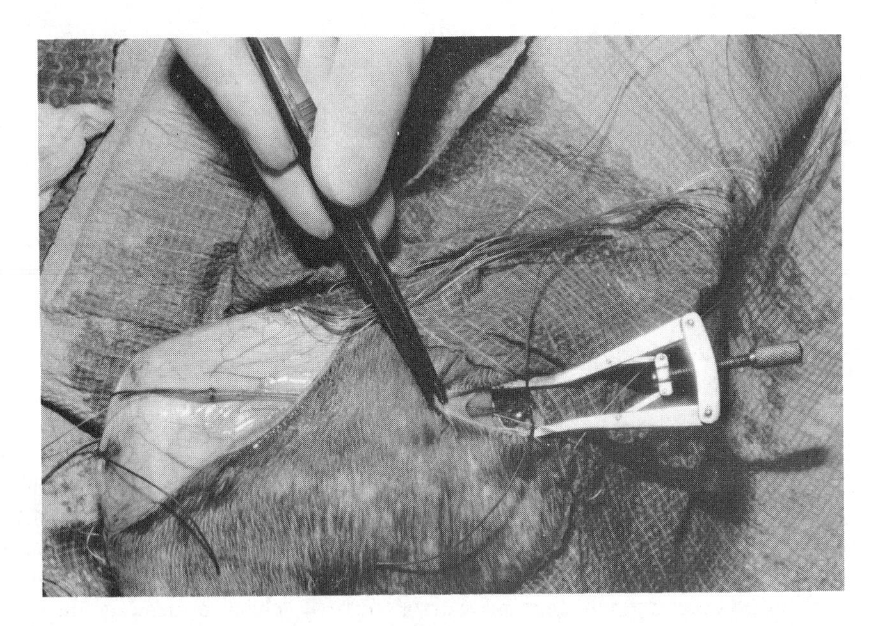

16.3. Surgical installation of a length gauge across the vasti muscles of a cat hindlimb, from a bone screw in the greater trochanter of the femur (*right*, between jaws of retractor) and tendon sutures through the patellar tendon (*left*).

mission time (and hence the distance) between ultrasonic source and detector across air- and fluid-filled spaces. It is likely that these will begin to find more common use in kinesiological measurement systems, particularly as increased power and efficiency improve immunity to noise and as the high-speed support electronics can be mass produced (see also chapter 15, section E).

3. Accelerometers

Another way of tracking displacement indirectly is by the use of accelerometers. It is critical that these be suitable for low-frequency detection. The high-frequency models used for vibration analysis often do not respond to the slow movements of interest in functional studies and tend to be useless for displacement analysis.

The sensitivity of accelerometers is apt to differ with the direction of movement. Some accelerometers are designed to be unidirectional; they are most sensitive for movement along a single axis (remember that movements along other axes will still produce variable signals at low levels). Placement of two such units adjacent to each other, but with crossed direction, may indicate movements more or less at right angles to each. Remember that an accelerometer will indicate not just the movement of the terminal element, but also the aggregate acceleration of the entire animal relative to the gravitational system. For instance, a set of accelerometers placed on the lower jaw of a goat records the movement of the jaw relative to the skull, plus the effect of movement of the entire head as this is flung in feeding; the accelerometers also will respond to accelerations caused by walking as well as any movements of the surface on which the animal stands. Hence, the outputs of multiple accelerometers may have to be compared to establish the movements of particular elements (fig. 16.4). However, acceleration is presumably what muscles induce by exerting forces; consequently, we may expect EMG signals of some muscular actions to be more closely related to the output of accelerometers than to the magnitudes of muscular displacement. Furthermore, accelerometers bypass the often unacceptable noise introduced by taking the second derivative of what appear to be clean displacement or position records.

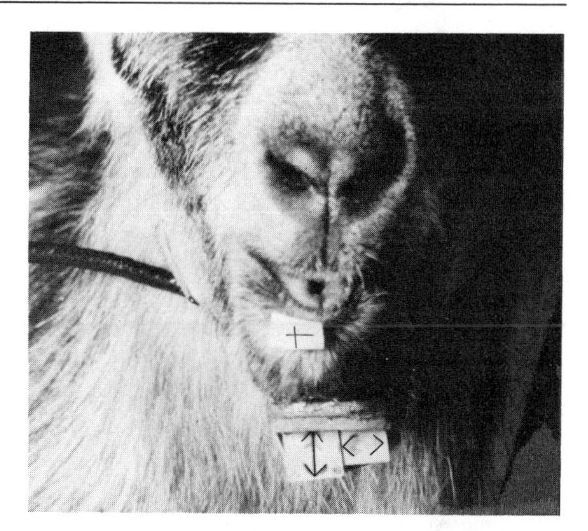

16.4. View of a goat bearing two accelerometers on its lower jaw (each responding, respectively, to horizontal and vertical movements). A patch of leather was glued to the shaven skin, and stout rubber bands attached to it held the accelerometers during mastication. (Courtesy Prof. F. de Vree)

D. FORCE TRANSDUCTION

1. Introduction

Most measurements of deformation utilize strain gauges (which are also integral to some accelerometers). Such gauges reflect the property that the resistance of certain materials changes as they are deformed. Various bridge circuits permit quantification of the resistance changes and, hence, strain values. Consequently, strain gauge circuits normally require a power supply and active amplification (fig. 16.5). Output voltage is best calibrated in units of force by application of known forces in the form of weights; this is usually performed in the absence of internally generated force. Care must be taken to keep track of gain and baseline values, both during experimentation and during calibration sessions. Most DC amplification devices have the problem that an increase of sensitivity will cause changes in baseline (zero force) output voltage. This imposes problems in calibration unless one has a modern system that incorporates reference voltages internally.

2. Force Gauges

A force gauge really consists of a highly elastic element that senses deformation (the strain gauge). This is bonded to a substrate so that the mechanical properties of the substrate determine the sensitivity of the whole to applied stresses and forces. Usually, the sensor responds to a given amount of stretching or bending by a small change in its electrical resistance, although piezoelectric elements ac-

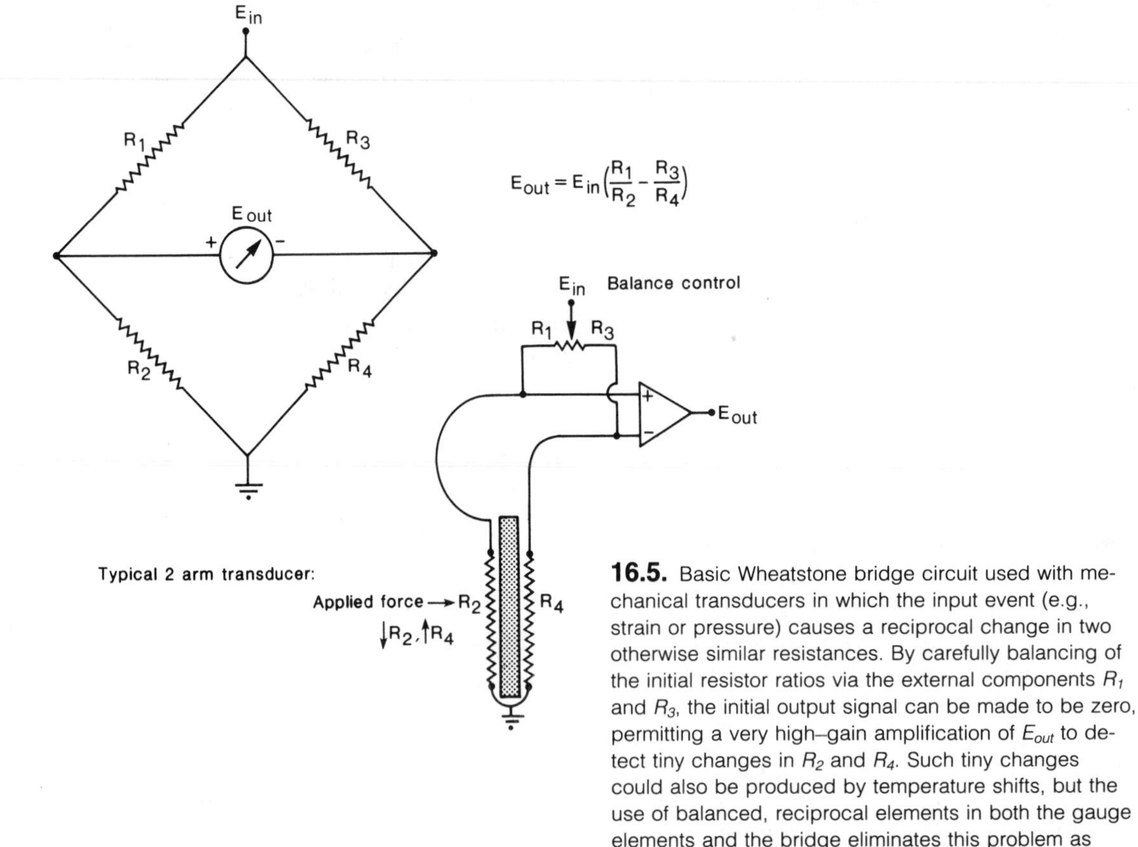

16.5. Basic Wheatstone bridge circuit used with mechanical transducers in which the input event (e.g., strain or pressure) causes a reciprocal change in two otherwise similar resistances. By carefully balancing of the initial resistor ratios via the external components R_1 and R_3, the initial output signal can be made to be zero, permitting a very high–gain amplification of E_{out} to detect tiny changes in R_2 and R_4. Such tiny changes could also be produced by temperature shifts, but the use of balanced, reciprocal elements in both the gauge elements and the bridge eliminates this problem as long as the temperature effects are similar in both members of each pair.

tually emit voltages. The sensitivity in terms of percentage of resistance or voltage change for a given amount of mechanical strain depends on the composition of the sensing element, and must be weighed against many properties including ruggedness, temperature sensitivity, and cost.

This is a complex and highly developed field of engineering, and most suppliers of sensor elements have well-documented instructions and explanations in their sales literature. Users frequently custom mount strain sensors on the structures that are to bear the stresses. Complete kits, including adhesives and encapsulants for particular substrate materials, are often available. The problem is relatively simple as long as the gauges may be bonded to dead or inorganic materials. They become more complex for gauges to be bonded to living substances, such as bone. The bond must then be with the surface of the intercellular matrix, as screws or pegs likely affect the stress distribution within the bone. Preparation of the surface generally involves skinning off of the periosteum and the surface cells. Even so, the bond lasts only days, barely long enough to permit the animal to overcome the effects of the operation (see also below).

There are many suppliers of bridge amplifiers, and chart recorders often include modules for what are essentially strain gauge arrays, such as blood pressure monitors. Once you know the input impedance and sensitivity of such an amplifier channel, you can replace the standard transducer head with a similar electrical configuration of elements on transducers of your own design. There are two common bridge circuits in use, two-arm and four-arm circuits. Strain elements are always used in identical but reciprocally mounted pairs to compensate temperature changes, to which the elements are usually quite sensitive. If your design does not have room for nor require four elements, you can still use a four-arm bridge by installing two precision, matched resistors equal to the sensor values in lieu of the two missing arms.

3. Implantable Tendon Force Gauges

Simple and surprisingly linear devices can be attached to intact tendons (fig. 16.6). This permits one to monitor continuously the tension transmitted by the tendon in a freely moving animal. The two shapes in common use are the belt buckle and the E type, as illustrated in figure 16.6. In both cases the exact shape and dimensions need to be carefully tailored to the tendon in question and the thickness and strength of the substrate (usually spring-type stainless steel) tailored to a sober calculation of the anticipated maximal forces. Be sure to pay particular attention both to strain relief and to the electrical insulation of the connection points to lead wires. The excitation voltage of the bridge circuit will be DC, so it will rapidly corrode exposed metal and will prove to be very irritating to the animal. Even the best designed gauges become unreliable after implant periods of more than 1 or 2 months because of reorganization of the chronically distorted connective tissue in the tendon. Occasionally, a gauge will constrict the cross-section of the tendon or bend it into a tortuous path; this is likely to lead to sudden tendon rupture, usually secondary to sharp edges or constricted blood supply.

Despite their complex modes of flexion, the output of E-shaped and belt buckle–type gauges are often surprising linear in terms of applied tendon tension. However, their absolute calibration depends entirely on the thickness of the inserted tendon and any deformation or reorganization that occurs over time. The gauges must be calibrated within 1 or 2 days of recordings at the most; this must proceed in situ by application of calibrated forces to the tendon at a point some distance from the gauge (e.g., by application of various weights via a suture to a cut end).

4. Bone and Carapace Strain

Strain gauges have recently been applied to the surface of bones, a method that induces a certain amount of damage to the tissue and permits relatively short-term recording. This method has been accorded substantial attention because it is likely to be affected directly by muscular action. These measurements rely on the fact that solid bodies loaded in various ways will deform and that the deformation

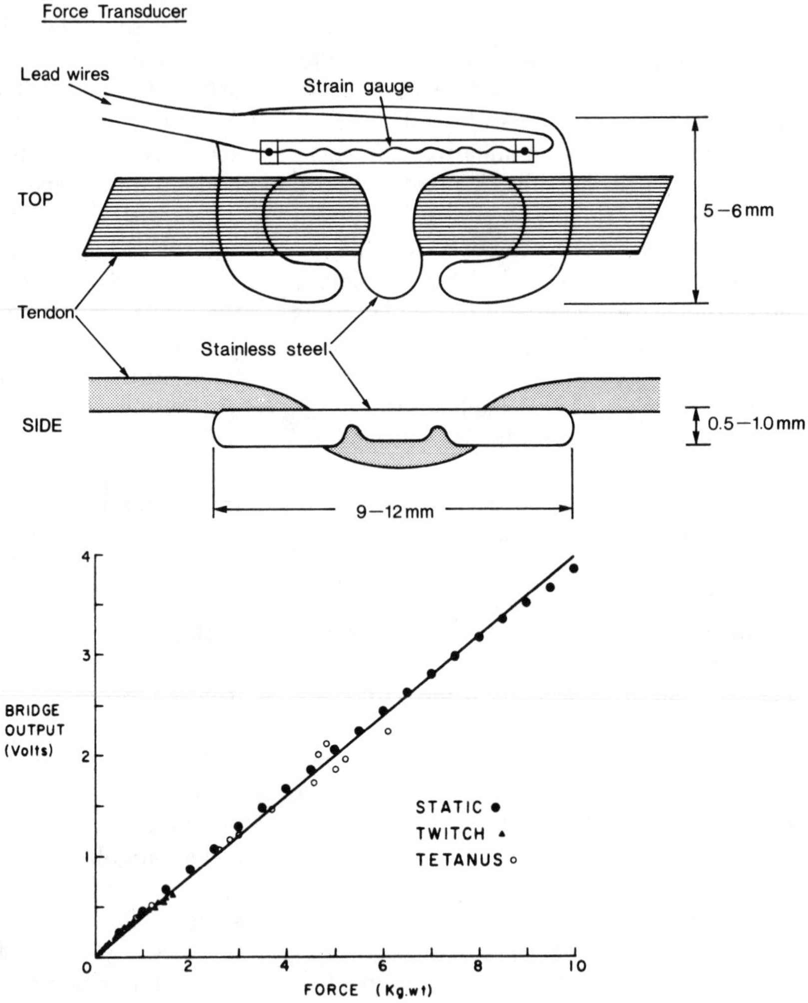

Force Transducer

Lead wires
Strain gauge
TOP
5–6 mm
Tendon
Stainless steel
SIDE
0.5–1.0 mm
9–12 mm

BRIDGE OUTPUT (Volts)
FORCE (Kg.wt)
STATIC •
TWITCH ▲
TETANUS ○

16.6. Tendon strain gauge showing course of tendon and placement of leads and strain sensing elements. Calibration curve in situ shows bridge output from a tendon strain gauge versus force output from a calibrated, conventional strain gauge for various hanging weights (static) and active isometric contractions of the medial gastrocnemius muscle of a cat. (From Walmsley, Hodgson, and Burke, 1978.)

measured on their surfaces in some ways reflects the distribution of stresses throughout the element. Most modern approaches use so-called strain gauge rosettes, which measure strain in three axes and thus allow determination of the maximal and minimal directional deformation. This allows determination of the axes of maximal and minimal deformation and of their magnitude.

Such gauges provide an indirect measure of load. First of all, they indicate only part of the tension generated by a muscle. In a dynamic situation the attenuation produced by the tendons of attachment may be significant. Also, even the smallest strain gauges and their attachments occupy a relatively large area on the surface of the bone so that they average the strains over a substantial part of the bony cortex. Furthermore, such gauges only measure the conditions within the local surface layers of an intrinsically nonhomogeneous system that is subject to strain from adjacent muscular attachments. The output of such gauges must, therefore, be calibrated against known forces applied to the bony insertion sites.

5. Force Plates

The ground forces of a locomoting animal often provide a great deal of information about the dynamics of the motor program. In conjunction with information about the mass and acceleration of body segments (the latter obtainable by taking the second derivative of the observed motion), the laws of Newtonian mechanics permit a complete reconstruction of the internal net torques operating around each joint.

There are several commercial devices commonly used in human kinesiology laboratories that simultaneously and automatically measure all or most of the three axes of ground force (vertical, longitudinal, and lateral). They may even calculate the center of pressure as a point on an extended platform surface (important to consider if the foot is not a point contact). Two such devices used together can keep track of the center of mass of the body and the distribution of weight on the two feet. Unfortunately, many such platforms are sized only for humans and tend to be quite expensive.

A single-axis force plate (usually for vertical force) is produced relatively easily by means of an appropriately machined and oriented support arm and two of the strain-sensing elements described above. Four such independent devices under the limbs of a standing quadruped can provide much useful information regarding postural set and response to perturbations. Homemade, multiaxis force plates are fraught with problems regarding linearity, compliance, and nonorthogonality (the tendency to pick up forces perpendicular to the axis supposedly being measured). A good machinist, careful design, and a lot of tinkering and calibration will be required if you try to build your own. Use of force plates in a free gait situation can be very frustrating. It is amazing how infrequently an animal will set one, and only one, limb down on a force plate of any chosen dimensions.

E. VOLUMETRIC SENSORS

1. Pressures

Several kinds of muscular activities may best be correlated with pressures. The muscles involved may act circumferentially upon spaces and affect the pressure of their fluid or gas contents. Multiple devices are available for pressure recording. The most expensive and currently most desirable are miniature pressure cells, normally placed in the ends of small tubes. These permit direct recordings of the pressure in fluid-filled spaces. The critical issues in selecting such cells for biological applications are the time required for the system to reach equilibrium in response to applied pressure, the system's response to changes in temperature, its resistance to moisture, and the possibility of hysteresis.

A less elegant but cheaper method is to connect the fluid-filled spaces to the pressure gauge by tubes of rubber, plastic, metal (particularly lead), and glass. The former two are flexible but deform in response to imposed loads; the latter two are stiffer but show minimal signal-dampening deformation. Whenever filled with gases, such systems also are subject to dynamic delays as a result of compression effects. Thus, it is advantageous to keep the volume of gas to a minimum. However, reduction of the diameter of gas-filled tubing has the major disadvantage that condensate may form in the tubes, dampening dynamic pressure waves even before intermittent plugs occur. Hence, the tube diameter should be at least 160 PE (polyethylene tubing).

Liquid-filled systems do not tend to show compression effects, although the often higher pressures may permit the tubes to deform; however, the system only avoids compression effects as long as the liquid-filled tubing does not contain gas bubbles. If EMG signals are to be correlated with values of an aerated liquid, it is best to fill the connecting tubing with a fluid, such as boiled water, that will have minimal dissolved gases.

One way of avoiding the bubble problem is to utilize closed, fluid-filled systems—for instance, small balloons at the end of a plastic

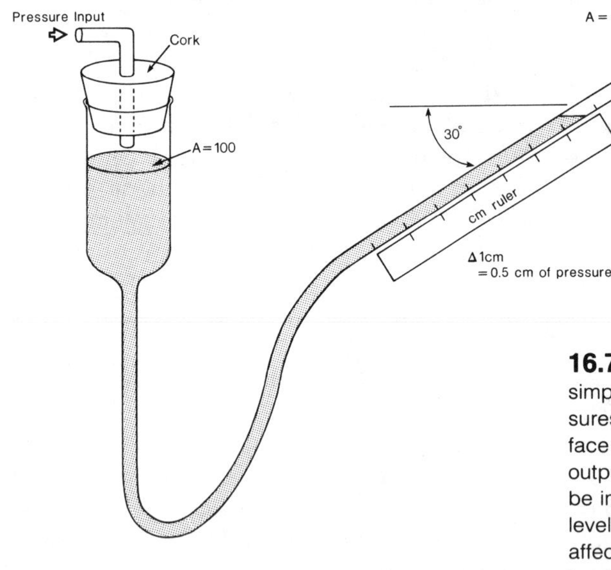

16.7. Bourdon tube is one example of a long-known, simple manometer for measuring slowly changing pressures in fluids. As the area (*A*) of the manometric surface on the input side is 100 or more times that of the output side, any depression of the input level is likely to be insignificant in comparison to the rise of the output level (the ratio may be modified as desired in order to affect the performance). Furthermore, as the output leg is arranged at 30°, 1 cm linear movement of the manometric fluid also corresponds to 0.5 cm of pressure, providing a doubling of reading sensitivity. Of course, the increase of manometer volume (and the diametric reduction of the manometric tube) also affects the response time, so such systems are best used for static calibrations.

or flexible metal tube. However, readings from these are indirect and likely to be affected by the shape of the space into which the balloon is placed. Assuming that the balloon is under some level of pressure, so that its surface deforms elastically, it will only provide an accurate indication of the pressure of the surrounding space as long as the space is also filled with a fluid. Mechanical deformation of the balloon caused by the movement of solids against its elastic surface is apt to increase the apparent pressure even though the real pressure does not change.

Static values of pressure recording systems can always be determined by connection of the gauge to a simple manometer. In this, the difference in level of the two legs of the recording fluids represents the pressure; however, slight differences may be difficult to read. Various special gauges, developed in engineering practice prior to the electronic age, ameliorate these difficulties. The Bourdon tube shown in fig. 16.7 represents a simple and useful example.

2. Flow

Two additional correlation systems perhaps deserve attention. The first are flowmeters, which can now be miniaturized remarkably. For gases, they commonly involve thermistors—small, heated units the resistance of which changes as a function of their temperature. If two such units are placed in series within a gas stream, the upstream one will shade the downstream one, and the differential readings of the thermistors can be calibrated for flow as long as the thermal properties of the gas do not change significantly (fig. 16.8). Another approach that is more expensive but has a better time constant consists of placing two pressure gauges at right angles to one another within the air stream. One measures the velocity pressure

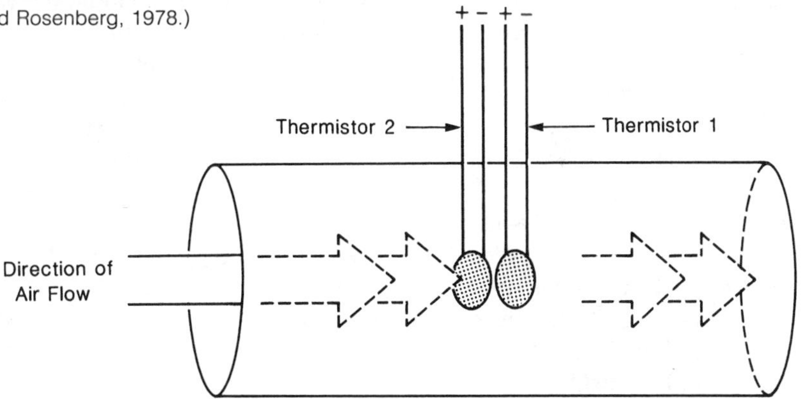

16.8. Use of thermistors to generate a flowmeter. Thermistors have a termperature-dependent resistivity and the current flowing through them generates heat. If two thermistors are placed in series within an air stream, the upstream thermistor will shield the downstream one and hence cool more rapidly. A simple electrical circuit will show zero resistance difference for zero flow and differential response as the flow increases in either direction. (Described in Clark, Gans, and Rosenberg, 1978.)

and the second the static pressure. The Bernoulli relations indicate that the difference between these readings then rises with increased velocity.

3. Sound

Finally, one notes that various flow and deformation events produce sounds that can be picked up with microphones. These sounds have the further advantage that they incorporate almost no time delay beyond that introduced by the excitation-contraction coupling of the muscle itself and that involved in the production and transmission of the sounds. Hence, closely placed small microphones provide obvious event recorders and are clearly the correlation tool of choice for muscular systems involved in sound production.

However, the sounds produced by muscle-controlled syringes and larynges are relatively far removed from the direct action of the musculature. They should not be used as synchronizing devices unless forces, pressures, and displacement associated therewith have already been independently determined. Here, and for many other modes of indirect recording, it may become of some interest to stimulate the various muscles as a first step in calibration of mechanical transducers. However, stimulation should not occur through the EMG recording electrodes, as this may involve other problems and uncertainties (chapter 8).

F. TEMPORAL CORRELATION

We have seen that EMGs may be correlated to each other or to various electrical or mechanical events. Thus the temporal relationships among data stored on the several channels of a tape or chart recorder will be critical. It is important to consider not only the time course of recorded events but also the time between their occurrence and the consequences they generate in the electromechanical record. This interval is apt to differ for each event being recorded.

The EMG represents a voltage measurement. If the several EMGs to be compared are handled by approximately equivalent preamplifiers, the raw signals will be recorded close to instantaneously (although many envelope detection schemes, such as integration, introduce considerable time delays). In chapter 15 we dealt with the relation of EMGs to images recorded at the time they are observed. Whereas calculation of velocity and acceleration involves delay, there need be little such for the establishment of position at the instant of image formation (although many transducers use bridge circuits, the integrating and filtering components of which impose potentially significant delays).

However, various delays do occur between the electrical event recorded as an EMG signal and the effect of the resultant muscular activity. First there is the excitation-contraction (E-C) coupling, meaning the time difference between the electrical events sweeping over the surface of the muscle fiber and the start of force generation and mechanical deformation. Obviously, there is the complication that not all parts of all fibers of a motor unit may be excited simultaneously. Also, there are the differences in E-C interval among motor units with different fiber types. Beyond this, one must be clear whether one defines the excitation-contraction interval as that between the appearance of the electrical event and the first indication of force production, or that between the former and the generation of maximum force by the fiber or some intermediate value, such as 80% of peak force.

Differential delays may become even more of a problem whenever EMGs are compared with the output of electromechanical transducers, such as strain gauges; these linkages may incorporate substantial delays, so that the EMG signal produced by the muscle appears on the records much earlier than does the signal of the transducer recording the mechanical event.

Although there is a theoretical basis for calculating the time delay for each record, it is better to measure it. Various simple devices allow one to make such direct comparisons. A signal may be imposed on multiple channels of EMGs by simultaneous shorting of all inputs or by deliberate production of large voltage artifacts as time markers. The delay times of EMG and pressure-detecting circuits may be compared, for instance, by attachment of the pressure transducer to the nozzle of a plastic syringe and placement of two EMG wires so that slight movement of the plunger will induce shorts. Suddenly unloading a strain gauge by cutting a string attached to a weight will produce a dampened ringing, the cycle time of which indicates mechanical resonance in the transducer. Such simple schemes allow check of the delay times within the electromechanical transducers. However, calibration of the transducer itself for delay and signal magnitude will not correct for linkage problems to a particular signal source (e.g. bubbles in a fluid pressure tube). Therefore it is important to introduce the calibrations as close as possible to the in situ source of the signal.

The complications caused by linkages may be explained by means of an example. Contraction of the intercostal muscles will be reflected in the pulmonary pressure. However, the force generated by muscular contraction must first stretch their intrinsic and extrinsic connective tissues, then move and deform the ribs rotating them about their articulations (and in the process deforming the cartilaginous surfaces). The change in the proportions of the rib cage affects the pulmonary surfaces and tends to change the volume occupied by the lung. This, in turn, modifies the pressure within the lung, the change depending on such factors as the degree of glottal opening and the compressibility and flow directions in the gas mixture being breathed and the site at which the pressure is being recorded.

The changes in gas pressure will theoretically be detected by the surface of the pressure gauge (which will have its own time constant). However, unless the gauge consists of a membrane directly exposed to the intrapulmonary space, it must be connected by tubing. If the bore of the tubing is large, the flow resistance will be minimal, but the tubing may then contain as much gas as does the lung, introducing a physiologically significant dead space. If the tubing diameter is narrower, the flow resistance is increased, in turn decreasing the rate at which pressure waves pass down them. Furthermore, accumulation of mucus and condensate may restrict the passages, and more work then will have to be done to overcome viscous and capillary forces. Whatever the diameter of the tubing, its walls will be deformed by pressure waves; this may introduce a significant delay if the tubing is formed of rubber or thin plastic.

All of these factors cumulatively attenuate the magnitude and delay the occurrence of peak strain levels. Consequently, systems are unlikely to achieve a mechanically steady state until after the electrical events in the muscle have decayed; this is often after reverse forces have started to act. All of these aspects emphasize that we should not expect that the EMG record and the mechanical products thereof will have equivalent delay times. They also emphasize the complexity, even of mechanically simple systems. It is best to measure the delay times in order to ensure that their magnitudes are equivalent to those calculated or assumed.

17 Signal Processing and Display

A. OVERVIEW

The rapid passage of signals across the screen of an oscilloscope can provide a quick impression of biological events. Although such signals may be scanned repeatedly, they only tend to give a very superficial impression of the nature of the events thus portrayed. For more detailed analysis it is best to stop the signal, converting it into a hard copy that may be analyzed at leisure and may be published or otherwise submitted to the scrutiny of colleagues.

The EMG signal is recorded as a relatively wideband AC signal, typically spanning the frequency bandwidth of 50 to 3000 Hz. However, much of the detail in the instantaneous ups and downs of the signal amplitude reflects only random processes of summation and occlusion of the thousands of individual, brief action potentials of the various muscle fibers. Also, the apparatus used for processing and generating hard copy displays of high-frequency AC signals is expensive and demanding, examples being oscillograph cameras, electrostatic chart recorders, ink jet recorders, and computers capable of high-speed digitization.

Frequently, the investigator is trying to interpret the EMG signal as an indicator of the mechanical force being generated by the muscle or of the net depolarization received by the spinal cord motoneuron pool. Both of those processes are much slower changing positive-going events that are more or less proportional to the envelope amplitude of the AC EMG signal. This chapter is concerned with various techniques for extracting and displaying just the low-frequency envelope information from the AC signal, a process called demodulation in the broadcast industry (see the discussion of demodulation, below). It includes brief descriptions of somewhat more exotic techniques for detecting and summarizing repeating patterns buried within the apparently random EMG signal. Before beginning a discussion of demodulation, we should review the characteristics of typical hard copy equipment found in the laboratory. After all, it is far more economical to obtain an integrator for your old, reliable chart recorder than to buy a new display system of initially unknown reliability.

B. HARD COPY SYSTEMS

1. Moving Pen Chart Recorders

Traditional chart recorders rely on magnetic coil galvanometers to deflect a stylus along one axis, as the paper is moved along another axis at right angles to this. The relative motion of stylus and paper leaves a trace generated by an ink pen on regular paper or by a thermal reaction between a heated tip and a special heat-sensitive paper. Depending on the linkage, the tip of the pen will move in a curve or a straight line. The former may make comparison among traces more difficult; however, some straight-line pens may deform the signal variably with changing amplitude. Frequency response is obviously limited by inertia and resonance of the stylus deflection system, typically to a high-frequency limit of about 100 Hz, even for servo-assisted ones. This may attenuate the signal so that the pen response will not be linear; thus, the peaks of high spikes will be attenuated more than those of low ones and very rapid events may disappear altogether. Furthermore, both ink and thermal marking systems suffer from fading or irregular trace darkness at the higher frequencies (if the density is adjusted to high rates, the wide trace tends to obscure small signals). One can improve the apparent response characteristics of a chart recorder by playing back a tape-recorded signal at a slower rate (assuming that this is permitted by the reproducing amplifiers). Halving the playback rate should double the signal rate that will be transcribed by a given recorder. Obviously, raw EMG signals cannot be applied directly to the chart recorder input (although some systems have built-in rectification and integration circuits similar to those described below). However, multichannel chart recorders offer an economical and photographically reproducible record of several simultaneous signals once these have been demodulated to suitable bandwidth.

The decision of which response characteristics are adequate will of course be substantially affected by the biological question. Sometimes the concern is whether the EMG is coincident with a slow mechanical event; for instance, one may ask whether a muscle participates in the slow breathing of a reptile or the slow locomotion of a sloth. If only the onset and cutoff of the EMG signal are of interest, a poor response characteristic may still be acceptable. Attenuation is more of a problem if one desires to obtain a quantifiable estimate of the magnitude of the signal—for example, if the question regards the magnitude of muscular activity when a particular function is performed. Then one must establish the number and magnitude of electrical events, at least of their aggregate. If the task is to recognize the subtle interactions of multiple muscles in a complex movement, one probably needs to establish not only the sequence in which the muscles act, but also the time-varying magnitude of their action.

The records of a chart recorder may provide an adequate representation for onset or cutoff of signals; its representation of signal amplitude will always be suspect. Regardless of whether the performance of a chart recorder appears to be adequate, it is still useful occasionally to check onset and cutoff times of the EMG onset versus those provided by a recorder with better response characteristics. Sometimes patterns emerge from temporally compressed long records that cannot be seen in the details of expanded ones.

2. Oscilloscope Films

Several different recording schemes dispense with the mechanical limitations of a moving stylus, further delayed by the need for frictional contact with the paper. They use light beams to mark a photosensitive film or xerographic transfer device or they mark the chart with ink sprayed through a low-mass vibrating capillary. The simplest such recorder is a fixed oscilloscope camera with a shutter that remains open during the sweep time of a single trace. To cope with the limited picture width, electronic rastering circuits are available that add a slow ramp to the AC signal so that multiple sweeps are automatically stepped successively downward on the face. All these systems have the disadvantages of the expense of the medium and the nonimmediacy of the output.

Hard copy printouts of the oscilloscope record are most simply achieved by photographing of the screen—for instance, by means of a 35 mm camera and Plus-X film or perforated AZO paper with or without an enhancing filter. Polaroid film backs are also available for most oscilloscope cameras, although such film is relatively expensive and it may be difficult to keep track of loose prints during a hectic recording session. However, such records will show only a single sweep of the beam, and one may have some difficulty coordinating camera, shutter, and sweep rate. A storage oscilloscope, either digital or storage tube, is useful for this application.

One may obtain a much better hard copy record by turning off the sweep signal so that each beam or split beam becomes a single dot on the face of the cathode ray tube. The signal is then expressed as vertical displacements of the dot on a single line and multiple spikes appear coincident. A special oscilloscope-mounted camera will use this signal to produce a high-resolution strip record. The image of the cathode ray tube screen is focused onto the film plane, and movement of the film (rather than the oscilloscope sweep) then reproduces a straight line on the emulsion. Whenever the dot moves vertically, it traces a line on the film, as would the light pen of a recorder.

Several manufacturers provide simple transport systems for 35 mm film, and it is even possible to build them from old 35 mm cameras. It is always useful to mark the film with a time signal independent of the motor drive, as the speed of the film past the gate is likely to be affected not only by the speed of the motor, but also by the amount of film on the takeup reel (hence, pin feed control is highly desirable). A combination of a flashlight, fiberoptics, and a spinning disc or a pulsed light-emitting diode will generate such a time signal directly on the film emulsion without occupying one of the instrumentation traces on the oscilloscope. (The marker spacing should also be checked against absolute time markings on the start and end of the tape.)

The records obtained from such a recording system have the same frequency response as does an oscilloscope, and the system can produce records equal to those of oscillographs and better than those of all but the best chart recorders. However, they do have several problems. One is that the intensity of the oscilloscope beam and the exposure of the film have to be adjusted to each other; otherwise, the baseline will tend to be overexposed and the vertical movement of the dot underexposed. (Remember that the number of photons incident on the film will be constant per unit of time; the greater the distance that the dot travels along its zigzag path, the weaker the image.) Also, the record tends to be rather compressed, particularly when one is "writing" with multiple beams on a single strip of 35 mm film. Furthermore, it will take from 15 to 30 minutes to develop the film. Still this is a relatively inexpensive way of obtaining records from which to calibrate the response characteristics of other equipment.

3. Oscillograph Recorders

Recently, several types of continuous high-response chart recorders have been developed, some using light processes. These recording schemes provide what is effectively a built-in version of the old-fashioned oscilloscope camera. The Visicorder uses ultraviolet light–sensitive paper and a multichannel linear cathode ray tube to improve oscillograph performance. Bandwidth per channel is 5 kHz, and the continuous trace provides a smooth and constant linewidth regardless of paper or beam speed. However, contrast is rather limited and the exposed paper requires fluorescent lights to develop its traces, which then fade if left exposed to typical ambient light or sunlight. Some xerographic copiers can generate high-contrast copies of the traces, but the copier flash tube overexposes and destroys the original within seconds.

Electrostatic processes such as used in xerography have been adapted to continuous records, with excellent, albeit expensive, results. Digital preprocessing gives the records a somewhat less continuous appearance, but with a 15 kHz per channel bandwidth, this is not noticeable for EMG signals. Contrast, while not as high as a good ink pen trace, is quite constant and reproduces well photographically. The records are fully stable and immediately visible (although there is a distance lag between the point of photon impression on the electrostatic drum and the point at which the paper bearing the traces leaves the recorder).

A unique chart recorder system sprays pressurized ink through low-mass capillaries that do not touch the moving paper. The output is permanent and looks very nice, but the dimensions of the system make it sensitive to mechanical disturbance.

4. Computer Output

The EMG signal can be digitized, after which a computer can operate on the data to compress, expand, integrate, or perform just about any imaginable transformation. It can then generate either a direct picture on its own hard copy system (such as x-y plotter, video copier, and graphic printer) or a set of output voltages with a bandwidth suitable for an existing chart recorder. The catch is that the raw records must be converted from analog to digital at a minimum of two samples per cycle of the highest frequency of the bandwidth of the input signal. Several channels with 3 kHz bandwidth will overwhelm all but the fastest digitization systems, and the volume of sample points that must be saved as fast as they are generated is truly awesome, even in these days of cheap microprocessor memory. A 10-second stretch of four such channels would consume 240,000 bytes of storage. Although this is less than the capacity of modern floppy discs, the data accumulate far faster than the computer could transfer them to this medium. Hard disks and other expensive options then become necessary; even then it is still useful to reduce the problem to a more manageable version.

Although computers may be very useful, there are several reasons why they may not be the most desirable storage medium. Apart from cost, there is the problem that digitization removes the output yet one stage further from a version that can be checked quickly and comprehended intuitively (as can a chart record). Something may go wrong and continue to do so for some time before the problem is recognized. If the hard disc fails prior to being backed up (on tape or floppy discs), the experiments will have to be repeated. Hence, do not use the computer as a spare closet, but reserve it for its most appropriate functions—calculation and comparison. Magnetic tape provides a much more stable and effective storage medium that retains more of the detailed information. Paper chart strips (run at a relatively slow speed) are cheap and permit a long experiment to be summarized on a single roll.

The demands upon computer memory decrease sharply and computer utility increases even faster if the raw data remain stored on tape and the computer is asked only to extract and summarize. Multiple kinds of demodulation are now available (see sections C and D, below), but you have the task of deciding which of them are truly useful for the job at hand. Is the change of spike frequency critical for explaining biological events? Or is the amplitude of the spikes, or perhaps the envelope touching their peaks, the most significant aspect? The computer lets you rapidly generate such values for selected segments of many records. It allows you to scan long segments and to compare the values from multiple muscles. Thus, it can elevate EMG from the anecdotal characterization of individual phenomena by fostering the application of statistical tests of significance. All of these aspects have long been with us, but the computer at the least can remove the drudgery, so that we can concentrate on what the animal is doing.

5. Off-line Transfer

One simple way of coping with limited bandwidth or sampling rate in the display system is to record the signal on tape and then play it back drastically slower. Also, one may play and analyze only selected short segments at normal speed using an oscilloscope camera. This can be a very economical, if tedious, way to examine high-frequency detail in AC signals with simple systems. However, many instrumentation tape recorders require a special set of channel filters for each playback speed, and the capacity to play back the signals at a varied speed can be a significant expense for multichannel systems.

C. DEMODULATION

1. Concepts

The activity of the muscle that is measured as the EMG signal is the end product of a chain of events that starts with the centrally generated "command signal" to the motoneuron pool. At the level of the motoneurons, this signal appears as the relatively gradually changing level of net postsynaptic depolarization from all of the neural inputs, both excitatory and inhibitory. This is turned into an FM rate of action potentials in each of the many motoneurons innervating a particular muscle.

At the level of the muscle, the electrical activity of the many motor units tends to sum into an AM envelope of AC signals with a fairly high bandwidth, looking somewhat like the output of an AM carrier frequency. Various techniques can be used to quantify the envelope of this modulation signal, which should be related in some way to the command signal that originally modulated the recruitment and firing rates of the motoneuron pool. Thus, we here group these techniques under the rubric of "demodulation."

In choosing among techniques, one must consider unique characteristics of the particular EMG signal on which they will be used (e.g., smoothly modulated versus spiky and irregular) as well as the purpose of the study and the properties of the display medium. In implementing any particular method, one must consider just how the envelope is to be defined. An EMG trace from a smoothly performed behavior displayed at slow sweep on the oscilloscope appears to the envelope detector of the human eye as a smooth, spindle-shaped envelope; all the sudden peaks and valleys will be ignored.

It is relatively simple electronically to rectify and integrate such an input signal, thus smoothing out the output to this imagined form (fig. 17.1). However, the eye readily picks out important but brief departures from the smooth behavior in the raw EMG trace. In contrast, the electronic integration system indiscriminately blurs such deviations into a smooth output. Obviously, the use of demodulation techniques carries implicit assumptions about exactly how big and how long a fluctuation must be before it is considered real datum as opposed to random variability to be smoothed away. In the discussion that follows we emphasize the characteristics and trade-offs of multiple techniques in making this compromise (fig. 17.2).

2. Area-under-the-curve Methods

a. Approach. All area-under-the-curve methods are based on the notion that no part of the signal should be disregarded and that the total amount of energy in the envelope is the best indicator of the level of effort represented by the source EMG signal recorded. We have already discussed how the random reinforcement and occlusion of asynchronous, brief action potentials will sum in a volume conductor (see chapters 4 and 5). Obviously, there are conceptual problems as to how far this notion can be extended quantitatively.

However, the absolute area under the curve of such random summations should be a valid (if somewhat noisy and perhaps nonlinear) indicator of the real sum of the mechanical activity. Furthermore, there is recent empirical evidence that such integrals are closely correlated with the frequency modulation of individual motoneurons and, by implication, with the net synaptic depolarization impinging on motoneuron pools. We will here adopt these methods as the standard with which other envelope estimation methods can be compared for validation.

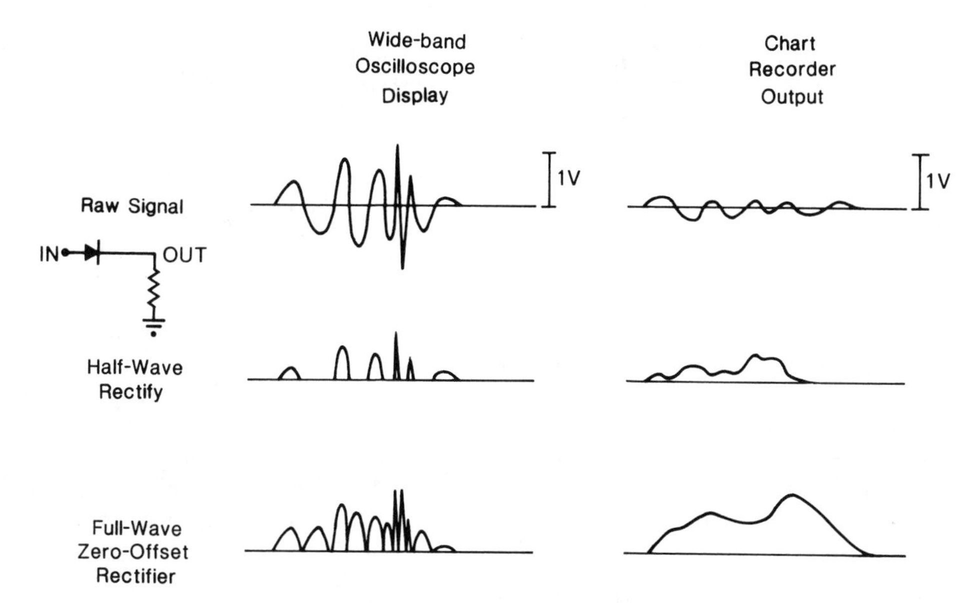

17.1. Chart recorder with a limited bandwidth will respond poorly, if at all, to the high-frequency AC signals typically present in raw EMGs. A half-wave rectification can be accomplished by means of a simple diode and resistor, but this works only for those parts of the signal that are above the threshold voltage of the diode (typically 0.6 V for common silicon diodes). The slow response of the chart recorder acts as a leaky integrator, producing an output somewhat like the envelope of the rectified signal. If the input signal can be amplified to 10 V or more before rectification, this may be acceptable. The best response is given by a full-wave rectifier in which internal biasing of the diodes is used to keep them near threshold for both positive and negative deflections, which are then recombined as all-positive deflections. This response is readily integrated into a smooth envelope by the chart recorder dynamics, although it may be desirable further to smooth the signal electronically with a controllable time constant.

17.2. Various filtering and rectification effects on raw, wideband EMG signal (*second trace*) of a cat splenius muscle. High-pass filtering emphasizes brief peaks, whereas low-pass filtering removes them. Full-wave, zero-offset rectification inverts the negatives to provide a dense plot going up from the baseline only; in contrast, half-wave rectification by means of only a simple diode passes only the positive peaks, missing low-level activity entirely.

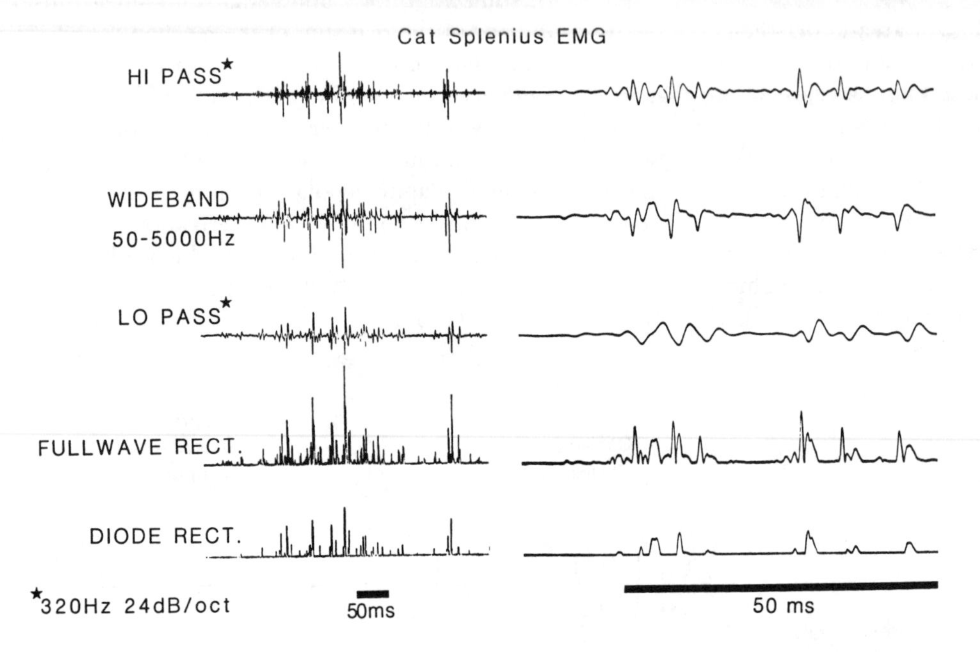

As we mentioned above, the eye is an excellent discriminator of events that lie within an envelope as well as those that represent significant or artifactual deviations. By selecting the sweep speed of the trace, you can subtly influence your eye to be more or less aware of brief fluctuations. You can then divide the trace into pieces of equal temporal duration or events of equal behavioral significance (e.g., cycles of gait, breathing, or chewing) and take the area subsumed by the traces (individually or sorted by some experimental criterion) as an indication of level of effort.

b. Paper Weighing. If you have the ability to generate a high-fidelity AC tracing of the wideband EMG signal (e.g., to use an oscillograph recorder or slowed-down galvanometric recording), you may planimeter this to establish its area. Another very simple and quite accurate method for quantifying its envelope area is simply to cut out the paper containing the tracing and weigh it. There is experimental evidence that this method correlates surprisingly well with the mechanical action of muscles.

c. Analog Filter Integration. Once you have a rectified version of the input signal (see fig. 21.1 and below), there are both simple and complex circuits for smoothing it into an envelope. The simplest is a two-component low-pass filter, in which a series resistor and a capacitor smooth out the fast fluctuations of the rectified signal by means of the charge storage properties of the capacitor. The rate (in seconds) at which the capacitor charges and discharges depends on the product of its capacitance (farads) times the resistance (ohms), a value which is known as the time constant τ (tau). The inverse of τ is the frequency in radians per second (ω), which must be divided by 2π to give cycles per second (hertz). This so-called first-order RC (resistance to capacitance) filter tends to change its output voltage (as measured across the capacitor) at a constant rate; the output only increases and decreases significantly whenever the mean value of the input signal changes for longer than the time constant (fig. 17.3).

One problem with the simple RC filter is that it has a fairly slow rolloff for frequencies above its cutoff time constant. This means that one has to use rather long time constants to produce significant smoothing out of brief, high-amplitude transients, such as occur in fairly spiky EMG records. Long time constants mean very sluggish response to the onset and offset of the EMG envelope as a whole, which may actually start and stop abruptly. Because the investigator is often interested in the exact times of onset and offset of the envelope, this phase lag can be quite misleading.

One important advance has been the introduction of three-pole Butterworth filters to the

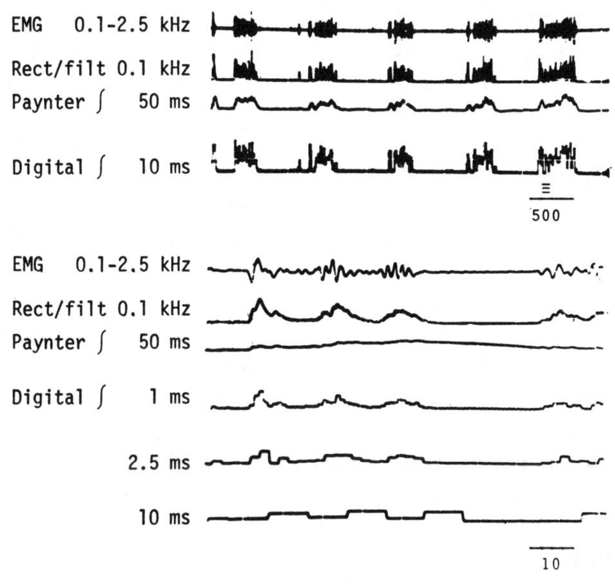

17.3. Various methods for rectification and integration of EMG signals to reveal envelope modulation by area-under-the-curve method. *Top,* Raw wideband EMG (from the hindlimb of a cat walking on a treadmill) at low sweep speed (500 msec calibration). *Bottom,* expansion of the sequence shown by the *three short bars* above (10 msec calibration). The traces marked *Rect/filt* show full-wave, zero-offset rectification with a relatively low time constant RC filtering (corresponding to 100 Hz cutoff). The *Paynter* filter trace shows long time-constant, high-rolloff filtering (50 msec three-pole filter, corresponding to 4 to 20 Hz). This is useful for perceiving the shape of the envelope but completely obscures short-term detail. The *Digital* traces show the output of a commercially available Pulsed Sample-and-Hold Integrator (Bak Electronics, Inc.) set at the bin integration times of 1, 2.5, and 10 msec and designed primarily to simplify the further processing of these data by computer sampling once per bin. (From Bak and Loeb, 1979.)

analysis of EMG; these are called Paynter filters in this application. Paynter filters roll off at 18 rather than 6 dB per octave and are often a satisfactory (and commercially available) compromise with a fixed time constant of around 10 to 50 msec depending on the application. Their main disadvantage is that it is difficult to incorporate variable time constants; the circuit of the Paynter filter contains several critical components that must be simultaneously adjusted to change this constant.

d. Mechanical Integration. The Paynter filter modules consist of a rectifier and a low-pass filter. We have already noted that the stylus of a typical chart recorder cannot respond to frequencies greater than about 60 Hz. The 3 msec time constant to which this corresponds ($\tau = 1/2\pi f$) happens to be a convenient low-pass filter setting for slightly smoothing rectified EMGs.

The simplest half-wave rectifier is a diode that is placed in series with the input signal. It does pose the problem that diodes have a biasing voltage, typically 0.6 V for silicon diodes. Below this voltage, any diode has a very high resistance to both signal polarities, with essentially no rectifying properties. Thus, the single-diode rectifier circuit acts like a threshold device, eliminating signals of low amplitude. This threshold effect can be varied usefully to reject baseline noise by adjustment of the overall signal amplitude to a percentage of the ignored input. However, simply ignoring this property courts disaster, because you will not see any low-amplitude signals and may assume that they do not exist. Remember also that a half-wave rectifier throws away half of the signal, aggravating integration problems. You can easily construct more complex full-wave rectification circuits with zero offset with an operational amplifier and a few simple components; however, once you add such an active device, you might as well incorporate a real integration circuit with a controllable time constant, as described above.

e. Pulsed Sampling Integrators. All of the analog filtering techniques cause temporal distortions of the input, such as frequency-dependent phase lag and hysteresis. These distortions are difficult to describe or correct quantitatively. They pose a particularly difficult problem in situations in which exact timing of EMG responses is needed, such as during fast movements and reflexes. A pulsed sampling integrator may overcome this problem. It can be used to generate accurate integrals of reproducible, brief, discrete time intervals of the input signal and to facilitate temporal comparison of events on several channels with each other and with external events.

The device performs a full-wave, zero-offset rectification of the wideband EMG signal and then integrates it in a very long time constant RC circuit. At intervals specified by the user, the voltage in the integration circuit is read out and the integrator circuit is reset to zero volts for the next period of integration. A sample-and-hold circuit maintains the previous integral readout as the output signal during each successive integration period, so that the output appears as a steplike signal that is always exactly one interval behind the current EMG signal value. Sudden onsets and offsets of EMG signals are reflected as instantaneous step increases and decreases in output subject to the temporal resolution of the bin width selected. The integration period is conveniently varied by a simple timing circuit and may be synchronized among several channels or with an external event such as a stimulus. If played out on a chart recorder, the steplike appearance will be somewhat smoothed, resulting in a signal that is both an accurate envelope measure and reasonably pleasing to the eye.

This form of integration is particularly useful for preprocessing wideband EMGs prior to digitization, because the degree of integration is easily adjusted to the desired sampling rate of the computer and because the digitization can be timed to occur at stable values of the integrated output. This overcomes the difficulty of "aliasing" errors to which digitization of continuously fluctuating analog signals is prone. In the latter, the sampled value is severely influenced by vagaries of the point of digitization on rapidly changing waveforms and poorly represents true fluctuations of the source signal.

3. Zero-crossing Method

Zero-crossing counts represent a more popular and simpler form of digital quantification of EMGs than area integration. It is based on the notion that a more active muscle will be generating more action potentials and that there will tend to be more zero crossings (or peaks) in the signal. Of course, few spikes in the waveform correspond to individual unitary action potentials (unless the record has very few units active), but the effects of summation and occlusion are assumed to be random. A simple electronic circuit produces a single digital pulse of fixed amplitude for each zero crossing detected (following a peak that must be above a threshold level to prevent triggering on baseline noise). These pulses can then be processed by an average rate-meter, displayed as dots in a raster, or counted by a computer. One variation of this is inflection counting, usually accomplished by counting of the zero crossings of the differentiated signal. This avoids the problem of low-frequency signals (e.g., motion artifact) causing "dead periods" when none of the EMG spikes actually go through zero volts.

The properties of this transformation vis-à-vis our hypothetical pure integral are somewhat dependent on the particular nature of the muscle and its EMG signal. Some muscles consist of large numbers of homogeneous units, generating similarly sized action potentials all recruited at similar levels of effort. In this case increments in the level of effort at high baseline levels may be more accurately reflected in the integral rather than the zero-crossing count, which will tend to saturate (i.e., increased firing rate will ever more affect the area under the envelope, rather than the number of zero crossings). Conversely, EMG records consisting of low rates of widely varying spikes (e.g., small numbers of recruited motor units or closely spaced differential electrodes) may generate poorly fused and noisy integrals, whereas zero-crossing counts may reflect more accurately the level of effort.

4. Amplitude-weighted Inflection Counting

A variant of the inflection-counting approach not only counts the peaks but also establishes the amplitude of the value at which each inflection (peak) occurs (fig. 17.4). (Such amplitude values may be expressed as mean per sampling interval.) The sum of these amplitudes per sampling interval includes information about large-amplitude spikes, which are counted as equals of smaller spikes in pure counts of inflection or zero crossing. This would represent important information whenever the mixture of events of large and small amplitude reflect a heterogeneous population of motor units. However, it may also be quite misleading whenever it reflects a random biasing as a result of a few fibers that lie very close to an electrode of small surface area, a

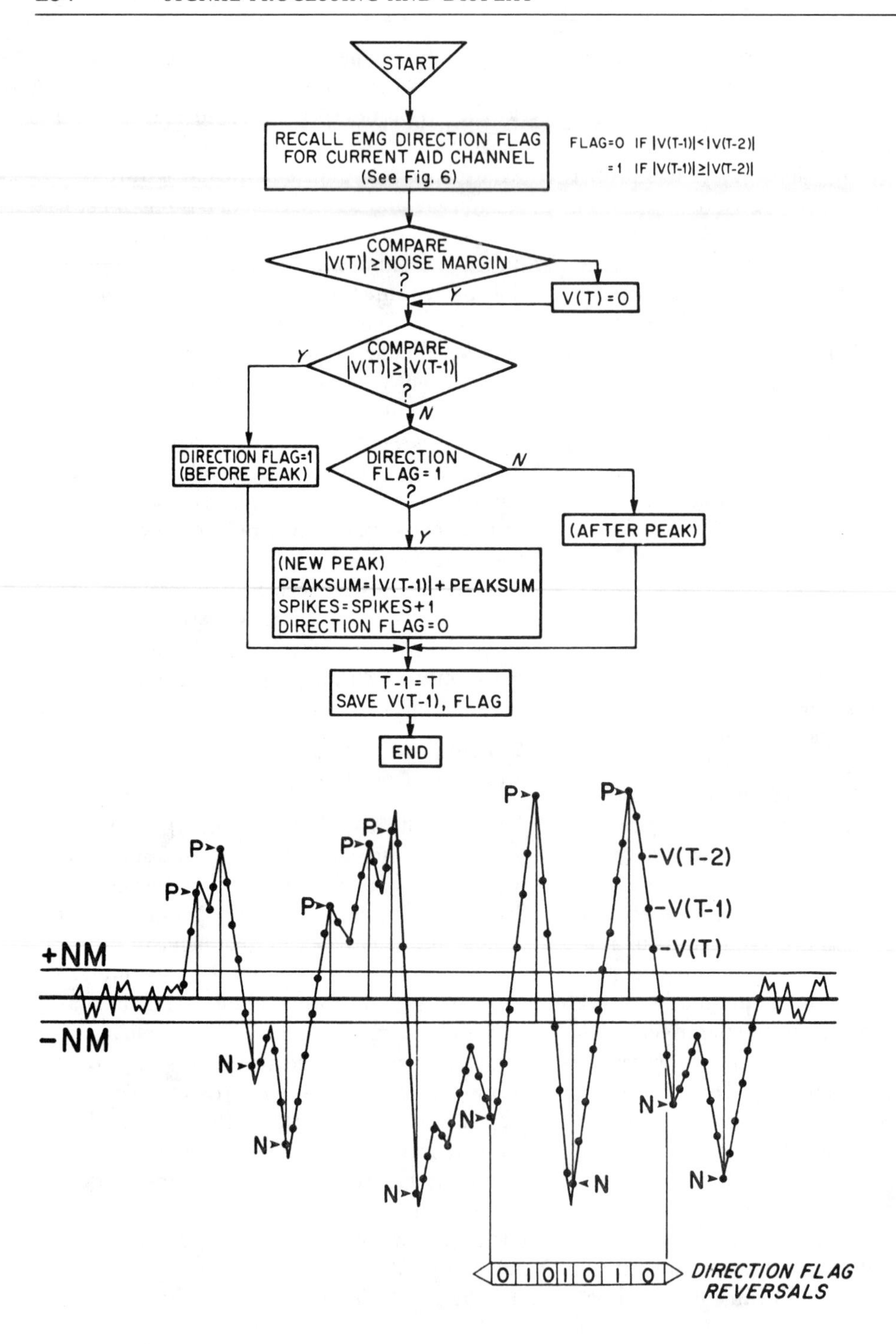
START
RECALL EMG DIRECTION FLAG FOR CURRENT AID CHANNEL (See Fig. 6)
FLAG=0 IF |V(T-1)|<|V(T-2)|
=1 IF |V(T-1)|≥|V(T-2)|
COMPARE |V(T)| ≥ NOISE MARGIN ?
V(T)=0
COMPARE |V(T)|≥|V(T-1)| ?
DIRECTION FLAG=1 (BEFORE PEAK)
DIRECTION FLAG=1 ?
(AFTER PEAK)
(NEW PEAK)
PEAKSUM=|V(T-1)|+PEAKSUM
SPIKES=SPIKES+1
DIRECTION FLAG=0
T-1=T
SAVE V(T-1), FLAG
END
+NM
-NM
V(T-2)
V(T-1)
V(T)
0 1 0 1 0 1 0
DIRECTION FLAG REVERSALS

17.4. *Top,* Algorithm for computer analysis of EMG waveforms. The number of spikes and their mean amplitude are determined for each time interval. Signals are subdivided into regular intervals and in this idealized sketch (*bottom*), the *dots* indicate the points sampled. Those sampling points falling within the noise margins (+NM to −NM) are ignored. Those outside the noise margins and nearest the peak of each wave or wavelet are counted and their amplitude is calculated from the baseline (zero). *N,* negative peaks; *P,* positive peaks thus recorded. A time-base generator establishes the duration for which points are sampled; alternatively, the sampling intervals may perhaps be set by signals from an event recorder or a camera shutter. For each interval, such a system can easily determine the number of peaks and their mean or aggregate amplitude) over such an interval. Consideration of the sketch documents why the sampling rate must be several times the highest frequency to be acquired to avoid biasing the results (i.e. obtaining results that do not reflect the spike amplitude). (After Beach, Gorniak, and Gans, 1982, copyright 1982, Pergamon Press, Ltd.)

possibility that is easily tested for as it would produce quite different signals from other electrodes in the same general space. This variant of inflection counting probably produces results most like area-under-the-curve methods and is suited to computer processing without analog preprocessing of EMG signals.

5. Peak Detection

Peak-detecting circuits are readily available and frequently used with analog-to-digital converters to avoid the aliasing problem. The measurement of peaks is frequently used in clinical work, and the method of paper weighing tends to concentrate on peaks, particularly if slow sweeps are used. The measurement of peaks obviously throws away much information inherent in the signal waveform and concentrates on events that may depend on the rare random events of constructive addition of small individual action potentials or on the firing of a few fibers close to a small electrode. Furthermore, the peaks are likely to be unduly influenced by the infrequently recruited, large, fast motoneurons that innervate large numbers of fast-twitch, fatiguable muscle fibers, whereas the contribution of smaller, less fatiguable units is undervalued.

6. Peak-resettable Integration

Another variant that combines both inflection counting and amplitude information is to use a resettable integrator in which the resets are triggered by each deflected peak.

D. SPECIAL PURPOSE ANALYSIS

1. Signal Averaging

a. Properties. General purpose signal averagers are becoming increasingly powerful and cost effective for the analysis of transient biological signals. They are a particularly powerful means of identifying fine structure within a signal that may be time locked to an external event even when the signal is highly unpredictable and random, as is the EMG signal. However, their use requires a sound understanding of the statistics of signals and the manner in which variously processed signals tend to cancel and reinforce each other when summed.

In its simplest application the averager accumulates and thereby produces a mean of several trials of performance of the same motor act. Because the AC signals have a net zero value, they must be rectified so that they will add constructively. The duration of the EMG bursts to be averaged must be similar to avoid false spreading out of the resultant. It is absolutely critical that each sweep of the averager be triggered at the same point in the movement. Randomness in the synchronization signal used to initiate each sweep will wipe out real fluctuations of duration briefer than the trigger uncertainty. With enough sweeps, the overall envelope of effort will be readily apparent, even if the individual trials are quite irregular. Calibration of amplitude can be difficult to keep track of, given variables such as analog to digital (A/D) gain, number of sweeps, and display gain. It is often advisable to add a small, known calibration pulse at the input of the amplifier.

b. Treatment of Variations. The usual problem is how to know whether a given wiggle in the shape of the average is "real" or whether it results simply from not having averaged enough sweeps to eliminate (smooth out) random fluctuations. This is a recurring problem in the use of signal averagers, for which there are two general considerations, both of which should be used whenever possible.

First, one can consider the probability that a given wiggle is significant, based on its shape and size and the properties of the underlying single sweeps. This is a complicated problem to address rigorously. Few averagers automatically compute the standard deviation of the composite trace; even this only gives the significance of single point fluctuations, whereas one is usually interested in trends lasting many bins. The following factors should be kept in mind.

1. The contribution of a given amount of noise decreases as the square root of the number of sweeps (i.e., it will take four times as many sweeps to reduce noise by half).

2. Fluctuations lasting much longer than the lowest frequency components present in the input signal are quite likely to be real, because there is no way for individual, brief, random events to produce such an artifact. Judiciously applied high-pass filtering can be helpful.

3. Fluctuations that are three times larger than the fluctuations of background noise of similar duration have about a 10% chance of being caused by noise.

4. The more independent information you can bring to bear on the precise timing of a fluctuation, the more certain you can be of its presence or nonpresence. A fluctuation that has a 10% probability of occurring at a particular point in a record has a 50% chance of occurring at random within a record that is 10 times as long as the fluctuation itself.

The second test is simpler and more convincing. Repeat the average with an independent data set and check whether the fluctuation in question is still in the same place. This is a far better use of a given volume of data than is running all the sweeps in a single average. Doubling the number of sweeps will produce only a 40% improvement in the signal (square root of two); furthermore, if the fluctuation was caused by a large, random event in just one sweep, it is unlikely to disappear completely with further averaging. However, two independent runs of the averager permit many detailed comparisons between the two resultant traces. One can then look for similarities in the exact timing, shape, and amplitude of the fluctuation in question, all of which add powerful arguments to the determination of significance. Just be absolutely certain that the two data sets are independent, particularly when working from short stretches of tape-recorded data.

c. Special Uses. The signal averager also may be used to identify fine timing structure that may correlate with external signals, such as the discharge of a single neuron believed to be presynaptic to the motoneurons responsible for the EMG signal. Such uses require a careful estimation of the degree of synchronization expected between the two signals and of the frequency spectrum of the EMG. For example, the discharge of a particular motoneuron will be time locked to the action potentials in its synaptically driven muscle fibers with a jitter of less than 1 msec (as a result of variability in synaptic transmission and propagation of the action potential along the muscle fiber).

Because the frequency components of the EMG are mostly at or below such a waveform duration (1 kHz), the unrectified EMG signal can be averaged with each sweep time locked to the action potentials of the motoneuron. The mean so generated will be zero regardless of the level of use of the muscle. Any significant fluctuations from zero are likely to reflect fairly accurately the composite biphasic action potential generated by the motor unit in question. On the other hand, the activity of a premotoneuron (e.g., a monosynaptically linked pyramidal tract cell in motor cortex) will be reflected probablistically in the composite activity of all of the motor units, with a jitter equal to the duration of the excitatory postsynaptic potentials of the trigger unit, perhaps 10 msec long. The much shorter individual action potentials making up the EMG signal are likely to occlude each other completely over time, because each unrectified unit waveform has zero net area. Thus, the rectified EMG signal will have to be used, and the averaged response will be seen as a small, relatively broad increase in the rather large overall sum (or a decrease if the synaptic relationship is inhibitory).

d. Programmable Triggering Criteria. Once the EMG data have been reduced to a table of numbers and stored in the memory of a microcomputer, a procedure that is ever more likely and less costly, one can easily write programs that permit the signal itself to be used for the generation of synchronized averages. For instance, one may instruct the system to scan the record for any pattern in the output of a single muscle. To use muscular activity as the trigger, signal levels that do not exceed a "set point" (noise level) are ignored and the system proceeds to the next datum. Once a significant value is observed, a certain number of points are stored in the first row of a table, with each subsequent such trigger initiating the filling of a following row. Such programs should incorporate subprograms that permit one to select criteria, such as how many "zero" values must pass before a new event is recognized and printed out as the next row of the table and the number of zero values that may occur within a data string without triggering a new one.

Computer programs are easily modified to generate multiple comparisons; for instance, they permit one to test whether the start (or stop) of action in one muscle will indeed predict a constant action pattern in a second one or to correlate duration of activity among various muscles. Also, it may be interesting to test for correlation among features that occur in the middle of each EMG string. A microprocessor can be programmed to try sequentially many different, perhaps complex criteria for starting each sweep. More than just confirming and documenting relationships that are already known to be true, they permit "fishing expeditions"—that is, the search for correlation among events that have not previously been considered to be associated but that incorporate the potential for biologically significant aspects. However, you must remember that the more fishing you do, the more likely you will find the 5% (or 1%) chance (artifactual) correlation implicit in the $p < .05$ (or .01) probability test of significance.

e. Computer-based Statistical Analysis. Microcomputer-based analyses also permit more complex comparisons. The descriptors of spike number or area under the curve may be tabulated for multiple muscles (columns) and sequential time intervals (rows). Descriptor values generated by particular muscle(s) may be summed (horizontally) per event (i.e., bite, foot placement, inhalation) or by temporal segment (i.e., first quarter of duration). The values obtained for multiple replicates (i.e., steps in a sequence) may be treated statistically (mean, standard division) for each of the defined categories. As indicated in more detail below (section D3, this chapter), this also permits the test whether the variance differs among muscles and among categories, with the possibility that aspects showing lower variance are more tightly constrained and perhaps more significant to the animal.

Microcomputer-based comparisons also permit comparisons with events that occur after the EMG sequence so that triggering may be used with a reverse scan of the sequence. Finally, there is the possibility that EMG signals associated with successive events may differ in their time scale, without varying in the shape of the envelope. Simple software tests permit one to evaluate such possibilities and even to rescale time to correct for them.

2. Fourier Analysis

There has been much interest in spectral analysis of EMG signals during the past few years, although one suspects that this is related much more to the general availability of the Fast Fourier Transform Algorithm in signal averagers than to the suitability of this particular form of signal analysis to biological questions. Since the dawn of electricity and signal theory, it has been known that any time-varying signal, however complex, can be described in terms of the relative energy present in each of a harmonically related series of sinusoidal frequencies encompassing its bandwidth. As the level of recruitment of muscles changes, certain processes may occur that will change the frequency distribution of this energy as well as its overall amplitude. The recruitment of higher threshold, fast-twitch muscle fibers may be associated with higher frequency components in the signal because the larger diameter muscle fibers conduct action potentials more rapidly past the recording electrodes.

On the other hand, intensely activated muscles frequently demonstrate synchronization of the motoneuron activity, presumably as a result of complex interneuronal linkages. Such synchronizations show up as an increase in the low-frequency energy. Somewhat lower frequency components are present in the various forms of tremor, which are associated with finely controlled movements in both normal and pathological subjects. Very low values are also observed in the truly slow (tonic) muscle fibers. Obviously, many complex and often poorly understood factors contribute to the spectral content of EMG signals, as revealed by Fourier analysis. For a well-defined experimental paradigm and given strong empirical confirmation, this form of analysis can provide a powerful insight into changes in the mechanisms of motoneuron recruitment. However, it is best used in parallel with an accurate method of quantifying overall EMG activity (such as integration), so that the interdependence of spectral content and overall recruitment can be examined carefully.

3. Variance Analysis

Another way of looking at mechanisms of recruitment concerns the smoothness of control at any given level of effort. In the analysis of animal behavior it may provide an indicator of the relative advantage of different motions to the animal, as indicated by the relative tightness of the control exercised. Various disorders and disabilities manifest themselves as an inability to perform smoothly modulated muscle contractions and movements. Digitized EMG signals can be computer processed to quantify the instantaneous variability of the

EMG signal amplitude (integrated over a short period) with respect to the mean level of effort over a longer period.

This sort of analysis can be approximated on-line by means of analog signal processing such as the Paynter filter.

The Paynter-filtered signal is subtracted from the full-wave rectified signal, producing a balanced AC signal that contains only the instantaneous fluctuations around the mean rather than the mean itself. This signal is, in turn, rectified and Paynter filtered, giving a smoothed version of the variance. One can also correct this variance for level of effort by subtracting out the original Paynter-filtered EMG signal, giving a signal that reflects the degree of incoherence of the EMG signal independent of the level of effort.

E. INFERRING FORCE FROM THE EMG RECORD

1. Statement

Variously processed EMG signals are frequently offered as a direct indication of the force output of the muscle, which is most unlikely at best (fig. 17.5). The conversion of EMG records into an accurate dynamic picture of muscle force currently is the subject of intensive research and involves issues so specific to each particular preparation that no general solution is ever likely to emerge. The discussion below assumes that we have already obtained an accurate representation of total muscle fiber recruitment from a suitably designed and placed sampling electrode (chapters 5 and 6) and its suitably processed signal (e.g., area-under-the-curve integral, this chapter). However, it should always be kept in mind that even the best such sampling and processing will have been subject to significant nonlinearities.

We here consider the various factors that can be expected to influence the degree to which the electrical activity recorded from a muscle will be reflected in force acting on a body part. These should help the investigator select experimental paradigms in which these factors are suitably constrained or in which their effects may be empirically identified. Only after this is done may one begin to consider correlation of EMG and mechanical actions.

2. Trigonometry and Physics

In most cases, muscles convey their forces to body parts via leverlike structures such as bones. In addition to the length of this lever arm from its real or implied fulcrum at a joint, one must know the angle of pull of the muscle with respect to the axis of the lever. This is because only the component of the force vector that is normal to the lever arm contributes to the torque; the perpendicular vector component that is parallel to the lever arm acts only to compress the joint, usually assumed to be a rigid structure. Usually the angle of pull and even the distance to the instant center of rotation change significantly over the physiological range of muscle length. Furthermore, the angle of pull may differ within the muscle as a whole in variously pinnate muscles.

3. Length/Tension Curve

The tension-generating capabilities of a single muscle fiber are strongly influenced by the well-known inverted-U length/tension curve (fig. 17.5). It must always be remembered that the fiber architecture of the muscle can cause significant deviations in the proportionality between changes in the length of the entire muscle and that of its individual fibers, so that a calibration of force output for a fixed amount of activation needs to be performed empirically at all physiological muscle lengths. A further complication is the possibility that the length/tension relationship will differ with fiber type (particularly in muscles with sarcomeres of various lengths) and with activation frequency.

4. Velocity/Tension Curve

The tension generated by an active muscle depends strongly on the velocity imposed on the muscle by the compliance of the external system to which it is linked. Typically this linkage (and the length/tension curve) has been studied by measurement of the velocity achieved during isotonic contractions against various loads, resulting in the typical exponential curve from maximal isometric force to zero force at V_{max}, the maximal shortening velocity. However, muscles frequently operate under conditions of imposed lengthening, for which they can generate more than the maximal isometric force, particularly for short excursions (fig. 17.5). For longer excursions and

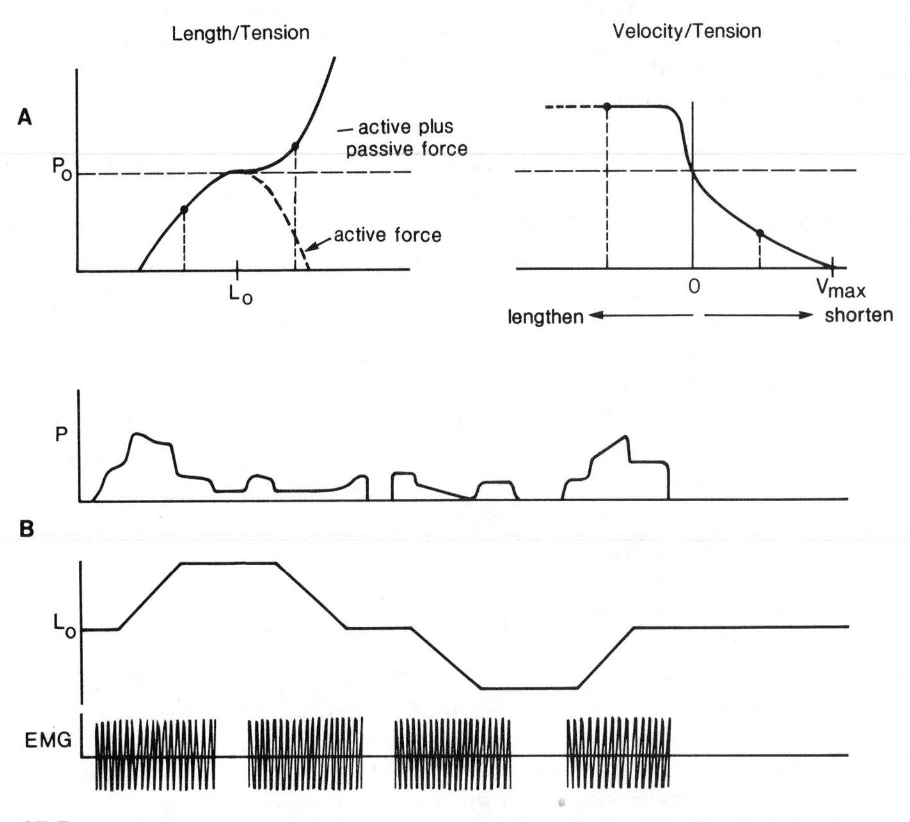

17.5. Amount of tension generated by a given rate of muscle activation (as measured by EMG) will depend complexly on the length/tension and velocity/tension properties of the muscle. Four similar EMG activity periods are shown with four different states of motion, including both lengthening and shortening velocities (*dotted lines in velocity/tension curve*) transiting between L_o (muscle length associated with peak active force [P_o]) as well as lengths less than and greater than this (*dotted lines in length/tension curve*). V_{max}, Maximal velocity; *P*, tension.

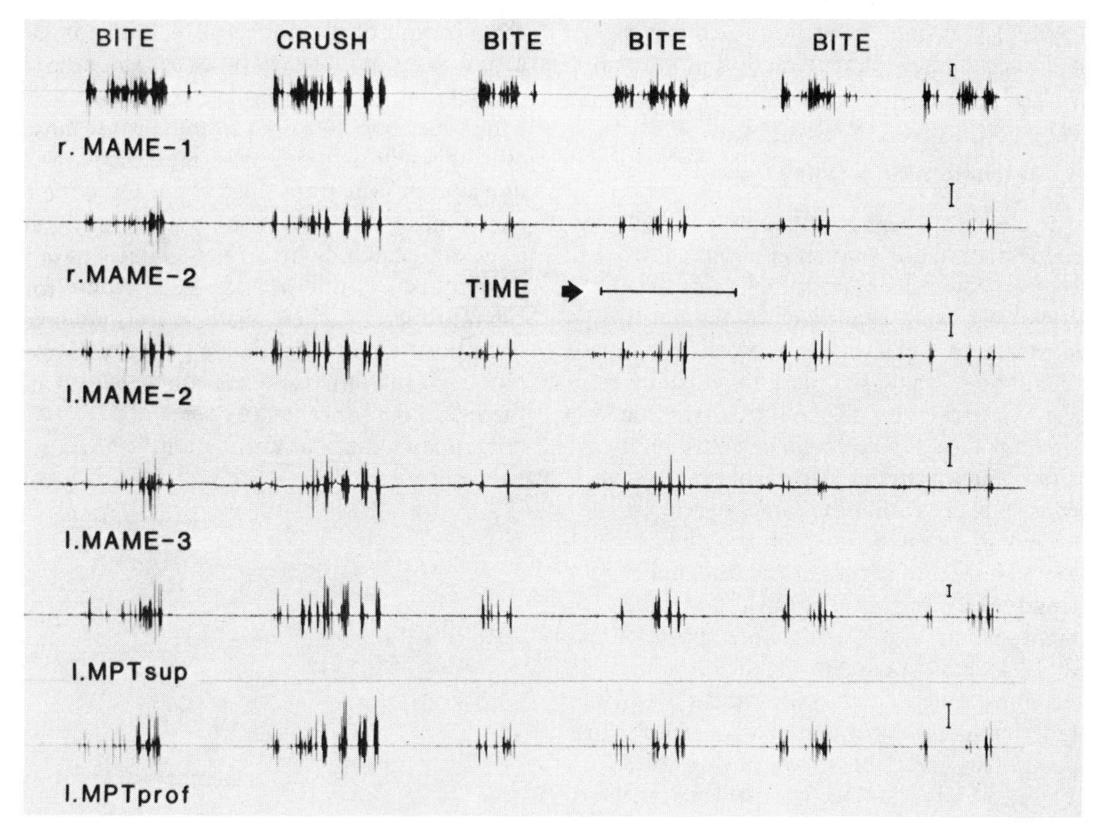

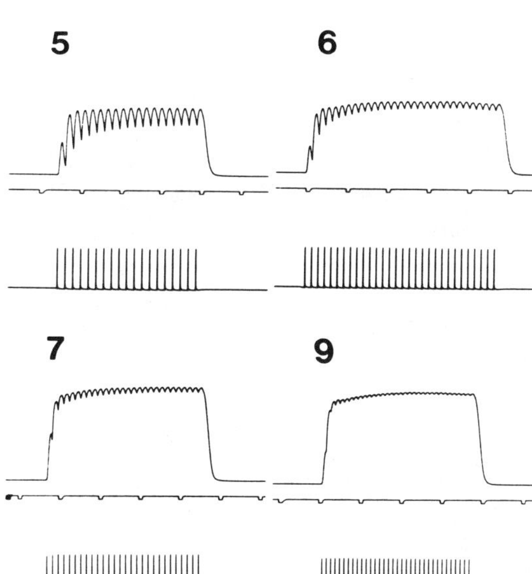

17.6. Pulsatile muscle activation by the shingle-back lizard, *Trachydosaurus rugosus*, apparently allows it to sum some of the adductor musculature in unfused tetanus, thus generating forces sufficient to crush heavy snails. Top, Note the pulsatile pattern in the six adductor muscles sampled here. (The individual spikes within each pulse apparently reflect incomplete synchronization of multiple motor units. The vertical scale bars equal 0.1 mV, and the time line equals 1 second) *Bottom,* Results of low voltage stimulation of the major adductor mass. The numbers give the stimulation pulse rate. Note that the major pulsation frequency observed agrees with the values of unfused tetanus (staircase region). The animal apparently utilizes a minimum number of pulses at the frequency optimum for stepwise increase toward the tetanus plateau. (Courtesy C. Gans and F. de Vree.)

higher lengthening velocities, force may rise or fall depending on the interaction of internal viscosity terms with cross-bridge kinetics.

5. Activation Time Delay

The electrical event recorded by the EMG signal is the sodium ion action potential proceeding down the outer sarcolemma. The contractile event is initiated by the much slower release and diffusion of calcium ions from the internal sarcoplasmic reticulum to the adjacent myofibrils, and it is terminated by reuptake of the calcium at active binding and pumping sites in this membrane system. Muscle fibers within the same muscle are known to differ markedly in the rate at which they release and sequester calcium and the length of the diffusion pathway for these highly bound ions. Such factors appear to underlie at least in part the classic division of mammalian muscle into fast- and slow-twitch types, although most mammalian muscles are composed of variable mixes of these fiber types; the situation becomes still more complex as muscles of lower vertebrates are considered. Typical time delays between EMG and force range from less than 10 to more than 100 msec, although a single integration time constant is usually employed to approximate this effect. Furthermore, the delay usually has a shape quite different from an RC time constant and may be affected by activation history and the temperature at which the muscle is acting.

6. Time-varying Contractile Properties

The mechanical output of a muscle fiber is complexly influenced by the entire recent and not-so-recent history of activation. The highly nonlinear relationship between single twitch tension and tetanic tension is only one class of these effects; depending on the rate and synchronization of physiological recruitment, the actual force output may be closer to one end or the other of this typically large range. Sometimes one notes the tetanic peaks; however, this will only be true if multiple motor units act in synchrony (fig. 17.6). Other effects that fit this pattern are potentiation, fatigue, and catch. Some of these effects have time courses of milliseconds (e.g., mammalian catch property), whereas others last minutes (e.g., posttetanic potentiation) to hours (e.g., fatigue). Only after one has controlled for these and the other factors listed above, may one attempt cautious correlation of EMG signals and force.

18 Graphical Conventions and Preparation of Illustrations

A. PRESENTATION OF RESULTS

Papers or research reports are the mechanism by which our results are transmitted to our colleagues. They must serve two purposes. First, they should document the work and indicate that it has been done properly. Second, they should explain the results of the study and indicate that the conclusions are significant and well founded. In each case they should provide the reader with data that permit appropriate decisions on such issues. Some of the data will be given as text and other portions in tables. However, man recognized form long before the development of writing or even speech. Presumably, this underlies the proverbial equality between the picture and the thousand words.

Assuming that one has obtained acceptable EMG records, the question remains how to illustrate convincingly that they explain something useful about the activity of the muscles. In this chapter we make suggestions about the kinds of pictures that may illustrate functional analyses and about ways of increasing their effectiveness. We will try briefly to indicate some conventions and approaches, incorporating rules of what must be done and samples of what can be done. Remember that some of your readers will read only your abstract and scan your figures. The key aspects of your message should be discernible thereby, and the quality of the picture should motivate the scanner to read your text. Ideally, your illustrations should represent a fine blend of the extreme detail and accurate perspective of Albrecht Dürer's water color of a rabbit (fig. 18.1*a*) and the terse but elegant simplicity of line in a Japanese brush drawing (fig. 18.1*b*). Finally, remember that different rules apply to illustrations to be published and those to be presented as slides. The former will be available to be perused at length, whereas the latter must transmit their message in seconds.

In the same way as an anatomical drawing will be designed to provide an immediate impression of three-dimensional structure, illustrations of EMG records should provide an immediate concept of the general function of the muscle. They should not only document the characteristic properties of the particular signal but should be designed to facilitate comparisons among muscles and among animals. Several different kinds of information must be presented. First is the nature of the mechanical system. Next are the kinematic events of the motor behavior. Then there are the time course and magnitude of the EMG signals. Finally, there are combined pictures generally designed to synthesize concepts and facilitate discussions. As past discussion should have made clear, the nature and quality of EMG output will vary, depending on the place of recording. Consequently, the illustrations should indicate how and where the records were taken. For this, description is good; figures are best.

All figures should be labeled to a common

18.1. *A,* Albrecht Dürer was justly renowned for the extreme detail of his illustrations. Even in this watercolor there can be no doubt about the kind and characteristics of the animal depicted. (Photo taken from Fonds Albertina, from The Photo Archive of The Austrian National Library.) *B,* Some schools of art specialize in illustrating a subject by a minimum of lines, expressing the essential characteristics and leaving the remainder for the imagination. Whereas this kind of cartooning is not to be recommended for illustration of anatomical detail, it may be useful to show characteristic behavior in parallel with EMG measurements.

A

B

final size after reduction (rather than buying one size of letter and pasting it on large and small pictures so that the final format differs widely). If you initially plan your figures for a standard reduction you avoid this problem. Remember that a simple letter style will be scanned more easily and will hence communicate your message better than a complex version, such as old English or the copperplate curls dear to the makers of eighteenth century engravings. Use of a reducing lens will indicate something of the appearance of your final figure.

Plan your labeling scheme carefully. It is better to mount the labels on the periphery and use leaders (with a white countershade if necessary) than to glue lettering where it will obscure details of the figure. Keys to all abbreviations used belong in the figure caption, but use a mnemonic scheme and remember that written out labels (even of a shortened version) are better than codes, memorization of which may interfere with immediate understanding of the message and comprehension of structure.

There is no ideal standard for data presentation. The nature of the structures being compared and the properties of events and signals being generated should indicate the kind of description and concept that are most important.

B. ILLUSTRATION OF STRUCTURE

Animal bodies are complexly three dimensional and the integument often spans multilayered structures. Indeed, we have seen that the complexity does not stop at the surface of the muscles but involves their internal subdivisions as well. Understanding animal topography is critical to understanding the possible range of actions. Good illustrations are needed to communicate it, but of what kind?

Initial description should characterize the fundamental anatomy. Good photographs have depth, which means that one can tell deep from superficial structures. Some (generally expensive!) illustrators can render a complex structure and even improve this effect. One way of documenting form might be by illustrating dissections layer by layer from skin to bone (fig. 18.2). The effect may even be enhanced by printing of stereo pairs—views taken from slightly different angles; these require that the reader use a stereo viewer or learn to combine images without it. This approach may be both costly and often inappropriate in functional settings. Furthermore, fine detail is likely to be lost as a result of the vagaries of reproduction by all but the finest grained printing methods. Also, not everything visible is necessarily significant; indeed, the best illustrations may be those that provide only enough information to let one understand the movement patterns and the forces that induce them. Thus, "keep it simple" may be the first message.

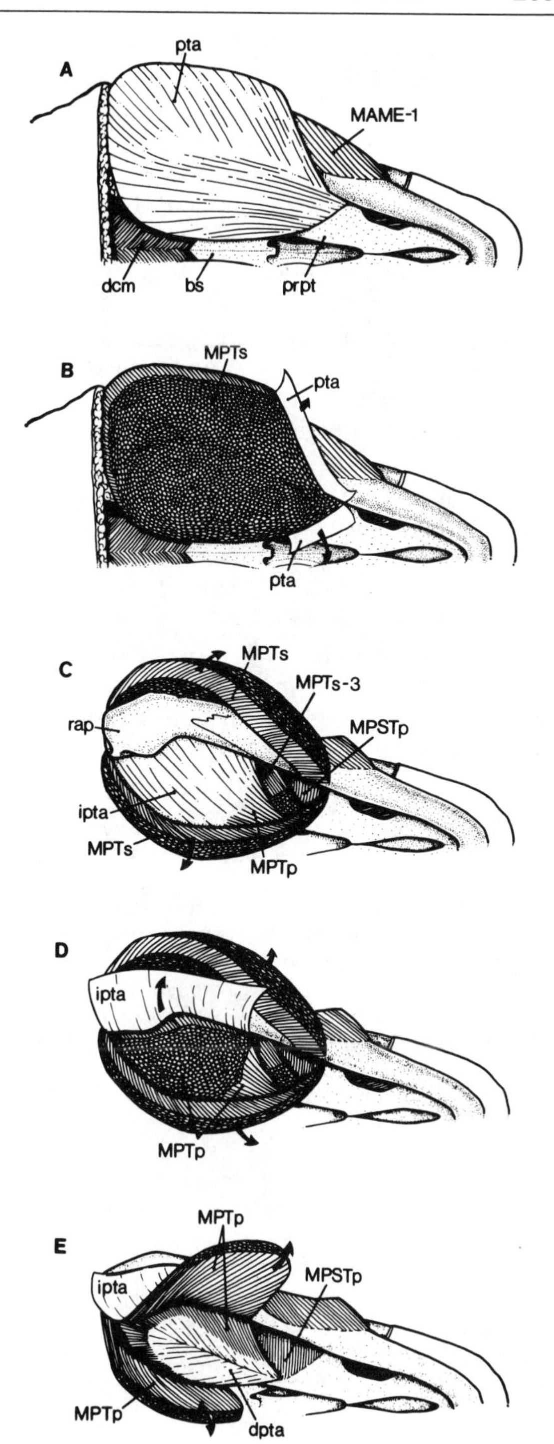

18.2. Series of drawings illustrating fiber architecture selected to show that muscle placement changes with depth. Often the details shown by successive muscle layers, such in this pterygoid muscle of the stump-tailed lizard *Trachydosaurus*, differ so drastically that sequential illustration of the several levels is the best way of communicating the nature of the anatomy. (By J. Dimes, from Wineski and Gans, 1984.)

Whereas simple may be fine, accuracy is also critical. The best way of ensuring accuracy is by illustrating complex structures from their appearance in multiple directions with all views drawn to the same scale (magnification may be given in the caption or by illustration of a 1 mm or 1 cm line to scale). This pattern, exemplified in engineering drawing as descriptive geometry, shows the object (animal or its part) as suspended in a polygonal box with plane sides. The image of the object is traced on each side (avoiding parallax with all projection lines at right angles to the surface (fig. 18.3). The box is then unfolded along lateral edges so that the contacting sites lie flat. Only the minimal number of sides is actually retained; those providing duplicate information are omitted, although the set must consist of contiguous sheets. The rules of descriptive geometry then allow one to derive the position in space of any point from such illustrations; the conventions of this branch of mathematics are usefully reviewed here.

Naturally, one should try to orient the box so that its sides parallel important surfaces (fig. 18.4). It is hardly useful to orient the top at 15° from the horizontal unless this corresponds to a natural line, such as the slope of the snout. Instead, use vertical and horizontal surfaces at right angles to each other, unless there is reason for deviation. These diagrams also have the major advantage of facilitating check of the proportions of the individual views, a self-checking process that is lacking for the seemingly more elegant views in single perspective. Members of the paired dorsal and ventral, paired lateral, and anterior and posterior views must have equivalent outlines and renderings; structures such as eyes or ears must lie on lines that are parallel to each other. Breach of these simple rules is immediately obvious and indicates inaccuracy.

Many illustrators accustomed to the production of artistic renderings have initial difficulty with the discipline imposed by descriptive geometry, and it may be useful to

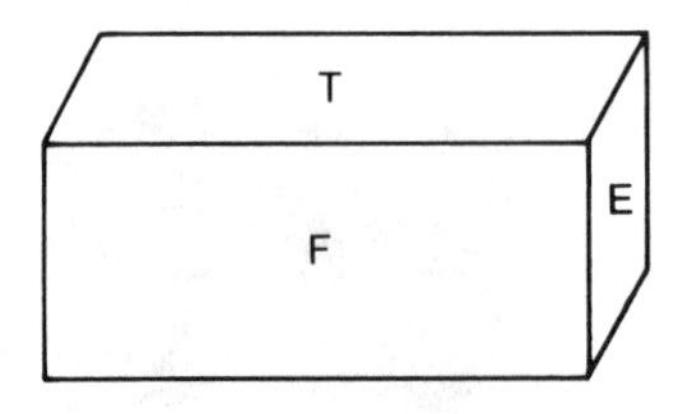

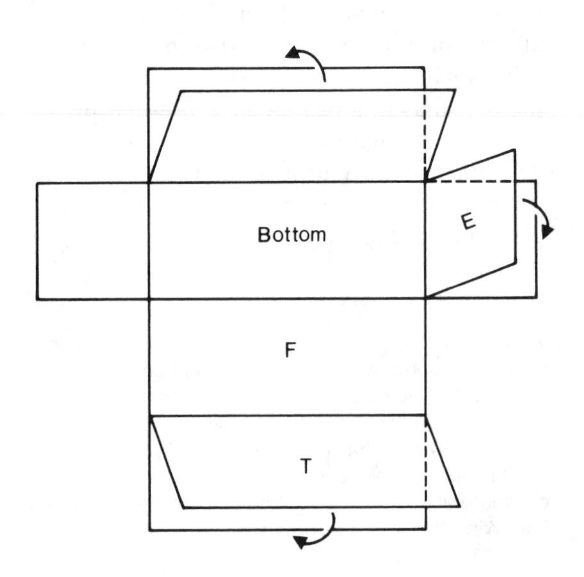

18.3. Principles of descriptive geometry show all views to the same scale and oriented adjacent to each other, as if the object had been placed into a box and illustrated on its faces, the box then being opened up. The ventral view then lies adjacent to the lateral view, and this lies adjacent to the dorsal view. If this convention is followed (and the drawings are prepared for the same magnification), the position of the individual points may be compared among views. Consequently, the approach facilitates comparisons and also lets the observer recognize the degree of curvature shown at any level.

18.4. Skull of the lizard-like, burrowing reptile *Pachycalamus.* This illustration indicates why such an elongate, complexly shaped skull may be difficult to visualize should this only be illustrated in dorsal, ventral, or lateral view (drawn to slightly different scale and oriented at random). Consequently, the views and angles were selected along axes at right angles to the surfaces to be illustrated. The posterodorsal and anterodorsal views illustrate the sutures, respectively, of the posterior and anterior portions of the dorsal surfaces of the skull without introducing foreshortening. The insert shows the position of the views. (By S. B. McDowell, Jr., from Gans, 1960, Bull. Am. Mus. Nat. Hist., with permission from the publishers.)

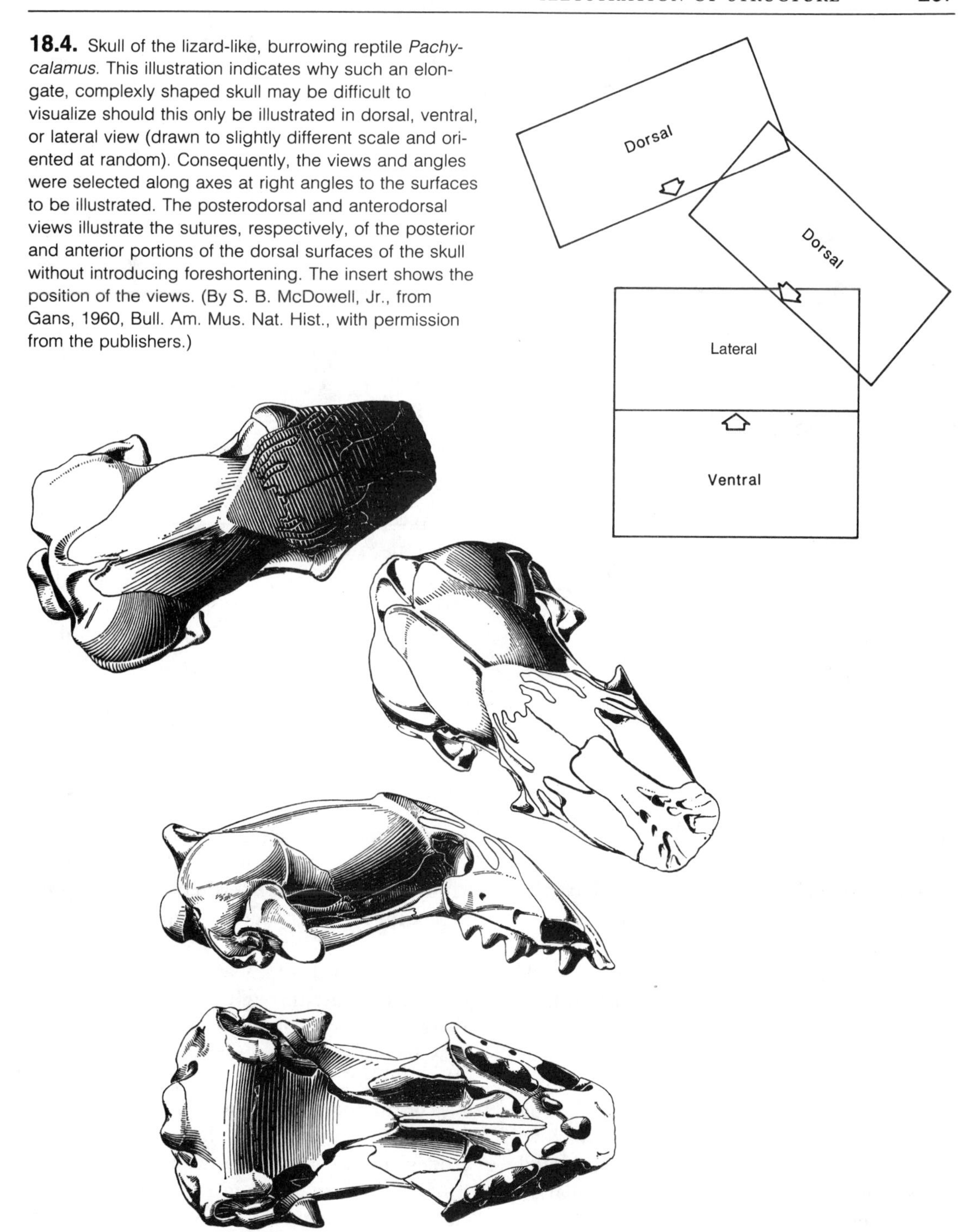

18.5. Cat hindleg in cross-section. Although a series of muscles may pass from roughly parallel origins to roughly parallel insertions, their actual topography over the distance may vary markedly. Here we see a set of four sections illustrating how the proportions and areas occupied by the soleus and gastrocnemius muscles change along the tibia between knee and ankle of a cat.

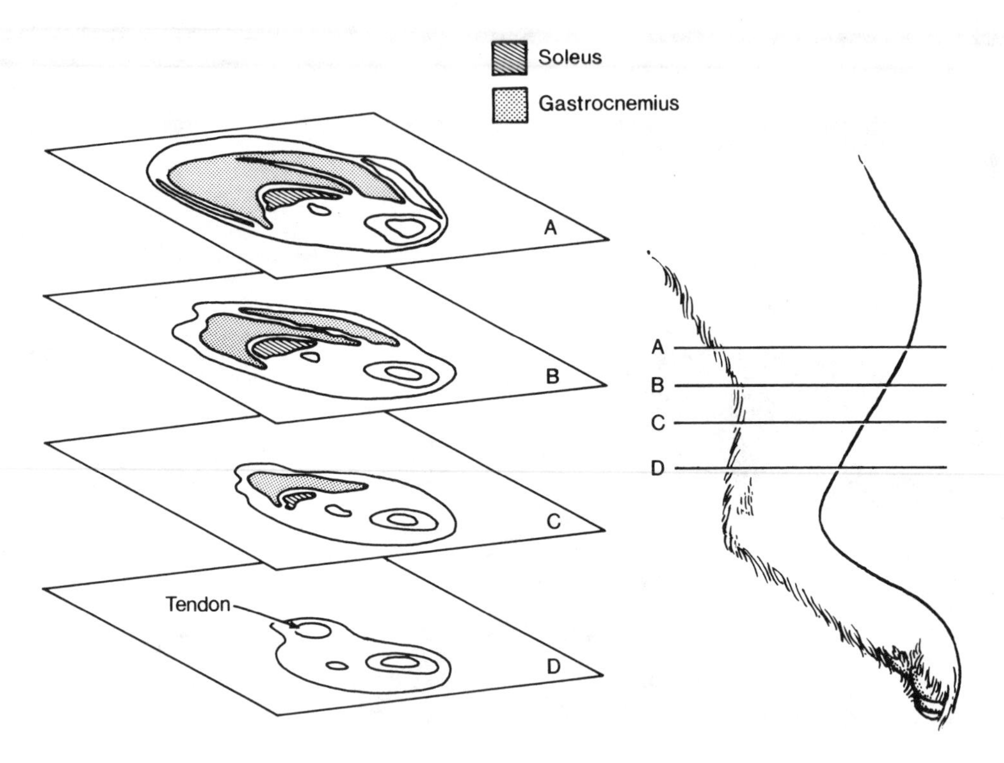

have them generate the initial outlines with a camera lucida or from photographs (taken from far enough away to avoid parallax). Also, some editors object to having the figures placed contiguously. However, the approach deserves attention as it fosters accuracy and documents that the structural pattern was fully understood.

Drawings of the muscles in situ may also be combined with exploded views showing them displaced and perhaps rotated. Unusual views are useful, and multiple views may be necessary to show the connective tissues and muscle fibers separately. This is particularly important in complex muscles in which the internal architecture may be difficult to understand. Sections are desirable; although they do require some skill in interpretation, they provide a much better index of three-dimensional structure than do surface views by themselves. The arrangement of muscles with central tendons that rotate through their length so that there is no obviously desirable external view may be visualized as a series of stacked sections (fig. 18.5).

The anatomical drawings, or parallel overlays, should indicate the placement of electrodes and transducers in detail (e.g. fig. 10.5). Even the insertion track for the wire may usefully be indicated. Numbers in figure or caption may remind the reader of the actual number of repeated samples from each particular site.

One critical aspect of anatomical rendering is to remember that joints have a range of positions. For instance, whereas masticatory muscles are normally shown with the jaws closed and teeth in occlusion, the lines of action of the muscles are markedly different when the jaws are open. The position of par-

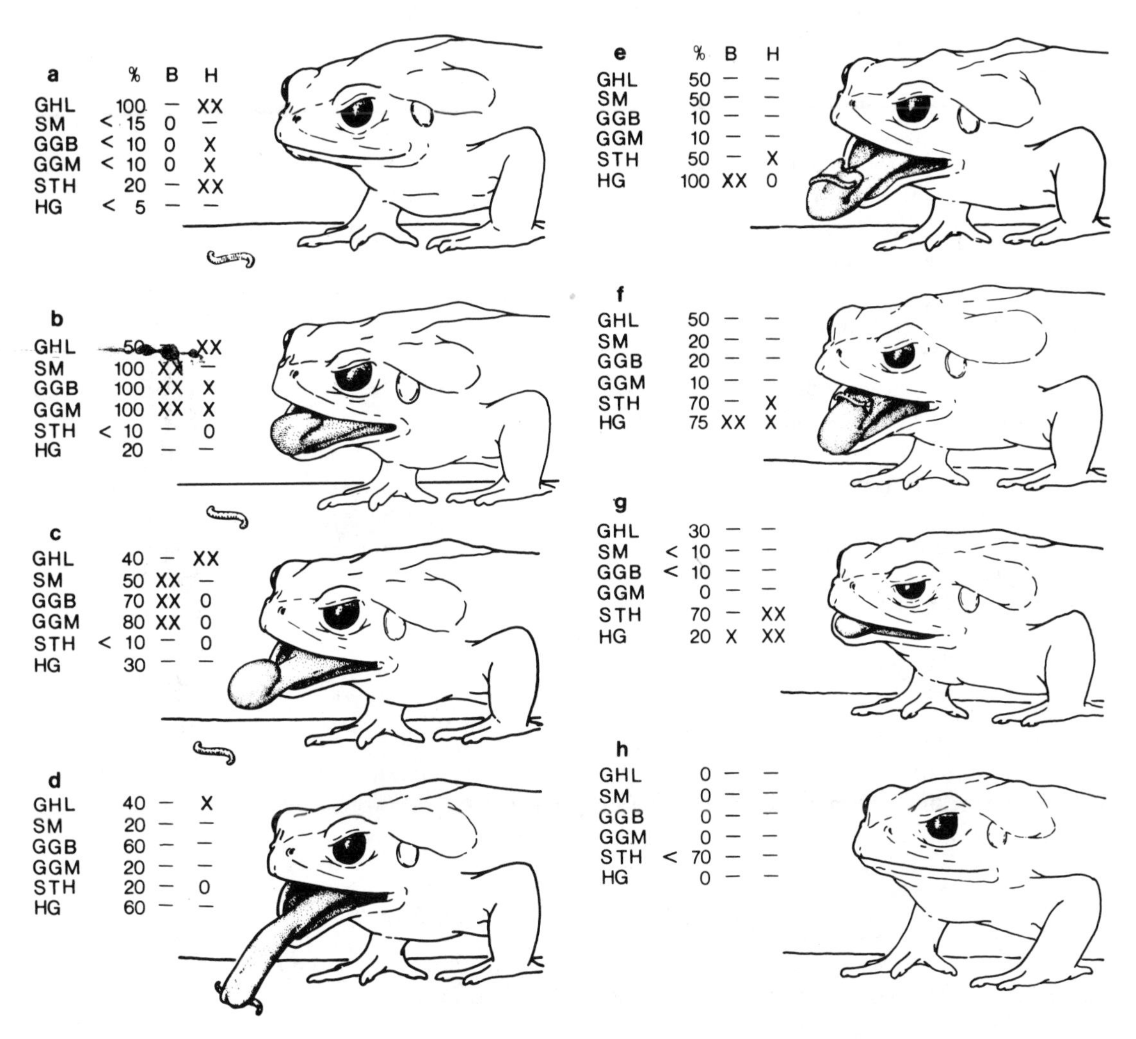

18.6. Views of a toad flipping its tongue. The activities of the major muscles determined by EMGs are indicated as percentages of maximal observed activation (expressed as number of spikes times amplitude). The illustration incorporates a single test of two hypotheses (*B* and *H*). For each, the expected muscle activity is coded from *0* (least) to *XX* (greatest). The simple set of sketches makes it easy to understand that, based on these criteria, *B* appears to have a greater predictive power than *H*. (*B* more closely matches the observed values.) (By L. Trueb, from Gans and Gorniak, 1982b. Copyright 1982 by the AAAS.)

allel flexors and extensors of limbs may shift between joint angle extremes, and the relative cross-sectional area of muscles seen in a transverse section may differ profoundly. Whenever such aspects bear upon the analysis, they should be illustrated.

The elegant rendering of bone and muscle architecture may be usefully supplemented by a series of cartoons offering only the key morphological features and showing how these are displaced during activity (fig. 18.6). Solid arrows may show forces and open arrows may show velocities (use of a single type of arrow for both may confuse the issue). It is better to use multiple sketches to show successive positions, rather than to note all possible states in a complex picture bristling with arrows like a porcupine. This also avoids the abomination of arrowheads on both ends of an unbroken line (see also fig. 15.1).

Even simpler cartoons may later be added to composite summaries of the EMG record.

C. PRESENTATION OF THE EMG RECORD

1. The Raw Signal

The best way to convince the reader that the signals that have been analyzed are really EMGs is to illustrate a portion of the raw record taken from some major muscles. The illustration should be of a portion of the record played on a fast enough time scale to permit resolution of the individual spikes of the EMG (see frontispiece of this book). If signals from adjacent muscles are illustrated side by side, the possibility of cross-talk may be visually evaluated. Furthermore, the illustration should include several events and an inactive (silent) portion. A vertical scale should indicate the voltage referred to the electrode tips for each signal (rather than the amplification of the chart recorder!). Finally, it is necessary

18.7. Various ways of illustrating the activity of muscles, in this case for a series of tongue-protruding muscles in toads. *A,* Mean spike number for intervals of 7 msec plotted above the baseline and the mean spike amplitude below the baseline as compared with a series of mechanical events *(dash-dot lines).* Note that time axes and base lines run vertically in this figure. *B,* Separate values are combined, thus the plot gives percentage of spike number times amplitude, normalized to the maximal value observed for each muscle. Note the indication of the actual maximal value, without which one could assume that the maximal output of the several muscles might be equal. *C,* Plot of the variance calculated for each interval. Theoretically, one may posit that the variance will be least during the interval of biological interest for the particular muscle. (From Gans and Gorniak, 1982a.)

to include a time base (perhaps as a bar measuring a certain number of milliseconds in length) and to note the direction of paper travel if time does not run simply from left to right (e.g. fig. 10.5).

Given these aspects, the record will give an immediate impression of the signal-to-noise ratio and of the level of baseline noise, and it will confirm that there is neither obvious clipping nor other deformations. If you are going to document the correlation of the EMG signal and a mechanical event, you may well include an analog or similar record. Note critical events with vertical arrows, perhaps crossing multiple traces to provide a synchronization reference. Also note whether the records have been offset to compensate for transducer delays.

2. Block Diagrams

Illustrations of raw data indicate signal quality and little else. The spikes depicted likely will differ even among the successive repetitions of a cyclical sequence. Hence, they are often abstracted into various kinds of "histograms," the length of which shows the duration of activity with the height of each bar showing instantaneous magnitude. These by

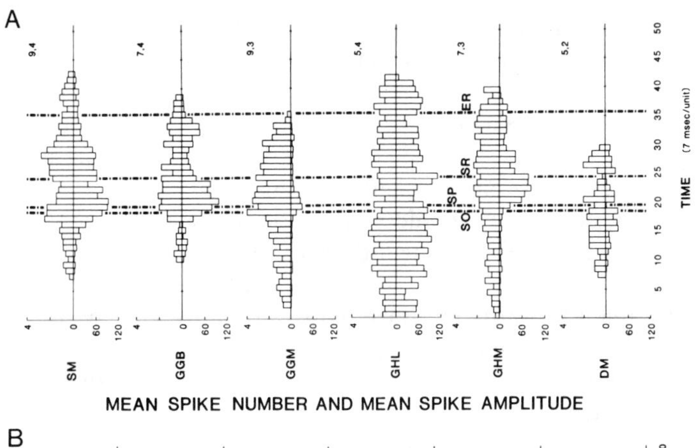

MEAN SPIKE NUMBER AND MEAN SPIKE AMPLITUDE

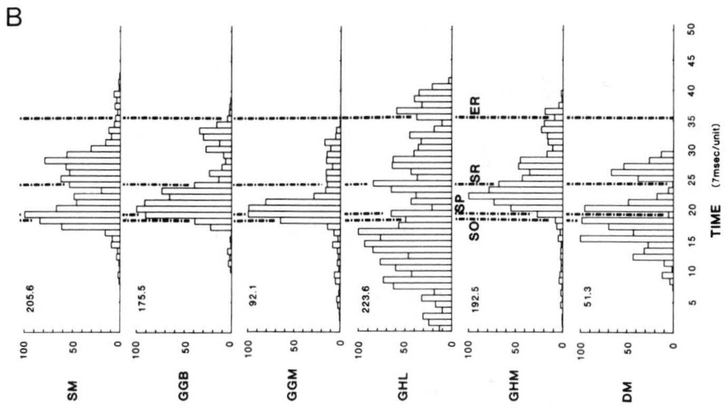

PERCENTAGE OF SPIKE NUMBER TIMES AMPLITUDE

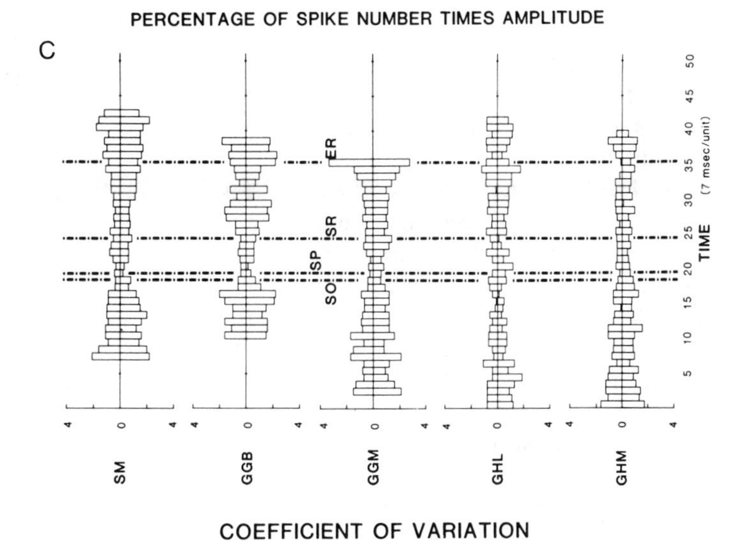

COEFFICIENT OF VARIATION

themselves represent a bare minimum of signal illustration. Implicit within them may be a statistical reduction of data; if so, it must be described mathematically.

In general, the envelope of such a histogram represents an abstraction; means and various descriptors should be potentially available for any point although there can be no assumption that the data will be normally distributed. Obviously the start and cessation of activity will be both independent and important, although in some situations the significant factor may be the duration of the event (or of the gap between events) rather than either of these. Hence, bar diagrams of duration should indicate the range and measure of deviation for each extreme, as well as those for the duration and interval, if deemed significant.

As seen in chapter 17, the vertical dimension of the histogram may represent any one of a number of attributes of the EMG tracing. It may be useful thus to plot the maximal spike height per unit of interval, the mean number of spikes, or their average height. In some applications spike number and amplitude change independently (fig. 18.7); this may be illustrated by plotting these as a flip diagram with one aspect above and the other below the line. Sometimes, the EMG record contains what may be described as biological noise. The activity of interest may then be included in a string of seemingly random action, produced by the overlap of activation processes resulting from other activities of the animal. In this case the recorded EMG signal may reflect all of these. It then may be useful to plot the variance of the particular attribute of the EMG signal for each interval; this is likely to be lowest in the zone of biological interest.

We have already seen that the electrical output of muscles, even if recorded with standardized electrodes, differs depending on the cross-sectional area of the active fibers and other factors. This is often ignored in illustration. High- and low-voltage signals are shown as equivalent bars; indeed, some automatic graphing systems now make it facile to express instantaneous muscular activity as a percentage of the maximal value within the string of interest, or of the maximal recorded value for any activity. This may be an informative way of illustrating real changes, but should always be combined with the inclusion of the actual voltage values.

The availability of automated spike and amplitude counters, which can sample long strings of EMG activity, now allow one the option of having the system automatically present the data as a scatter diagram—for instance, of spike number versus spike amplitude. Should the region sampled contain two populations of units that have different properties but act independently during overlapped periods of activity, one can sometimes sort these out because they cluster in different portions of the field.

D. GROUPINGS AND COMPOSITES

Groupings and composites include all those figures that involve more than the simple demonstration of the EMG signal and an analog trace; thus, one may combine EMG with cine and multiple analog records to resolve the complex associations studied. The EMG normally is not studied by itself; as it is an indicator of muscular activity associated with particular mechanical events, illustrations combining maps of the mechanical events with those of EMG magnitudes will often be informative.

The kinds and formats of combined illustrations that will ultimately be useful depend on the innovativeness of the investigator (and of the artist). Most important, although here mentioned last, remain the questions of what is most significant about the observations and what does one desire to communicate. Once these critical matters have been resolved, the remaining problem is how these phenomena are best illustrated. Figures are only one part of the message, so one must decide how they are to interface with tables of raw and processed data.

One fundamental rule is to keep things

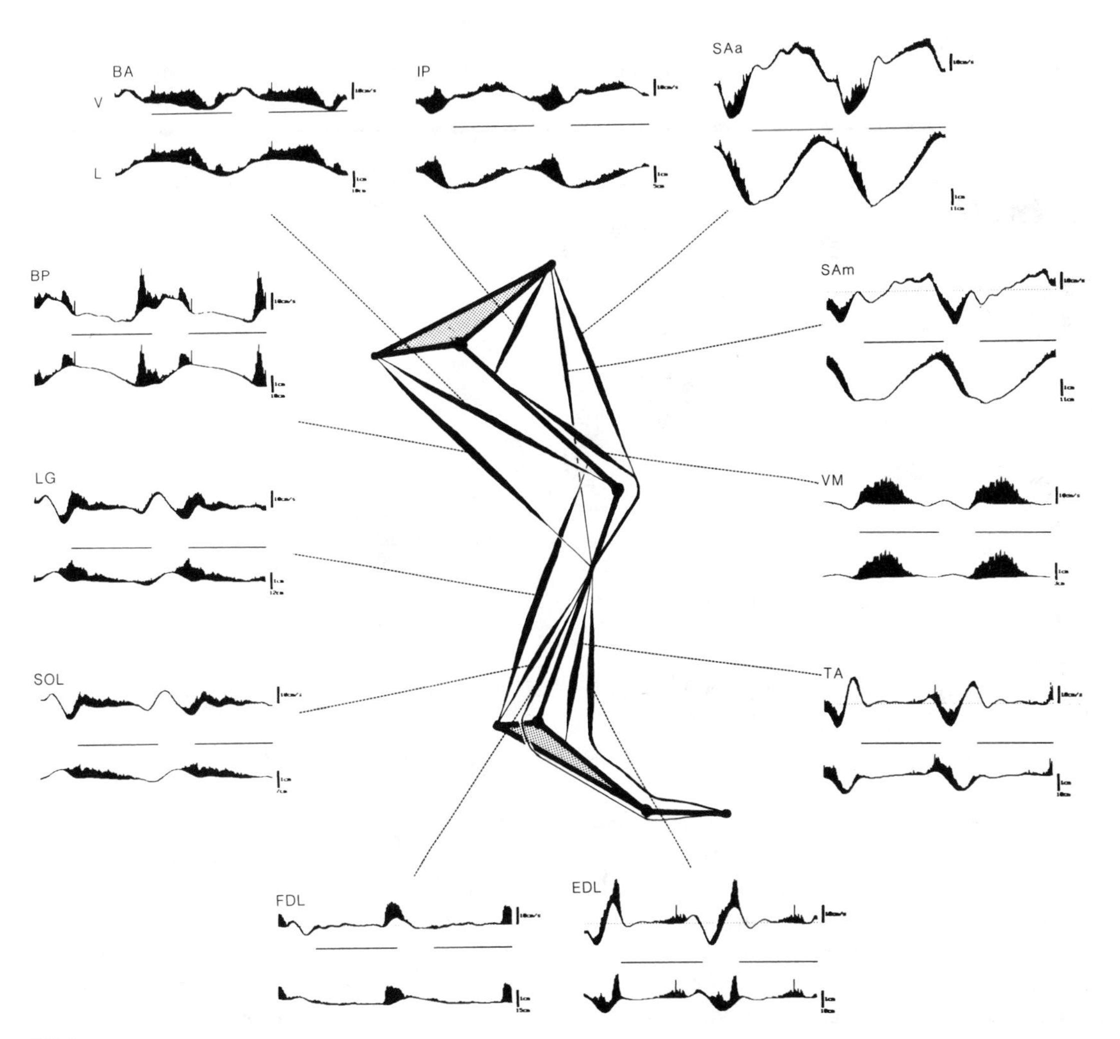

18.8. Computer-generated family of tracings showing average EMG activity from many step cycles of a cat hindlimb superimposed on muscle length (*L*) and velocity (*V*) for two cycle sequences. Length and velocity of each muscle were calculated from reconstructed video stick figures, based on a computer model of the gross anatomical pathway of each muscle in the limb. *BA*, Biceps anterior; *IP*, iliopsoas; *SAa*, sartorius pars anterior; *SAm*, sartorius pars medialis; *VM*, vastus medialis; *TA*, tibialis anterior; *EDL*, extensor digitorum longus; *FDL*, flexor digitorum longus; *SOL*, soleus; *LG*, lateral gastrocnemius; *BP*, biceps posterior. (Courtesy G. E. Loeb, W. B. Marks, and W. S. Levine, work in progress.)

18.9. Raster (generated by a computer) of reflex responses to a sequence of sural nerve electrical stimuli delivered as a cat walked on a treadmill at a constant speed. Activity in each trace of the tibialis anterior EMG was rectified and integrated into 2 msec bins prior to analog-to-digital sampling for 20 msec before and 100 msec after each stimulus (at *0*). Stimuli occurred at 2 second intervals as the cat walked at about 1.3 steps per second, so they were essentially random with respect to phases of the step cycle. Point of stimulation in each step cycle was derived from an implanted length gauge across the ankle by identification of the inflections in the digitized record and calculation of the phase of the stimulus. The traces were then reordered by phase of the stimulus (Phillipson phase divisions, shown at *left*). Prestimulation control EMG levels were integrated and smoothed into the histogram shown at the *right* of the raster. All traces were summed in synchrony with the stimulus to provide the histogram at the *top,* showing mean temporal patterns. *Diagonal dashed lines* indicate phase delay in normal expected EMG level after stimulation as a result of latency being later in the step cycle than at the point of stimulation. (From Abraham, Marks, and Loeb, 1985.)

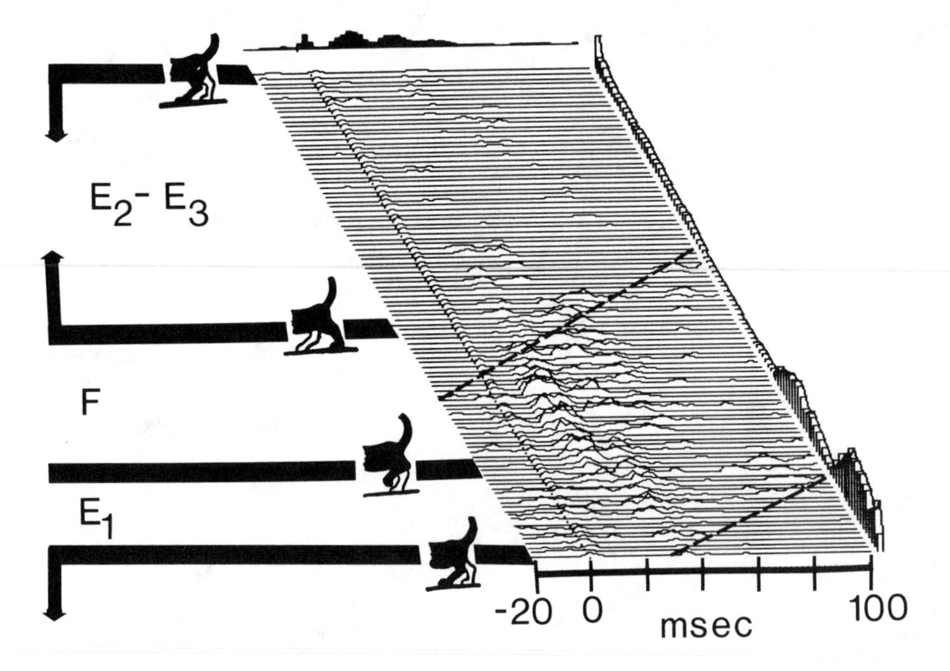

simple. If two events are associated, show them on the same diagram. If the mechanism is complex, break it into components and show each separately. A series of sketches, each showing one phase of a complex activity, is preferable to a single combined diagram highly decorated with code letters.

The simplest such correlation diagrams combine the EMG with the output of a device such as a length gauge (see fig. 18.8) strain gauge, pressure recorder, or microphone. Should the event be regularly cyclic, the multiple steps of a cycle may be demonstrated in sequence in a single diagram showing the EMG magnitudes correlated with different phases of a cycle. Such diagrams should include more than a single cycle to show intercycle relationships. For more complex cyclic movements one may wish to show the positional states occupied by the animal during each phase as a series of sketches. These may be placed lateral to or above the EMGs and coded by letters. Rather than cramming sketches into a horizontal bar graph, one may show them as a vertical sequence, presenting the EMG magnitudes for each muscle as voltage levels or percent of maximal EMG at the position indicated (see fig. 18.6).

Other attributes of the EMG record permit plots that allow immediate check of EMG magnitude with that of the effects produced—for instance, the inhalatory effort of a snorkeling turtle inhaling against a variable overburden of water. Such illustrations can indicate not only force and EMG signal correlation but also its possible dropoff near the ends of the range of force and displacement.

Computer analysis and intermuscle correla-

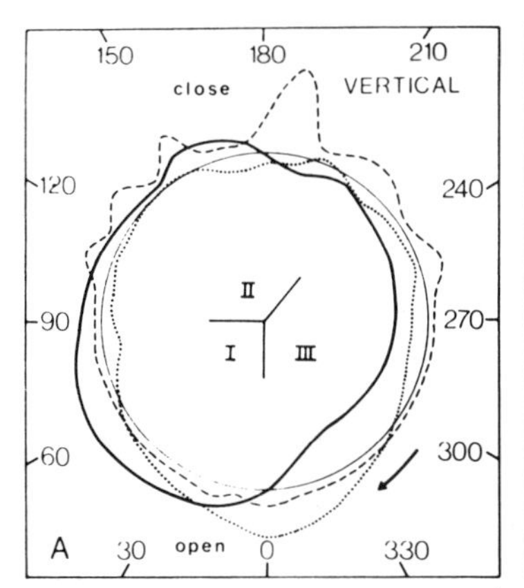

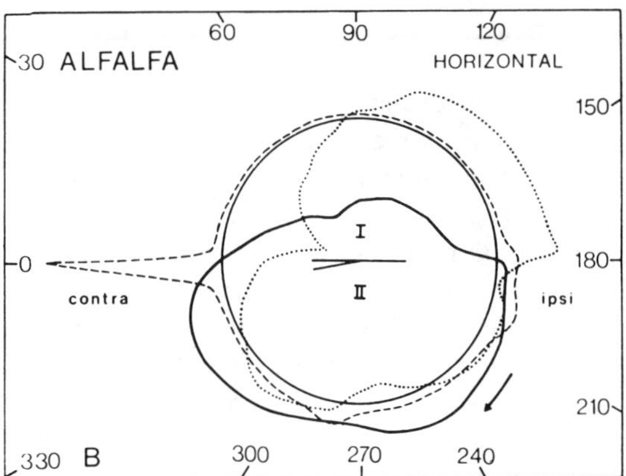

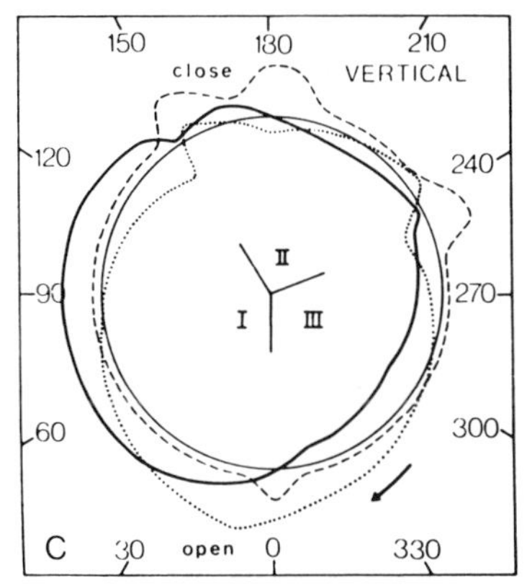

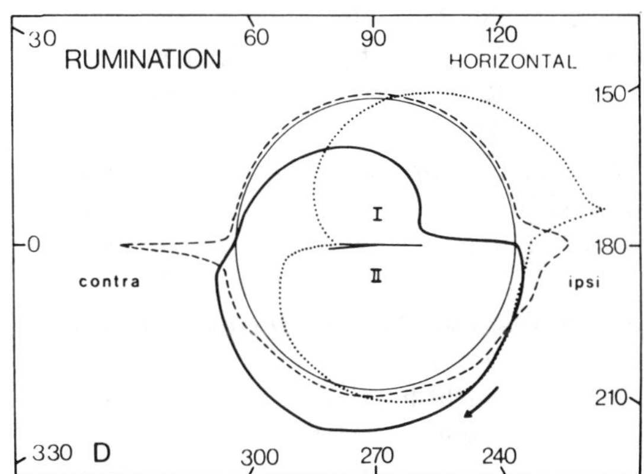

18.10. Regularly cyclic events may be plotted on polar diagrams rather than linearly, in this case to compare the movements of the mandible of a goat masticating alfalfa and cud. Polar coordinate graphs show the distribution of velocity (*heavy line*), acceleration (*dots*), and time (*dashes*) during which the jaw is at a particular phase of mastication. Opening and closing velocities and accelerations are indicated as displacements toward and away from the center. For each food type there is a plot in the horizontal (left and right, or contralateral and ipsilateral positions of the mandibular tip) and vertical planes. The same method of plotting may be applied to the descriptors for EMGs of the driving muscles. These sketches not only indicate that the animal masticates these foods quite differently but also allow conclusions regarding the nature of the differences. (From de Vree and Gans, 1976.)

tion techniques make it possible to survey activation times or intervals among those of different muscles and, if informative, to graph these in illustration. A tight histogram—for instance, one resulting from stimulus-triggered rastering and averaging—is likely to provide an immediate impression of association of responses to triggering events (fig. 18.9). Always consider including data from an uncorrelated event to document the differences within the series.

Whenever displacements of animal structures have been recorded, it is useful to reduce these to their time derivatives and illustrate velocity (fig. 18.8) and acceleration. After all, acceleration requires force; as this is the end result of muscle activation, the acceleration (second derivative of position) may yield the most obvious correlations. However, the demonstration of visually correlated acceleration values does not in any case obviate statistical treatment of the data.

More than one complete cycle of a cyclic behavior should be shown, so that all transitions can be easily seen. Regularly cyclic movements may also be shown as polar diagrams in which the central 360° turn represents the time or the aggregate displacement of a single cycle. On concentric circles around this will be mapped the EMG values, or derivatives of other attributes of the cycle, such as the acceleration then exhibited and the force measured at a tendon (see fig. 18.10). A sequence of such diagrams permits one to visualize the changes exhibited among the mastication or locomotion series shown by different animals. Alternately, such diagrams may be used to compare the behavior of animals of a single species in response to varied behavioral conditions (food types in mastication) or of different species under equivalent conditions. Naturally, the selection of the parameter mapped and the scale of mapping will be critical to any successful explanatory scheme. If you have averaged data from many such cycles, be sure to include the reference point in the cycle, as the temporal uncertainties increase the further one looks from such a reference.

19 Medical and Surgical Techniques

A. HUSBANDRY

The care of experimental animals is a science and an art. It must be an integral rather than an incidental activity. Hence, the arrival of animals in the laboratory should follow rather than precede some careful planning. Each species likely has some peculiar requirements in terms of space, diet, illumination, humidity, and temperature; remember that the latter values cycle daily and seasonally so that continuous maintenance of animals at mean "optimal" values is unlikely to produce optimal conditions for them. Also remember that some species are solitary and others are gregarious. Introduction of many solitary individuals in a group enclosure will produce a skewed mortality curve; in that case, the last individual is likely to survive for a very long time.

You might not know about all of the behavioral characteristics of the species you propose to study for some esoteric reason; presumably someone does, and more information about exotic species often may be found in the literature about their natural history. The closer the animal-holding facilities are to natural conditions, the more comfortable the animals are apt to feel between experiments and the greater the likelihood that your results will indeed have biological applicability. Quite apart from the fact that disturbed animals provide inadequate data for most of the studies utilizing EMG, it remains necessary for humanitarian reasons to minimize trauma and stress on any living creature.

We know that EMG is a technique that involves conscious animals. Consequently, the psychological conditions of the beast remain a significant and critical component of the study. Many investigators who never consciously practice cruelty forget that wild animals will naturally be disturbed when encountering other animals, or other unexpectedly moving objects, that are 20 or more times their size. Hence, shields and blinds are often useful adjuncts to holding and experimental situations, as are nest boxes, hollow logs, and other devices that let animals isolate themselves from people, indeed from each other.

Different species, particularly predators and prey, are usefully kept separate, and you should note that they may recognize each other not by visual but by olfactory or auditory cues. Also remember that some species can be trained or conditioned; however, the response of animals to such actions differs among species and individuals. A useful technique for conditioning some species is to provide food or similar positive reinforcement whenever an animal must be moved or otherwise manipulated. Finally, note that animals have a remarkable capacity to tell individual people apart; even some snakes learn to recognize particular keepers. Consequently, it is an excellent idea to keep the areas in which the experimental animals (even lower vertebrates!) are housed accessible to only the few staff members actually involved in the training and care regimens. Discourage visits by strangers (particularly those who like to tap the cages to make the animals move). Check that the custodial staff has been informed of the reasons for these rules and is responsible; one of us once lost some months of work because a janitor would enter late at night to play with the instrumented animal.

There should be staff responsible in the care of each animal, and weekend and holiday rosters of staff should be made early enough to ensure that no last-minute problems arise. Rules regarding those responsible should be firmly established; there must be a backup available should a keeper suddenly become ill or be detained. Also, the issue of who is to be notified if the animals show signs of illness should be determined early in the program.

Restriction of access to the animal facility has the further advantage of maintaining security, which can become a rather serious problem. Months of work may be lost and experimental animals be exposed to severe trauma if they are removed from the holding area by "visitors." These include persons who steal animals from zoos and laboratories for a dare or to add specimens to their private collections. It applies as much to persons who chose to liberate animals because of a commitment in opposition to any research on animals. Once animals have been removed from the wild they can no longer fend for themselves, and the experimentalist has the responsibility of protecting them.

Various agencies of the United States Government and several of the professional societies supporting electrophysiological research have recently issued guidelines for standards of animal facilities and care. Even though these probably represent only an initial basis for minimal levels of animal care, we include three of them (appendix 3) as samples of issues to be considered in the specific case.

B. THE EXPERIMENTALIST AS SURGEON: APPROACH TO SURGERY

Even the insertion of percutaneous electrodes represents an invasive technique and has the potential for generating medical problems. In other approaches it becomes necessary to cut the skin. With this the experimentalist enters the realm of surgery.

Any student of laboratory biology frequently has recourse to surgical instruments to perform anatomical dissections and to prepare terminal physiological experiments. Thus, it is not surprising that most graduate investigators feel comfortable, even proud, of their manual dexterity at using these special tools. However, any clinically trained surgeon or veterinarian can tell you that the techniques required to expose living tissue successfully and then put it all back together again share little, besides those instruments, with laboratory dissection techniques.

Unfortunately, students of physiology have few opportunities to come into contact with surgeons or to work with chronic animal preparations during their training. This chapter is intended at least to raise their consciousness to this yawning chasm that has the potential for causing distress to the experimental animals and has swallowed so many otherwise well-designed experiments. Although acknowledging that manual techniques cannot be mastered without guided on-the-job practice, we would point out that such practice will result in improvement rather than frustration only if the student has been alerted to the fundamental processes at work.

C. SURGICAL PRINCIPLES

1. Plan

It is amazing how often one finds oneself staring vacantly into some macerated tissues and wondering what to do next. Sometimes this is because Mother Nature has ignored the carefully constructed plan of attack, but too often it is because the investigator is mentally window-shopping. Just going in and having a look is wasteful enough in the gross anatomy laboratory, but it is disastrous in the operating room. Here, the number of surgical complications (such as anesthetic death, hypovolemic shock, and postoperative infection) rises exponentially as time passes.

We have already emphasized the importance of designing EMG experiments and devices based on a careful study of the gross anatomy of the subject. While conducting such preliminary dissection, you should be sure to make careful notes and sketches regarding the best location for incisions and to note landmarks that guide the dissection quickly to the target. First, try out the device implantation and fixation techniques in a cadaver to make sure that the devices fit and the proposed surgical fixation actually can be executed without unduly traumatizing the subject, or requiring more hands than you possess.

During and just after actual implant operations, repeat and annotate your plan sketches for the surgical approach and the devices. Force yourself to write or dictate a postoperative note while your ideas and experiences are fresh in your mind, no matter how late the hour. Changes in surgical approach may be needed to accommodate anatomical variation, or you may think of special instruments to facilitate tricky procedures.

Be sure to consult these notes and take their advice. There is a tendency to avoid spending a few hours on procedural changes or special devices in the hope that things will simply turn out better next time. Whenever there is little invested in each procedure and animal, such an approach may be efficient. However, each chronic physiological experiment usually entails many hours and even days of training animals, building devices, and setting up recording equipment. The success of it all cannot rest on a 50% chance that, in one brief moment, you will grab the wrong thing with a clamp.

Also plan on having an assistant available to deal with unpredictable contingencies. This person should not be sterile but should be prepared to reach for and sterilize or replace that dropped instrument. Finally, have a good general anatomical atlas in the operating room. After all, you would not drive cross-country without a road map, no matter how much you trusted your memory or the advice of gas station attendants.

2. Workspace

A good scrub nurse in an operating room will frequently point out to the surgeon or the circulating nurse that the lights should be adjusted or a retractor repositioned. Few physiologists have the luxury of a scrub nurse on hand when they operate, so it is important to ask why surgeons themselves so often fail to make such observations.

The incision is your work space. It has to be in the right place and of adequate size in which to work, and it has to be properly lighted. It is not a constant space, but changes in size, shape, and depth as you delve deeper into the various layers and reposition retractors. The intensity of concentration on the goal in mind can blind you to the very things hindering your achievement of that goal. Fixing the light or extending the incision will interrupt and delay the work at hand, but often these actions will more than return the invested time and effort. A surgeon's halogen headlamp and fiberoptic illuminators are useful options for illuminating dark corners, but only if you remember to use them.

3. Tools

Surgical instruments are often found lying loosely in poorly marked drawers, if they can be found at all. It is one thing to have to stop fixing the lawnmower in order to search for the right wrench and another to interrupt a surgical procedure on an anesthetized animal to look for a decent fine forceps. The best way to limit the natural migratory tendency of surgical instruments is to use cataloged and wrapped packs. If your species and procedures require aseptic technique, this is easily combined with the need to sterilize the instruments. Once you have constructed a list of the instruments actually needed for your procedures (perhaps with some sublists for specialized variations), you will be amazed how easy it is to keep one or more such sets complete and serviceable.

In making up these lists, you should think

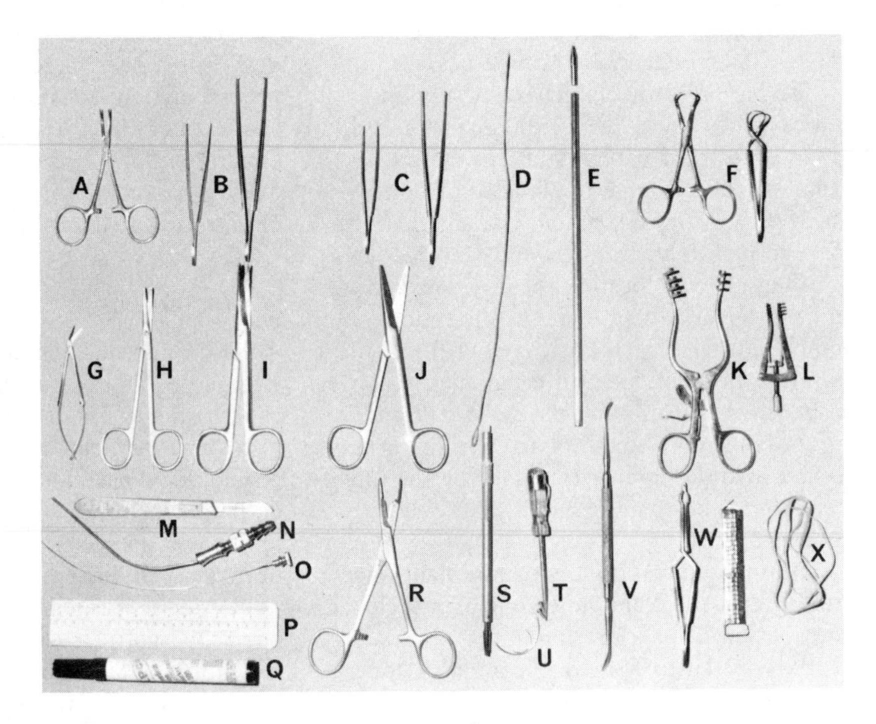

19.1. Basic instrument set for general surgical procedures including installation of long-term electrodes and passage of percutaneous leads in mammals. This set should be supplemented with an ample supply of sponge gauze, sterile saline in a bowl, drapes, electrocautery forceps, and suction line tubing. *A*, Mosquito hemostat; *B*, toothed forceps; *C*, serrated forceps; *D*, malleable lead-passing probe; *E*, hollow lead-passing probe; *F*, towel clips; *G*, Castro-Viejo scissors; *H*, fine curved blunt scissors; *I*, curved Metzenbaum scissors; *J*, straight suture scissors; *K*, Weitlaner retractors; *L*, microretractors; *M*, scalpel; *N*, suction tube; *O*, suction trochar; *P*, ruler; *Q*, indelible marker; *R*, Olson-Hegar needle holder; *S*, screwdriver, screw holding; *T*, screwdriver, magnetic with bone screw; *U*, curved needle; *V*, small periosteal elevator; *W*, Wachenfeld wound clips and applier; *X*, rubber bands.

about the stages of various procedures (fig. 19.1). There should be one general instrument pack containing the basics: scalpel handle, forceps, towel clips, clamps, and needle holders. Having such a sterilized set on hand can be a real lifesaver (at least for the subject) if a device suddenly needs repair in the middle of a chronic experiment. It may be useful to have a sterilized pack of just prepared items, such as sponges, towels, suction tubing, fixation clamps, and pins. Specialized instruments, such as bone cutters or very fine microsurgery instruments, often deserve their own packs, which will help keep these critical instruments in good repair by isolating them from abuse and by facilitating their being sent out for sharpening and adjustment.

Some instruments deserve special attention and handling. Very fine cutting instruments such as iris scissors and nondisposable scalpels should not be autoclaved because the steam will dull their edges. "Instrument milk" or similar nonirritating lubricants may protect instruments after cleaning. However, for most animal work, careful cleaning and dry storage bins that permit the tools to be dipped in 70% ethanol and air dryed at the time of operation are likely to be adequate asepsis. (It is a little known fact that 70% alcohol kills bacteria by dehydrating them during the air drying process rather than by direct action of the alcohol itself). If this is considered inadequate, gas sterilization with ethylene oxide is the next best choice, although the outgassing time makes this a considerably slower procedure that cannot be done in the middle of operation to cope with a dropped instrument. Avoid surgical antiseptic solutions; they are not that superior to ethanol and the instruments will have to be rinsed carefully to avoid contact of the tissues with their toxic residues. Most important, keep your surgical instruments separate from dissecting tools. Remember that formalin and glutaraldehyde are apt to leave toxic coatings that persist even after the tools have been scrubbed.

All types of scissors will be seriously dulled if they are used to cut either electrical wires (obviously) or even cotton gauze and various suture materials. One useful rule is to keep two pairs of general use scissors in the instrument set—one straight and one curved. Then convince everyone to use the curved scissors only on tissue and sacrifice the straight scissors for general use.

Fine forceps require protection for their tips while being knocked about with other instruments. A short piece of snugly fitting, autoclavable rubber tubing slipped over the closed tip end is the best bet. Keep lots of short pieces of tubing on hand in a nearby drawer, because these always seem to get lost during cleanup.

One very handy tool is the glass probe or "nerve hook." Such tools are easily shaped and polished in the flame of a Bunsen burner to whatever sizes and shapes may be desired. Distributors or manufacturers of glass-blowing supplies can provide opaque glasses that melt at low temperatures and are easily seen in the surgical field. Clear glass probes can be coupled to fiberoptic illuminators to provide a convenient source of light at their tips.

4. Understanding the Properties of Animal Tissues

Living tissues have two essential properties as far as the surgeon is concerned: extensibility and metabolic rate. Interestingly, these properties tend to be inversely related. The tensile properties of tissue are obviously critical to the method of blunt dissection, the placement of sutures, and the anchoring of devices. The metabolic properties are less obvious but even more critical for chronic procedures, because they dictate the response of the tissue after its blood supply has been compromised, which is an inevitable consequence of any surgical intervention. Tissues, such as bone and tendon, that tend to have the lowest metabolic rate are the best for strong anchoring. Those that have the highest metabolic rate, such as muscle, nerve, and skin, are least useful and rapidly necrose if their blood supply is interrupted by surgical transection or by applied pressure.

Yes, skin really was included under least useful tensile materials in the paragraph above. As noted in chapter 11, the active dermal layer itself is easily compressed, which shuts off its dense capillary network; if continued, this causes necrosis of the skin. Skin does not hold sutures well under constant tension. The word "skin," like many other loosely used structural names, actually refers to a composite of many different tissue types present in varying proportion in different locations on the body. The strength of skin frequently resides in collagenous subdermal layers, which can be utilized specifically with some of the many elegant suturing techniques known to surgeons (see below). Similar considerations apply to muscles, which are usually invested with variable thicknesses of fascia, an excellent substrate for sutures. However, muscle fibers themselves have virtually no strength perpendicular to their axes, and their high metabolic rate causes them to turn to necrotic mush if they are sutured.

5. The Germ Theory Revisited

It is generally believed that infections are caused by bacteria and that antibiotics can prevent and cure them. Although both statements have some basis in fact, they are very misleading in their implications for the prevention and treatment of surgical wound infections.

Ubiquitous and opportunistic bacteria will make a home in culture media provided for them by a technically incompetent surgeon, resulting in wound infection. All the aseptic technique in the world cannot prevent a few bacteria from lodging in a wound, having been carried there via airborne dust, remained embedded in the pores of the skin, or transferred from the gums into the bloodstream during a hearty meal. If they land in well-perfused healthy tissue, the natural immune system makes quick work of them. If they land in a hematoma, a bone chip, or a frayed tissue fragment, they will set up housekeeping and multiply until their numbers permit an assault on adjacent healthy tissue. If there is a foreign body present, such as an electrode and its associated lead wires, the bacteria will travel along these surfaces, protected by the niches and spaces created by surface irregularities and the relatively avascular encapsulation layer of connective tissue. These infections can rapidly spread far; they may be very difficult to eradicate because of their chronic, recurring nature.

The best protection is the elimination of the culture media and a reduction of the initial inoculum to a manageable quantity. The best way to achieve this is to practice good techniques of surgery, rather than dissection, and to irrigate, irrigate, irrigate. A number of specific pointers regarding frequent mistakes of surgical technique are given later, but the main rule is to work quickly and efficiently. Make every movement count, so that tissue is not unnecessarily poked, stretched, clamped, burned, cut, or air dried. Keep a bowl of saline or Ringer's lactate solution handy at all times and frequently wash tissue debris and blood from the surfaces on which you are working, as well as from all cul-de-sacs. Irrigation with fluids at room temperature will not damage mammalian tissue and may even improve it by temporarily lowering its metabolic rate and causing some local vasoconstriction. Just be sure that body core temperature is being adequately maintained, with a heating pad if necessary, and avoid physiological measurements until the structures have warmed.

Be particularly thorough with irrigation during and after drilling through bone, both to prevent overheating at the drill bit and to wash out bone chips, which make particularly attractive culture media because of their mechanically inaccesssible niches. As you close each layer, vigorously irrigate around, among, and under sutures to flush out dead tissue and blood and to reveal any shortcomings in your hemostasis. Finally, roll a tightly wadded sponge across the sutured wound along the axis of the incision to force out any remaining free fluid. If the emerging liquid shows more than the faintest pink tinge, consider carefully going back in to control bleeding; at least apply a pressure dressing.

Specialists in the field of infectious disease control have spent the past 15 years coming up with scientifically sound and empirically effective regimens for the prophylactic use of antibiotics by surgeons, supplanting a long and hotly debated tradition of spreading them around like amulets against evil spirits. It behooves anyone calling himself or herself a scientist to take note of these quite sensible findings.

First, prophylactic use of antibiotics does not prevent wound infections. It only guarantees that possible infections will not be treatable with the antibiotics that were used prophylactically. Second, antibiotics circulating in the bloodstream or sitting on the surface of the wound are ineffective against bacteria sitting in devascularized deep structures.

If you do use antibiotics prophylactically, you must administer them at least 30 minutes before operation to guarantee that they will be in the tissue that you propose to traumatize. If your animals have had wound infections in the past, the best bet is to use whatever antibiotic was effective against that organism, because it is likely to be endemic to your animal quarters. There is no point in giving the antibiotic for more than one or two doses before and again after operation; if you do not eliminate the bug completely, you will encourage replacements that are resistant to your best weapon. If infection develops, use a broad-spectrum antibiotic of a different basic chemical type, on the presumption that the infection will be resistant to the antibiotic you administered prophylactically. Chloramphenicol is a good choice, because it has been out of favor for human use for many years. Do give the antibiotic systemically rather than topically. Pink and yellow powders and ointments painted over the skin may be decorative, but organisms generating the pus underneath will not be impressed.

If all else fails, treat surgical wound infections surgically. Few abscesses are ever cured by antibiotics alone, and never when they include foreign bodies. If a suture line breaks down (dehisces) a week or more after operation, it is because the skin is necrotic and incapable of supporting itself, much less healing in the face of a microbial army. All necrotic or even marginal tissue must be surgically debrided (cut away), including the entire perimeter of the dehiscence. This is called "ellipsing." Use scissors on subcutaneous tissue and a scalpel on the skin to get to the tissue that will bleed when cut, indicating a viable blood supply, always your best ally. Close carefully, in several layers, to distribute as best you can the inevitable tension caused by the loss of skin.

If you do decide to nurse a small, superficial infection along for a few days with systemic antibiotics, probe the area carefully with a cotton-tipped applicator to be certain of its real extent and irrigate it frequently into those depths with dilute hydrogen peroxide (the standard 3% drugstore solution). The mechanical foaming action plus the high free oxygen levels are ideal for cleansing the wound without inhibiting the granulation tissue (metabolically active vascular bed) that must gradually fill in the defect in the time-honored process of "healing by secondary intent." You can instill dilute antibiotic solutions into the wound if it makes you feel better to do so, but be sure to avoid anything toxic to the healing tissue, such as alcohol, iodine, or nonisotonic solutions. Be vigorous when irrigating, as granulation tissue has no nerve supply, so the animal will not mind.

If the nature of your subjects and procedures makes wound infections more than a rarity with even the best techniques, it is advisable to look into some form of microbiological testing of wound cultures. Consider the microbiology laboratory of an affiliated hospital or a consulting veterinarian. There are several very inexpensive and easy to use commercial lines of bench top incubators and prepared culture media, complete with sets of test disks for antibiotic sensitivity. Even if you never become adept at identifying gram-stained enemy organisms, a little intelligence about their weak spots will vastly improve your choice of weapons.

6. "Inner Surgery"

There are several sports education books regarding activities such as tennis and skiing that admonish the reader to attend to his or her mental attitudes and emotional set at least as much as to the mechanical actions required. Although no one has written such a text for surgeons, it is worth asking what is going on in one's head as one is rearranging a living animal.

We have already noted instances in which a third-party observer may notice important technical handicaps such as lighting or poor choice of instruments. One of the most important aspects of the clinical training of surgeons is to enable them to act as their own efficiency expert, mentally analyzing every motion and every procedural decision in terms of its purpose, its efficacy, and its risks. For example, if you make the incision smaller to minimize trauma and suturing time, you may encumber the work space and lose the anticipated gain. If you cut quickly through a bit of obstructing connective tissue, you may overlook a blood vessel in need of cautery or ligation, which will be much harder to deal with after it has been cut. Obviously, the more often you do the same procedure, the more secure you can be in electing certain shortcuts, based on familiarity with the local anatomy and its variants.

The emphasis on surgical speed may seem to be a "macho" obsession among clinical surgeons, but it is well founded in mortality and morbidity statistics. Good, objective studies indicate that operating time is closely correlated with the level of complications, such as wound infections and prolonged convalescence. The meticulous approach toward anatomical dissection demands patience and rewards the time spent with accuracy and completeness of observation. However, the surgeon must remember the job to be done, in this case the implantation of electrodes as atraumatically as possible so that the real study of physiological properties can proceed. Therefore, the objectives are to know what you are trying to do, to do it quickly, and to get out neatly.

D. COMMON ERRORS OF SURGICAL TECHNIQUE

1. Anesthesia and Vital Support

a. General. There are no general cross-species rules for the use of anesthetics. Always consult the literature and a veterinarian before proceeding with a new species. Also label your syringes, particularly if multiple drugs are available for the procedure.

b. Lack of Pharmacological Control. Maybe the procedure was only supposed to take 10 minutes, and maybe the intraperitoneal anesthetic should have lasted that long, and maybe significant bleeding was quite unlikely. However, it only takes a few minutes to start an intravenous line, and it will be your only handle on the situation if things go wrong.

c. Misuse of Drugs. Some aspects of human pharmacology, such as routes of administration, half-lives, and tissue perfusion, tend to be unchanged in animals. Others, such as dosages based on body weight, involve special considerations. Veterinarians and veterinary pharmaceutical houses have lots of information about these things. If the need is great enough to warrant use of the drug in the first place, it is worth reading the package enclosures and considering the appropriateness of the choice beyond its availability in the refrigerator. Remember that each new species involves a new challenge. Mice, rats, and guinea pigs may all be rodents, but they differ drastically in their responses to drugs. Anesthetics that work well on vipers may kill some cobras (e.g., halothane). Always do a bit of library research and then run some trials if a new animal is to be approached.

d. Bad Posture. When a human or an animal is anesthetized, it cannot tell you that its peroneal nerve is being compressed against the metal table or that its airway is half occluded. A competent anesthesiologist must think about these things for the patient and consider how the patient is going to feel after lying in the chosen position for several hours. Consider the architecture of your subject; a back downward position may cause difficulties for some animals.

e. Lack of Temperature Control. The usual thermoregulatory mechanisms of birds and mammals are often severely compromised by general anesthesia and the prolonged recovery process. Exposed tissues lose heat rapidly via evaporation and surgical irrigation. On the other hand, an excessively warm animal will attempt to lose body heat by vasodilation, increasing bleeding in the surgical site. Be sure to provide a gentle amount of warmth by wrapping the unneeded parts of the body loosely, perhaps with a warm water circulating pad. An incubator for premature infants can be very handy for providing both warmth and humidity for the first few hours of the recovery period, but remember that the animal must have a reasonable gradient to dissipate excess body heat by convection and evaporation; 37° C at 100% humidity is rapidly fatal. Cold-blooded animals (ectotherms) pose different problems, and here cooling may be an excellent aid to anesthesia. However, be careful that your spot lamps do not cause local overheating.

f. Hypovolemia. Blood loss is extremely hard to estimate accurately and may only represent a small part of the loss of effective circulating volume, which includes respiratory evaporation, wound evaporation, serous discharge, tissue edema, and unreplaced natural secretions. Animals such as cats and dogs often have quite variable postoperative periods during which they are unable to drink liquids. As long as there is an intravenous line already in place, nothing is lost by slow administration of isotonic saline, perhaps 1% of body weight or more if particular losses have been observed. This will usually improve tissue perfusion, antibiotic uptake and circulation, and anesthetic agent clearance, as well as reducing greatly the chances of sudden death.

g. Poor Attention to Cardiac Rate. Particularly in studies on poorly understood ectothermal species, you may find that the second dose of anesthetic has less effect. The tendency is then to supply a third dose, and this may result in the subject's death. Cardiac rate may be a good indicator of general condition and thus of rate of perfusion, absorption, and clearance. It may well be monitored during prolonged and innovative approaches to previously unstudied organisms. Just hook up an EMG amplifier with good low-frequency response to a bipolar electrode consisting of one contact on each forelimb.

h. Use of Antidotes. Everything went well, the epidermal embroidery is beautiful, and then the breathing remains shallow and the pulse faint. Before you start, plan to have antidotes available for your anesthetics, as well as proven stimulants. Consult a knowledgeable colleague before rather than during the procedure. Remember to match the procedure to the species. Reptiles often respond well to a few forced breaths of a high-oxygen mixture directly into the glottis.

2. Tissue Quality

a. Overdissection. If you cannot find the structure which you seek, maybe you are looking in the wrong place. Stop and look for other landmarks. Consult your anatomical atlas. Do not keep looking around in the same place simply because the light is good there. Once you do find your target, remember that you just wanted to get an electrode into it. There is no need to have it cleaned and mounted like a trophy.

b. Devascularized Debris. Debris is the consequence of overdissection. Anything that does not look viable should be removed. Sometimes you must severely lacerate a particular structure to gain access to your goal. Take it out. If it was vital, the preparation is doomed anyway. You are kidding yourself and setting up the subject for an infection if you sweep the mess under the edge of the incision and close over it. Bits of subcutaneous fat and fat pads frequently are extraneous and are best removed, as long as vital nerves and vessels aren't coursing through them.

c. Hesitant Incisions. You simply have to know how much pressure to apply to a firmly held, vertically oriented, fresh, sharp scalpel blade to get through the skin in one cut. And you have to know exactly where and what is under the skin before deciding whether the margin of error is to be on the shallow or the deep side. Incising the skin in multiple slashes leaves multiple ridges of bleeding and generates nonviable dermal tissue.

d. Use of Scissors on Skin. The use of scissors on skin is a no-no. Scissors will crush hard, keratinous epidermal layers, compromising both the scissors and the regenerative power of the skin. Of all the forms of trauma, few heal more slowly than crush wounds.

e. Parallel Incisions. Much of the blood supply of skin travels laterally within the skin. There is a rule in skin grafting that a vascular pedicle (skin flap with blood supply) should never be longer than the width of its base. An incision that runs parallel and close to another incision or even to an old surgical scar is risky, indeed.

f. Inadequate Incisions. We have already emphasized that the full extent of the incision should be made in one continuous stroke. However, suppose that, despite your best intentions, you find yourself working in a long tunnel under the skin to get to your target.

Consider how long it will take you to extend the incision and then close it versus how much time you will waste mucking about in the dark.

g. Inadequate Strain Relief. Biological tissues are great at resisting brief stresses and nearly worthless for surviving constant stresses. Preventing the latter requires careful planning of incisions, the routing of lead cables, provision of adequate slack, and making the best use of stronger tissues, such as bone and tendon, to protect weaker ones, such as muscle and nerve.

h. Dried Meat. Water evaporates from warm surfaces, leaving behind dissolved solids. This is the equivalent of throwing salt crystals on the open wound, which you probably are unlikely to do intentionally. Frequent irrigation with isotonic saline solution, flooding all exposed tissues and skin edges, is absolutely necessary.

i. Use of the Scalpel on Subcutaneous Tissues. There are a few, rare circumstances in which the sideways motion of the scalpel can be used like a sharp periosteal elevator. However, cutting down onto subcutaneous tissue is a definite no-no. The instruments of choice for combined blunt/sharp dissection (but not for skin) are the curved, rounded-end Mayo scissors. Deft spreading motions are used to slip the closed tips under tissue needing to be cut and to evaluate the thickness, texture, and architecture (does it contain nerves or blood vessels?). Then the exquisitely sharpened tips, carefully pointed upward, are used to cut the elevated tissue with a single snip.

j. Serrated Forceps. You can distinguish the dissectors from the surgeons by their response to the choice between a pair of finely serrated forceps and vicious-looking, rat-toothed forceps. The serrated ones look so gentle, until you realize that they work by crushing the tissue in their jaws, ever more tightly as the operator concentrates on what the other hand is doing. The rat-toothed forceps will leave just three tiny puncture wounds no matter how much retraction is needed. Serrated forceps are great for handling implanted devices prone to puncture damage (e.g., insulated wires and length gauges); they should be banned from general surgical use.

k. The Sponge as Sandpaper. The ubiquitous cotton gauze sponge is very absorbent, making it ideal for dabbing away excess fluid and blood. It is not a washcloth. The microscopic surface consists of razor-sharp cellulose filaments capable of scratching glass, so just think what happens each time you wipe it across the raw, exposed surfaces of living tissues. Dab, but do not wipe.

l. Underuse of Cautery. The radio frequency electrocautery machine is a real boon to both surgeons and patients when properly used. The fire stick, as it is colloquially known, is fast and leaves little dead or foreign material behind. However, these advantages are often missed by those who do not understand its principle of operation and, therefore, miss out on the benefits. The underlying idea is that fluids such as blood conduct electricity better than tissues. In a dry operating field, the applied radio frequency current will concentrate on the bleeding blood vessel and coagulate its contents, stopping all but the worst arterial flow. As long as the cautery is on, any tissue adherent to its surface is rapidly burned away, leaving a charred residue that must be scraped off the tips from time to time. If you turn off the cautery while it is still in contact with the tissue, the coagulated clot will probably stick to the tips and then be pulled away, causing the bleeding to resume. With good suction, even a freely flowing leak can often be stopped, but the device is useless in a pool of blood or irrigation fluid. Cauterizing a vessel before it is cut is also fast and easy, leaving much less debris than even the neatest ligature.

A tiny soldering iron may be considered the poor man's cautery. However, it lacks the focal cauterizing effect of selective conduction. Use it with great care.

m. Overuse of Cautery. The fire stick is not a magic wand. Some tissues such as nerves and skin edges should never be cauterized, because it greatly retards their healing. Remember that the best hemostatic agent is tincture of time. Surgeons usually stop going after those little oozing capillaries once they are "down to a dull roar." The clotting time of most animals tends to be 2 to 4 minutes, so a little benign neglect or perhaps compression under a moist sponge for several minutes is preferable to constantly knocking off the forming clot. Major arteries must be tied with a ligature and not cauterized at all, lest they burst loose as the coagulated material is phagocytosed away.

3. Closure

a. Incision over the Device. If the incision must be closed over a protruding device or coil of lead wires, the skin will have its blood supply compromised at precisely the point that is most vulnerable and already compromised. Usually this leads to a wound dehiscence over a foreign body, a certain setup for a noneradicable chronic infection. Proper location of the incision represents a trade-off between surgical accessibility of the implant site and the need to have the closure away from the implant (cf. fig. 11.3).

b. Poorly Aligned Epidermal Edges. Epithelium grows until it meets other epithelium. If the cut epithelial edges of mammalian skin meet each other, this healing process can take as little as 2 days. If they are everted, with a short gap of upturned, healthy basal layers, the spread is almost as rapid. If the edges are turned inward, the epithelium will start growing downward into the incision, forming an ugly, slow to heal, and intrinsically weak scar. Furthermore, sebaceous glands and hair follicles are likely to be trapped on the inside of the wound, providing both a source and culture media for infections. Learn how to evert the skin edges by always orienting the course of the needle obliquely through the skin so that you grab a wider margin of the deep layers than of the superficial layers. Also, start at the middle and work by successively halving the suture line, rather than starting at one end and leaving a "dog ear" at the other whenever the amount of skin on opposite sides fails to match.

The skin of other vertebrates may pose quite different problems and encourage distinct approaches. Whereas healing may take longer, ectothermal vertebrates are remarkably resistant to bacterial infections, although fungal attacks may reflect an inappropriate moisture regimen. The keratinous external cover of reptilian skin responds well to some contact adhesives; for instance, a thin coating of Nobecutane Spray (Duncan, Flockhart and Evans, Ltd.) quickly stabilizes incisions. This has some merit (despite the caveats of the next section) as the skin of some reptiles and amphibians is remarkably loose. Folds and wrinkles occurring during the early stages sometimes heal as such. Not only should they be smoothed during the postoperative stage, but incisions should be planned along rather than across lines in which the skin is likely to be in tension. In any case, it is best to learn something about the natural history of your subject and consult the volumes written for zoo veterinarians who have faced similar problems for quite different reasons.

c. Goop on the Wound. Mother Nature learned how to cope with healing lacerations a long time ago. The proteinaceous effusion that comes from the cut skin edges soon dries into an antibacterial sealing layer that breathes freely and eventually biodegrades and is sloughed as a scab. Nothing you can buy from the pharmaceutical houses is anywhere near as good for mammals, and many products cause distinct problems by trapping fluid pockets underneath or irritating the skin, particularly if hair follicles are trapped by the layer of synthetic goop. This will cause many mammals to start chewing and scratching at sutures they would otherwise have left alone.

d. Knots Are for Boy Scouts. Face it, you probably do not really understand the physics and topology of the square knot, the granny knot, and the slip knot (two half-hitches). Making one and not the others with speed and proper tensioning in awkward positions

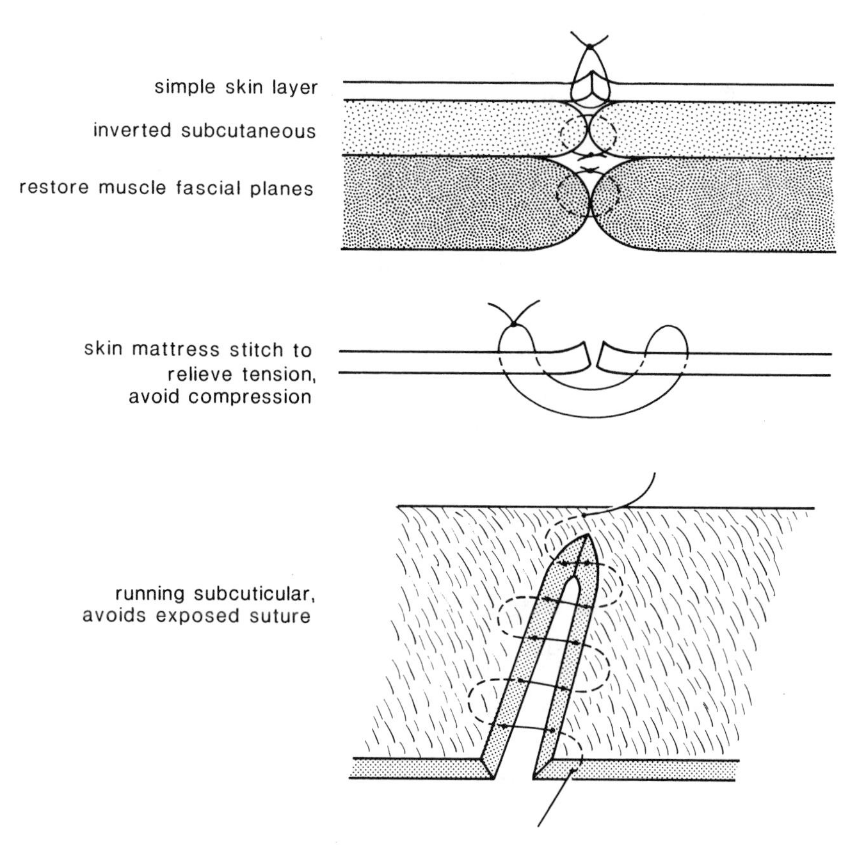

19.2. Common surgical techniques for closing clean wounds. Layered closure (*top*) emphasizes restoration of surgical planes and avoidance of pathways from skin surface to deep tissue. Note the inverted knot in the subcutaneous layer to prevent suture tails from poking upward into the skin layer. The mattress stitch (*center*) provides a way to improve strain relief of skin and to approximate edges under some tension without compressing wound edges. The running subcuticular stitch (*bottom*) eliminates surface-exposed sutures, loops, and knots, on which the animal could chew.

just takes practice. Furthermore, modern synthetic suture materials require rather different knots and handling techniques than plain old silk, which has little place in the implantation of long-term devices. Beyond these rudimentary knots, there are all sorts of specialized suture placements that are ideally suited to special problems, such as distributing tension evenly, promoting hemostasis, hiding skin sutures from prying teeth and claws, and dealing with stellate corners (fig. 19.2).

Instead of just hoping for the best with your mother's advice about "left over right," take some shoelace size strings and examine carefully how loops formed in different manners draw down when pulled in different directions. Learn some reliable habits that tend to result in successful knots of an acceptable form, and keep your eyes open while you work. The companies that manufacture sutures make some very nice brochures and booklets about the tying of surgical knots. Most third-year medical students spend the first few weeks of their surgery rotations practicing sleight-of-hand maneuvers that they are usually proud to show off to any interested party.

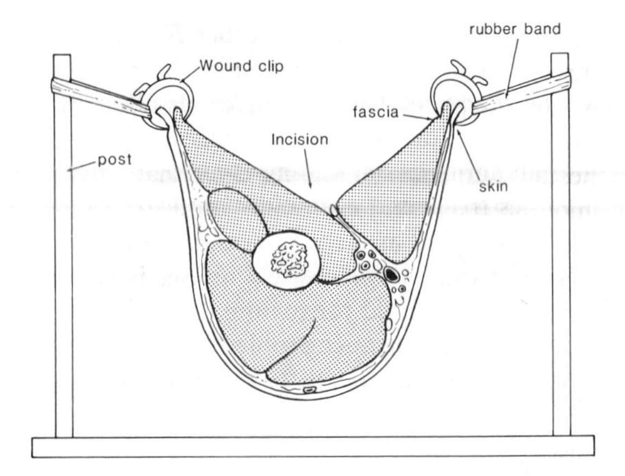

19.3. Constant tension retraction scheme to hold wound edges apart. This improves surgical exposure without requiring an assistant; also, retractors need not be repositioned to accommodate small amounts of motion. Note that wound clips incorporate a tuck of fascia as well as skin, which prevents the exposed undersurface of the skin from drying out.

E. USEFUL TRICKS OF THE TRADE

1. Rubber Band Retractors

Most clinical surgeons work with one or more assistants to apply just the right amount of tension on the retractors to provide an ideal exposure. For those who work alone, there are all sorts of self-retaining retractors, most of which sag and fall out repeatedly during the operation and all of which tend to get very much in the way, constantly snagging wires and sutures. What you usually want is a gentle, constant tension applied radially away from the surgical field in several directions. If you have some sort of anchoring sites around the preparation (parts of the stereotaxic frame or tie-down points for the limbs on the operating table), hook rubber bands over some small wound clips or towel clips as you attach them to the tissue to be retracted (usually skin or fascia) (fig. 19.3). Then hook the free end of each rubber band over anchors located in the desired direction (or tie the ends with lengths of suture if the distance is too great). As the tissue relaxes during operation, the rubber bands will take up the slack and maintain retraction. You can also hang weights over the edge of the table to get the same effect, provided no one bumps into them.

2. Handling of Subcutaneous Leads

The site at which the leads are to leave the skin and the site at which they are to be inserted into muscles are often widely separated. In chapter 11 (section A4/5) we dealt with the formation of a narrow subcutaneous tunnel for the passage of the electrical leads from the implanted device to the connector site, as well as with simple techniques for marking the leads from each of multiple devices. Both of these approaches permit you to limit the amount of trauma and still ensure that the leads are properly connected.

3. Suture Drag Lines

Suppose that you must anchor a device deep in some narrow crevice between two bulky muscles. You can just about get your needle holder down there, but you cannot suture the device once it has been placed there. The trick is to place all of the sutures into the bottom of the hole by themselves, using however many separate sutures are needed. String the device onto them outside the surgical field, being sure to keep multiple sutures properly oriented and not crossing. Then tie a simple loop into each suture and use it to drag the device into place as you draw down the loop. You can then easily throw the other loops down into the hole, although you should be sure to use several loops including one granny pair (adjacent loops in the same direction) to prevent the poorly controlled square knots from degenerating into double half-hitch slip knots, the topological sister form of the square knot.

4. Needle Holder/Suture Scissors Combination

One of the things that delays skin closures significantly when one is working alone is the need to set down the needle holders to pick up the suture scissors at the end of each knot. One trick is to hold the suture scissors tucked into the palm of the hand that is holding the needle holders farther out on the same fingers. This takes a little practice but is a cute trick often performed with a little flourish by experienced surgeons. A more direct solution is the Olson-Hegar needle holder, which incorporates a short section of scissors blades into the arms of the needle holder just behind the gripping surfaces. Remember to check that your electrode wires are out of the way before you reset the needle.

5. Closure of Skin in Two Layers

Closing skin in two layers takes just a little more time, but the resultant decrease in the incidence of wound infections and dehiscences is spectacular. Using an inverted knot placement as shown in figure 19.2, make a small number of fairly widely spaced, loose, individual sutures that reach under the skin perhaps 1 cm on either side of the incision edges. Often only one or two such subcutaneous sutures near the center of the incision are needed, whereupon—voila! the skin edges just fall into place with no tension or gaping. Then all you need are a few very superficial dermal sutures near the edges to keep them approximated. These minimize compression of the skin, which cuts off capillary flow. Plastic surgeons regularly use this trick with the buried subcuticular stitch to leave scar-free wounds, and you can use the same suture technique to eliminate exposed sutures. Like a plastic surgeon, you too will leave the table muttering "a thing of beauty is a joy to behold."

20 Single-Unit Electromyography

A. DEFINITION

In this chapter we provide techniques for recording from only one or a few motor units. This is a special situation, the fundamental biophysical considerations of which dictate very distinct designs of electrodes and electronic equipment. This situation can arise from two rather different experimental needs. In one case it may be desirable to record the activity patterns of single motor units in a large muscle. These may be of interest in order to correlate activity with unit type identification, to study details of recruitment order among motor units, and to consider the nonlinear effects of frequency modulation on probable force output. An entirely different situation is the need for EMG measurements in very small muscles, which may be composed of only a few motor units or may be so small that use of the bipolar macroelectrodes is simply impractical.

In both of these cases the experimentalist will need to use monopolar semimicroelectrodes or microelectrodes for which the amplitude and the selectivity of the recordings are based on entirely different phenomena. These are the physical relations that predominate in regions very close to individual active muscle fibers rather than the field effects that arise some distance from the surface of the fibers (figs. 20.1 and 20.2). Most of this chapter concerns the attainment of single, discriminable unitary action potentials. However, their biophysical considerations are essentially the same as those from multiunit recordings of small muscles. The major consideration is the much larger amplitude and much higher frequency spectrum of the action potentials that can be recorded when one is so close to the active cell surface that the action currents have not yet spread out into the volume conductor surrounding them.

The recording of activity from single motor units (sets of muscle fibers innervated by a single motoneuron) has undergone relatively little exploration in research on vertebrate animals. However, it is a well-developed method

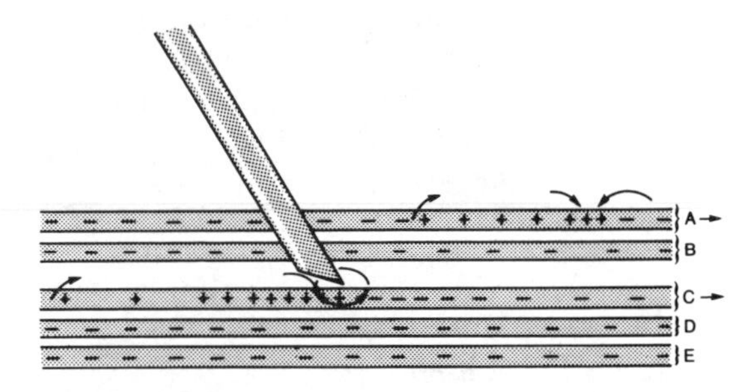

20.1. Tip of a typical single-unit EMG electrode shown to scale among a group of parallel muscle fibers. The peak of the single-unit waveform from *C* will occur at the instant shown, at which the edge of the active depolarization front produces a large, fast transition in the extracellular potential in the negative direction. The rest of the electrical events associated with repolarizing ion fluxes are more diffuse in space, producing slower signal components that are also more similar in amplitude and shape to the analogous events in more distant fibers, such as *A*.

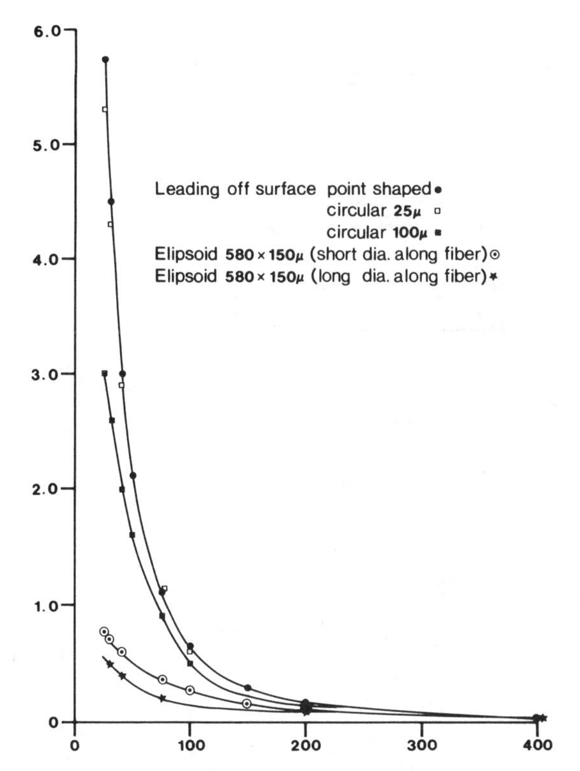

20.2 Amplitude of recorded potential (arbitrary units on ordinate) versus distance from surface of muscle fiber (microns on abscissa) for various monopolar microelectrode surfaces. Small surface area contacts that lie very close to the fiber can pick up very large potentials. Farther than 200 μ from the fiber, all electrodes record potentials of similar amplitude, although the small surface electrodes will be much more noisy because of their higher impedance. (From Ekstedt and Stålberg, 1973, reprinted in Stålberg, 1976.)

at both extremes of the phylogenetic spectrum of research subjects, being used routinely in awake humans and in the small muscles of insects, arachnids, and other invertebrates.

This seeming paradox can be understood by consideration of the requirements for successful isolation of unitary all-or-none potentials—namely a recording milieu in which an electrode (antenna) can be positioned so close to one source of such signals that it records a potential that is significantly larger than that coming from any other source in the preparation. In humans this is achieved by careful manipulation of a percutaneous needle with a tiny recording surface into a muscle that the subject recruits under controlled, isometric conditions intended to minimize motion with respect to the muscle fibers. In invertebrates it is achieved by selection of muscles with so few spiking motor units that the action potentials can be separated on the basis of individual shape, with little probability of overlap and occlusion of the unitary waveforms. Usually, these are small muscles requiring fine wires for which the exposed contact areas of the tips approach the dimensions of microelectrodes (linear dimensions of 5 to 50 μ).

B. MECHANICS OF MICROELECTRODES

1. Characteristics

Whenever the dimensions of the contact area of an electrode begin to be measured in microns rather than millimeters, a whole class of electrical and mechanical considerations that would usually be considered insignificant suddenly take on importance. In previous chapters considerable attention has been paid to maximizing the surface area of electrode contacts in order to minimize both their impedance and their tendency to pick up motion artifact as they rub over charged cell membranes. Obviously, an electrode that must be significantly closer to one particular muscle fiber with a diameter of 50 μ than to its neighbors must have contact dimensions smaller than 50 μ.

To maintain the stability of the recording, in terms of magnitude and shape of the unit potential, the recording configuration must remain constant in size and shape throughout the time course of the muscular action to be studied. Not only must the tip of the electrode be small, but its motion must be constrained, not more than a few microns relative to the muscle fiber in question. In humans this is achieved by careful selection of the muscle and the percutaneous insertion of a rigid needle, along with strong dependence on the cooperation of a steady subject. In small animals this is usually achieved by the use of highly flexible wires and the judicious selection of anchoring sites such as fascia and exoskeletal structures.

2. Flexible Fine Wires

The properties of metals that make them flexible (ductile) rather than springy (elastic) are exactly the properties responsible for stress fatigue and tensile failure. Fortunately, most studies in invertebrates rarely last more than a day, permitting use of fine gold or copper monofilament wire with very limited flex life. Typically, a fine-gauge, commercial armature wire with enamel insulation can be cut obliquely with a scissors, thus producing a reasonably sharp point that is simply inserted into the muscle, using the exposed cut end as the electrode contact. Such measures are unlikely to succeed in most vertebrates, as these show movements of much greater amplitude. Also, the leads must often course through several planes of skin, muscle, and connective tissue, all moving in different directions. Even if a fortuitous recording site can be found, the experimentalist interested in a study lasting more than a couple of days will have to contend with the tendency of any foreign body to be walled off by connective tissues. The resulting capsule tends to displace the contact surface from all active muscle fibers, making it impossible for the electrode to be much closer to one fiber than to another, defeating the critical condition for single-unit recording. Furthermore, even small amounts of motion stimulate growth of the connective tissue capsule and tend to kill the closest muscle fibers.

3. Percutaneous Needles

Studies in humans require electrodes that are sufficiently stiff and slender to be inserted through the skin with a minimum of damage to either the electrode or the cooperation of the subject. Usually, these electrodes are based on small-caliber hypodermic needles with fine-gauge, insulated electrode wires in their lumens. A monofilament wire (usually a biocompatible stainless steel or platinum alloy) is embedded in the lumen by means of epoxy or any low-viscosity, water-resistant material that can be cured in place. The needle is then reground at the tip so that the cross-section of electrode wire is exposed flush on the face of the embedding agent. Hypodermic tubing alone (without a typical hub) is usually used to minimize the mass and inertia that might cause the needle to drift after insertion. Similar considerations apply to the very flexible leads forming the core conductor and to the tubing itself, which acts as a grounded shield or as the negative input to a differential amplifier. The fine wires usually terminate in a more rugged shielded cable that is strain relieved at a convenient site on the subject.

4. Muscle Cuff Electrodes

One new technique currently being explored by one of us (G.E.L.) is the use of cuff electrodes on long, thin muscles, such as the mammalian tenuissimus muscle (see fig. 10.1). If the muscle has a long stretch (1 cm or more) along which it receives no innervation, blood supply, or important lateral attachments, it is possible to record unusually large, clean potentials by surrounding a muscle with a cuff electrode, such as commonly used for chronic nerve recordings. Such a cuff is made up of silicone rubber tubing with an inside diameter slightly larger than the diameter of the muscle, longitudinally (or spirally) slit to allow it to be slipped around the intact muscle. The lumen is equipped with three (or more) circumferential electrode contacts equally spaced along the length of the tube. The tube is tied around the muscle with a few external circumferential sutures. The two end contacts are connected together to the negative input of a differential amplifier with the positive input taken from the central contact. This tripolar electrode configuration records only potentials generated by the tissue in the cuff, to the almost complete exclusion of cross-talk from external sources, such as other muscles. Externally generated signals will cause a uniform voltage gradient along the constant-diameter lumen; consequently, the potential in the middle is exactly equal to the sum of those at the two ends and will be completely canceled in the differential amplifier. By varying the contact spacing in relationship to the dipole lengths of the muscle fibers, the recorded potentials can be altered in terms of amplitude, frequency spectrum, and discriminability. Because each contact is large and has low impedance, noise levels are very low. In muscles that have only a few motor units active at any time, or those that have a few of unusually large units composed of many large, synchronously active fibers, it may be possible to use good unit discrimination techniques to pull out one or more single-unit signatures.

5. Partial Denervation

Most EMG practitioners have encountered occasional macroelectrode recordings in which the signals were unusually spiky, apparently dominated by one or a few unit potentials with little smooth modulation. These usually signify some failure of technique, such as damage to the innervation, with loss of most of the normal activity. If sufficient time has passed since the clumsy surgical intervention, the remaining or reinnervating motoneurons may take over unusually large and tightly packed numbers of muscle fibers, giving rise to giant polyphasic unit potentials (see fig. 5.3). This suggests a way to obtain discriminable single-unit activity from at least part of a muscle with relatively large, stable electrodes suitable for long-term studies.

C. ANTENNA PROPERTIES OF MICROELECTRODES

We have already mentioned the essential principle of getting closer to one source than to any other. We use the term "close" to refer to circumstances in which the recording will be dominated by the action currents from one or a few muscle fibers.

If several adjacent muscle fibers all belong to the same motor unit, their action currents may add constructively or destructively, depending on the exact times they arrive in the vicinity of the electrode. At the very least, they are likely to cause complex and somewhat prolonged signatures. The complexity is good in that it allows discrimination of any particular signature; the prolongation is bad in that it increases the length of this distinctive shape, increasing the probability that a potential from another unit will add to and thus distort the apparent waveform (fig. 20.3). The complex interactions among numbers, distributions, and innervation points of fibers making up the various units in a given muscle probably account for the well-known variation in the ability to pick up single units in different muscles. Judicious selection of the preparation and a few pilot experiments may save a great deal of trouble.

It is important to remember that distance from a dipole source is best expressed in units

20.3. Formation of occluded EMG waveforms as a result of the asynchronous activity in two muscle fibers belonging to two different motor units. Depending on the exact timing of the two spikes, the resulting waveforms may be normal, smaller, or larger in amplitude; the number of countable events will always be less than the simple sum of the individual action potentials before occlusion.

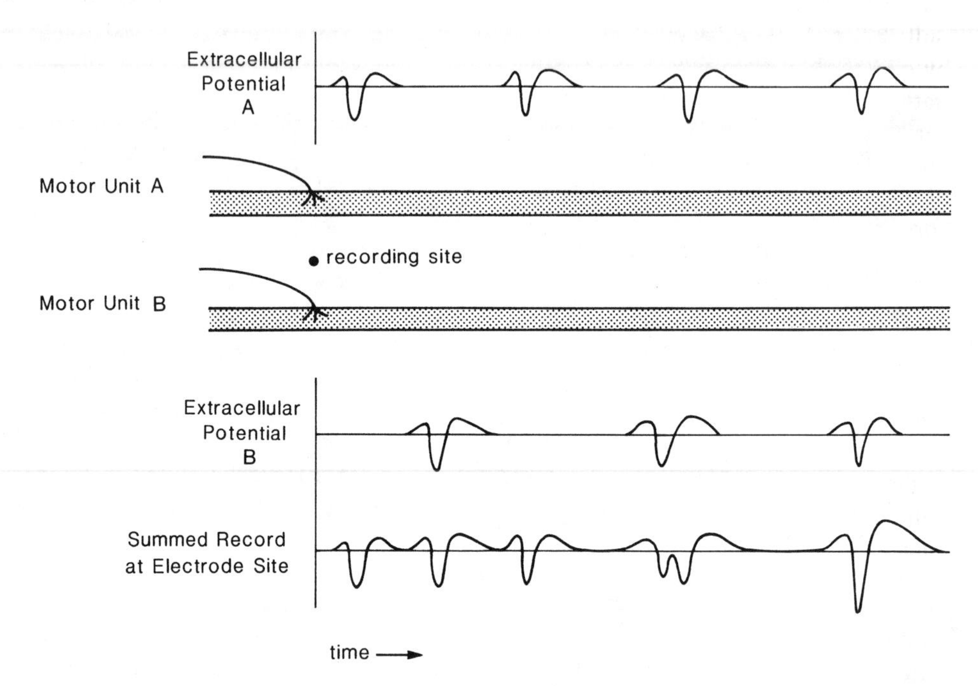

of the dipole separation. After all, the amplitude of the recordable potential decreases much more rapidly as the distance of a perpendicular to the long axis of the dipole is increased past the dipole separation (see chapter 4). At first glance, it seems not to matter whether one expresses the distances between the electrode and the muscle fibers in microns, dipoles, or furlongs. However, the equivalent dipoles differ among the various frequency components of a wideband EMG spike. The lowest frequency components are generated by the slow repolarization currents, coming from extended regions of previously depolarized membrane. The middle frequencies constituting the bulk of the gross EMG come from the active depolarization region, perhaps a few millimeters long. The highest frequencies come from the sharp boundary between open and closed sodium gates at the leading edge of the actively propagating potential (see fig. 4.2).

The latter high spatial and temporal frequencies contribute almost nothing to gross EMG recordings, which are often made at bandwidths that may not even include them. However, they are critical to single-unit work, for which the bulk of the energy spectrum may be in the 1 to 5 kHz band. High-frequency signals are selectively recorded by electrodes that are located within a few microns of the membrane surface and hence are able to respond to the passage of the narrow boundary that constitutes the high-frequency dipole source. The use of coaxial bipolar electrodes, such as hypodermic needles, further enhances the selectivity for these local potential gradients. Judicious high-pass filtering can pull pearls out of a sea of overlapping signals by effectively performing a temporal differentiation of the recording. Remember that their very localized nature makes the high-frequency signals particularly labile in the event of even small shifts in the position of the recording electrode.

D. ELECTRONIC CONSIDERATIONS

For most macroelectrodes the large surface area of the contacts results in source impedances that are low enough to be considered negligible in the equivalent circuit of the amplifier and cables. This is definitely untrue for microelectrodes. Tip impedances are generally inversely proportional to surface area and can easily reach megohms rather than the usual kilohms. At these levels there is significant possibility of losing most of the signal to stray attenuation and noise.

The most obvious consideration is the input impedance of the amplifier itself. Remember that the input impedance and the source impedance act like a pair of series resistors to form a voltage divider of the source signal. If the electrode impedance is 1 MΩ and the amplifier impedance is 10 MΩ, you lose 10% of the signal. If the electrode impedance is 1 MΩ and the amplifier impedance is 100 kΩ, you lose 90% of the signal. If they are equal, you lose 50%. Electrode impedance is like the laws of thermodynamics; one always loses, but one can minimize the losses (see fig. 12.4 for a suitable preamplifier).

The next most important consideration is stray capacitance. Any electrical conductor has a certain amount of capacitive conductance relative to any other electrical conductor in the vicinity (see fig. 12.2). The larger the surface area and the smaller the distance, the greater the capacitance. Capacitive shunting depends on signal frequency, which may be fairly high, as discussed above. The two major sources of stray capacitance are the amplifier itself and the cables from the electrode contact to its input. Usually, the input impedance of the amplifier is specified as an equivalent resistance in parallel with an equivalent capacitance. The capacitance of any lead cable can be added to that of the amplifier. The true input impedance of the amplifier for a given signal frequency is the inverse of the vector sum of the parallel resistive and capacitive conductances. That mouthful is better understood from the example that an input impedance of 10 MΩ paralleled by 48 pF represents an equivalent impedance of just 1 MΩ at 3 kHz. One foot of the common, small-caliber coaxial cable, type RG-174, has 31 pF of capacitance. If the electrode impedance is over 1 MΩ, you can kiss your single-unit potentials goodbye.

Last, but often not least, you have to be worried about noise (see fig. 11.1). Whenever you record an EMG potential of 1 mV from a 1 kΩ electrode, you have a source with an equivalent power of 1 nanowatt (10^{-9} W). That may not sound like much, but the same 1 mV from a 1 MΩ electrode can put out only 1 femtowatt (10^{-15} W). This means a drastic increase in the number of potential sources that will introduce enough power into your single-unit recording configuration to generate significant noise. To make matters worse, most single-unit electrodes are monopolar; hence, they are used with single-ended amplifiers or with the negative (inverting) input of a differential amplifier connected to a low-impedance reference, such as the outer tubing of a hypodermic needle. That means that one is stuck with whatever noise the electrode picks up.

The general strategy for microelectrode recording is to keep the electrode impedance as low as possible (but still consistent with the needed selectivity for units) and keep the cabling to the first stage of amplification as short as possible. The latter will minimize both the size of the capacitive shunt and the amount of antenna for noise pickup. Often you may best accomplish this by having the first amplification stage right on the preparation itself, as a miniature, voltage-follower, junction-field-effect-transistor amplifier with low current noise. You can make such a device using an operational amplifier in a miniature dual-in-line package (DIP), requiring only a couple of leads for power and few additional components (see fig. 12.4). The output of a follower will be precisely the same voltage as the input; however, its output impedance is very low (under 1 kΩ). Such a package provides the advantage that you may use long pieces of coaxial cable and place the main amplification and filtering stages at a more convenient location. With a few extra components, you may be able to put in a gain of 10. This will prevent the amplifier noise in the second stage from becoming a significant factor. Do not get greedy and try to put too much gain in one stage on the preparation, or you risk feedback and cross-talk into the input of this and any other low-power channel in the vicinity.

E. PROCESSING SINGLE UNITS

1. Aims

The main reason for recording single units is usually to obtain detailed information about the exact temporal pattern of activity of a single motor unit, perhaps contrasting it to the activity of other units in the same or other muscles. The usual objectives of gross EMG processing are to obscure such instantaneous fluctuations and to obtain a perspective on the overall envelope of activity over a longer term. Obviously, an expedition into the unit recording jungle will require some special outfitting.

2. Discrimination

The most important electronic instrument you will need is a unit discriminator. This is a device that electronically identifies the unitary events the waveforms of which sound so distinctive to the investigator whenever EMG signals are played over an audio monitor. Unfortunately, these waveforms are often not nearly so distinctive to the eye. They are easily confused with similar waveforms from adjacent units and frequently distorted by summation with the random background fluctuations from large numbers of more distant ones. Both the sophistication of the discriminator and the quality of the data it produces rest heavily on the detailed nature of these little annoyances. Improvement of the distinctiveness and stability of the unit potentials on the recording side is likely to pay for itself many times over.

If the experimental paradigm requires only the identification of one single unit, one obviously picks the largest one from a multiunit recording. Theoretically, discrimination requires only a voltage comparator that puts out a digital pulse whenever the input goes over a certain threshold value. This is essentially what an oscilloscope trigger circuit does, usually with the added advantage of being able to pick positive or negative levels whenever approached from either a positive or negative slope. In practice, this will work reliably only if the unit amplitude is several times larger than the background activity and larger than the sum of the next largest unit plus twice the background noise. This is because the random summation of the units plus background potentials will frequently reduce the desired unit by that much and increase its competitors equivalently. Also, this simple threshold device will happily accept any large artifact as a real event.

Separation of unit waveforms that are similar in overall amplitude but that differ in shape requires a window discriminator. Most of the commercially available devices work on a common principle. They first detect all the units that cross a certain threshold, and then they test each of these for the presence of a particular amplitude at some time interval after the threshold crossing. All waveforms that pass this amplitude window are considered to belong to a single unit; this places considerable responsibility on the setting of the window criteria by the operator. The window discriminator generates two kinds of output signals. One is a standard digital pulse signifying the acceptance of a waveform by the window. The other is a complex signal that can be displayed on an oscilloscope to show the exact time and amplitude of the window boundaries with respect to the various unit potentials crossing the intial triggering threshold.

The efficacy of a window discriminator is just as dependent on signal-to-noise ratio and the size of the differences between units as is the voltage comparator. However, it lets units be separated if they differ by an adequate amount at any point on their waveforms, rather than only at the peak. EMG potentials frequently have complex multiphasic shapes that are distinctive for each unit, although their overall amplitude is similar.

The biggest problems in discrimination of EMG units usually reflect changes in amplitude and waveform resulting from movement of the muscles. These changes are caused both by the motion of the electrodes with respect to the muscle fibers and by the changes in propagation of action potentials and their summation that occur as the muscle fibers change in length and cross-sectional area. Such changes in geometry may cause artifactual dropout and false acceptances that are systematically correlated with the ongoing muscle activity and may produce a completely unacceptable situation. Extreme care and frequent inspection of raw records as well as acceptance pulse records are absolutely essential.

The literature is full of ingenious circuits and computer programs for pulling single-unit potentials out of garbage. Some of these even claim to be able to compensate for or track fluctuations in unit size. Unfortunately, the very few biological signals that can benefit from such approaches would probably produce equally good results when run through a carefully set window discriminator. What is worse, many of these devices are extremely complicated and obscure in their internal machinations. This may make it impossible to be confident that they are not turning out wholly artifactual results. Even a window discriminator in the wrong hands is likely to generate rather convincing-looking "data" out of a signal composed of nothing but amplitude-modulated wideband noise.

3. Displays of Instantaneous Interval Frequency

The simplest display of instantaneous spike rate is the interval of time between two spikes. One can easily turn this into a proportional voltage by having each successive event reset a ramp generator. However, this produces an upside-down view of intensity of motor unit recruitment, with high firing rates appearing as minima in the output. Also, nonphysiologically short intervals caused by spuriously identified spikes are hard to discern because they cluster near the bottom of the graph.

One of the most useful tools for both the

display and confirmation of validity of discriminated units is the frequency gram (fig. 20.4). Commercially available devices or simple programs for a digital computer can be used to display against time each accepted spike as a dot or an increment of a bar, with the ordinate of the graph being the instantaneous frequency given by the inverse of the time interval from the previous spike. Usually such a display automatically resets to zero between events so that it does not continue to register indefinitely the last frequency at the end of an activity period. This kind of display is particularly useful for motoneuron activity because the firing rates often are expected to modulate smoothly between narrow physiological boundaries such as 10 to 50 pps. If a single action potential is missed by the discrimination system, the contour of the displayed curve will drop immediately to exactly half the amplitude of the surrounding events. Generation of a single false acceptance pulse will produce an immediate jump up to at least twice the prevailing frequency; more often it will produce a very high and nonphysiological firing rate, because the extra spike is completely unsynchronized to the unit under study.

In designing or selecting a system for computing the frequencygram, it is important to keep in mind the somewhat unusual resolution characteristics caused by taking the inverse of interpulse interval time. Usually this time is obtained by a digital count of a clock rate. Obviously, the highest frequency that can be identified is that given by the inverse of the clock rate—for example, 1000 pps for 1 msec. However, the frequency resolution in the region of these high rates is very poor. The next lower output that is possible is for two ticks of the clock—for example, 500 pps for the example given. It is only when the intervals get to 20 clock ticks that the resolution of the output reaches even 5% error levels, corresponding to 50 pps for the example of a 1 msec clock. Obviously, a fast clock is an asset, but that risks dynamic range problems for very long intervals. In a cyclic activity, such as walking, each burst of the motor unit begins with a spike of very low instantaneous frequency—namely the inverse of the period of time since the last activity period of the unit. If the pause between recruitment periods is 1 second, even a 1 msec clock will rack up 1000 ticks, needing about 10 bits for the counting register. If this device is implemented in conventional 8-bit microprocessor architecture, it may be necessary to make some special provisions for handling overflow in order to prevent spurious output. Commercial frequency-gram devices use a voltage ramp that is read by an A/D chip whose output bits switch an array of precision current sources into a summing amplifier to calculate the inverse. Again, resolution is limited by that of the A/D, which must usually be 12 bits or so.

20.4. Two types of signal processing: one suitable for single-unit recordings (*left*) (here from ventral root axons but similar for single-unit EMG) and one for whole muscle sampling (*right*). Multiple-unit record (*left*) is first discriminated into single-unit spike events, which trigger a frequency-gram generator that shows the instantaneous frequency from the inverse of the time interval between two successive spikes. The whole muscle EMG is rectified and integrated into bins prior to computer digitization and smoothing into the demodulated envelope. The two representations can be fitted and overlaid to indicate the degree to which unit rate modulation reflects whole muscle activity modulation. (From Hoffer et al., in press.)

4. Rasters and Histograms

The field of single-unit neurophysiology has provided a wealth of display techniques for showing patterns of activity of one or more individual sources of all-or-none spikes. One

SINGLE MOTOR UNIT | **WHOLE MUSCLE**

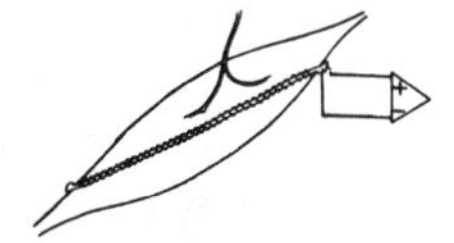

SIGNAL RECORDING

hatpin microelectrode
1,000-10,000 Hz filtering

spiral bipolar EMG electrode
50-5,000 Hz filtering

ANALOG PROCESSING

unitary spike discrimination

EMG rectification, integration and sampling: 4 ms binwidth

DIGITAL PROCESSING

instantaneous firing rate computation

analog to digital sampling: 4 ms; 15-point moving average

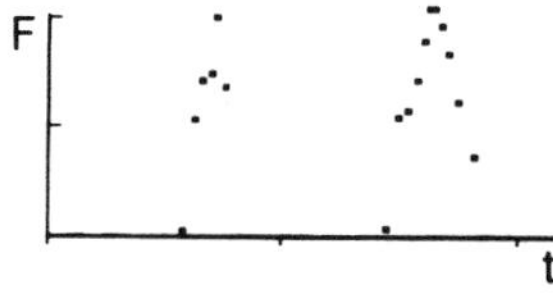
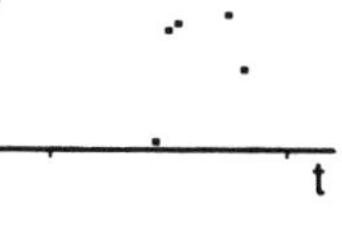

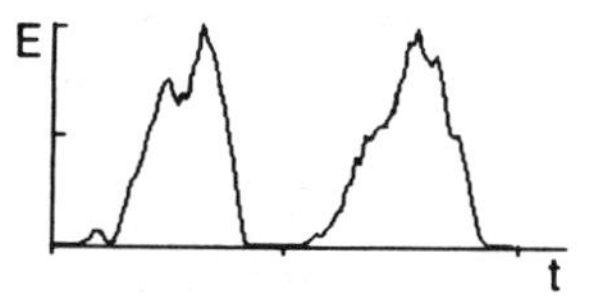

FIT AND DISPLAY

$F(t) \doteq K * E(t)$

least-squares fit; K adjusted to maximize correlation coefficient r.

Unitary recruitment threshold Th: mean calculated from all steps.

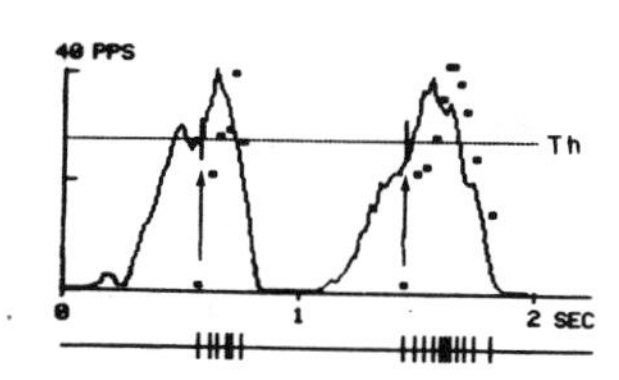

of the most popular techniques is the dot raster, in which each unit spike is shown as a single intensified dot along a sweep against time, usually synchronized with some external event such as a behavioral response. This display has the advantage that each line takes up almost no vertical space, allowing large numbers of repeated sweeps with slight vertical offset to be examined for systematic or random changes in the exact timing of every single spike recorded. Such condensed displays may permit an oscilloscope camera to take the place of a chart recorder. If the synchronizing event is of interest, it is common to display a summary histogram. The trace time is divided into a number of discrete bins. A bar graph then shows the number of recorded events as the amplitude of each temporal bin.

Another way to examine trends in frequency of activity is to construct an interpulse interval histogram. In such a display the abscissa consists of bins of time corresponding to the various intervals between two successive spikes in a long train of activity. The ordinate of the bar graph indicates the number of spike intervals falling into each bin.

Whenever more than one source of unit activity is available, more sophisticated cross-correlation techniques can be used to ask questions regarding the commonality of the inputs driving such activity. Even slight degrees of correlation between two apparently asynchronous spike trains can be discovered by generation of histograms from the intervals between the occurrences of two separate unit spikes.

21 Quick, but Not Too Dirty

A. APOLOGIA

Although most of the earlier discussions in this book have noted minimal requirements, the accounts have generally proceeded far beyond these to characterize complex, multichannel approaches. Well-planned complexity of approach may have the benefit of simplified analysis; still, it does this at the cost of rather involved and often expensive instrumentation that must be checked and maintained. In many situations this initial expense will be unacceptable. For instance, some readers may wish to perform pilot experiments in order to establish whether the problems of interest to them are amenable to solution by EMG. Other readers may be faced with the absolute limitation that the expenditure of very substantial funds, even for a guaranteed solution to their biological problem, may be impractical or impossible.

Finally, there are the particular needs of those who must apply for initial research support. Reviewers of grant proposals often inquire whether the applicant has ever performed the kind of work described. Also, some granting agencies may not award funds for equipment and supplies until after an investigator has successfully carried out the kinds of experiments listed in the application. The required pilot experiments are often carried out in the laboratory of established investigators. What follows should allow some pilot experiments to be performed at home. In any case, it should shorten the duration of training required. We propose to describe the minimal effort and equipment that will generate acceptable EMG records, which in turn will permit the investigator to report results with confidence (table 21.1). Frequently, this approach will provide the experience necessary to justify and efficiently plan for more quantitative experiments.

What kinds of approaches should be recommended for persons facing such problems? It is easy to argue for compromise, for accepting any electrode that gives a signal whatever the event, any amplifier although it is known to distort this signal, and any chart recorder no matter what its frequency response. Two reasons suggest that this advice is inappropriate. First, as we have tried to show, nothing is to be gained by recommending approaches that carry a very high risk of generating artifact. Second, we cannot imagine that the espousal of inadequate techniques will ultimately achieve our aim of convincing colleagues that EMG is indeed a serious approach to any problem.

However, we have demonstrated that it is possible to generate acceptable EMGs with relatively simple equipment. Much of the necessary equipment may even be assembled by judicious loans from colleagues and from instructional facilities (the undergraduate physiology laboratory is often a remarkable source of potentially useful components). These loans may have to be implemented at times conve-

Table 21.1 Costs for a Minimal Laboratory Configuration

Equipment	Cost
Electrodes	$ 50
Spring clip connectors	10
Differential amplifier and power supply	200
Oscilloscope monitor	600
Impedance testing and calibration equipment	200
Stereo cassette tape recorder	50
Super-8 sound home movie system	300
Frame count synchronization equipment	10
Total	**$1420**
Chart recorder (probably can be borrowed)	($ 900)

nient to the owners, perhaps at night, away from working hours. Experiments may have to be restricted as only a few events may be recorded simultaneously; consequently, the experimentalist may have to be willing to face a somewhat more complicated analysis. Such compromises being accepted, the use of simple approaches may yield results that allow useful interpretation.

B. PLANNING AND ANATOMY

The first concept that must be recognized in the selection of pilot experiments is that planning remains critical. If the question is poorly defined, the answers will lack definition. For instance, it may be possible or desirable initially to ignore differences among fiber types. However, it is undesirable to postpone characterization of musculosketetal anatomy and the fiber architecture of the target muscles. In short, the word "quick" in the title of this chapter applies only to the EMG approach and not to the anatomical or behavioral portion of the investigation. Both anatomy and behavior must be characterized, procedures for dissection and filming (chapters 3 and 15) must have been utilized, and the results of these must be available before the start of EMG analysis.

The anatomical approach will probably have to be carried out more carefully for simple experiments than for multichannel ones. The fewer the electrodes from which one may record simultaneously, the greater the problem of ultimate comparison. Consequently, knowledge of anatomy must aid experimental design by providing the basis of decisions about the order in which muscles will be recorded and the sequence in which recordings will be compared. When the signals from 8, 10, 12, or 14 muscles may be recorded simultaneously with a single tape recorder, the experimentalist may defer decisions about the sequence of ultimate comparisons, whether of left versus right side, of openers versus closers, or of abductors versus adductors. If the tape recorder has only four, three, or two channels, the decisions regarding signals that must be recorded simultaneously become much more critical. Remember the useful rule of thumb that the closer in time the activity periods of two muscles are, the less suitable it is to compare them on the basis of their temporal relation to a third muscle and the more important it becomes that they be recorded simultaneously on the same tape. Fine discrimination requires direct comparison. Also, remember that your experiments may not succeed on the first trial. Plan to carry out pilot experiments with a common and readily available species before running tests on the rare exotic form of ultimate interest.

C. MINIMAL LABORATORY CONFIGURATION

1. Basics

Single-channel EMG is a reasonably simple approach. The number of problems rises exponentially with the number of simultaneous channels being recorded. This suggests the fundamental rule: keep things simple! Remember the earlier comments about reliability of complex systems and that this is likely to be lowest during the initial phase. Start out by trying each component separately and combine components only after each has been shown to work by itself. First film the animal without implanted electrodes, then record separately from each muscle, then film the animal with one electrode in place, and finally proceed to filming while recording from multiple muscles.

It is unlikely, but possible, that the attempt to record simultaneously from multiple muscles will by itself change the response. The noise may increase or oscillations may appear on the screen of the oscilloscope. Although the geometrical shapes of harmonics sometimes seen have a certain aesthetic beauty, you are unlikely to appreciate them at this point. Instead, start out again to check your cabling connections. Look for ground loops. Ask which component is responsible (if you turn it off, will the noise stop?). Check that all of your cables are really grounded. A supply of 2- or 3-foot lengths of heavy-gauge (number 14 or lower) insulated copper wire, fitted with a large alligator clip at each end, will come in handy.

It is always best to start out by making some initial records from a few, large, superficial muscles. Once you have demonstrated to yourself that it is possible to record from such muscles and have carried out the necessary tests (chapter 14) to confirm that the wriggles on the screen are indeed EMG signals, there will be time to consider recording from more deeply placed or smaller muscles. The relatively simple EMG signals from large superficial muscles are also ideal to check out the effects and success of anesthesia and other procedures. Remember that few things are more likely to discourage the neophyte than to perform an "incredibly" (although sometimes unnecessarily) complex series of instrumentations only to find the animal uncooperative and avoiding precisely those activities one set out to study in the first place. Only after the postoperative response has been considered is it appropriate to implant multiple electrodes simultaneously and to consider a formal recording session.

2. Electrodes and Connectors

a. To Start. The first question in selecting an electrode and insertion method will obviously require information about the kind of animal to be studied, its size, its tolerance of anesthesia, the depth to which the electrodes are to be inserted, and the availability of operation.

b. Percutaneous Bipolar Wire Electrodes. For pilot experiments the twisted bipolar hook electrode made of Teflon-coated stainless steel wire may be near optimum (see chapter 10, section C3). The main difficulty is that the shape and placement of the electrode must be well standardized and may have to be checked before the experiment is started. Materials for these electrodes are available off the shelf. They can be constructed in advance, tested, sterilized, and stored for prolonged periods. With training, the experimentalist will learn to achieve reasonable precision in inserting these electrodes stereotactically through relatively small nicks in the skin.

For the pilot experiments described, designed to test whether EMG is an appropriate technique, the twisted bipolar hook electrodes are best inserted via hypodermic needles with continuous electrode wire used as subcutaneous leads and avoidance of any joints within the animal. Whereas open insertion of electrodes has obvious advantages in terms of precision of placement and orientation, it is counterindicated until the investigator has mastered all of the other procedures and unless trained surgical assistants are available.

c. Implanted Bipolar Patch Electrodes. If you have to work on a shoestring, you will probably be doing without many of the features that provide extra assurance about the quality of the behavior and the EMG signals, such as multichannel tape recorders, high-quality film records, and implanted transducers. For this reason, it is all the more important that you pick electrodes that have the highest a priori probability of providing clean, selective records from undamaged muscles. Both theory and experience now favor the relative newcomer, the bipolar patch electrode (see chapter 10, section C5). These electrodes require just a little more in supplies and preparation, but they are much more likely to work in difficult situations, such as posed by relatively small, fragile muscles or large, adjacent sources of cross-talk.

The most expensive supply is the Dacron-reinforced silicone rubber sheet, which usually comes in 5 by 7 inch sheets used for neurosurgical repair of defects in the dura mater. One sheet used carefully will make 30 to 50 electrodes; you may be able to beg one from a friendly operating room supervisor. The medical grade silicone adhesive (usually Dow Corning Medical Adhesive A) is also expensive, but clear silicone rubber bathtub caulk from the hardware store works just as well and is quite biocompatible; however, both give off acetic acid while they are curing. The usual stranded stainless steel wire with the Teflon jacket serves as the contact and the leadout. The fabrication procedure in figure 10.6 is simple enough with a little practice, and a bubble tester (see fig. 14.1 and below) will keep it all honest. If you are feeling insecure, there is now a commercial supplier of both standard and custom sizes of single- and double-sided (four contact) patch electrodes (see appendix 2).

d. Spring Clip Connectors. It is next necessary to transmit the signals from the electrode wire to the differential preamplifier. For the initial experiments it may be worthwhile to utilize some kind of spring clip connectors, an otherwise outdated technique. Although not really appropriate for prolonged experimentation, spring clips permit rapid connection of fine wires to a harness. They allow a neophyte investigator to concentrate on the signals rather than to begin by intense training on how to solder fine wires in tight quarters before the animal emerges from anesthesia or otherwise reacts to the vicinity of a hot iron. The simplest spring clips consist of short segments of a helical spring. A separate coil is attached to each of the lead wires of the flexible cable that connects the animal to the preamplifier. The clips are best formed of stainless steel and must be kept clean and tight. To use them one removes the insulation from the last 1 cm of the electrode (lead) wire. One then bends the spring separating the adjacent wires on one side of the coil and threads the bared end of the electrode between them. As the spring is released, it closes the gap, providing for mechanical and electrical contact with the electrode wire.

Unfortunately, spring clips are bulky, tend to get dirty, and lose their tension when they are bent too far. They may contact each other, thus producing shorts; also, they may act as antennas during recording sessions, so that they may need to be used in a properly grounded Faraday cage. Consequently, these spring clips cannot be recommended as connectors except for testing purposes and pilot experiments. However, there are now tiny pin connectors that consist of a spring-loaded metal pin that slides in a hollow tube. A hole perforates tube and pin, and it is only necessary to depress the pin until the two holes align, slide the bare end of the electrode wire through it, and release. Such pins can be mounted closely to each other and represent a useful approach to temporary connection.

e. Better Connectors. Once individual components have been tested, the electrode wires should presumably be attached permanently to a harness involving a connector matching the one on the major cable to the preamplifier. At this point it may be useful to see if one can borrow two or three pairs of matching connectors from an electrophysiologist; as long as only a few units are required for a pilot study, the procedures described in chapter 11 are likely to make purchase of only a few connectors rather costly in time and money. On the other hand, your friendly neighborhood electronics shop may stock one of the types of connectors recommended in appendix 2.

If one takes proper precautions and compensates for their difficulties and limitations, one can utilize nonstandard materials in pilot experiments. Earphone wire represents a good example. It is cheap and easily available, tends to be flexible, and usually has very low noise properties; its internal stranding makes it particularly resistant to repeated bending. When such wire is used outside of the animal, its attachments and composition are not likely to lead either to heavy metal poisoning or infection. However, other problems remain. For instance, earphone wire is very difficult to solder, and the solder joints tend to have little strength as the conductive combination of fine wires and foil is supported only by the thermoplastic coating, which must be stripped. Attempts to load such wire mechanically after attachment of electrodes will demonstrate the fallacy of combining an electrical and a mechanical connection. However, should the next store be hundreds of miles away but a machine shop handy, one can easily machine small clamps of spring steel that will provide mechanical support for the electrical joint by pressing against the intact portion of the thermoplastic coating.

Whenever such nonstandard arrays are built, they must be tested before use. This starts with the isolation of the individual channels and check of the grounding of the cables thereof with an ohm-meter. Connector cables also must be connected into a low-current, low-voltage circuit (see fig. 14.3) and the transmitted voltage checked with an oscilloscope. The signal should not vary even though the cable and the homemade connector are moved or vibrated.

3. Amplification and Monitoring.

a. Borrowing Equipment. The minimal equipment package for establishing the utility of EMG would seem to consist of a combination of a variable number of differential amplifiers, an oscilloscope, and a multichannel AM tape recorder.

This equipment had best be borrowed so that the pilot experiments can be performed with minimal investment. This means knowing what to borrow and how to test whether the things that can be borrowed, are adequate and functional. Some simple test and calibration routines and various precautions are hence emphasized. If at all possible, start by obtaining the instruction book for any borrowed equipment.

Remember that all of those neat little knobs and buttons on the front (and back) of the various gadgets tend to have a meaning. Thus, start by making up a master record sheet for future experiments, copy one for each run, and write down the setting of every switch and knob, giving the amplification, filter setting, carrier voltage, and so on, regardless of whether you intend to use them. Few other compulsions are as likely to be helpful in the long run (see also chapter 12).

If the equipment you borrow is old, it probably contains vacuum tubes rather than transistors; this means that it should be allowed to warm up for 20 minutes or so in order to achieve a stable output. Be sure that the perforations in the equipment housings are left open, as such equipment generates much heat and may fail unless cooling air can circulate. If there are filters on the integral cooling fans, clean these. If there are circuit boards or other plug-in devices in the borrowed equipment, reseat them for good contact. Contact cleaner may be useful, but apply it sparingly and do not squish and spray it randomly over the equipment as some solvents may degrade components. Avoid cleaners containing "conditioners" that leave residues.

The preamplifiers selected should be differential, as the significant aspect of the EMG record is the local voltage gradient, rather than the absolute voltage relative to ground. Amplification should be close to 1000 (be sure to record its level and calibrate it with an oscilloscope). Many differential preamplifiers have high- and low-cutoff filters that allow elimination of stray signals. Make sure that these are working and set them away from the frequency range in which you hope to acquire useful information. Finally, the preamplifier may have a carrier voltage setting; this establishes the voltage level, relative to ground, on which the output signal will be transmitted. (This is known as chopper stabilization and is a somewhat old-fashioned strategy to improve low signal fidelity.) You will have to check and adjust this in order to make sure that your signals will indeed record on tape with minimal distortion.

Some preamplifiers have very limited final driver stages so that they produce too little power to support your oscilloscopes plus a tape recorder plus that odd chart recorder. Test the driver capacity by connecting the output to an oscilloscope and then placing another unit, whether tape recorder or chart recorder, into the circuit. If the voltage drops substantially, things are overloaded and you may have to monitor the output with one unit at a time. Remember that you are trying to obtain permanent records; thus, the tape recorder must be the first unit on-line, with the oscilloscope next and the chart recorder perhaps relegated to the output of the tape system. (Does its driver have adequate capacity?)

Whatever kind of equipment you borrow or buy, be sure to organize it in a simple pattern and to arrange the cables logically, so that noise is kept to a minimum (see chapter 13).

b. Differential Amplifier and Power Supply. When working with low-impedance electrodes that detect relatively large signals as do EMG contacts, the more important requirement is common mode rejection. Fortunately for the budget conscious experimentalist, the best way to achieve high common mode rejection is to keep the head stage preamplifier very simple. Figure 12.4 shows a very simple circuit easily powered by a couple of transistor batteries. The dual, matched field-effect transistors provide automatic temperature compensation. The circuit is intended to provide just the preamplification and conversion to a single-ended signal suitable for direct input to an oscilloscope, tape recorder, or chart with adequate sensitivity. Any bandpass filtering can probably be done by appropriate setting of filters on the second stage device or by default if the bandpass of the second stage device is reasonable (as it often is with audio frequency devices such as cassette tape recorders). The cost for parts is under $20. A similar device is commercially available for under $200.

c. Oscilloscope Monitor. Most oscilloscopes, even those sold for instructional purposes 20 years ago, are more than adequate for monitoring of EMG signals. Start by noting the regularity of the sweep pattern; the beam should not accelerate noticeably (or jerk) while it shifts from the left to the right side of the screen. Use a calibration voltage, sometimes supplied within the instrument, to ensure that the vertical deflection achieved for a particular voltage is equal across the full vertical range of the screen. Check whether the unit deflection distance differs between 0 and 1 V, 1 and 2 V, 2 and 3 V, and so on, and do the same for the negative portion of the scale.

Whereas single-channel oscilloscopes are sufficient to permit checking whether a particular signal is acceptable, they permit intersignal comparison only with great difficulty. Either a double-beam or a split-beam oscilloscope will allow you to make direct comparison on a single screen. For the split-beam scope check the sampling rate and for the double-beam scope be sure that both beams are sweeping at the same rate. Sometimes you may be faced with a storeroom filled with single-channel oscilloscopes but no multichannel ones. Consider the placement of two single-channel units, side by side or on top of one another. This will achieve some preliminary indication of possible coincidence (but check the time base calibration and trigger synchronization relative to each unit). Whenever the number of muscles of interest exceeds the number of channels that may be displayed, you should consider establishing a quick switching system by taping down a single cable leading to the input of the recorder or display, opposite a mating set of cables running from the several differential preamplifiers carrying the signals that you wish to check.

4. Impedance Testing and Calibration

There are two basic requirements for testing. First, you must be sure that the recording surfaces of the electrodes are where they are supposed to be, which means not shorted together and that the insulation has no pinholes or other defects. This can be certified with a bubble tester consisting of a transistor radio battery and a bowl of saline (see fig. 14.1).

Second, you must have some way of verifying that the complete circuit consisting of electrodes, leads, and amplifiers is actually intact when the animal is hooked up (see fig. 14.2). This requires a signal generator and a couple of resistors and capacitors in conjunction with your existing amplifier. Figure 21.1 shows a system that permits check of the differential electrode impedance in one configuration and the amplifier calibration in the other. The signal generator is ideally a sine-wave generator with a frequency in the middle of the bandpass of your recordings. However,

21.1. Simple circuit for in situ testing of electrode impedance and amplifier calibration as well as high-pass filtering of the recorded EMG to reduce lower frequency line noise and motion artifact. The abbreviations refer to source impedance (*Z*), signal potential (*E*) and signal frequency (*F*).

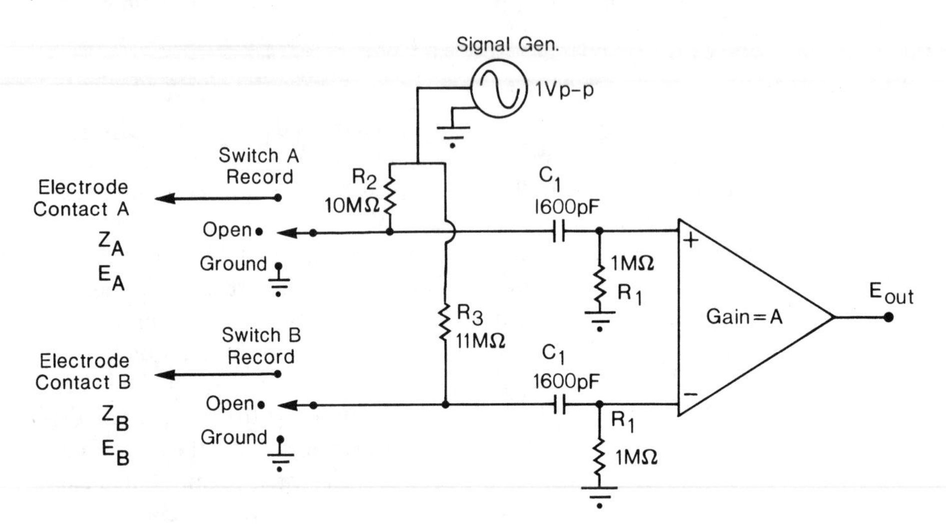

Assumptions for correct function:

$Z_A \approx Z_B \ll R_1$

Output Characteristic:

for $F_{in} > F_{cut\text{-}off}$, $E_{out} = A(E_A - E_B)$

$$F_{cut\text{-}off} = \frac{1}{2\pi R_1 C_1} = \frac{1}{2\pi(1M\Omega)(1600pF)} = 100\text{ Hz}$$

Operating Conditions:

Mode	Sw. A	Sw. B	Gen. Signal	Sig. (ref. to input)
Record EMG	Record	Record	OFF	High Pass EMG
Test Impedance A	Record	Ground	ON	0.1 mV/kilohm
Test Impedance B	Ground	Record	ON	0.1 mV/kilohm
Calibrate Gain	Open	Open	ON	10mVp-p

for rough estimations (all that is usually needed), any convenient source such as the square-wave calibrator built into many oscilloscopes can be used.

A very simple and useful substitute for generators of test signal pulses is to borrow them from another laboratory. Use your AM tape recorder, and store the various signals of interest on clearly defined segments of a cassette tape. During recording, monitor pulse rate and waveform on an oscillosope to determine the kind of distortion that one thus introduces. Be sure that the signal tracks are sufficiently long and well defined on the tape. Separate them by blank periods, and perhaps include records of your voice describing what comes next. Desirable patterns to record are first of all a series of sinusoidal wave frequencies of 10, 30, 100, 300, 1000, 3000, and 10,000 Hz. Also, include a 100 Hz square-wave. Finally, it may be nice to have a recording of an archetypical clean EMG signal that shows a reasonable dynamic range and lacks noise, motion artifact, and any of the faults against which we have warned.

5. Stereo Cassette Tape Recorder

The most useful component to any simple setup for pilot studies is a multichannel tape recorder. It lets the signals be stored for later playback and facilitates their check and comparison. Even small, single-channel AM cassette recorders tend to have an adequate frequency response for recording the fast rise time spikes of the EMG signal. The critical region is between 80 and 3000 Hz, which is well within the range desired for the audio response for which the recorders are designed. Indeed, the audio capacity of the system permits one to switch one channel back and forth between the output of an EMG preamplifier and a microphone. This provides a simple system for identifying the records of individual EMG recording sessions. Use of a speaker also allows the investigator to listen to the signal determining its quality and its coincidence with observed movements of the animal.

A stereo recorder (two channels) represents a minimal experimental unit and the standard tape cassette an excellent medium for recording. It is often possible to obtain quadraphonic recorders and thus have four records on a single tape. (Remember that many such recorders use the full width of the tape in one direction only. Turning over the tape and attempting to record the other side will erase the records.) The mass production of AM tape recorders for the audio market has made them affordable to students and considerably less expensive than FM instrumentation recorders. The major disadvantage remains the inability simply to record slow signals, such as the gradually changing voltages obtained from pressure or strain gauges. Sometimes one can compensate for this disadvantage by building a simple circuit that will transform the slow signals into a high frequency that can be recorded but must be demodulated during playback. There are also multiplexors that combine several low-frequency signals on a single channel of wide frequency range, which may be recordable on a home video tape recorder. Most people buy a multichannel FM instrumentation recorder about the time that they start to encounter such problems and thus bypass them.

6. Rectified Chart Records

The wideband AC signals of the raw EMG record must be written out onto some form of hard copy for examination of the time course, correlation with other records such as film, and eventual incorporation into publishable figures. The most common apparatus of this type is the pen galvanometer chart recorder, which uses ink or thermal paper. The pen motion is limited to about 100 Hz frequency response, much lower than the EMG bandwidth, so that it acts like a low-pass filter. This can be turned to advantage once the signal is rectified because filtering action will then produce the smooth envelope of changing DC level that actually represents the overall intensity of the EMG signal over time. If you can make your preamplified signal large enough so that the baseline noise represents the first 0.6 V of a signal at least 10 times larger during muscle activity, you can simply use a diode in series with the chart recorder input to half-wave rectify the signal. (Put a 10 kΩ resistor across the chart recorder input to lower its input impedance to terminate the rectifier circuit.) However, remember that the biasing voltage of the diode (0.6 V for the garden variety silicon diode) means that the diode has a threshold, so that signals below 0.6 V do not get through at all. This will clean up the baseline but you will risk overlooking low-level activity whether or not it is noise.

A battery and some resistance can be connected (fig. 21.2) to bias the rectifying diode up to its cutoff voltage so that any positive voltage, regardless of amplitude, is passed to the amplifier while all negative voltages are blocked. Whenever rectified signals are provided to the chart recorder, it should be set DC coupled to prevent scalloping of the baseline. If the chart recorder has additional high-frequency cutoff filtering, this can be used further to smooth and integrate the records as desired.

7. Super-8 Home Movie Systems

a. Filming. There are literally hundreds of devices that record motion for correlation with the EMG signal. Cine cameras are likely to be the easiest and often the least expensive of these, although video cameras may soon rate a close second. Super-8 is presumably the cine format of choice, as there is a vast selection of cameras, reasonably inexpensive film, and multiple editing, splicing, and projection devices. Whenever selecting a camera, you should pick the one with the sharpest optics rather than the greatest zoom range (which makes the camera more expensive, but not necessarily more useful for experiments). Buy good-quality, screw-threaded, add-on (close-up) lenses; these cost much less than an integral zoom lens and achieve nearly the same effect. Saving a dollar or two on their quality may be poor economy. Do look for a camera with an integral shutter synchronization; this may be used to drive an inexpensive stroboscope for sharp exposure of the pictures or to generate a direct signal for synchronizing the cine camera and EMG signals. Do invest in a

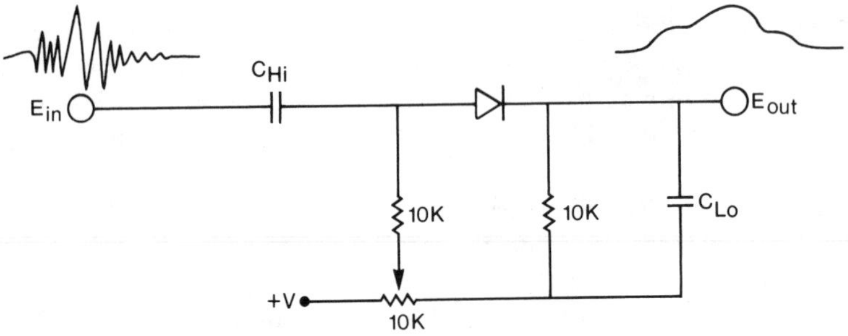

$$F_{Lo} = \frac{1}{2\pi C_{Lo} \cdot 10K}$$

$$F_{Hi} = \frac{1}{2\pi C_{Hi} \cdot 10K}$$

e.g. F = 100 Hz, C = 0.15 μF

21.2. Simple circuit consisting of passive components and any source of DC-positive voltage (e.g., small battery) that can be used to eliminate the diode turn-on offset potential when half-wave rectification is performed. The potentiometer should be turned up until an applied AC signal appears on the far side of the diode as only and all of its positive-going components. In this configuration the optional capacitor (*CLo*) is shown to provide filtering of low-frequency components such as motion artifact that would produce large-output signals. The optional capacitor (*CHi*) is shown providing some smoothing of the rectified signal. Typically both filters might be set to 100 Hz corner frequency with 0.15 μF capacitors.

sturdy tripod; always avoid hand-held cameras. Remote shutter releases are desirable as they tend further to reduce possible tremor when the camera is being turned on. Vibration produces unsharp images and complicates analysis. If necessary, mount a 1/4″-20 (1/4 inch—20 threads per inch) bolt on top of a small crate and fasten your camera to it.

Super-8 cameras have the advantage that they represent a widespread commercial technology, so that much support equipment is readily available. For instance, there are numerous kinds of inexpensive camera-mounted photoflood lights, automatic exposure controls, and remote controls for the cameras. All have variable benefits when applied in experiments. The lights may need to be remounted for close-up photography, and care must be taken that the intense heat does not affect the behavior of the experimental animal. The automatic exposure devices with which most such cameras are equipped tend to ensure that the overall field is well illuminated, but the mean exposure value may be inadequate for the portion of the image that is of particular interest. Thus, it is often desirable to install lateral reflectors outside of the field in order to illuminate the sides of the animal; this will improve resolution of all parts of the final images. (Also consider opening the aperture one or two stops beyond the "mean" value, particularly if the animal is darker than the background).

Most modern Super-8 cameras have a zoom feature that lets a single lens generate images with variable magnification. Even should this feature remain unused, it imposes special requirements for calibration. Thus, the first portion of each film should show a standard grid, and this should be filmed again at the end of the roll to prove that the zoom setting has not been changed accidentally during focusing. The beginning and end of each experimental film had best be defined by filming of a record number; this may be combined with the grid. The image may also contain additional information—for instance, the specimen number, date, film number, and a footage indication for the tape on which the EMG signals are being recorded in parallel.

When selecting floodlights, you should select the cheapest and remember that as long as the camera is fixed on a tripod, the lights need not be mounted on the camera. Consider life, cost, and availability of replacement bulbs when selecting a model. Certain brands of floodlights tend to generate electrical interference that appears as noise on the EMG traces. Generally you can reduce or eliminate such noise by connecting the floodlights into a different power circuit and by inserting a bit of grounded copper screening between the floodlights and the animal or adjusting the opening of the Faraday cage.

b. Synchronization and Analysis. It is critical that the demands of future analysis be kept in mind during filming. The problem may be much reduced by recording of the EMG waveform directly on the film or video tape recorder (fig. 21.3). If this is impractical, there are two ways of achieving future synchronization of films and EMGs. The first is to generate a synchronization signal with the camera shutter and to record this on one track of the chart records or one track of the tape. An alternative is to introduce periodically a simultaneous event, whether signal or interruption, both on the film and on one or more channels of the tape system. Well-defined voltage spikes coincident with filmed movements are obviously appropriate for this. It may be possible to mark the film with a time signal, which may be generated by a combination of a flashlight, fiberoptics, and a spinning disc or a pulsed light-emitting diode. This marker spacing should also be checked against absolute time markings and recorded on the start and

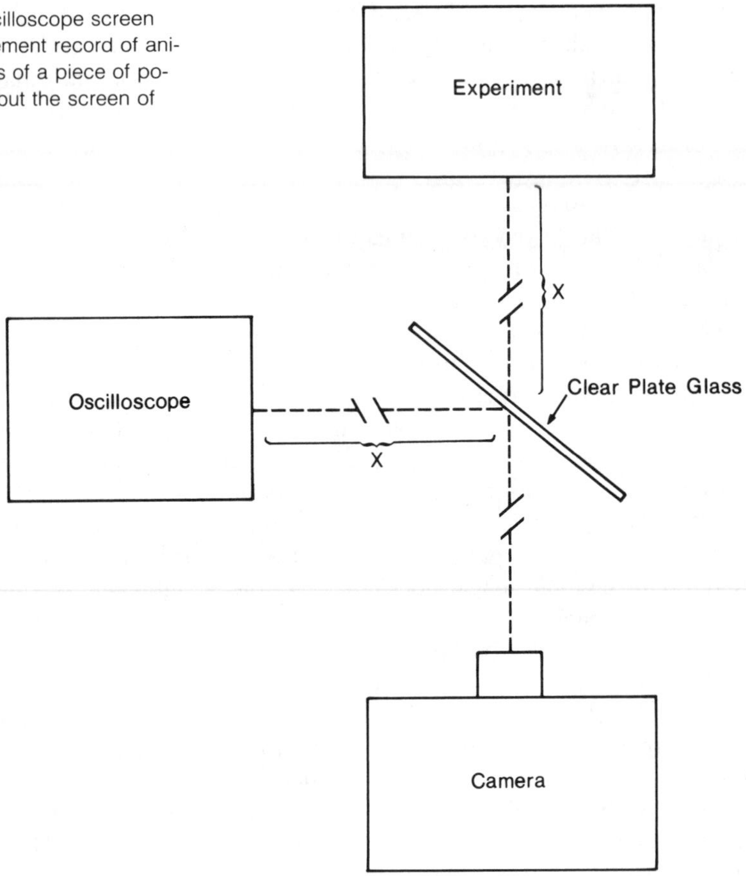

21.3. One can superimpose an oscilloscope screen on the cinegraphic (videotape) movement record of animal by using the reflecting properties of a piece of polished plate glass and by having all but the screen of the oscilloscope darkened.

end of the instrumentation tape.

Ordinarily, do not start an experimental sequence until the film is definitely being exposed (and has progressed beyond the portion that will be trimmed during processing). Stop a recording sequence before the film has run out. It is then possible to check the synchronization of events by counting both forward and backward in a sequence. One of the key mistakes made by many beginning investigators is to run the camera for long sequences. Not more than 10 feet of film should ordinarily be exposed, even when digitizing equipment is available for analysis. Unless the EMGs are to be part of a major behavioral analysis, a larger number of short sequences is better than a few long ones.

Consideration of analysis illustrates why short sequences are desirable, as even 10 feet of Super-8 film aggregates more than 400 frames. Accurate frame counts are essential, as a miscount even by a single frame is likely to have a substantial effect on the synchronization of mechanical events. We have already noted that the count must proceed independently, both from the beginning and from the end of each film sequence. One way of ensuring proper count is to mark the edge of the film, perhaps piercing every tenth frame with a small pin in one corner and every hundredth frame in another one.

The film has to be projected for analysis, most simply onto a sheet of white cardboard covered with acetate or tracing paper. Useful is a simple light box consisting of a horizontal sheet of ground glass, perhaps 30 by 30 cm, with a mirror placed at exactly 45° below it (fig. 21.4). For analysis, the plane of the glass should be absolutely normal to the beam of light incident from the mirror. Thus, the centerline of the lens will point at the middle of the field, and the image will incur minimal deformation (and will not be projected as a parallelogram rather than a square). Projection of the calibration grids filmed at the beginning and end of each film will quickly disclose possible distortion. In such a system the magnification is determined by the distance between

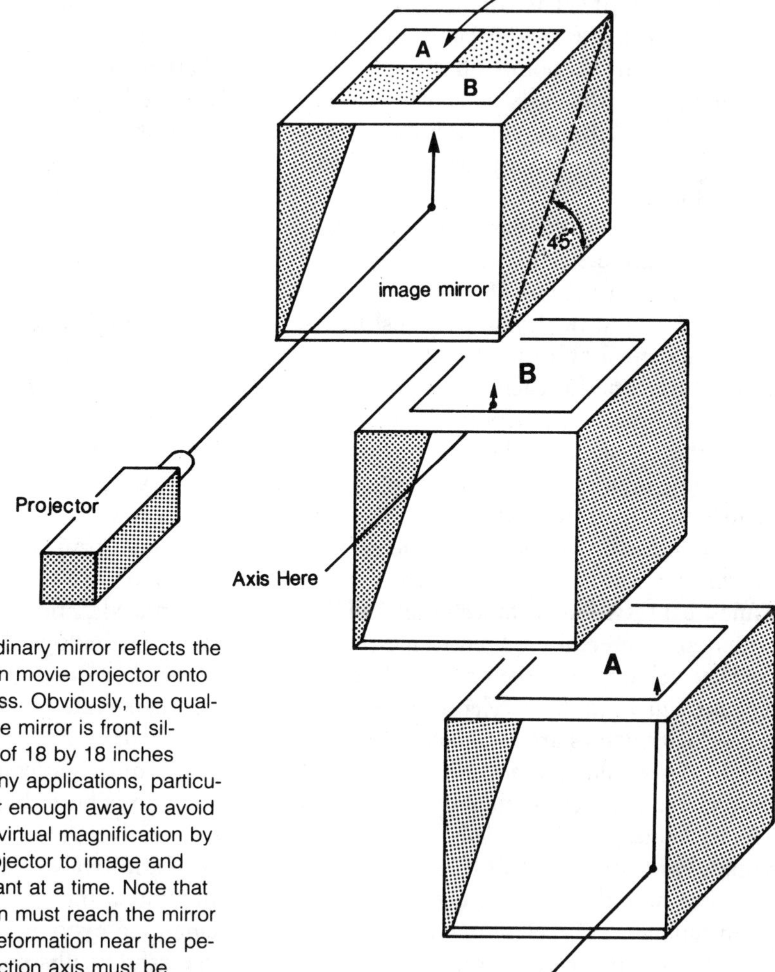

21.4. In this simple box, an ordinary mirror reflects the image provided by a stop-motion movie projector onto a horizontal piece of ground glass. Obviously, the quality of the image is improved if the mirror is front silvered. However, an image area of 18 by 18 inches provides adequate detail for many applications, particularly if the projector is placed far enough away to avoid parallax. One may increase the virtual magnification by increasing the distance from projector to image and projecting the image one quadrant at a time. Note that the incident axis of the projection must reach the mirror at 45° in order to avoid image deformation near the periphery. Consequently, the projection axis must be shifted whenever only part of the image is to be shown.

the projector and the ground glass. This distance will vary between the center and corner of the image, and the relative difference will increase as the projector is moved closer to the screen.

Once the basic rules of projection have thus been established, one must decide what aspects of the animal are important. The potential record may be conceptualized as a large data table, each line of which represents one slice in time derived from a single frame of a film and separated by a time determined by the filming rate. The columns in turn contain the coordinates (and perhaps angles of rotation) of various points on the animal, in two- (or three-) dimensional space. This suggests the desirability of constraining the animal so that it moves in a plane parallel to that of the image.

It is critical that the baselines for particular movements are defined within the image. If the interest is in the movement of portions of a limb relative to the body, one must be able to view the position of the body axis as a constant baseline for each frame and thereafter record the new position of the limb relative to this. A picture only of a moving foot is likely to be uninformative unless ankle movements only are being studied.

Once linear or rotational displacements are known, the change of displacement between frames, expressed as distance (angle) divided by temporal interval between frames, will yield the velocity, and the change of velocities per unit will yield the acceleration. Remember that velocity values are established for the interval between film frames, whereas the values of acceleration, representing an average of the velocity values, are coincident with the film frames. For purposes of EMG synchronization, they may thus be considered to represent the instantaneous acceleration at the time of exposure. In any case, the rules of synchronization given in chapters 15 and 16 deserve review at this point.

D. KEEPING IT CLEAN

The steps thus far delineated involve the determination of the EMGs of perhaps four muscles, the recording of the signals on tape and its representation as a hard copy, as well as the parallel analysis of mechanical events on film and correlation with the EMGs. The costs listed in table 21.1 could be less than $5,000, even if nothing may be borrowed.

Of course, the test procedures remain to be discussed. First, postmortem dissection must confirm that the electrodes were indeed in the correct muscle. Next is the question of whether the recorded EMG signal was indeed produced by the muscle under consideration. Some of the testing methods described in chapter 8 should certainly be used here. Most require relatively inexpensive equipment, and all of them tend to promote valid decision.

A much more critical issue is the overall nature of proof in studies such as these. When EMG is used, one should test not only for correlations between mechanical actions and the function of particular muscles, but also for their possible correlation with the activation of diverse other systems. More is needed than a simple correlation. All alternate possibilities must be eliminated. Will the action be performed equivalently if the muscle is incapacitated? Are the movements of skeletal levers indeed critical for the execution of particular biological roles? Can the animal still feed or jump although the muscles have been denervated or the skeletal linkages stabilized?

The most successful experimentalist is the one who is best able to discover and serve as devil's advocate for opposing hypotheses. The more effectively alternate possibilities are eliminated, the greater the probability that the remaining hypothesis is correct. The price of successful results remains the willingness to weigh alternatives and to search for errors in one's own conclusions.

APPENDIXES

1 Anatomical Techniques

This appendix provides an introduction and initial guide to some anatomical techniques found useful and important for characterizing muscle. They derive from our laboratories and those of colleagues who kindly made them available. As in most histochemical approaches, there is a certain amount of idiosyncrasy. Methods reported in research papers and later in guidebooks tend to be modified slightly or markedly to take into account the nature of the histologist, the conditions in the particular laboratory, and the nature of the animal. Hence, we present these formulas as starting points subject to modification and have noted places where the techniques may need adaptation to meet special circumstances. Some reference to the literature is provided (see References, section F), but the list is not inclusive and we stress that this account only includes the first steps toward anatomical and histochemical techniques.

Certain other procedures would obviously generate data desirable to the electromyographer, indeed to any student of muscle. Among these are laser techniques for establishing filament lengths and their overlap in muscles (correlated with joint angles), for instance expressed as relative widths of A and I regions (Rüdel and Zite-Ferenczy, 1979). Also there is the burgeoning field of immunohistochemistry for identification of fiber types by means of myosin isoenzymes (Cuello, 1984; Williams, 1983). However, these are not procedures derivable from a simple cookbook appendix and had best be approached via their primary literature.

A. CHEMICAL DISSECTION OF MUSCLE

1. General Principles

Serial cross-sections of whole animals represent a poor basis for muscle reconstruction as the knife is likely to intersect the fibers at odd angles and the original fiber architecture is often extremely difficult to interpret. However, careful dissection of muscle will give a good indication of fiber placement and may thus serve as a guide to determining sites for potential implantation of electrodes. The main problem is that of separating fascicles from their overlying connective tissues, the fibers of which generally run at an angle to those of the muscle, wrapping around the latter. Three simple techniques (surface treatment, staining, and total digestion) facilitate visualization.

The first technique is the chemical procedure of digestion of superficial connective tissues. Alternate application of full-strength sodium hypochlorite (used in household bleach such as Clorox), gentle washing, and surface drying removes the superficial connective tissues. For various reasons, the bleach attacks connective tissue fibers more rapidly than it attacks the muscle, so that the fascicles stand out and become obvious. This method is particularly useful for analyzing the surface architecture of preserved muscle as it makes few demands on previous preservation techniques.

Once the surface of the muscle fiber has been freed from connective tissue, other methods let one further differentiate the architectural pattern of the muscle. The best way of doing this is to stain the muscle, perhaps by use of a methylene blue surface stain that permits visualization of discontinuities, such as the clefts between adjacent fibers and the disposition of collagen strands. Even more useful for the present purpose is to use an iodine stain, such as Lugol's solution suggested by Bock and Shear (1972), although plain tincture of iodine also works well. The iodine solutions specifically stain the sarcoplasm so that individual muscle fascicles appear dark brown against a paler background of connective tissue. These approaches provide a rapid way of checking fiber architecture at different levels in the tissue. These staining methods are effective for aqueous and alchohol-stored materials; both of these methods have the further advantage that the stains will leach out once the material is returned to its liquid preservative.

Total digestion involves placement of the muscle into dilute nitric acid (see below). This slowly destroys the connective tissue septa and permits the fascicles to be moved about, separated, and if necessary stored in glycerin. The fascicles then may be counted and measured and their sarcomere lengths can be determined under a light microscope. Assuming that the sarcomere length for the species is known and that the sarcomeres of all its muscles have equal length (an assumption that does not hold for some lower vertebrates), one can then use the relative length of the sarcomeres to generate a correction factor that will permit calculation of the length that the fiber would have occupied if all sarcomeres were at the plateau region.

Total digestion is a relatively rough, but rapid approach, which incorporates the assumption that the muscle was fixed at appropriate length and that shortening of sarcomeres was equivalent throughout the muscle—matters that are easy to test unless the muscle has been digested too extensively. With this method, it is possible to obtain a standardized fiber length for the system and combine this with estimates of fiber position to engage in first-level comparison of muscle architecture.

Reagent

Lugol's solution

Iodine crystals	1 gm
Potassium iodide	2 gm
Distilled water	12 ml

Other formulas recommend weaker solutions; just add more distilled water.

Methylene blue surface stain may also be useful (cf. Romeis, 1948).

2. Nitric Acid Dissection

Prepare three beakers containing, respectively, at least 0.5 L of 30% nitric acid in saline, 0.5 L of pure glycerin, and 0.25 L of 30% nitric acid with 0.25 L of glycerin.

The previous fixation history of the specimen has a major effect on the way nitric acid dissection will work. Also the position at which the muscle is fixed has an obvious bearing on sarcomere lengths. Remember that this technique will decalcify bone and then dissolve it as well as the connective tissue. Consequently, it is useful to dissolve sample muscles to check the state of preservation; this is particularly important if the specimens have been deep-frozen for some time. Also, note that nitric acid does very well at dissolving fingers and metallic objects (gloves are needed). It is useful to have a supply of glass fingerbowls and plastic Petri dishes and to draw out a series of fine glass probes to manipulate material in the fluid. Precious dissection tools (even stainless steel ones) should be kept far away. Also a disposable plastic pad over the table surface helps to prevent indelible stains and scars. (We use plastic bags from the dry cleaner and a layer of newspaper over this. Laboratory supply houses will gladly provide a more convenient protective material.)

Procedure

a. Expose the block of fixed (not fresh) tissue, dissecting it out so that the origins and insertions of all the muscle fibers are included.
b. Place the specimen into the center of a dish and flood it with fresh 30% nitric acid in saline. (If the tissues are very fragile, it may be desirable to reduce the concentration of nitric acid to 15%, 20%, or 25%).
c. It is often useful to maintain the specimen in position and to photograph or to draw the original and subsequent stages (layers) using camera lucida methods. If the position of the several portions of a muscle is critical, you may find it useful to coat the bottom of the dish with perhaps 5 to 10 mm of paraffin wax and to fix the muscle onto this, perhaps by means of stainless steel minipins (the "minuten" pins used in entomology). This will keep the parts of the muscle from floating about during digestion. The fixed specimen may also be sutured to polyethylene mesh.
d. Cover the dish with a fitting lid and label the dish by a method that will not be affected by the acid (the ink in certain laboratory marking

pens is acid fast; adhesive paper labels may hold a plastic number to the outside).

e. Check the specimen every hour or so for the first 6 hours, until you are certain that digestion is not proceeding too rapidly. If digestion is too fast, substitute a weaker solution; if digestion is too slow, wait.
f. Thereafter it is often desirable to check every 6 or even 24 hours depending on the rate of digestion. Look for the moment when the fibers begin to fall apart of their own accord, separating into sets of pinlike units that are easily moved about by means of the glass needle.
g. Try not to disturb the specimen too much once this stage is reached, but gradually decant the fluid with an aspiration pipette applied to the side of the dish. Replace the nitric acid, first with a 50% glycerin (in water)/nitric acid mixture and then with only 50% glycerin (in water). This will arrest breakdown.
h. Digestion proceeds from the outside inward. If there is an expectation that the muscle is heterogenous, it may be useful to dissect away and remove the surface layer of fibers before letting digestion continue to the next layer. If more than a general topographical arrangement is to be determined, it may be critical to examine fiber lengths in several layers of the muscle.
i. Select a series of fibers, making sure that they are taken from a region that has not been damaged by dissection. Gently float them onto a standard microscope slide and place a coverslip over the top.
j. Examine the fibers under a light microscope. First check that the fiber ends are complete, showing a rounded or pointed architecture, rather than being truncated sharply as if they were broken. It may even be useful to break a couple of fibers in order to acquaint yourself with the difference. Then measure the length of the fiber, either by using a calibrated microscope stage or by examining the specimen under a dissecting microscope.
k. For measurement, place the slide on top of a graduated scale on the stage of a dissecting microscope. To avoid parallax, note the position of each end by shifting scale and slide together, so that the position of each end can be determined in the center of the microscopic field.
l. The specimen can also be examined to determine sarcomere length. For this, an ocular micrometer is best and it may be most useful to use dark-field, polarized, or similar contrast-enhancing optics. Measure approximately 10 sarcomeres and scan the sarcomere length along the fiber to note the range of deviations from a mean value.

The observed sarcomere length in microns should give a ratio when divided by the value of 2.3 μ assumed to be standard for mammals. The ratio can be used to standardize the fiber length. If the measured sarcomere length is shorter than normal, the resting length of the intact muscle fiber is likely to have been greater than observed; if the sarcomeres are longer than normal, the resting length of the fiber is likely to have been shorter. However, a different approach would set aside expectations and describe the fiber length and number and length of sarcomeres in terms of the position of the joint moved by the muscle. This approach reduces the number of preconceived notions and may increase the potential for discovery of previously undescribed phenomena.

An inevitable limitation of nitric acid digestion is due to the possible occurrence of very thin muscle fibers that may lie interspersed among larger ones. If these thin fibers adhere to the thicker ones, the technique may generate fascicles, giving the impression that they are really fibers. For this reason, at least some specimens should be stained with silver or gold reagents; this approach permits visualization of the connective tissues and motor end plates and thus ensures that fibers and fascicles can be distinguished.

3. Gold Reagent Fiber Marking

Gold reagent fiber marking stains muscle fibers in various gradations of color so that they can be more easily differentiated at sites of overlap. It has the further advantage that it can in favorable cases stain motor end plates, vascular networks, and sensory receptors. It does have the disadvantage of requiring fresh material (Swash and Fox, 1972).

Reagents

25% Formic acid

0.2% Gold chloride solution (prepare fresh)

Gold chloride	0.2 gm
Distilled water	100 ml

Glycerin

Procedure

a. Cut the fresh muscle into longitudinal slices 3 to 5 mm wide and parallel to the fibers. If the fiber directions are complex, try to cut some slices parallel to each set. Small or flat muscles can be stained directly.
b. Fix in formic acid solution for 10 minutes to 4 hours, depending on muscle thickness (use glass or plastic instruments).
c. Blot tissues on filter paper.
d. Place tissues in gold chloride solution for 20 to 90 minutes.
e. Blot tissues on filter paper.
f. Place in fresh formic acid solution and keep in dark for 8 to 12 hours.
g. Wash thoroughly in running tap water. Transfer to glycerin. Stir occasionally. Change the glycerin as it becomes diluted by fluids from the stained muscle.
h. Wait 2 or more days—the specimen often improves with storage.
i. Tease the preparation apart using a dissecting microscope.

Nerve endings should appear brown-black; muscle fibers range from purple on the surface to pale gold in deeper regions.

B. TIPS FOR HISTOCHEMISTRY (FROZEN SECTIONING)

1. Preparation of Muscle

The histochemical techniques here described rely on frozen sections for the study of succinic dehydrogenase (SDH), adenosine triphosphatase (ATPase), and nerve acetylcholinesterase (AChE) enzyme activity. The tissue is excised from the animal and immediately cooled to a temperature at which enzymatic reactions are essentially stopped. Storage time is minimized and the material sectioned while it remains frozen. Staining and reaction for particular compounds follow without prior fixation.

Cold isopentane (2-methylbutane) is usually used for freezing the muscle (cold acetone also gives good results). Fill about half of the Dewar flask (260 ml capacity) with isopentane that (preferably) has been precooled. Break dry ice in small chunks. Place one chunk at a time in a spoon and dip it slowly into the isopentane. Do not let the chunks fall into the isopentane, otherwise it will boil over. Keep adding dry ice until the isopentane no longer boils. Add a few more chunks, which will settle at the bottom. These will keep the liquid cold. Close the flask with a perforated cork to let gas escape. A CO_2 syphon tank with attached dispenser provides a useful source of finely powdered dry ice, avoiding the storage problem with intermittent usage.

Prepare as many tongue depressors as there will be muscles to be cut. Label each depressor in advance. Anesthetize or kill the animal and take out the muscle as quickly as possible. Trim the chunk of muscle to the size you eventually wish to section. Then orient the chunk on the tongue depressor in the plane you intend to cut—that is, transversely or longitudinally. (Some investigators prefer to mount samples longitudinally and reorient them later for transverse sections, otherwise samples may "lean.") Cover the muscle completely with a cryo-embedding compound. Place the muscle on one end of the tongue depressor and dip it in cold isopentane, keeping the H Dewar flask tilted so that the tongue depressor stays horizontal (to keep compound from dripping off). Keep it in this medium for 15 to 30 seconds, and then transfer it quickly into the cryostat. In some laboratories, the cryo-embedding compound is omitted.

Let the specimen equilibrate for about 30 to 60 minutes to the temperature of the cryostat. Specimens may be stored at $-20°$ C or colder for up to 1 week, if adequately sealed with embedding compound. ATPase, SDH, and nerve AChE all survive such storage; however, storage never improves the preparation. Still, some colleagues report on successful storage at $-70°$ C for more than 6 months.

Muscles may also be frozen directly in liquid nitrogen. This minimizes artifact (but there is a tendency for some blocks to split). The muscle may be oriented on the chuck in a layer of embedding compound and sprinkled liberally with talcum (to prevent formation of an insulating gaseous film). The chuck is then lowered into the N_2 on a metal plate with a long handle.

2. Sectioning

For very large muscles, it may be necessary to section local samples. Orientation will then be critical.

Be sure that the cryostat contains a single-edged razor blade, a forceps, and a camel hair brush and that all of them are cold. The chunk of muscle usually has become attached to the tongue depressor, so scrape it off with the blade. Trim it further if necessary. Remove the block holder from the cryostat and warm it a little with your thumb. Pour a drop of Cryoform (International Equipment Co., Boston, Mass.) (or other such matrix) on it. Return the block holder to the cryostat. Pick up the chunk of muscle in the cold forceps and be ready to fix it onto the block holder as soon as the Cryoform starts to freeze at the edges. Place the muscle in the middle of the drop of Cryoform and press it a little until its base is firmly embedded. Let the Cryoform solidify, which takes about 5 to 10 minutes. Sometimes, the chunk of muscle is very small, and you may have to embed it completely into the Cryoform. In such cases you will have to trim the block as for normal paraffin histologic examination.

The technique and precautions adopted during sectioning are the same as in normal microtomy, except that it is now necessary to cut and stretch one section at a time, on a coverglass (for fiber typing) or a microscope slide (for nerves and AChE). After properly aligning the muscle to the blade, start cutting the block slowly until you get an even, complete section of the muscle. Longitudinally oriented muscle is best cut when mounted at a slight angle horizontally (approximately 10°).

Removing the section from the blade is a delicate procedure. Normally the cut section pops up as the clean coverglass or microscope slide approaches; this makes it easy to attach it to the coverglass. If this does not happen, you may have to place the glass on top of the section, leave it there for a few seconds, and then pick it up. (This operation tends to warm the blade. Use Cryokwick [International Equipment Co.] for cooling it.) Stretch the section by rubbing the index finger on the *bottom* of the coverglass beneath the section. Permit the section to dry for 30 to 60 minutes. No moisture should remain in the section as this would interfere with the staining. Some cryostats are equipped with antiroll plates, which are useful for sections that tend to curl.

3. Staining

Staining procedures vary. Sections to be stained for SDH and ATPase should be processed on the same day. Overnight storage in the cryostat or a freezer is possible but may result in some loss of enzymatic activity, especially of SDH. (Be sure to redry the sections after removing them from the refrigerator.) Sections for nerve cholinesterase may be stored in sealed containers in a freezer (nondefrosting type!) for several months.

4. Subbing

Coverslips do not generally require subbing, but some brands allow sections to adhere better than others, so it may be necessary to test different brands. Microscope slides intended for collection of frozen sections had best be precleaned in acid alcohol or hot soapy water followed by a long rinse in running water. Slides are then subbed so that the muscle does not become distorted and the sections will not float off during staining. Dipping the slide in 3% EDTA just prior to use is often sufficient. The EDTA is also useful because it minimizes artifacts (i.e., shrinkage, holes) in the section.

Alternate Subbing Solution

Distilled water	100 ml
Pure gelatin	0.5 gm
Chromium potassium sulfate ($.24H_20$)	0.05 gm

The solution is sufficient for subbing 200 to 250 slides. Add the gelatin and chromium potassium sulfate to the water. Warm the solution slowly to 70° C. Shake well at this temperature. *Filter* and let the solution cool gradually to room temperature. If desired, add 3% EDTA and adjust to pH 7.0 to

7.4. At 30° C the solution is ready to use. Clean the slides by dipping them in acid alcohol (1% HC1 in 70% ETOH) or 100% ETOH and air dry them. Dip the slides into subbing solution and dry in a 36° C oven. Keep them dust free. Subbed slides should be dried for 1 hour before use and should be used the same day, if possible. Prolonged storage (a few weeks) is sometimes feasible, but such slides should be tested before use. Rinse used subbed slides and reclean them in EDTA.2H_2O = 3.32 gm per 100 ml (3%).

C. FIBER TYPE AND METABOLIC STAINS

1. Histochemical Demonstration of Myofibrillar Adenosine Triphosphatase

The times here listed are for some mammalian tissues. Experimentation will adjust these procedures to other organisms.

Reagents

0.1M Sodium barbiturate
1.0M $CaCl^2$
2% $CoCl_2$
0.1M NaOH
1% Ammonium sulfide
ATP
2,4 Dinitrophenol

Reaction Medium

Prepare immediately before use. Mix 5 ml 0.1M sodium barbiturate, 2.5 ml 1.0M $CaCl_2$, and 14 ml distilled water. Then add 30 mg ATP. Adjust pH to 9.4 with 0.1M NaOH; keep track of volume added. Filter into Coplin jars.

Procedure

a. Air dry 10 to 14 cryostat sections for 20 to 30 minutes at room temperature prior to incubation. Immerse sections in preincubation medium for 7 minutes at 37° C.
b. Transfer to reaction medium for 20 to 30 minutes at 37° C.
c. Remove to 1% $CaCl_2$, three changes of 2 minutes each.
d. Transfer to 2% $CoCl_2$ for 2 minutes.
e. Wash thoroughly with distilled water. Dip in three to four changes of water, blotting between each change. *Important*: This removes excess Co and prevents formation of scaly precipitate during the next step.
f. Immerse in 1% ammonium sulfide (yellow) for at least 30 seconds. (Leave longer if convenient.)
g. Wash well in tap water. (Hold under gentle stream of water for 10 to 15 seconds.)
h. Air dry for at least 30 minutes. Adequate drying is crucial for the preservation of the ATPase activity. *Alternative:* Dehydrate.
i. Mount with glycerin jelly. *Alternative:* Permount.

Preincubation Media

Preincubation media of different pH (acid or alkaline) may provide a further subclassification of fiber types (Brooke and Kaiser, 1970a, b).

Alkaline Reaction Medium

Same as for incubation, but with ATP omitted and the pH adjusted to 10.2 to 8.4. May be reused. Preincubate for 15 minutes at room temperature prior to ATPase incubation.

Acid Reagents

Stock solution

Sodium acetate	1.94 gm
Sodium barbital	2.94 gm
Water to make 100 ml	

For each dish, take 2.5 ml stock sollution, 4 ml of water, and 0.1M HCl to desired pH. To prepare buffer series at pH 4.6, 4.3, and 3.8, mix 7.5 ml stock solution, 12 ml water, and 0.1M HCl to pH 4.6 (approximately 12 to 14 ml). Place approximately 8 ml into staining jar number 1; 0.1M HCl to pH 4.3, 8 ml to staining jar number 2; 0.1M HCl to pH 3.9, 8 ml to staining jar number 3. Incubate sections for 5 minutes at room temperature prior to routine myofibrillar ATPase incubation.

2. Histochemical Demonstration of Succinic Dehydrogenase Activity (Nachlas et al., 1957)

Reagent

Buffered succinate stock—equal volumes of 0.2M phosphate buffer (pH 7.6) and 0.2M sodium succinate. Dissolve 10 mg nitroblue tetrazolium in 10 ml distilled water. Add 10 ml buffered succinate stock.

Procedure

a. Cut 10 μ sections on cryostat, mount on coverslip.
b. Air dry for ½ hour. (Sections stored overnight before incubation lose much of their activity. It is best to incubate immediately following air drying. Useful results can be obtained up to about 4 hours after the section is cut.)
c. Incubate at 37° C for 20 to 60 minutes depending on activity. (With very red muscles it is best to incubate for shorter times because cell boundaries are difficult to determine in heavily stained sections; 30 minutes is a good compromise whenever dealing with a muscle of unknown composition.)
d. Rinse in distilled water.
e. Quickly dip (1 to 2 seconds) in safranin O (0.1% aqueous solution) (optional step.).
f. Rinse thoroughly in distilled water and blot excess moisture lightly.
g. Air dry for about 30 minutes or dehydrate.
h. Mount in Permount.

3. Histochemical Demonstration of Unsaturated Lipids and Carbohydrates, especially Glycogen (Lillie, 1965)

Glycogen-rich fibers will stain deep pink or red; glycogen-poor fibers will be pale. Use extreme caution as glycogen may stream in formalin-fixed material. Use of frozen sections is safest.

Fixative

10% Buffered neutral Formalin

Technique

Paraffin sections of 5 μ

Solutions

0.15N Hydrochloric acid	
Hydrochloric acid	0.55 ml
Distilled water	99.45 ml
0.25N Hydrochloric acid	
Hydrochloric acid	0.92 ml
Distilled water	99.08 ml

(Use 1N HCl made up on shelf and dilute for .15N or .25N with distilled water.)

0.5% Periodic acid	
Periodic acid	0.5 gm
Distilled water	100 ml

Schiff's reagent ("cold method") (may be purchased ready-made)

Basic Fuchsin	1.0 gm
Sodium metabisulfite	1.9 gm
0.15N Hydrochloric acid	100 ml
(for doubling the recipe, use 0.25N HCl)	
Activated charcoal	500 mg

Procedure for Mixing Reagent

a. Combine basic fuchsin, sodium metabisulfite, and 0.15N HCl. Shake at intervals or put on a mechanical shaker for 2 hours. The solution should be clear and yellow or light brown at the end of this time.
b. Add the charcoal and mix for 1 to 2 minutes.
c. Filter into a graduated cylinder, washing the residue with a little distilled water to restore to the original 100 ml volume.
d. The solution should be as clear as pure water. If it has a yellow tint, repeat the decolorization process using a fresh lot of activated charcoal.
e. Store at 0° to 5° C. The solution is stable as long as it is colorless.

Procedure for Staining (use control slides; if sections separate from slides, consider use of celloidin-coated slides)

a. Deparaffinize and hydrate to distilled water.
b. Place in 0.5% periodic acid for 10 minutes.
c. Rinse in distilled water.
d. Place in Schiff's reagent for 30 minutes.
e. Rinse in running tap water for 10 minutes or longer.
f. Dehydrate, clear, and mount using Permount.

4. Periodic Acid–Schiff Method for Frozen Sections (procedure modified from Chayen et al., 1973)

The periodic acid–Schiff (PAS) method may be better for frozen sections. Considerable controversy has been generated over the problem of how to fix polysaccharides in tissues. Diffusion of glycogen in a block of tissue during fixation and dehydration is unimportant if frozen sections can be used. The picric acid–formalin fixative will then fix the polysaccharide material in the section, without diffusion from its site. More intense results can often be obtained if 5% acetic acid is added to the fixative, or if the sections are fixed in absolute alcohol; both these procedures intensify by partially solubilizing and redistributing the carbohydrate material.

Solutions

Picric acid–formalin fixative

40% Formaldehyde (technical)	10 ml
Ethyl alcohol	56 ml
Sodium chloride	0.18 gm
Picric acid	0.15 gm
Distilled water	100 ml

Periodate solution (prepare just before use): Dissolve 400 mg periodic acid in 15 ml distilled water. Add 135 mg crystalline sodium acetate ($CH_3.COONa.3H_2O$) dissolved in 35 ml of absolute ethyl alcohol.

Reducing rinse

Potassium iodide	1 gm
Sodium thiosulphate	1 gm
Distilled water	20 ml

When dissolved, add (while stirring the solution) 30 ml alcohol. Then add 0.5 ml 2N hydrochloric acid. A precipitate of sulphur may form; this should be allowed to settle out.

Procedure

Fresh cryostat sections should be fixed for 5 minutes in the picric acid–formalin fixative. (For an estimate of the total glycogen, it may be preferable to fix in absolute alcohol or in a picric acid–formalin solution to which 5% acetic acid has been added.) Red or purple-red indicates a positive PAS reaction.

a. Wash in 70% alcohol.
b. Immerse in the periodate solution for 5 minutes at room temperature.
c. Wash in 70% alcohol.
d. Immerse in the reducing rinse for 5 minutes. *Note:* Because periodate is a very strong oxidizing agent, even traces of it left in the section will recolor Schiff's reagent. Consequently, it is necessary to remove all adsorbed traces of periodate with this reducing solution. If no reducing solution is used, as unfortunately recommended by some workers, there is no reason to believe that the staining reactions of the tissue are indeed due to polysaccharide and not to adsorbed periodate.
e. Wash in 70% alcohol.
f. Immerse in Schiff's reagent for 30 minutes.
g. Wash in three changes of SO_2-water.
h. Dehydrate (stain the nuclei with Ehrlich's hematoxylin if required); mount in Euparal or in DePeX.

D. HORSERADISH PEROXIDASE METHODS

1. General

Horseradish peroxidase (HRP) techniques fill neurons and make all of their processes stand out. They are ideal for definitive tracing of neuronal or neural branches and connections. However, the procedures are dangerous as the 3′3′-diaminobenzidene (DAB) generally used for incubation is very carcinogenic; the procedures must be carried out under a properly vented hood and with appropriate precautions.

After the reaction, liberally add neutralizing bleach (sodium hypochlorite used to deactivate the DAB) to the DAB solution used for the reaction and flush all glassware or instruments that have contacted the DAB. *Never* pour the DAB down the drain; send it to a toxic waste disposal facility.

The dangerous DAB methods are useful if electron microscopy is to be used and for special applications. Tetramethyl benzidine (TMB) (see section 7) is preferred by some workers.

There are several methods of applying HRP. It is often injected into muscle as a 20% to 50% solution in saline, but this risks spreading. Better, the crystalline material may be placed immediately into the biological tissue. Novel methods include preparation of an HRP paste that is applied by means of a grooved dental appliance or large hypodermic needle, which inserts it into the muscle. In addition, cut peripheral nerves can be exposed to HRP, which may be applied as crystals, solutions, and Gelfoam implants. Whatever the application, the animal must survive for hours or days prior to sacrifice. Another important consideration with the use of any method of HRP application is the specificity of the approach. The tracer should be taken up only by the tissues of interest and not by neuronal elements in nearby tissues that are not being studied.

There are in essence two major techniques of visualizing the cells that have been filled with HRP.

The first is to fill the cells and then mount, section, and examine the tissue under the microscope. However, this requires a later, three-dimensional reconstruction of the system in order to trace the path of each cell through the animal's body. A different and often preferable technique is to fill the cells and then clear and stain the tissues.

Clearing and staining techniques have been developed for demonstration of skeletal elements such as bone and cartilage. The tissues are first bleached and then cleared in oil of cedar wood or glycerin. Present clearing and staining techniques allow one to follow the paths of neurons relative to these, either inward toward the spinal cord or outward toward sense organs and muscles. The method provides rapid visualization. For large animals it may be combined with a sectioning technique except that the cuts made are measured in millimeters rather than microns. Finally, it is possible to open up the specimen under the dissecting microscope and to trace the neuronal paths in this fashion. Whatever the approach, it would also be desirable to control the amount of clearing and to counterstain the specimen in order to obtain an impression of the paths of the neuron relative to such other tissues as bone and cartilage.

2. Preparation of Gels (Griffin et al., 1979; Mesulam, 1982)

Spreading of introduced HRP will be limited by introduction of it in gel form. The following procedure will make 0.05 ml of 15% HRP–15% polyacrylamide gel.

Reagent

Acrylamide-bis solution (44.4% acrylamide, 1.2% N,N′-Methylene-bis-acrylamide, w/v)	0.01675 ml
Distilled water	0.005 ml
0.1M Tris buffer (pH 7.2)	0.025 ml
TEMED (N,N,N,N-tetramethylethylenediamine)	0.001 ml

To produce this mixture, dissolve 7.5 mg HRP (Type VI, Sigma Chemical Co.). Then add 0.0025 ml ammonium persulfate (60 mg/ml). Mix, quickly transfer to a container, and allow to polymerize for 45 minutes at room temperature.

Sources (Sigma Chemical Co. 1983 catalog)

Acrylamide: Sigma A8887
N,N′-Methylene-bis-acrylamide: Sigma M7256
TEMED: Sigma T8133
Ammonium persulfate: Sigma A6761

The amounts given in the protocol make enough gel for many experiments and the gel has been kept for 1 week with no apparent ill effect. After the ammonium persulfate has been added, the fluid can be quickly drawn inside 1 ml disposable syringes. Such syringes provide sterile and airtight containers so that the gel remains hydrated. Drawing the gel into a syringe, as suggested above, always involves some loss of the gel.

Pieces of the wet gel also can be used for implants. However, the polymer is more easily handled if slices of the necessary size and shape are cut from the wet gel and are allowed to dry to a hard form. The wet 15% HRP–15% acrylamide gel loses approximately 60% weight upon drying. The ingredients may be combined in a V-shaped vial. Once they have gelled, the V-shaped mass may be removed and cut into usable pieces with a sharp scalpel. The dried pieces can be stored in a covered container in the freezer until ready for use.

The gel becomes difficult to handle if wetted (i.e., as soon as one tries to implant it) because it gets slippery and sticky. You will have to develop your own set of motor skills. The following procedure works well: Perform the implantation under a dissecting microscope using two pairs of fine forceps, one (A) to hold open the area (CNS or muscle to be implanted) and the second (B) to place the pellet into the opening and hold it there. Then release the tissue with forceps A so that it closes around the tips of forceps B (and the gel), and continue to use forceps A to scrape the pellet off of forceps B. If you lose control of a pellet, it probably will not help to keep poking around trying to get it into the opening. Put it aside (it will dry out and be reusable) and start with a dry piece, which will be much easier to handle. A large-gauge hypodermic needle with a plunger may also be used to implant pellets: press the pellet onto the end of the needle (it will stick), insert the needle into tissue, and push the plunger.

HRP pellets give a very restricted uptake site,

which has advantages and disadvantages. This technique obviously reduces the danger of leakage to adjacent muscles, but it also makes it very difficult to achieve an HRP concentration high enough for uptake and transport when used on large muscles. It works best with small, compact muscles. Usually several pellets must be implanted in a single muscle—the bigger the muscle, the more pellets. You can help offset this difficulty by using a sensitive procedure for demonstration of the HRP (see chapter 1 of Mesulam, 1982).

3. Fixation

Ideally, the tissue intended for sectioning should be fixed by perfusion, first with saline under a hood (add amyl nitrite to act as a vasodilator), and then with buffered glutaraldehyde or paraformaldehyde. Overfixation of tissue will reduce the enzyme reaction. The initial saline perfusion should remove the red blood cells as these display endogenous peroxidase activity. If saline perfusion is not feasible, add a preincubation step, using H_2O_2, to eliminate staining by the endogenous peroxidases. Many workers advocate flushing a 20% to 30% sucrose solution through the animal after fixation with aldehydes.

4. Sectioning

Muscle can usually be sectioned directly as frozen sections, 40 to 50 μ. Successful sectioning of brain and spinal cord tissue may require the use of agar support or a preembedding procedure similar to that below.

Procedure

a. Remove the brain or spinal cord, then wash in buffer with 30% sucrose for some hours or overnight. (30 gm sucrose, with enough Ringer's lactate solution to make 100 ml).
b. Prepare a 10% gelatin–30% sucrose solution with distilled water. To dissolve the gelatin, heat the solution to at least 60° C.
c. Pour the gelatin into a mold, place it into an ice bath to cool until it is tacky, and then insert the tissue and orient it properly.
d. Allow the gelatin to harden in the bath or refrigerator, then trim the block to the usual trapezoidal shape used for sectioning blocks.
e. Fix the block in 2% glutaraldehyde in 30% sucrose and saline for 3 to 4 hours or overnight. Keep it in the refrigerator.
f. Wash the block in 30% sucrose and saline and store in this solution in the refrigerator until sectioning.
g. Freeze the block in the cryostat and cut 40 to 50 μ thick sections. Mount the sections on gelatin-subbed slides and air dry. *Alternative*: The sections can be reacted while they are floating, transferred to phosphate buffer, and then mounted on the slides after the reaction.

5. Whole Mounts (limbs or entire muscles)

Procedure

a. Prefix the tissue with 2% glutaraldehyde or a combination of paraformaldehyde and glutaraldehyde. Either cacodylate or phosphate buffer may be used.
b. Incubate in 0.8% DAB in phosphate buffer for 2 to 4 hours in the refrigerator.
c. Remove from the refrigerator, add 0.5 ml of 0.3% H_2O_2 (fresh) per 5 ml of incubation medium. Incubate for equivalent time as in step a.
d. Rinse in phosphate buffer, put in 70% alcohol. Tissue can then be prepared for paraffin sections, whole mounts (clear in methyl salicylate or cedar wood), or plastic sections.

6. Alternate Method For Whole Mounts or Frozen Sections

Procedure

a. Place specimen into 0.1M Tris-HCL, at a pH of 7.6 for 5 minutes.
b. Switch into 0.5% cobalt chloride in tris buffer (1 gm $CoCl_2$/200 ml Tris) for 10 minutes with agitation.
c. Wash in Tris buffer (use the same solution from step one) for 5 minutes.
d. Wash in phosphate buffer for 5 minutes.

Remember: DAB is very carcinogenic. Use it under a hood with the hood fan on at all times (Adams, 1981). React the sections in a solution containing 1 mg DAB per 1 ml phosphate buffer and about 5 drops of 30% hydrogen peroxide per 200 ml solution. First dissolve the DAB in a small amount of distilled water (about 30 ml for 200 mg). Next add buffer to the final volume. Then add the peroxide, mix, and immediately add the sections. Allow the sections to react for about 20 minutes with agitation, then remove them and wash them in distilled water a few times.

The DAB concentration used here is about double that used elsewhere and the peroxidase is more than is usually used as well, but the combination works. You can tell that the reaction is going properly if small brown crystals form in the reaction vessel. (Follow Mesulam, 1982, to visualize fine connections.)

7. Tetramethyl Benzidine Procedure for Development of Horseradish Peroxidase

Two additional procedures for visualizing HRP may be considered—the Hanker-Yates method (Hanker et al., 1977), and the TMB method (Mesulam, 1978; Mesulam et al., 1980). The former is now available in kit form from a number of biological supply houses. The latter (Mesulam, 1978) is described here.

Prereaction Media (note time limit)

Solution A

Sodium nitroferricyanide	100 mg
0.2M Acetate buffer, pH 3.3	5 ml
Distilled water	92.5 ml

Solution B

TMB	5 mg
Absolute ethanol	2.5 ml

Carefully heat the solution to 40° C to dissolve the TMB.

Stabilizing Reagent

Absolute ethanol	50 ml
Sodium nitroferricyanide	9 gm
0.2M Acetate buffer, pH 3.3	5 ml
Distilled water	45 ml

Procedure

a. Perform all operations on floating sections. Keep cut sections for 9 to 48 hours at 4° C in the phosphate buffer.
b. Prepare solutions A and B within 2 hours of the staining procedure. Get ready to mix them.
c. Briefly wash sections in three changes of distilled water.
d. Add solution A to solution B in a reaction vessel only seconds before introducing the tissue into this prereaction medium. Soak for 20 minutes at 19° to 23° C. (Keep vessels free of oxidizing agents. Unless color remains pale reddish brown, suspect contamination.)
e. Add 2.0 to 4.0 ml of 0.3% H_2O_2 per 100 ml of medium removing the sections while adding this. Use trial procedures with 2.0, 2.5, 3.0, 3.5, and 4.0 ml to establish the optimal amount (i.e., that which stains the neurons but limits artifacts).
f. Transfer sections to the stabilizing solution at 0° to 4° C without prior washing. Maintain here for 20 minutes.
g. Counterstain (see next section). Wash, dehydrate in alcohol, clear in xylene, apply Permount, and cover.

8. Neutral Red Counterstain

To counterstain slides, add 40 ml acetate buffer at pH 4.8 to 1 L of a filtered 1% aqueous solution of neutral red. The acetate buffer is 500 ml of 0.1N acetic acid solution plus 750 ml of 0.1M sodium acetate.

Procedure

a. Counterstain in neutral red for approximately 1 to 3 minutes depending on the degree of staining that is desired.
b. Wash in distilled water for 15 seconds.
c. Clear in graded alcohols to toluene or xylene (15 seconds per step).
d. Mount.

E. IRON STAIN FOR ELECTRODE LOCALIZATION

If it proves impossible to dissect out the electrodes immediately, or if their leads are to be trimmed off and the tips left in situ while a second experiment is performed, one needs to ensure that they will not have shifted prior to check by dissection. An easy technique consists of electrolytically corroding the electrodes and subsequent staining for the ionized iron. This generates obvious spots with a dark blue center.

Procedure

a. Connect the recording electrodes to a 3 to 4.5 V battery for 30 to 60 seconds. To avoid stimulating the muscles electrically, gradually increase the current at the beginning and gradually decrease it at the end, using a series potentiometer.
b. Transfer unfixed tissue to a solution of 10 parts of 4% potassium ferrocyanide, 10 parts 4% acetic acid, and 1 part 40% formaldehyde (100% formalin solution). Maintain here for 24 hours.
c. Wash carefully with distilled water.
d. Counterstain for cell nuclei if desired.
e. Look for obvious spots with a dark blue center.

(Note: hemosiderin deposited in a hematoma will also react similarly.)

F. NERVE CHOLINESTERASE STAINS

There are many techniques for staining nerves and combined staining of nerve cholinesterase and AChE. Instructions for many of them may be found in the report by Zacks (1973). A few variations are listed here.

1. Cholinesterase Motor End Plate Stain for Gross Identification of End Plate Bands

This method works for both fresh and formalin- or glutaraldehyde-fixed tissue. It can also be used for frozen sections, but it does not work for paraffin-embedded samples. The method has the disadvantage that the end plates appear white.

Solution

20% Na_2SO_4 (34 gm anhydrous)	170 ml
$CuSO_4.H_2O$	0.3 gm
$MgCl_2$	1.0 gm
Glycerine	0.375 gm
Maleic acid	1.75 gm
4% NaOH	30 ml

This solution must be adjusted to pH 6.0 and heated to 37° C prior to use. Add 20mg acetylthiocholine iodide in a few drops of distilled water for every 10 ml of this stock solution.

Procedure

a. Incubate tissue in cholinesterase medium for 8 minutes at 37° C.
b. Rinse in distilled water for 2 minutes.
c. Treat in a 1% ammonium sulfide solution for 2 minutes.
d. Rinse in distilled water 3 to 5 minutes.
e. Repeat steps a to e if necessary.
f. Individual muscle fibers or small fiber bundles may be teased out and mounted for microscopic examination.

2. Cholinesterase Motor End Plate Stain (Karnovsky and Roots, 1964)

Solutions

0.1M Acetate buffer, pH 5.0

1M Acetic acid (5.81 ml glacial acetic acid made up to 100 ml)	36.5 ml
1N NaOH (1.0 gm/25 ml)	25 ml

0.6M Na citrate (17.64 gm/100 ml)
30 mM $CuSO_4$ (0.75 gm/100 ml)
5 mM $K_3Fe(CN)_6$ (0.16 gm/100 ml)
(The above solutions may be stored for some weeks at 4° C.)
1% Ammonium sulfide
(This solution should be stored in the fume hood.)
10% Formalin

Incubation Medium.
The following ingredients are added in order, stirring between each addition. Note that the concentration of acetylthiocholine iodide may be increased by a factor of 10 for optimal results with whole muscle preparations.

Acetylthiocholine iodide	12.5–125.0 mg
Acetate buffer, pH 5.0	16.2 ml
0.6M Na citrate	1.3 ml
30 mM $CuSO_4$	2.5 ml
5 mM $K_3Fe(CN)_6$	2.5 ml
Distilled water	2.5 ml

This should be made up fresh and is clear green. The solution should be discarded if it becomes cloudy and gold.

Procedure

a. Freshly dissected whole muscles or muscle pieces should be sutured at physiological lengths to coarse nylon mesh. This prevents the muscle from shrinking and allows an even distribution of stain.

b. The muscle (attached to the mesh) is completely immersed in an adequate volume of the incubation medium. The container of stain should be large enough to allow the muscle to lay flat during incubation. Whole muscles should be incubated overnight in the refrigerator. A brown precipitate will form, but this will not affect staining and may be washed off with distilled water.
c. Drain the incubation medium and rinse the muscle in distilled water. Motor end plates can be distinguished as white spots on the muscle surface.
d. Place the tissue in 1% ammonium sulfide. Motor end plates should appear dark brown after 1 to 2 minutes.
e. Fix muscle in 10% formalin. The muscle can be stored indefinitely in formalin.

3. Localization of Motor End Plates with Acetylcholinesterase Staining (suitable for electron microscopy)

Solution: Modified Karnovsky's Medium

Stock solution

0.1M Na acetate, pH 5.0	32.5 ml
300mM Na citrate	2.5 ml
30 mM $CuSO_4$	5 ml
5 mM $K_3Fe(Cn)_6$	5 ml
distilled water	5 ml

Incubation medium: For reaction for AChE, add 25 mg acetylthiocholine iodide (Sigma Chemical Co.) to the stock solution. For preincubation, add 0.1 mM ISO-OMPA (tetraisopropylpyrophosphoramide, M.W. 342, Koch-Light Laboratories, Ltd.

Procedure

a. Cut fixed muscle into thin strips or, ideally, use a "tissue chopper" to obtain slices 100 to 150 μ thick. Rinse in buffer.
b. Incubate in modified Karnovsky's medium (without acetylthiocholine) containing 0.1 mM ISO-OMPA for 20 minutes at 4° C. (This step may be omitted if distinction between other cholinesterase and AChE activity is not a factor of interest.)
c. Incubate in medium containing acetylthiocholine and ISO-OMPA for 4 minutes at 4° C.
d. Rinse at least twice with 0.1M Na acetate, pH 5.0.
e. Store in cacodylate buffer until ready to dissect end plates.
f. Place small pieces of muscle in drops of buffer on a microscope slide. The end plates may be visualized at low magnification in a compound microscope and precisely sliced out with a small scalpel. *Note:* If longer incubation times are used, the end plates can be easily visualized under a dissection microscope. However, the reaction product will not be well localized for electron microscopy. If the AchE is being stained simply to locate the motor end plate band, without a specific need for a precisely located reaction product in 1 μ or thin sections, use a dissecting microscope.
g. Toluidine blue stain may be used for 1 μ sections. The reaction product will appear brown. An electron-dense precipitate will be observed in thin sections.

4. Holmes Silver-staining Method for Axons (Lillie and Fullmer, 1976)

Solutions

Boric acid buffer

Boric acid	12.4 gm
Distilled water	1000 ml

Borax buffer

Boric acid	19 gm
Distilled water	1000 ml

20% Silver nitrate in distilled water (350 ml per staining dish)

1% Silver nitrate in distilled water

0.2% Gold chloride in distilled water (may be used repeatedly providing there is no brown precipitate)

2% Oxalic acid in distilled water

5% Sodium thiosulphate in distilled water

10% Pyridine in distilled water

Reducer

Hydroquinone	2 gm
Sodium sulphite	20 gm
Distilled water	200 ml

(may be used 2 consecutive days before discarding)

Impregnating solution

Boric acid buffer	110 ml
Borax buffer	90 ml

(Use only once)

Mix all the buffers in the 1 L cylinder and dilute to 988 ml with distilled water. Then add 2 ml 1% silver nitrate and 10 ml 10% pyridine. Mix thoroughly.

Stock Solutions

Saturated picric acid solution (filter before use):
1% Acid fuchsin

Picrofuchsin Stain

Saturated picric acid (filtered)	237.5 ml
1% Acid fuchsin	12.5 ml
Concentrated HCl	0.65 ml

Procedure

a. Deparaffinize the sections and rinse in running water, then distilled water.
b. Place in 20% silver nitrate, in the dark, at room temperature for 1 hour. Prepare the impregnating solution.
c. Rinse slides in three changes of distilled water for a total of 10 minutes.
d. Place the slides in impregnating solution at 37° C and leave them overnight.
e. The following morning, prepare the reducer. Remove the slides from the impregnating solution, shake off the excess, and place the slides in the reducer for 2 minutes (longer if necessary).
f. Wash the slides in running water for 3 minutes.
g. Rinse well in distilled water.
h. Tone in 0.2% gold chloride for 3 minutes (tone longer if the sections are still brown).
i. Rinse well in distilled water.
j. Transfer the slides to 2% oxalic acid for 4 minutes (2 to 10 minutes).
k. Rinse well in distilled water.
l. Transfer the slides to 5% sodium thiosulphate for 5 minutes.
m. Wash in running water for 10 to 15 minutes.
n. Counterstain in picrofuchsin for 3 minutes.
o. Dehydrate and differentiate in three changes of 95% ETOH (rinse quickly in the first two changes of ETOH and leave the slides in the third change for 3 minutes).
p. Continue dehydration in three changes of 100% ETOH (3 minutes each change).
q. Apply a coverslip. The tissue will stain as follows: myelinated axons, purple-black; muscle, brown-black; connective tissue, pink; collagen, red; nuclei, black.

5. Modification of Palmgren Stain for Nerve Fibers and Motor End Plates in Frozen Sections (Palmgren, 1960)

This technique can also be adapted for nerve staining of paraffin sections, but the acetylcholine actually does not well survive the paraffin embedding process.

Procedure

a. Mount and dry 10 to 50 μ frozen sections using slides dipped in 3% EDTA (pH about 7.2). Sections should be dried at least overnight but may be kept at room temperature or in the freezer longer.
b. Fix sections for 20 to 30 minutes at 4° C in the following solution:

Formalin	10 ml
Calcium chloride	1.0 gm
Sucrose	11.5 gm
Cacodylic acid	1.6 gm
Distilled water	90 ml

(Solution may be refrigerated and stored for months. It can be reused unless it becomes cloudy, then it should be discarded.)

c. Wash for 1 to 2 minutes in distilled water.
d. Prepare a cholinesterase medium (for preparation see section F1).
e. Incubate sections in cholinesterase medium for 4 to 8 minutes at 37° C.
f. Wash in distilled water for 2 minutes.
g. Treat for 2 minutes in a dilute ammonium sulfide solution (1 ml ammonium sulfide to 99 ml distilled water.)
h. Wash in distilled water for 3 to 5 minutes. (At this stage, the sections can be dehydrated and mounted unless nerve staining is not desired). Sections can be stained first with eosin.

i. Dehydrate through graded alcohol (70%, 95%, 100%) and then hydrate back to distilled water. (Dehydration is necessary for uniform staining.)
j. Incubate slides in 10% $AgNO_3$ for 30 minutes at 37° C. ($AgNO_3$ can be reused. Store in sealed container in 37° C oven.)
k. Transfer slides without rinsing to a 2% sodium borate solution (made fresh no longer than 24 hours before staining) and gently agitate for 3 to 5 seconds.
l. Rinse in distilled water until all the white precipitate formed in the preceding step is removed.
m. Develop for 10 minutes in the dark at 37° C in a freshly prepared solution of 100 ml 2% sodium borate (add first), 5.0 gm sodium sulfite, and 0.05 gm hydroquinone (add last). (For unknown reasons, this step frequently gives varied results.)
n. Rinse through three 1.5 minute changes of 50% alcohol.
o. Tone with 0.5% gold chloride (aqueous) for 3 to 3½ minutes. (The gold chloride solution may be stored in the dark and is often better after aging. Always do a test run whenever a new solution of gold chloride is prepared, as the timing may change.)
p. Rinse in distilled water.
q. Darken tissues *slightly* in freshly prepared 0.5% oxalic acid in 50% alcohol for 15 to 30 seconds. (The gold chloride step may be omitted, but if it is, increase the time in oxalic acid to darken the nerve fibers.)
r. Rinse in distilled water.
s. Fix in 5% sodium thiosulphate for 10 seconds. Rinse in distilled water, dehydrate through alcohols, clear in xylene, and apply a coverslip. (Sections may be counterstained with eosin for approximately 30 seconds prior to dehydration.)

6. Demonstration of Innervation in Teased Preparations by Bulk Staining

Silver Stain (modification of de Castro's [1925] version of Cajal's method, courtesy Dr. D. C. Quick). Stain may be capricious and often will give uneven staining of sensory axons. All solutions should be prepared fresh.

Solution 1

Chloral hydrate	1 gm
95% ETOH	45 ml
Distilled water	50 ml
Concentrated nitric acid	1 ml

Mix 95% alcohol and distilled water, add chloral hydrate, and stir until dissolved. Slowly add nitric acid while stirring.

Solution 2

95% Ethyl alcohol	1000 ml
Concentrated ammonium hydroxide	0.9 ml

Solution 3

Silver nitrate	1.5 gm
Distilled water	100 ml

Solution 4

Hydroquinone	2 gm
Distilled water	75 ml
88% Formic acid	25 ml

Procedure

a. Fix in solution 1 for 4 to 6 days.
b. Wash in running tap water for 1 day.
c. Incubate in solution 2 for 1 to 2 days (keep covered).
d. Blot surplus fluid.
e. Incubate in solution 3 for 5 days at 37° C (keep in dark and well covered).
f. Reduce in solution 4 for 2 days (keep in dark and well covered).
g. Rinse in distilled water for 1 day (use a couple of fresh changes).
h. Mounting options
   (1) Store in glycerin for at least 2 days and make squash mounts.
   (2) Freeze and section on a cryostat microtome.
   (3) Dehydrate and embed in wax or other medium.

G. ELECTRON MICROSCOPY/PLASTIC-EMBEDDED MATERIAL

1. Fixation of Tissue

Ideally, muscles destined for electron microscopy (EM) should be perfused with an oxygenated physiological solution, followed by perfusion with glutaraldehyde (at room temperature). However, in many cases perfusion is not feasible. An alternative procedure is to remove the muscle or muscle segment, pin out at a "resting length" on dental wax, and cover with glutaraldehyde (see below). As soon as the tissue starts to turn slightly darker (within 5 minutes), carefully subdivide the sample *longitudinally* into as many thin (1 to 2 mm) strips as possible to allow adequate internal fixation. (Use a very sharp razor blade, and cut with single, downward strokes rather than a "sawing" motion.) After 15 to 30 minutes, the strips may be subdivided into 1 to 2 mm^3 cubes or left as strips (if motor end plates are to be detected). The tissue should remain in the cold fixative 2 to 24 hours and then be rinsed at least twice with buffer. The samples may be stored in the refrigerator, but prolonged storage will cause them to deteriorate.

Fixatives. Glutaraldehyde (2% to 2.5%) is the most commonly used primary fixative for EM. After fixation in glutaraldehyde and adequate rinsing, the tissue is postfixed with 2% osmium tetroxide for 1 to 2 hours. This postfixation step is often incorporated in the process of dehydration and embedding.

Buffers. Either phosphate or cacodylate buffer may be used. However, the same buffer should be used for both fixation steps and all rinses.

Millonig's phosphate buffer (stock to make 100 ml)

$NaH_2PO_4.H_2O$	2.26 gm
NaOH	2.52 gm
Glucose	5.4 gm
$CaCl_2$	1.0 gm

To mix add 41.5 ml $NaH_2PO_4.H_2O$ and 8.5 ml NaOH. Mix thoroughly, test 5 ml for pH 7.3, and then add 5.0 ml glucose and 0.25 ml $CaCl_2$.

Cacodylate buffer (100 ml)

"Fixing" buffer (for use with glutaraldehyde and $OsSO_4$)

0.2M Na cacodylate

Hydrous	4.28 gm per 100 ml water
Anhydrous	3.2 gm per 100 ml water

Check pH (7.3 to 7.4), adjust with 0.1N HCl.

"Rinsing" buffer

0.1M Na cacodylate

Hydrous	2.14 gm per 100 ml water
Anhydrous	1.6 gm per 100 ml water
Sucrose	7.3 gm
0.1N HCl (adjust to pH 7.3)	4–8.5 ml

If desired, add $CaCl_2$ to rinsing and fixing buffers as this improves preservation of membranes in muscle: 0.555 gm anhydrous $CaCl_2$ per liter, 0.735 gm hydrous $CaCl_2.2H_2O$ per liter. (Will make 5M $CaCl_2$.)

2. Stains for Plastic-embedded Sections (0.5 to 2.0 μ)

Toluidine Blue Solution. The solution may be stored but must be filtered before use. It is convenient to fill a 10 ml syringe with stain and dispense through a Millipore filter attachment.

Toluidine blue	0.2 gm
Na borate	0.2 gm
Water	200 ml

Procedure

a. Be sure the sections on the slide have been dried thoroughly.
b. Place the section cover on a hot plate (medium heat) with a drop of stain.
c. Stain for 1 minute, then rinse. *Note:* Batches vary in staining time; hence, monitor slides visually.

Method of C. D. Humphrey and F. E. Pittman. This procedure stains muscle blue, whereas collagen and glycogen stain red.

Stock Solution A

Phosphate buffer pH 6.9 (add 9.08 gm anhydrous KH_2PO_4 and 11.88 gm anhydrous Na_2HPO_4 to make 1 L)	150 ml
Distilled water	250 ml
Methanol CP	50 ml
Glycerol	50 ml
Azure II Giemsa (Azure A)	0.1 gm
Methylene blue	0.65 gm

Stock Solution B

Basic fuchsin	0.5 gm
50% Ethanol	50 ml

To use, dilute 1 in 10 with distilled water. *Note:* Filter each stock solution just before using.

Procedure

a. Heat solution A to 65° C in a Coplin jar immersed in a beaker of water.
b. Stain sections for approximately 10 minutes.
c. Rinse two to three times in water and allow to dry thoroughly (usually overnight). *Do not dry in oven:* This causes fading.
d. Counterstain with solution B by putting a drop of the diluted stain over *unheated* sections for approximately 5 minutes.
e. Rinse thoroughly, air dry, and apply a coverslip.

3. Stains for Thin Sections on Grids

Staining of EM grids is accomplished by floating the grids with sections facing downward on a drop of stain solution. Individual drops are usually put on a layer of paraffin in an enclosed chamber, such as a Petri dish.

Solution

Uranyl acetate

Saturated solution of uranyl acetate in 50% ethanol. Store in dark bottle.

Lead citrate (add second)

Distilled water (boil first to remove CO_2)	10 ml
Lead citrate	0.25 gm
10N NaOH	0.1 ml

Store in dark bottle. Prepare frequently to avoid precipitates.

Procedure

a. Stain with uranyl acetate for 2 to 10 minutes. Rinse by dipping in at least three changes of water.
b. Stain with lead citrate, usually for 2 to 10 minutes. (To prevent precipitates, stain in presence of NaOH crystals or a few drops of NaOH solution on filter paper.)
c. Rinse. Add one to two drops of NaOH solution to the rinse solution.
d. Blot dry.

2 Suppliers

The following list of commercial suppliers of various goods and services is intended to provide only a starting place for would-be purchasers. It is not intended to be complete nor is it possible to keep such a list completely up-to-date. The listed companies are a mixture of those with which we have had personal experience, those recommended by colleagues, and those whose advertising has indicated potential interest to researchers pursuing techniques described in this volume. No specific recommendation or guarantee is intended by inclusion in this list, nor should omission from this list be taken as indicative of anything other than an oversight of the authors.

A. RAW MATERIALS

1. Electrodes and wire

Electrodes

Rhodes Medical Instruments
21044 Ventura Blvd.
Woodland Hills, CA 91364
818-347-3577

Description: Prefabricated multiwire and concentric EMG type semimicroelectrodes. Custom electrodes to specification.

MicroProbe Inc.
P. O. Box 87
Clarksburg, MD 20871
301-972-7100

Description: EMG patch and nerve cuff electrodes, custom configurations.

Microelectrodes, Conventional

Frederick Haer and Co.
4 Industrial Pkwy.
Brunswick, ME 04011
207-729-1601

Description: Epoxy insulated, also pipette glass.

Microelectrodes, Parylene-insulated

A-M Systems, Inc.
1220 75th St. S.W.
Everett, WA 98204
800-426-1306
206-353-1123

Description: SS and Tungsten, also pipette glass.

MicroProbe Inc.
P. O. Box 87
Clarksburg, MD 20871
301-972-7100

Description: SS and tungsten etched with Parylene-C.

Wire Sutures: Flexon, Teflon-coated Stainless Steel

Davis and Geck
Berdan Ave.
Wayne, NJ 07470

Description: Alternate gross long-term EMG electrode.

Wire, Many Alloys and Insulations

California Fine Wire, Co.
338 S. 4th St.
P. O. Box 446
Grover City, CA 93433
805-489-5144

Wire, Microelectrode

Sigmund Cohn Corp.
Medwire Division
121 S. Columbus Ave.
Mount Vernon, NY 10533
914-664-5300

Wire, Platinum: 30% Iridium, Pure iridium
Engelhard Industries
700 Blair Rd.
Carteret, NJ 07008
201-321-5721
Anaheim, CA 92807
714-779-7231

Wire, Platinum: 28 and 18 gauge
A. J. Thomas
Vine St. at 3rd
Philadelphia, PA 19105

Wire, BWR 3.48/7248 Teflon
Bergen Wire Rope Co.
P. O. Box 326
Lodi, NJ 07644
201-487-3521

Description: Stranded stainless steel (302) in Teflon jacket. Small, strong, flexible, biocompatible, perfect for EMG lead wires and very cheap in large quantities.

Wire, Biomed AS631
Cooner Wire Co.
9186-88 Independence
Chatsworth, CA 91311
213-882-8311

Description: Stranded stainless steel (316) as above, more expensive, but also great variety of multiconductor and shielded configurations, some very small. Also other alloys and jackets.

Wire
A-M Systems, Inc.
1220 75th St. S.W.
Everett, WA 98203
800-426-1306
206-353-1123

Description: Various special types for physiological electrodes.

2. Conductive Parts

Component Clips
Midland Ross
Electronic Connector Division
1 Alewife Pl.
Cambridge, MA 02140
617-491-5400

Description: 1 cm spring-loaded clips suitable for temporary connection of electrode wires. Thread mounts or board mounts.

Conductive Paint and Adhesives
E-Kote-3030, etc.
Acme Chemicals & Insulation Co.
P. O. Box 1404
New Haven, CT 06505
203-562-2171

Description: Complete line of silver-filled paints, lacquers, epoxies, etc. Great for unsolderable wires, shielding, and so on.

Epoxy Technology
14 Fortune Dr.
P. O. Box 567
Billerica, MA 01821
617-667-3805

Description: Conductive gold/silver insulation.

Conductive Polymers, CON/RTV-1, Nickel, Carbon, and Silver-filled sheet
Tecknit (Technical Wire Products)
129 Dermody St.
Cranford, NJ 07016
201-272-5500
320 N. Nopal St.
Santa Barbara, CA 93101
805-963-5811

Description: Silver-filled conductive equivalent of Medical Adhesive A plus prefabricated sheets and layered devices. Also wide selection of metal and cloth conductive meshes, shielding gaskets, etc.

Hypodermic Tubing, Stainless Steel
Small Parts, Inc.
Box 792
Biscayne Annex
Miami, FL 33152
305-751-0856

Description: Also brass tubing and rods.

Hyperdermic Needles
Vita Needle Co.
919 Great Plain Ave.
Needham, MA 02192
617-444-1780

Description: Source of strange sizes, e.g., 27 gauge by 3 inches

Plastic and Metal Stock
Small Parts, Inc.
Box 792
Biscayne Annex
Miami, FL 33152
305-751-0856

Description: Various extrusions, stampings, springs, etc.

Silver Flake
Alcan Metal Powders
P. O. Box 290
Elizabeth, NJ 07207
201-353-4600

Description: Fine, flake-shaped silver particles for filling your favorite polymer to make it conductive (works well in Insl-x and silicone rubber).

3. Polymeric Parts

Dacron mesh
No. 000887
USCI, Division of C. R. Bard, Inc.
Glens Falls, NY
617-667-2511

Description: Useful for reinforcing silicone rubber and providing tissue-invadable anchors.

Plastic Films: Teflon, Mylar, Kapton
Dupont
Film Department
Fabrics and Finishes
Wilmington, DE 19898

Silastic Parts: Tubing, Dacron-reinforced Sheet Sponge, Blocks, etc.
Dow Corning, Corp.
Medical Materials Business
Midland, MI 48640
517-496-4000

Description: Flexible, biocompatible, water-permeable raw materials for body tissue interfaces such as artificial dura, tendon sheaths.

Sil-Med Corp.
700 Warner Blvd.
Taunton, MA 02780
617-823-7701

Description: Great variety of tubing sizes.

Teflon Tubing, Medical Grade
Dixon Industries Corp.
386 Metacom Ave.
Bristol, RI 02809
401-253-2000

Description: Penntube Teflon, TFE, FEP, shrinkable, and nylon tubing, many sizes

4. Encapsulants and Adhesives

Alpha-cyanoacrylate
Ethyl Fast
Vigor Co., Division of B. Jadow Sons, Inc.
53 W. 23rd St.
New York, NY 10010-4275
212-807-3844

Description: Water vapor–catalyzed, rapidly setting, clear, rigid adhesive that sticks to body tissues.

Butyl Cyanoacrylate
Ethicon Ltd.
P. O. Box 408
Edinburgh EH11-4HE,
Scotland

Description: "Biocompatible nerve glue" available only in Europe.

Cyanoacrylate
Eastman 910
Permabond International Corp.
480 S. Dean
Englewood, NJ 07631
201-567-9494

Description: Similar to Ethyl Fast. Brochures have complete technical data.

Dental acrylic
L.D. Caulk, Co.
P. O. Box 359
Milford, DE 19963-0359
302-422-4511

Description: Old standby for rigid pedestals built up during operation and adherent to bone.

Epoxy

Expolite
Expolite Corp.
1911 E. Via Burton
Anaheim, CA 92816
714-956-9311

Description: Old standby rigid microelectrode insulation for dip coating.

Hardman, Inc.
Belleville, NJ 07109
201-751-3000

Description: Convenient packages of adhesives, differing in peel strength, work time, flexibility.

Insl-x-Products Corp.
115 Woodsworth Ave.
Yonkers, NY. 10702

Tra-Bond 2106
Tra-Con, Inc.
55 North St.
Medford, MA 02155
617-391-5550

Description: Complete line of epoxy types in convenient unit dose duopacks that mix without opening. Five-minute style particularly nice.

PU 61 putty
Abbeon Cal, Inc.
123-215A Gray Ave.
Santa Barbara, CA 93101
805-966-0810

Description: Like modeling clay that turns rigid in 1 hour at room temperature, comes in 1-pound bars, excellent adhesion.

Kerr Fast Cure

Stoelting
1350 S. Kostner Ave.
Chicago, IL 60623-1196
312-522-4500
Telex 314 176

Description: Rigid pedestals to be bonded to bone.

Neoprene Adhesive

Carboline Co.
350 Hanley Industrial Ct.
St. Louis, MO 63144

Description: Contact adhesive that sticks excellently (even to animal skin) and makes a metal-to-metal bond that resists autoclaving.

Parylene Deposition Services

Nova Tran Corp.
100 Deposition Dr.
Clear Lake, WI 54005
715-263-2333

Paratronix Inc.
129 Bank St.
Attleboro, MA 02703-1775
617-222-8979

Parmat, Inc.
905 Richmar Dr.
Westlake, OH 44145
216-835-0466

Description: Microprocessor-controlled Parylene coaters and deposition systems.

Viking Technology, Inc.
590 Laurelwood Rd.
Santa Clara, CA 95050
405-727-3057

Parylene Deposition Equipment

Union Carbide Corp.
Parylene Products Department
9474 Chesapeake, Suite 907
San Diego, CA 92123
619-569-8135

K.T.I. Chemicals
Parylene Products
2 Barnes Industrial Park
Wallingford, CT 06492
203-265-9242

Silicone Rubber

Silastic Medical Adhesive A (catalog no. 891)
Silastic casting compound (catalog no. 7982)
Dow Corning Corp.
Medical Products
Midland, MI 48640
517-496-4000

Description: Essentially highly purified bathtub caulk. Clear, single-component RTV adhesive, viscous enough for small parts and wire anchoring done in operating room, also two-part biocompatible molding compound.

Silicone Contact Cement
Silgrip-SR573
Silicone Products Dept.
General Electric Co.
Waterford, NY 12188

Description: Tack-free, pressure-setting contact cement, highly flexible, water and solvent resistant, probably biocompatible, wets even unetched Teflon.

Silicone potting compounds
RTV-602, RTV-615
Silicone Products Dept.
General Electric Co.
Waterford, NY 12188

Description: Low-viscosity, two-component, clear rubber potting and molding materials. See also Technical Data Book S-3C for complete line.

Tapes: Dielectric, Teflon, Conductive Foil
3M Industrial Electrical Products
Saint Paul, MN 55101
612-733-1110

Description: Generally available locally.

Teflon Coating
Phoenix Wire, Inc.
South Hero, VT 05486
802-372-4561

Comment: Will apply custom Teflon coatings to any size wire.

Temporary adhesives
Crystalbond 555
Aremco Products, Inc.
P. O. Box 429
Ossining, NY 10562
914-762-0685

Description: Water-soluble, low temperature–melting adhesive mounting material useful during fabrication steps. Also makes conductive adhesives and potting compounds.

Urethane Adhesives
Hardman, Inc.
Belleville, NJ 07109
201-751-3000

Description: Convenient packages of epoxy, urethane, and acrylic adhesives differing in peel strength, work time, and flexibility.

5. Solutions

Electroplating: Gold, Platinum, Nickel, etc.
Sel-Rex Corp.
Nutley, NJ 07110
201-667-5200

Orotemp No. 121, 24K gold
Technic Inc.
88 Spectacle St.
P. O. Box 965
Providence, RI 02901
401-781-6100

Kolrausch Solution
Hartman-Leddon Co.
Philadelphia, PA

Description: Pt Cl 3% dissolved in 0.025% lead acetate solution, for platinum black.

B. FABRICATION EQUIPMENT

Small Tool Vendors
Brookstone Co.
127 Vose Farm Rd.
Peterborough, NH 03458
603-924-9541
603-924-9511

Description: Odd tools.

Dri Industries, Inc.
11300 Hampshire Ave. S.
Bloomington, MN 55438-2498
612-944-3530

Description: Many tools and supplies, often in sets.

H&R Corp.
401 E. Erie Ave.
Philadelphia, PA 19134
215-426-1708

Description: Miscellaneous components. Electromechanical, optical, and electronic products.

Jensen Tools Inc.
7815 S. 46th St.
Phoenix, AZ 85040
602-968-6231

Description: Tool kits and individual tools, soldering and desoldering and test equipment.

Leichtung Inc.
4944 Commerce Pkwy.
Cleveland, OH 44128
800-321-6840
216-831-7645

Description: Woodworking and other tools.

Micro Mark
24 E. Main St.
P. O. Box 5112
Clinton, NJ 08809
800-225-1066

Description: Model maker's tools.

U. S. General
100 Commercial St.
Plainview, NY 11803
800-645-7077

Description: Hand and power tools.

Microwelders

Description: Essential for microwelding fine wires or small parts.

Ewald Instruments Corp.
Kent, CT 06757

Unitek
1820 S. Myrtle Ave.
Monrovia, CA 91016
818-574-7800

Description: Microwelding fine wires, small parts and thermocouples and strain gauges.

Rocky Mountain Orthodontics
P. O. Box 17085
Denver, CO 80217
800-525-6375
303-534-8181

Description: Inexpensive spot welder and resistance soldering unit.

Pressurized Dispenser

Portion-aire
Glenmarc Mfg., Inc.
330 Melvin Dr.
Northbrook, IL 60062
312-272-9030

Description: PMC-100-03-A is perfect for applying small dabs of adhesives, viscous encapsulants, etc., from disposable syringes; foot pedal air pressure control, no dribbling.

Soldering tip

Isotip
Wahl Clipper Corp.
2902 N. Locust St.
Sterling, IL 61081
815-625-6525

Description: Small tip, rechargeable battery, pen light, perfect for small wires in the operating room.

Soldering Flux

Johnson Manufacturing Corp., Inc.
Division of Serrmi Products
P. O. Box 43246
5290 Tulane Drive, S.W.
Atlanta, GA 30336

Description: Lloydes stainless steel soldering flux.

Solder

Indium Corp. of America
P. O. Box 269
Utica, NY 13503

Description: Various exotic solders.

Thermal wire stripper

Stripall
Teledyne Kinetics
410 S. Cedros Ave.
Solana Beach, CA 92075
800-344-4334
800-992-9988 (CA only)

Description: Hand-actuated insulation stripper, temperature safe for gold wire.

C. ELECTRONICS

1. Components

Cables

Malco, A Microdot Co., Inc.
306 Pasadena Ave.
South Pasadena, CA 91030
818-799-9171
12 Progress Dr.
Montgomeryville, PA 18936
215-699-5373

Connectors

Winchester, Amphenol, ITT, Cannon Electric

Description: Usual local suppliers. Small multipin and strip connectors to assemble your own pin configurations. Cannon Centi-loc strips are particularly reliable. Winchester 2024S and 2024P are crimpable with 107-0903-2A tool.

Trompeter Electronics
8936 Comanche Ave.
Chatsworth, CA 91311
213-882-1020

Description: More than 30 lines of connectors having mating options in bayonet (standard or keyed), threaded, or push-on patch.

Connector Parts

Malcon
306 Pasadena Ave.
So. Pasadena, CA 91030
818-799-9171

Description: Small, reliable, multipin connectors for transcutaneous use. Also many connector types.

Connector Systems

Scotchflex Systems
3M Electronic Products
Building 502
12209 Metric Blvd.
P. O. Box 2963
Austin, TX 78769-2963
512-834-6708

Description: Low noise, light weight, mechanically secure. Now standard on many computers. Automatic termination. Many local suppliers, also other manufacturers of similar types.

Crimping Tools

Daniels Manufacturing Corp.
6103 Anno Ave.
Orlando, FL 32809
800-327-2432
305-855-6161

Description: BNC and other crimpers and components. Connector tooling accessories.

Commutators

Litton Systems, Inc.
1213 N. Main St.
Blacksburg, VA 24060
703-552-3012

Description: (F2245) 30 Leadin/leadout gold slip ring commutator in small metal can suitable for relieving cable twisting caused by animal rotation.

Stoelting Co.
1350 S. Kostner Ave.
Chicago, IL 60623
312-522-4500

Description: Five- and 10-channel commutator with fluid paths.

Miniature Amplifiers

Bak Electronics, Inc.
687-L Loftstrand Ln.
Rockville, MD 20850
301-869-3700

Description: MMRS-1, 12-channel FET hybrid head stage and modular second stages.

MicroProbe Inc.
P. O. Box 87
Clarksburg, MD 20871
301-972-7100

Description: Inexpensive, low-noise transformer coupled head stage.

Miniature Telemetry

Midgard Harvard Apparatus
Pleasant St.
South Natick, MA 01760
617-964-4545

Description: Single-channel FM, high-impedance, wideband, battery-operated transmitter small enough for rat skull.

AMTI
Advanced Mechanical Technology, Inc.
141 California St.
Newton, MA 02158
617-964-2042

Description: Single-channel transmitter.

Biotelemetrics, Inc.
24650 Center Ridge Rd.
Westlake, OH 44145
216-835-5770

Description: Single-channel FM, high-input (100uv), DC-10kHz bandwidth, fully contained, 1.5–3.5 VDC operation. Smallest and lightest transmitter available.

Micro Probe Inc.
P. O. Box 87
Clarksburg, MD 20871
301-972-7100

Description: Single channel FM in inexpensive kit form with various transducer and electrode inputs and power configurations possible.

Matrix Switches

Co-Ord Switch Division
LVC Industries
Flushing, NY 11354
212-939-9777

Patch Panels

MAC Panel Co.
P. O. Box 5027
High Point, NC 27262
919-884-8111

Description: Low cross-talk programmable patch panels for infrequently reconfigured systems.

Switchcraft
5555 N. Elston Ave.
Chicago, IL 60630
312-792-2700

Description: Local electrical supply representatives. 2731-301 Modular telephone jack systems for quick, easy, on-the-fly equipment configuration. Accepts reliable floating shield, coaxial patch cords that are inexpensive (05AD05).

Trompeter Electronics
8936 Comanche Ave.
Chatsworth, CA 91311
213-882-1020

Description: Preassembled panels and multitype shielded systems, adapters, and components.

Ribbon Cable

Brand-Rex Co.
Willimantic, CT 06226
203-423-7771

Description: Standard and custom lightweight flexible copper foil ribbon cable that can be heat set into a self-retracting coil.

Deanco
2415 N. Triphammer Rd.
Ithaca, NY 14850
607-257-4444

Description: Good selection of multiconductor flat cable and 3M connectors. Will make flat cables with 3M connectors attached.

Special Electronic Components

AD521K Inst. Amp.
Analog Devices, Inc.
Rt. 1, Industrial Park
P. O. Box 280
Norwood, MA 02062-0280
617-329-4700

Description: True instrumentation amplifier, supplied in DIP form (i.e., totally integrated). Inexpensive (18 dollars), excellent CMMR, low bipolar noise, stable in true differential mode with input resistors of 2 MΩ, external gain control, and a gain band product of 40 MHz. Excellent for EMG and other recordings where source resistance is less than 100K and as a secondary differential stage amplifier. May be tricky to set up, i.e., may (not always) require compensation circuit and input configuration. Also consider their AD524 and AD625. (Manufacturer is willing to furnish a free copy of their extensive databook to readers who refer to this listing.

A07533LN
Analog Devices, Inc.
Rt. 1, Industrial Park
P. O. Box 280
Norwood, MA 02062-0280
617-329-4700

Description: Ten-bit multiplying digital-to-analog converter (DAC). Excellent general purpose DAC with good linearity low cost. Simple to apply using data sheet provided. AD7542 is a latched 12-bit microprocessor compatible version (see note above).

3522/K Op Amp.
Burr Brown
International Airport
Industrial Park
Tucson, AZ 35734
602-294-1431

Description: FET input operational amplifier. Low-noise, low-input offset voltage (no need for external nulling at low gains), low bias current, input capacity about 30 pF. Comes in TO-99 package. Excellent for extracellular recordings with metal microelectrodes.

2N4867A FET
Siliconix
2201 Laurel Wood Rd.
Santa Clara, CA 95054

Description: Low-noise FET transistor. Small TO-46 package. Operates at currents below 1 mA. Good as source follower for small head stage assemblies, inexpensive.

LH0002CN voltage follower
National Semiconductor
2900 Semiconductor Dr.
Santa Clara, CA 95051
408-737-5000

Description: Bipolar current amplifier, low output impedance (6 Ω). D.C. to 30 MHZ bandwidth, fast slew rate (200 V per μsec), low harmonic distortion. Excellent for driving analog signal through coaxial cables or multiple loads.

TL082CP, LF353N Op. Amp.
National Semiconductor
2900 Semiconductor Dr.
Santa Clara, CA 95051
408-737-5000

Description: Dual general purpose op. amp., no frequency compensation required, no offset nulling needed. Eight-pin DIP package and low power consumption. Good secondary amplifier and where high density packaging is required.

RAD 1024
Reticon
910 Benicia Ave.
Sunnyvale, CA 94086
408-739-4266

Description: "Bucket-brigade" delay line, 1024 stages implies delay = 1024/2F, where F is the clock frequency, which must be four times the upper frequency limit of the signal (see Bak and Schmidt, 1977a).

LE31H Comparator
Siliconix
2201 Laurel Wood Rd.
Santa Clara, CA 95054

Description: FET input comparator, low input currents, less prone to spurious oscillations and strobe input. Excellent as zero crossing detector and level detector. Used as level detectors in discrimination design (Bak and Schmidt, 1977b).

HCPL2530 Isolator
Hewlett-Packard
Optical Communication Division
640 Page Mill Rd.
Palo Alto, CA 94340
415-857-1501

Description: High-speed, dual-channel, optically coupled isolator. Excellent for both digital and analog information. Very easy to set up so that over two decades of gain may be achieved for isolating certain types of electrode recording preparations.

LF152 FET Instrumentation Amplifier
National Semiconductor
2900 Semiconductor Dr.
Santa Clara, CA 95051
408-737-5000

Description: JFET inputs. Rin = 10^{12} Ω, CMRR 110 db, low noise. Suitable as general purpose, high-impedance, inexpensive (8 dollars) instrumentation amplifier packaged as 16-pin DIP. Very easy to use.

2. Instruments

Bak Electronics, Inc.
687-L Lofstrand Ln.
Rockville, MD 20850
301-869-3700

Description: Complete line of amplifiers, stimulators, bridges, integrators, etc., for electrophysiological signals.

Dantec Electronics, Inc.
6 Pearl Ct.
Allendale, NJ 07401
201-825-3339
Telex 219 205 DANTC UR

Comment: Distributes DISA specialized, custom EMG systems for medical applications.

Stoelting Co.
1350 S. Kostner Ave.

Chicago, IL 60623-1196
312-522-4500

Description: General physiological instrumentation.

Medical Systems Corp.
One Plaza Rd.
Greenvale, NY 11548
516-621-9190

Description: Sophisticated electrophysiological systems.

Harvard Apparatus
22 Pleasant St.
South Natick, MA 01760
617-655-7000

Description: General physiological instrumentation.

Dagan Corp.
2855 Park Ave.
Minneapolis, MN 55407
612-827-5959

Description: Amplifiers, stimulators, averagers, voltage clamps, and patch clamp systems.

World Precision Instruments, Inc.
375 Quinnipiac Ave.
New Haven, CT 06513
203-469-8281

Description: Amplifiers, stimulators, averagers.

Carolina Biological Supply
2700 York Rd.
Burlington, NC 27215
800-334-5551

Description: General physiological instrumentation.

Frederick Haer & Co.
Brunswick, ME 04011
207-729-1601

Description: Amplifiers, stimulators.

A-M Systems, Inc.
1220 75th St. S.W.
Everett, WA 98204
800-426-1306
206-353-1123

Description: High-gain, differential AC amplifiers, DC Neuroprobe amplifiers.

Coulbourn Instruments
Box 2551
Lehigh Valley, PA 18001
215-395-3771

Description: Behavioral apparatus, interfaces.

3. Digital Data Processing

Comment: This area is exploding and any buyer should benefit from careful comparison shopping not restricted to the vendors listed.

Digitizing Systems

Comment: Generally available with signal analysis programs.

Analog Devices, Inc.
Route 1, Industrial Park
P. O. Box 280
Norwood, MA 02062-0280
617-329-4700
Telex 924 491

Andromeda Systems, Inc.
9000 Eton Ave.
Canoga Park, CA 91304
818-709-7600

Keithly Instruments, Inc.
28775 Aurora Rd.
Cleveland, OH 44139
216-248-0400

Description: DAS Series 500.

Interactive Microware, Inc.
P. O. Box 139
State College, PA 16804-0139
814-238-8294
Telex 705250

Data Translation, Inc.
100 Locke Dr.
Marlboro, MA 01752
617-481-3700
Telex 951 646

Gould Electronics
Recording Systems Division
3631 Perkins Ave.
Cleveland, OH 44114
216-361-3315

Description: Series 2000W.

MetraByte Corp.
254 Tosca Dr.
Stoughton, MA 02072
617-344-1990

Kirby Lester Inc.
P. O. Box 43
Riverside, CT 06878
800-243-2465
Telex 643 465

Description: Microlink.

Signal Processing Systems

David Computergraph Software System
J. D. Associates
275 Lodgeview Dr.
Oroville, CA 95965
916-589-2043

I.C.S.
8601 Aero Dr.
San Diego, CA 92123
800-621-0852 ext. 303
619-279-0084

Signal Technology, Inc.
5951 Encina Road
Goleta, CA 93117
805-683-3771
Telex 910-334-3471

TransEra Corp.
Suite 4
3707 N. Canyon Rd.
Provo, UT 84604
801-224-6550

4. Record and Display Systems

Motion Tracking Systems

Selcom Inc.
P. O. Box 250
Valdese, NC 28690
704-874-4102

Description: Selspot system. Three-dimensional motion analysis.

AMTI
Advanced Mechanical Technology, Inc.
141 California St.
Newton, MA 02158
617-964-2042

Description: Motion-monitoring systems.

Hamamatsu Systems Inc.
40 Bear Hill Rd.
Waltham, MA 02254
617-890-3440

Description: X-Y optical position detectors.

Northern Digital
415 Phillip St.
Waterloo, Ontario
Canada N2L 3X2
519-884-5142

Description: Three-dimensional motion analysis system.

Oxford Metrics, Inc.
11525 53rd St. N.
Clearwater, FL 33520-9990
813-577-4500
800-237-8923

Description: Vicon system.

Visicorder Oscillographic Recorder

Honeywell Test Instruments Division
P. O. Box 5227
Denver, CO 80217-5227
303-773-4700

Electrostatic and Pen Chart recorders

Gould Inc.
Instruments Division
3631 Perkins Ave.
Cleveland, OH 44114

FM Tape Recorder

A. R. Vetter Co.
Box 143
Rebersburg, PA 16872
814-349-5461

Time-base Generators

Chrono-log Corp.
2 West Park Rd.
Havertown, PA 19083-4691
215-853-1130

Datum, Inc. Timing Division
1363 State College Ave.
Anaheim, CA 92806
714-533-6333

Video Disc
Eigen Video
P. O. Box 848
Nevada City, CA 95959
916-272-3461

See also local representatives for large manufacturers of analog tape recorders: Ampex, Hewlett-Packard, Sangamo (Fairchild Weston), Honeywell; *and video equipment:* Sony, Panasonic, GYYR, Ikegami, JVC.

Film Projectors
Lafayette Instrument Co.
P. O. Box 5729
3700 Sagamore Pkwy. N.
Lafayette, IN 47903
317-423-1505
800-428-7545

Vanguard Instruments Corp.
1860 Walt Whitman Rd.
Melville, NY 11746
516-249-3031

High-speed Cameras
HYCAM
Red Lake Corp.
1711-T Dell Ave.
Campbell, CA 95008
408-866-1010

High-speed Video Equipment
Instrumentation Marketing Corp.
820 S. Mariposa St.
Burbank, CA 91506
213-849-6251

Stroboscopes
Chadwick-Helmuth Co.
4601-T N. Arden Dr.
El Monte, CA 91731
213-575-6161

D. ELECTROMECHANICAL EQUIPMENT

Accelerometers
Wilcoxon Research
2096 Gaither Rd.
Rockville, MD 20850
301-770-3790

Description: Miniature types.

Bruel & Kjaer Instruments, Inc.
185 Forest St.
Marlborough, MA 01752
617-481-7000
Telex 710-347-1187

Description: Sound and vibration analysis.

Kistler Instrument Corp.
75 John Glenn Dr.
Amherst, NY 14120-5091
716-691-5100

Description: General purpose.

Force Plates
Kistler Instrument Corp.
75 John Glenn Dr.
Amherst, NY 14120-5091
716-691-5100

Description: Complete systems.

AMTI
Advanced Mechanical Technology, Inc.
141 California St.
Newton, MA 02158
617-964-2042

Description: Motion-monitoring systems

Columbus Measurement Group
6952 Springhous Ln.
Columbus, OH 43229
612-422-0859

Pressure Sensing Catheter
Kyowa Electronic Instruments Co.
3-8, Toranomen 2-chome
Minato-ku
Tokyo, Japan
Telex 222-3854 Kyowat J

Description: Injectable pressure sensor. Millar PC-350

Strain Gauges
BLH Electronics
42 Fourth Ave.
Waltham, MA 02254
617-890-6700

Description: Elements and kits.

Micro-Measurements Division
Measurements Group, Inc.

P. O. Box 27777
Raleigh, NC 27611
919-365-3800

Comment: Product and technical information on request. Training programs and engineering support also available.

Omega Engineering, Inc.
1 Omega Dr.
Box 4047
Stanford, CT 06907
203-359-1660

Comment: Catalogs and handbook available.

Langer Biomechanics Group
21 E. Industry Rd.
Dear Park, NY 11729
800-645-5520
516-667-3462

Description: Foot force distribution.

Treadmill Belts

Potomac Rubber Co.
24 Kennedy St.
Washington, DC 20011
202-722-0800

Description: Custom-fabricated, endless belts. Neoprene multiply belts.

Treadmill Motors and Controllers

Graham Co.
P. O. Box 23880
8800 W. Bradley Rd.
Milwaukee, WI 53223
414-355-8800

Description: Electrically and acoustically quiet motors and reversing speed controls with conical rollers.

Zero-Max
2845 Harriet Ave. S.
Minneapolis, MN 55408-2291
612-872-6300

Description: Continuously variable speed controllers for synchronous electrical motors.

E. MEDICAL AND SURGICAL MATERIALS

Illuminators: Surgical Head Lamps

Welch Allyn Medical Division
State Street Rd.
Box 220
Skaneateles Falls, NY 13153-0220
315-685-8351

Description: 49003 Halogen head lamp, wall-mounted power supply. Excellent for small animal surgery.

Percutaneous connectors

Biosnaps
Bentley Laboratories, Inc.
17502 Armstrong Ave.
Irvine, CA 92714
714-546-8020

Description: Vitreous carbon forms that are sutured to skin for biocompatible exit point.

Suture Materials and Adhesives

Ethibond
Ethicon Inc.
Somerville, NJ 08876
703-281-4151

Vitallium Screws

Howmedica, Inc.
359 Veterans Blvd.
Rutherford, NJ 07070
201-935-2100

3 Animal Care Regulations

The following three statements represent minimal standards for the housing and experimentation of animals as adopted by some major professional societies. These standards represent the basic concerns. However, we note again that EMG generally involves the attempt to study muscular activity during natural behavior. This underlines that our standards of animal care should always be most stringent.

FASEB POLICY STATEMENT ON ANIMALS USED IN RESEARCH

The FASEB (Federation of American Societies on Experimental Biology) Board reaffirms the Federation's basic statement of 1913 on the issue of animal experimentation and its statement of "Guiding Principles in the Care and Use of Animals" approved in 1980.

Further, the Board has arrived at the following conclusions and supports the following resolutions:

Whereas, human and veterinary medicine have made great advances in relieving both humans and animals from disease and suffering; and,

Whereas, experimental use of animals has proven to be indispensable to the progress achieved in both human and veterinary medicine and product safety; and,

Whereas, continued progress is dependent upon the appropriate use of animals in research, education, and the testing of substances, techniques, and equipment; and,

Whereas, deterrence of appropriate use of animals for these purposes will impede continued progress in the search for treatment, cure, and the prevention of disease and suffering afflicting humans and animals;

Therefore, be it resolved that the Federation of American Societies for Experimental Biology supports the appropriate use of animals for scientific experimentation and education; and,

Also be it resolved, that animals obtained from local and municipal pounds are a valuable resource for medical and veterinary research and education and should be available for research and educational institutions with all appropriate safeguards to preclude inadvertent use of pet animals (i.e., identification, established holding periods, optional donation by owners, etc.); and,

Also be it resolved, that wide application of accreditation procedures for animal experimental facilities and well managed governmental inspection and regulation of such facilities and their supply channels are necessary and desirable protection against willful or inadvertent abuse of standards for animal care and management; and,

Also be it resolved, that continuing collection of appropriate data on the conditions and number of animals used in scientific research and education is necessary for development of legislative or administrative remedies in the field.

GUIDING PRINCIPLES IN THE CARE AND USE OF ANIMALS

Approved by the Council of the American Physiological Society

Animal experiments are to be undertaken only with the purpose of advancing knowledge. Consideration should be given to the appropriateness of experimental procedures, species of animals used, and number of animals required.

Only animals that are lawfully acquired shall be used in this laboratory, and their retention and use shall be in every case in compliance with federal, state, and local laws and regulations, and in accordance with the NIH Guide [see below].

Animals in the laboratory must receive every consideration for their comfort; they must be properly housed, fed, and their surroundings kept in a sanitary condition.

Appropriate anesthetics must be used to eliminate sensibility to pain during all surgical procedures. Where recovery from anesthesia is necessary during the study, acceptable technique to minimize pain must be followed. Muscle relaxants or paralytics may be used for surgery in conjunction with drugs known to produce adequate analgesia. Where use of anesthetics would negate the results of the experiment such procedures should be carried out in strict accordance with the NIH Guide. If the study requires the death of the animal, the animal must be killed in a humane manner at the conclusion of the observations.

The postoperative care of animals shall be such as to minimize discomfort and pain, and in any case shall be equivalent to accepted practices in schools of veterinary medicine.

When animals are used by students for their education or the advancement of science, such work shall be under the direct supervision of an experienced teacher or investigator. The rules for the care of such animals must be the same as for animals used for research.

Reference: *Guide for the Care and Use of Laboratory Animals*, DHEW Publication No. (NIH) 80–23, revised 1978, reprinted 1980, Office of Science and Health Reports, DRR/NIH, Bethesda, MD 20205.

GUIDELINES FOR THE USE OF ANIMALS IN NEUROSCIENCE RESEARCH

Published by the Society for Neuroscience

Introduction.
Research in the neurosciences contributes to the quality of life by expanding knowledge about living organisms. This improvement in quality of life stems in part from progress toward ameliorating human disease and disability, in part from advances in animal welfare and veterinary medicine, and in part from the steady increase in knowledge of the abilities and potentialities of human and animal life. Continued progress in many areas of biomedical research requires the use of living animals in order to investigate complex systems and functions because, in such cases, no adequate alternatives exist. Progress in both basic and clinical research in such areas cannot continue without the use of living animals as experimental subjects. The use of living animals in properly designed scientific research is therefore both ethical and appropriate. Nevertheless, our concern for the humane treatment of animals dictates that we weigh carefully the benefits to human knowledge and welfare whenever animal research is undertaken. The investigator using research animals assumes responsibility for proper experimental design, including ethical as well as scientific aspects.

The scientific community shares the concern of society at large that the use of animals in research should conform to standards that are consonant with those applied to other uses of animals by humans. While it is unlikely that any particular set of standards will satisfy everyone, it is appropriate for scientific societies to formulate guidelines that apply to the humane use of laboratory animals in particular areas of research. Ideally, such guidelines should also be acceptable to society at large as reasonable and prudent.

Most of the more specific sections of this document were formulated with respect to research using warm-blooded vertebrates. As a general principle, however, ethical issues involved in the use of any species, whether vertebrate or invertebrate, are best considered in relation to the complexity of that species' nervous system and its apparent awareness of the environment, rather than physical appearance or evolutionary proximity to humans.

Factors that Relate to the Design of Experiments.
The primary factor used to evaluate humane treatment in animal research is degree of distress or discomfort, assessed by anthropomorphic judgments

made by reasonable and prudent human observers. *The fundamental principle of ethical animal research is that experimental animals must not be subjected to avoidable distress or discomfort.* This principle must be observed when designing any experiment that uses living animals.

Although most animal research involves minimal distress or discomfort, certain valid scientific questions may require experimental designs that inevitably produce these effects. Such situations, while uncommon, are extremely diverse and must be evaluated individually. It is critical that distress and discomfort be minimized by careful experimental design. It is also important to recognize that there is no difference between distress and discomfort that may be inherent in a valid experimental design and that which may occur as an unintended side effect. It is therefore incumbent on the investigator to recognize and to eliminate all *avoidable* sources of distress and discomfort in animal subjects. This goal often requires attention to specifics of animal husbandry as well as to experimental design.

Invasive procedures and paralytic drugs should never be employed without benefit of anesthetic agents unless there is very strong scientific justification and careful consideration is given to possible alternatives. Advances in experimental techniques, such as the use of devices chronically implanted under anesthesia, can offer alternative approaches. If these are not feasible, it is essential to monitor nociceptive responses (for example, recordings of EEG, blood pressure, and pupillary responses) that may indicate distress in the animal subject, and to use these as signals of the need to alleviate pain, to modify the experimental design, or to terminate the experiment.

When designing research projects, investigators should carefully consider the species and numbers of animals necessary to provide valid information, as well as the question whether living subjects are required to answer the scientific question. As a general rule, experiments should be designed so as to minimize the number of animals used and to avoid the depletion of endangered species. Advances in experimental methods, more efficient use of animals, within-subject designs, and modern statistical techniques all provide possible ways to minimize the numbers of animals used in research. This goal is completely consistent with the critical importance of replication and validation of results to true progress in sciences.

Factors that Relate to the Conduct of Experiments. Research animals must be acquired and cared for in accordance with the guidelines published in the *NIH Guide for the Care and Use of Laboratory Animals* (National Institutes of Health Publication NO. 80–23, revised 1978). Investigators must also be aware of the relevant local, state, and federal laws. The quality of research data depends in no small measure on the health and general condition of the animals used, as well as on the specifics of experimental design. Thus, proper animal husbandry is integral to the success of any research effort using living animal subjects. General standards for animal husbandry (housing, food quality, ventilation, etc.) are detailed in the *NIH Guide*. The experienced investigator can contribute additional specifics for optimum care for particular experimental situations, or for species not commonly encountered in laboratory settings.

Surgery performed with the intent that the animal will survive (for example, on animals intended for chronic study) should be carried out, or directly supervised, by persons with appropriate levels of experience and training, and with attention to asepsis and prevention of infection. Major surgical procedures should be done using an appropriate method of anesthesia to render the animal insensitive to pain. Muscle relaxants and paralytics have no anesthetic action and should not be used alone for surgical restraint. Postoperative care must include attention to minimize discomfort and the risk of infection.

Many experimental designs call for surgical preparation under anesthetic agents with no intent that the animal should survive. In such cases, the animals ordinarily should be maintained unconscious for the duration of the experiment. At the conclusion of the experiment, the animal should be killed without regaining consciousness and death ensured before final disposition.

Certain experiments may require physical restraint, and/or withholding of food or water, as methodological procedures rather than experimental paradigms. In such cases, careful attention must be paid to minimize discomfort or distress and to ensure that general health is maintained. Immobilization or restraint to which the animals cannot be readily adapted should not be imposed when alternative procedures are practical. Reasonable periods of rest and readjustment should be included in the experimental schedule unless these would be absolutely inconsistent with valid scientific objectives.

When distress and discomfort are unavoidable attributes of a valid experimental design, it is mandatory to conduct such experiments so as to minimize these effects, to minimize the duration of the procedures, and to minimize the numbers of animals used, consistent with the scientific objectives of the study.

Ad Hoc Committee on Animals in Research, Robert E. Burke, Chairman. Neurosciences Newsletter, vol. 15, no. 5, 1984.

A newly revised edition of the "NIH Guide for the Care and Use of Laboratory Animals" may be obtained from the Office of Science and Health Reports, Division of Research Resources, Building 31, Room SB-10, NIH, 9000 Rockville Pike, Bethesda, MD 20205. It includes recommendations and references regarding animal care facilities and procedures.

References

This reference list is obviously selective and stresses introductory works that we have found useful in our own learning and in explaining particular approaches to colleagues. Various sources are intended as guides to areas that we approach but peripherally. We have included a series of sample studies in which EMG techniques were used on various animals and interesting details of presentation were offered. Some items might well have been listed under two or more headings, but we chose the one most relevant to the arguments presented in this book.

A. BIOLOGICAL STUDIES

1. Muscle Structure and Architecture, and Biomechanics

Akster, H. A. (1981). Ultrastructure of muscle fibres in head and axial muscles of the perch (*Perca fluviatilis* L.). *Cell Tissue Res.* 219:111–31. [Discusses architectural segregation and ultrastructural correlates of motor unit types.]

Alexander, R. M. (1969). The orientation of muscle fibers in the myomeres of fishes. *J. Marine Biol. Assoc. U.K.* 49:263–90.

An, K. N., Takahaski, K., Harrigan, T. P., and Chao, E. Y. (1984). Determination of muscle orientations and moment arms. *J. Biomech. Eng.* 106:280–82. [Presents methods useful for myoskeletal model development.]

Bock, W. J. (1968). Mechanics of one- and two-joint muscles. *Am. Mus. Novitates* (2319):1–45.

Buchthal, F., and Rosenfalck, P. (1973). On the structure of motor units. In *New Developments in Electromyography and Clinical Neurophysiology*, vol. 1, edited by J. E. Desmedt. Basel: S. Karger, pp. 71–85.

Cavagna, G. A., Heglund, N. C., and Taylor, C. R. (1977). Mechanical work in terrestrial locomotion: Two basic mechanisms for minimizing energy expenditure. *Am. J. Physiol.* 233:R243–61. [Considers elastic storage of energy.]

Dul, J., Johnson, G. E., Shiavi, R., and Townsend, M. A. (1984). Muscle synergism. II. A minimum-fatigue criterion for load sharing between synergistic muscles. *J. Biomech.* 17:675–84.

Dul, J., Townsend, M. A., Shiavi, R., and Johnson, G. E. (1984). Muscular synergism. I. On criteria for load sharing between synergistic muscles. *J. Biomech.* 17:663–74.

English, A. W., and Letbetter, W. D. (1982). A histochemical analysis of identified compartments of cat lateral gastrocnemius muscles. *Anat. Rec.* 204:123–30. [Discusses compartmentalization of muscles by fiber types.]

Galvas, P. E., and Gonyea, W. J. (1980). Motor-end-plate and nerve distribution in a histochemically compartmentalized pennate muscle in the cat. *Am. J. Anat.* 159:147–56.

Gandevia, S. C., and Mahutte, C. K. (1980). Joint mechanics as a determinant of motor unit organization in man. *Med. Hypothesis* 6:527–33.

Gans, C. (1980). *Biomechanics: Approach to Vertebrate Biology.* Ann Arbor: University of Michigan Press. [Reissue of 1974 volume. Contains a section on theory of structural analysis in animals.]

——— (1982a). Fiber architecture and muscle function. *Exerc. Sport Sci. Rev.* 10:160–207. [Reviews fiber architecture forms and their functional implications. See also Gans, de Vree, and Carrier (1985), section IC.]

——— (1982b). Functional analyses of muscular systems: Theory and an example. *Vertebrata Hungarica* 21:131–40.

Gans, C., and Bock, W. J. (1965). The functional significance of muscle architecture: A theoretical analysis. *Ergeb. Anat. Entwicklgesch.* 38:115–42.

Gaspard, M. (1965). Introduction à l'analyse biomathématique de l'architecture des muscles. *Arch. Anat. Histol. Embryol. (Strasb.)* 48:95–146.

Gaunt, A. S., and Gans, C. (1969). Mechanics of respiration in the snapping turtle, *Chelydra serpentina* (Linné). *J. Morphol.* 128:195–228.

Goldspink, D. F., ed. (1980). Development and specialization of skeletal muscle. *Soc. Exp. Biol. Sem. Ser.* 7:1–155.

Gonyea, W. J., and Ericson, G. C. (1977). Morphological and histochemical organization of the flexor carpi radialis muscle in the cat. *Am. J. Anat.* 148:329–44.

Goslow, G. E., Jr., Cameron, W. E., and Stuart, D. G. (1977). Ankle flexor muscles in the cat: Length–active tension and muscle unit properties as related to locomotion. *J. Morphol.* 15:23–38. [Considers preferred operating range versus force-generating capabilities.]

Grillner, S. (1972). The role of muscle stiffness in meeting the changing postural and locomotor requirements for force development by the ankle extensors. *Acta Physiol. Scand.* 86:92–108.

Henneman, E., and Olson, C. B. (1965). Relations between structure and function in the design of skeletal muscles. *J. Neurophysiol.* 28:85–99.

Herring, S. W. (1980). Functional design of cranial muscles: Comparative and physiological studies in pigs. *Am. Zool.* 20:283–93.

Herring, S. W., Grimm, A. F., and Grimm, B. R. (1979). Functional heterogeneity in a multipinnate muscle. *Am. J. Anat.* 154:563–76.

Hoyle, G. (1983). *Muscles and Their Neural Control.* New York: John Wiley & Sons. [An important, but idiosyncratic book that ranges widely through the literature on the muscles of many groups of animals.]

Huddart, H. (1975). *The Comparative Structure and Function of Muscle.* Oxford: Pergamon Press.

Jackson, K. M., Joseph, J., and Wyard, S. J. (1977). Sequential muscular contraction. *J. Biomech.* 10:97–106.

MacConaill, M. A., and Basmajian, J. V. (1977). *Muscles and Movements: A Basis for Human Kinesiology*, ed. 2. Baltimore: Williams & Wilkins.

Morrison, J. B. (1970). The mechanics of muscle function in locomotion. *J. Biomech.* 3:431–51.

Oguztoreli, M. N., and Stein, R. B. (1977). A kinetic study of muscular contractions. *J. Math. Biol.* 5:1–31.

Sacks, R. D., and Roy, R. R. (1982). Architecture of the hind limb muscles of cats: Functional significance. *J. Morphol.* 173:185–95.

Tueller, V. M. (1969). The relationship between the vertical dimension of occlusion and forces generated by closing muscles of mastication. *J. Prosthet. Dent.* 22:284–88.

van Mameren, H., and Drukker, J. (1979). Attachment and composition of skeletal muscles in relation to their function. *J. Biomech.* 12:859–67.

Wachtler, F., Jacob, H. J., and Christ, B. (1984). The extrinsic ocular muscles in birds are derived from the prechordal plate. *Naturwissenschaften* 71:379–80.

Willemse, J. J. (1977). Morphological and functional aspects of the arrangement of connective tissue and muscle fibres in the tail of the Mexican axolotl, *Siredon mexicanum* (Shaw) (Amphibia, Urodela). *Acta Anat.* 97:266–85.

Wineski, L., and Gans, C. (1984). Morphological basis of the feeding mechanics in the shingleback lizard *Trachydosaurus rugosus* (Scincidae: Reptilia). *J. Morphol.* 181:271–95. [Provides illustration of complex muscles.]

2. Muscle Fiber Physiology

Bendall, J. R. (1969). *Muscles, Molecules and Movement. An Essay in the Contraction of Muscles.* New York: American Elsevier.

Craig, R. (1983). Muscle *au naturel. Nature* 306:112–13. [Discusses properties of skinned muscle fibers.]

Gordon, A. M., Huxley, A. F., and Julian, F. J. (1966). The variations in isometric tension with sarcomere length in vertebrate muscle fibers. *J.*

Physiol. (*Lond.*) 184:170–92. [Classic study.]

Hess, A. (1970). Vertebrate slow muscle fibers. *Physiol. Rev.* 50:40–62.

Hill, A. V. (1938). The heat of shortening and the dynamic constants of muscle. *Proc. R. Soc. Lond.* (*Biol.*) 126:136–95. [Classic study of work output versus velocity of contraction.]

———(1970). *First and Last Experiments in Muscle Mechanics.* Cambridge: Cambridge University Press.

Magid, A., Ting-Beall, H. P., Carvell, M., Kontis, T., and Lucaveche, C. (1984). Connecting filaments, core filaments, and side-struts: A proposal to add three new load-bearing structures to the sliding filament model. In *Advances in Experimental Medicine and Biology: Contractile Mechanisms in Muscle*, vol. 170, edited by G. H. Pollack and H. Sugi. New York: Plenum Publishing, pp. 307–28.

Martonosi, A. N. (1984). Mechanisms of Ca^{2+} release from sarcoplasmic reticulum of skeletal muscle. *Physiol. Rev.* 64:1240–320.

McMahon, T. A. (1984). *Muscles, Reflexes, and Locomotion.* Princeton: Princeton University Press. [Includes history of muscle physiology plus biomechanics of scale among organisms.]

Morgan, D. L., and Proske, U. (1984). Mechanical properties of toad slow muscle attributed to non-uniform sarcomere lengths. *J. Physiol.* 349:107–17.

Pennycuick, C. J., and Rezende, M. A. (1984). The specific power output of aerobic muscle, related to the power density of mitochondria. *J. Exp. Biol.* 108:377–92.

Pollack, G. H. (1983). The cross-bridge theory. *Physiol. Rev.* 63:1049–113. [Presents a critical review of outstanding problems and questions in muscle physiology.]

Squire, J. M. (1983). Molecular mechanisms in muscular contraction. *Trends Neurosci.* 6:409–13. [Provides an overview of biochemistry of contractile coupling and its regulation.]

Wilkie, D. R. (1976). *Muscle*, ed. 2. Studies in Biology, No. 11. London: Edward Arnold.

Wood, J. E., and Mann, R. E. (1981). A sliding-filament cross-bridge ensemble model of muscle contraction for mechanical transients. *Math. Biosci.* 57:211–63.

3. Whole Muscle Properties

Bodine, S. C., Roy, R. R., Meadows, D. A., Zernicke, R. F., Sacks, R. D., Fournier, M., and Edgerton, V. R. (1982). Architectural, histochemical and contractile characteristics of a unique biarticular muscle: The cat semitendinosus. *J. Neurophysiol.* 48:192–201.

Close, R. J. (1972). Dynamic properties of mammalian skeletal muscles. *Physiol. Rev.* 52:129–97.

Granzier, H. L. M. Wiersma, J., Akster, H. A., and Osse, J. W. M. (1983). Contractile properties of a white- and a red-fibre type of the M. hyohyoideus of the carp (*Cyprinus carpio* L.). *J. Comp. Physiol.* 149:441–49. [Discusses physiological correlates of anatomical fiber types in fish.]

Hill, A. V. (1960). Production and absorption of work by muscle. *Science* 131:897–903. [Considers thermodynamics during active lengthening versus active shortening.]

Ismail, H. M., and Ranatunga, K. W. (1978). Isometric tension development in a human skeletal muscle in relation to its working range of movement: The length tension relation of biceps brachii muscle. *Exp. Neurol.* 62:595–604.

Joyce, G. C., Rack, P. M. H., and Westbury, D. R. (1969). Mechanical properties of cat soleus muscle during controlled lengthening and shortening movements. *J. Physiol.* (*Lond.*) 208:461–74. [Discusses tension increase during active lengthening.]

Komi, P. V. (1973). Relationship between muscle tension, EMG and velocity of contraction under concentric and eccentric work. In *New Developments in Electromyography and Clinical Neurophysiology*, vol. 1, edited by J. E. Desmedt. Basel: S. Karger, pp. 596–606.

Mackenna, B. R., and Türker, K. S. (1978). Twitch tension in the jaw muscles of the cat at various degrees of mouth opening. *Arch. Oral Biol.* 23:917–20.

Marechal, G., Goffart, M., Reznik, M., and Gerebtzoff, M. A. (1976). The striated muscles in a slow-mover, *Perodicticus potto* (Prosimii, Lorisidae, Lorisinae). *Comp. Biochem. Physiol.* (*B*) 54A:81–93.

Margaria, R. (1976). *Biomechanics and Energetics*

of Muscular Exercise. Oxford: Clarendon Press.

Maxwell, L. C., Barkley, J. K., Mohrman, D. E., and Faulkner, J. A. (1977). Physiological characteristics of skeletal muscles of dogs and cats. *Am. J. Physiol.* 233:C14–18.

Morgan, D. L., and Proske, U. (1984a). Vertebrate slow muscle: Its structure, pattern of innervation, and mechanical properties. *Physiol. Rev.* 64:103–69.

——— (1984b). Non-linear summation of tension in motor units of toad slow muscle. *J. Physiol.* 349:95–105.

Peachey, L. D., Adrian, R. H., and Geiger, S. R., eds. (1983). Skeletal muscle. In *Handbook of Physiology,* section 10. Bethesda, Md.: American Physiological Society.

Rice, M. J. (1973). Supercontracting skeletal muscle in a vertebrate. *Nature* 243:238–40.

Sallin, B., Henriksson, J., NyGaard, E., Anderson, P., and Janssen, E. (1977). Fiber types and metabolic potentials of skeletal muscles in sedentary man and endurance resources. *Ann. N.Y. Acad. Sci.* 301:3–29.

Spector, S. A., Gardiner, P. F., Zernicke, R. F., Roy, R. R., and Edgerton, V. R. (1980). Muscle architecture and force-velocity characteristics of cat soleus and medial gastrocnemius: Implications for motor control. *J. Neurophysiol.* 44:951–60.

Taylor, C. R., Weibel, E., and Bolls, L., eds. (1985). Design and performance of muscular systems. *J. Exp. Biol.* 115:1–412.

4. Motor Units and Neural Control

Baker, D., Hunt, C. C., and McIntyre, A. K. (1974). Muscle receptors. In *Handbook of Sensory Physiology,* vol. 3, part 2. New York: Springer Verlag.

Ballintijn, C. M., and Alink, G. M. (1977). Identification of respiratory motor neurons in the carp and determination of their firing characteristics and interconnection. *Brain Res.* 136:261–76.

Bixby, J. L., Maunsell, J. H. R., and Van Essen, D. C. (1980). Effects of motor unit size on innervation patterns in neonatal mammals. *Exp. Neurol.* 70:516–24.

Buchthal, F., and Schmalbruch, H. (1980). Motor unit of mammalian muscle. *Physiol. Rev.* 60:90–142.

Burke, R. E. (1967). Motor unit types of cat triceps surae muscle. *J. Physiol.* (*Lond.*) 193:141–60. [Discusses relationships between anatomy and physiology.]

——— (1977). Motor units: Anatomy, physiology and functional organization. In *The Handbook of Physiology: The Nervous System II,* Bethesda, Md.: U.S. Department of Health and Human Services, National Institute of Health, pp. 345–422. [Reviews correlations among motoneurons, muscle fiber types, and their function.]

——— (1978). Motor units: Their physiological properties and neural connections. *Am. Zool.* 18:127–34.

Burke, R. E., and Edgerton, V. R. (1975). Motor unit properties and selective involvement in movement. *Exerc. Sport Sci. Rev.* 3:31–81.

Burke, R. E., Levine, D. N., Salcman, M., and Tsairis, P. (1974). Motor units in cat soleus muscle: Physiological, histochemical and morphological characteristics. *J. Physiol.* (*Lond.*) 238:503–14.

Burke, R. E., Levine, D. N., Tsairis, P., and Zajac, F. E., III (1973). Physiological types and histochemical profiles in motor units of the cat gastrocnemius. *J. Physiol.* (*Lond.*) 234:723–48.

Burke, R. E., Rudomin, P., and Zajac, F. E., III (1976). The effect of activation history on tension production by individual motor units. *Brain Res.* 109:515–29. [Discusses nonlinear, time-dependent effects such as potentiation, fatigue, sag, and catch.]

Burke, R. E., and Tsairis, P. (1973). Anatomy and innervation ratios in motor units of cat gastrocnemius. *J. Physiol.* (*Lond.*), 34:749–65.

Clark, D. A. (1931). Muscle counts of motor units: A study in innervation ratios. *Am. J. Physiol.* 56:296–304.

Desmedt, J. E., and Godaux, E. (1981). Spinal motoneuron recruitment in man: Rank deordering with direction but not with speed of voluntary movement. *Science* 214:933–36.

Dum, R. P., and Kennedy, T. T. (1980). Physiological and histochemical characteristics of motor units in cat tibialis anterior and extensor digitorum longus muscles. *J. Neurophysiol.* 43:1615–30.

Eccles, J. C., and Sherrington, C. S. (1930). Numbers and contraction values of individual motor-units examined in some muscles of the limb. *Proc. R. Soc. Lond.* (*Biol.*) 106:326–56.

English, A. W., and Wolf, S. L. (1982). The motor unit: Anatomy and physiology. *J. Am. Phys. Ther. Assoc.* 62:1763–72. [Discusses fiber type and motor unit domain compartmentalization.]

Hoffer, J. A., O'Donovan, M. J., Pratt, C. A., and Loeb, G. E. (1981). Discharge pattern of hind-limb motoneurons during normal cat locomotion. *Science* 213:466–68. [Evaluates single-unit recording from ventral roots via floating microelectrodes.]

Hudson, R. C. L. (1969). Polyneuronal innervation of the fast muscles of the marine teleost *Cottus scorpius* L. *J. Exp. Biol.* 50:47–67.

Katz, B. (1966). *Nerve, Muscle and Synapse.* New York: McGraw Hill.

Loeb, G. E. (1984). The control and responses of mammalian muscle spindles during normally executed motor tasks. *Exerc. Sport Sci. Rev.* 12:157–204. [Includes notion of task groups of recruitment in kinematically heterogeneous muscle work.]

Olson, C. B., Carpenter, D. O., and Henneman, E. (1968). Orderly recruitment of muscle action potentials. Motor unit threshold and EMG amplitude. *Arch. Neurol.* 19:591–97.

Ridge, R. M. A., and Thomson, A. M. (1980). Polyneuronal innervation: Mechanical properties of overlapping motor units in a small foot muscle of *Xenopus laevis. J. Physiol.* (*Lond.*) 306:29–39.

Stephens, J. A., and Stuart, D. G. (1975). The motor units of cat medial gastrocnemius: Speed-size relations and their significance for the recruitment order of motor units. *Brain Res.* 91:177–95.

B. METHODOLOGICAL STUDIES

1. Biophysics of the EMG Signal

Andreassen, S., and Rosenfalck, A. (1981). Relationship of intracellular and extracellular action potentials of skeletal muscle fibers. *CRC Crit. Rev. Bioeng.* 6:267–306. [Includes analysis of dipole and tripole models of muscle fiber action potentials and effects of spatial filtering with distance from fiber surface to recording site.]

Barker, A. T., Brown, B. H., and Freeston, I. L. (1979). Modeling of an active nerve fiber in a finite volume conductor and its application to the calculation of surface action potentials. *Trans. IEEE-BME* 26:53–56.

Blinowska, A., Verroust, J., and Cannet, G. (1980). An analysis of synchronisation and double discharge effects on low frequency electromyographic power spectra. *Electromyogr. Clin. Neurophysiol.* 20:465–80.

Boyd, D. C., Lawrence, P. D., and Bratty, P. J. A. (1978). On modeling the single motor unit action potential. *Trans. IEEE-BME* 25:236–43.

Christakos, C. N. (1982). A linear stochastic model of the single motor unit. *Biol. Cybern.* 44:79–89.

Coggshall, J. C., and Bekey, G. A. (1970). A stochastic model of skeletal muscle based on motor unit properties. *Math. Biosci.* 7:405–19.

Dimitrova, N. (1974). Model of the extracellular potential field of a single striated muscle fiber. *Electromyography* 14:53–79.

Ekstedt, J. (1964). Human single muscle fiber action potential. *Acta Physiol. Scand.* 61 (suppl. 226):1–96.

Ekstedt, J., and Stålberg, E. (1973). How the size of the needle electrode leading off surface influences the shape of the single muscle fiber action potential in electromyography. *Comp. Prog. Biomed.* 3:204–12.

Fatt, P. (1964). An analysis of the transverse electrical impedance of striated muscle. *Proc. R. Soc. Br.* (*B*) 159:606–51.

Geddes, L. A., and Baker, L. E. (1967). The specific resistance of biological material: A compendium of data for the biomedical engineer and physiologist. *Med. Biol. Eng.* 5:271–93.

Griep, P. A. M., Boon, K. L., and Stegeman, D. F. (1979). A study of the motor unit action potential by means of computer simulation. *Biol. Cybern.* 30:221–30.

Guha, S. K., and Anand, S. (1979). Simulation linking EMG power spectra to recruitment and rate coding. *Comput. Biol. Med.* 9(3):213–221.

Gydikov, A. (1981). Spreading of potentials along the muscle, investigated by averaging of the summated EMG. *Electromyogr. Clin. Neurophysiol.* 21:525–38.

——— (1982). New investigations by averaging of the summated EMG on the propagation of the muscle potentials. *Electromyogr. Clin. Neurophysiol.* 22:89–103.

Gydikov, A., Gerilovsky, L., and Dimitrov, G. V. (1976). Dependence of the H-reflex potential shape on the extraterritorial potentials of triceps surae motor units. *Electromyogr. Clin. Neurophysiol.* 16:555–67.

Gydikov, A., Gerilovsky, L., Gatev, P., and Kostov, K. (1982). Volume conduction of motor unit potentials from different human muscles to long distances. *Electromyogr. Clin. Neurophysiol.* 22:105–16.

Gydikov, A., Gerilovsky, L., Kostov, K., and Gatev, P. (1980). Influence of some features of the muscle structure on the potential of motor units, recorded by means of different types of needle electrodes. *Electromyogr. Clin. Neurophysiol.* 20:299–321.

Gydikov, A., and Kosarov, D. (1972). Volume conduction of the potentials from separate motor units in human muscles. *Electromyography* 12:127–47.

——— (1972). Extraterritorial potential field of impulses from separate motor units in human muscles. *Electromyography* 12:283–305.

Gydikov, A., Kossev, A., and Christova, L. (1982). Influence of the interstimulus interval on the extraterritorial potentials of the motor units. *Electromyogr. Clin. Neurophysiol.* 22:563–77.

Håkansson, C. H. (1957). Action potentials recorded intra- and extra-cellularly from the isolated frog muscle fiber in Ringer solution and in air. *Acta Physiol. Scand.* 39:291–312.

Katz, B. (1948). The electrical properties of the muscle fiber membrane. *Proc. R. Soc. Br. (B)* 135:506–34.

Katz, B., and Miledi, R. (1965). Propagation of electric activity in motor nerve terminals. *Proc. R. Soc. Br. (B)* 161:453–82.

Kosarov, D., and Gydikov, A. (1975). The influence of the volume conduction on the shape of the action potentials recorded by various types of needle electrodes in normal human muscle. *Electromyogr. Clin. Neurophysiol.* 15:319–35.

Lago, P., and Jones, N. B. (1977). Effect of motor-unit firing time statistics on EMG spectra. *Med. Biol. Eng. Comput.* 15:648–55.

Lindström, L. H., and Magnusson, R. I. (1977). Interpretation of myoelectric power spectra: A model and its applications. *Proc. IEEE* 65:653–62.

Marks, W. B., and Loeb, G. E. (1976). Action currents, internodal potentials, and extracellular records of myelinated mammalian nerve fibers derived from node potentials. *Biophys. J.* 16:655–68.

Mobley, B. A., Leung, J., and Eisenberg, R. (1975). Longitudinal impedance of single frog muscle fibers. *J. Gen. Physiol.* 65:97–113.

Monster, A. W., and Chan, H. (1980). Surface electromyogram potentials of motor units: Relationship between potential size and unit location in a large human skeletal muscle. *Exp. Neurol.* 67:280–97.

Nanded Kar, S. D., and Stålberg, E. (1983). Simulation of single muscle fibre action potentials. *Med. Biol. Eng. Comput.* 21:158–65.

Perkel, D. H., Gerstein, G. L., and Moore, G. P. (1967). Neuronal spike trains and stochastic point processes. II. Simultaneous spike trains. *Biophys. J.* 7:419–40.

Perry, J., Easterday, C. S., and Antonelli, D. J. (1981). Surface versus intramuscular electrodes for electromyography of superficial and deep muscles. *Phys. Ther.* 6:7–15.

Plonsey, R. (1965). Dependence of scalar potential measurements on electrode geometry. *Rev. Sci. Instrum.* 36:1034–36.

——— (1974). The active fiber in a volume conductor. *Trans. IEEE-BME* 21:371–81. [Presents methods for modeling the effects of anisotropic tissue impedances.]

——— (1977). Action potential sources and their volume conductor fields. *Proc. IEEE* 65:601–11.

Ranck, J. B., Jr. (1975). Which elements are excited in electrical stimulation of mammalian central nervous system: A review. *Brain Res.* 98:417–40. [Surveys experimental data and relationship to biophysics of excitable cells and their size and orientation.]

Rosenfalck, P. (1969a). *Intra- and extracellular potential fields of active nerve and muscle fibers,* 117–33. Copenhagen: Akademisk Forlag.

——— (1969b). Intra- and extracellular potential fields of active nerve and muscle fibers: A physicomathematical analysis of different models. *Acta Physiol. Scand.* (suppl. 321):1–168

Schneider, M. F. (1970). Linear electrical properties of the transverse tubules and surface membrane of skeletal muscle fibers. *J. Gen. Physiol.* 56:640–71.

Stålberg, E., and Antoni, L. (1980). Electrophysiological cross section of the motor unit. *J. Neurol. Neurosurg. Psychiatry* 43:469–74.

Wani, A. M., and Guha, S. (1981). Synthesizing of a motor unit potential based on the sequential firing of muscle fibres. *Med. Biol. Eng. Comput.* 18:719–26. [Includes careful and readable accounting of factors that influence the generation of recorded EMGs.]

Weytjens, J. L. F., and van Steenberghe, D. (1984). The effects of motor unit synchronization on the power spectrum of the electromyogram. *Biol. Cybern.* 51:71–77.

2. Processing EMG Signals

Abraham, L. D., Marks, W. B., and Loeb, G. E. (1985). The distal hindlimb musculature of the cat: Cutaneous reflexes during locomotion. *Exp. Brain Res.* 58:594–603. [Presents rasters to show gating by phase of locomotion program.]

Agarwal, G. C., and Gottlieb, G. L. (1975). An analysis of the electromyogram by Fourier, simulation and experimental techniques. *Trans. IEEE-BME* 22:225–29.

Bak, M. J., and Loeb, G. E. (1979). A pulsed integrator for EMG analysis. *Electroencephalogr. Clin. Neurophysiol.* 47:738–41. [Discusses circuit to quantify area-under-the-curve and to avoid aliasing at low computer sampling rates.]

Bak, M. J., and Schmidt, E. M. (1977a). An analog delay for on-line visual confirmation of neuroelectric signals. *Trans. IEEE-BME* 24:69–71.

——— (1977b). An improved time amplitude window discriminator. *Trans. IEEE-BME* 24:486–89.

Basmajian, J. V., Clifford, H. C., McLeod, W. D., and Nunnally, H. N. (1975). *Computers in Electromyography.* Boston: Butterworth.

Beach, J. C., Gorniak, G. C., and Gans, C. (1982). A method for quantifying electromyograms (technical note). *J. Biomech.* 15:611–17. [Proposes a simple analytical scheme.]

Bigland-Ritchie, B., Donovan, E. F., and Roussos, C. S. (1981). Conduction velocity and EMG power spectrum changes in fatigue of sustained maximal efforts. *J. Appl. Physiol.* 51:1300–5.

Birö, G., and Partridge, L. D. (1971). Analysis of multiunit spike records. *J. Appl. Physiol.* 30:521–26.

Borchers, H.-W., and Pinkwart, C. (1983). A telemetry system for single unit recording in the freely moving toad (*Bufo bufo* L.). In *Advances in Vertebrate Neuroethology*, edited by J.-P. Ewert, R. R. Capranica, and D. J. Ingle. Series A, Life Sciences, vol. 56. New York: Springer Verlag.

Byrd, K. E., and Garthwaite, C. R. (1981). Contour analysis of masticatory jaw movements and muscle activity in *Macaca mulatta. Am. J. Phys. Anthropol.* 54:391–99.

De Luca, C. J. (1985). Towards understanding the EMG signal. In *Muscles Alive: Their Functions Revealed by Electromyography,* 5th ed., edited by J. V. Basmajian and C. J. De Luca. Baltimore: Williams & Wilkins.

Desmedt, J. E., ed. (1983). Computer-aided electromyography. In *Progress in Clinical Neurophysiology*, vol. 10. Basel: S. Karger.

Duysens, J., and Loeb, G. E. (1980). Modulations of ipsi- and contralateral reflex responses in unrestrained walking cats. *J. Neurophysiol.* 44:1024–37.

Gel'Fand, I. M., Gurfinkel, V. S., Kots, Ya. M., Tsetlin, M. L., and Shik, M. L. (1963). Synchronization of motor units and associated model concepts. *Biophysics* 8:528–41.

Kostov, K., Kossev, A., and Gydikov, A. (1984). Utilization of the stimulated electromyogram for estimation of the functional state of the muscles. *Electromyogr. Clin. Neurophysiol.* 24:387–99.

LeFever, R. S., and De Luca, C. J. (1982). A procedure for decomposing the myoelectric signal into its constituent action potentials. I. Technique, theory and implementation. *Trans. IEEE-BME* 29:149–57. [Discusses methods for single-unit EMG recording and discrimination.]

Lindström, L., Magnusson, R., and Petersèn, I.

(1970). Muscle fatigue and action potential conduction velocity changes studied with frequency analysis of EMG signals. *Electromyography* 10:341–56.

Lindström, L., and Petersèn, I. (1981). Power spectra of myoelectric signals: Motor unit activity and muscle fatigue. In *Neurology 1. Clinical Neurophysiology*, edited by E. Stålberg and R. R. Young. London: Butterworths, pp. 66–87.

Mambrito, B., and De Luca, C. J. (1983). Acquisition and decomposition of the EMG signal. In *Computer-aided Electromyography: Progress in Clinical Neurophysiology*, vol. 10, edited by J. E. Desmedt. Basel: S. Karger, pp. 52–72. [Discusses single-unit EMG techniques.]

Masuda, T., Miyano, H., and Sadoyama, T. (1982). The measurement of muscle fiber conduction velocity using a gradient threshold zero-crossing method. *Trans. IEEE-BME* 29:673–78.

McComas, A. J., Fawcett, P. R. W., Campbell, M. J., and Sica, P. R. E. (1971). Electrophysiological estimation of the number of motor units within a human muscle. *J. Neurol. Neurosurg. Psychiatry* 34:121–31.

Mize, R. R. (1983). Microcomputer applications in cell and neurobiology: An annotated bibliography. Paper presented to the American Association of Anatomists, Atlanta, 6 April 1983. [Presents basic subroutines.]

Person, R. S., and Kudina, L. P. (1968). Cross-correlation of electromyograms showing interference patterns. *Electroencephalogr. Clin. Neurophysiol.* 25:58–68.

Person, R. S., and Mishin, L. N. (1964). Auto- and cross-correlation analysis of the electrical activity of muscles. *Med. Electron. Biol. Eng.* 2:155–59.

Schmidt, E. M. (1984a). Computer separation of multi-unit neuroelectric data: A review. *J. Neurosci. Methods* 12:95–116.

——— (1984b). Instruments for sorting neuroelectric data: A review. *J. Neurosci. Methods* 12:1–24. [Discusses spike waveform discriminators.]

Stulen, F. B., and De Luca, C. J. (1981). Frequency parameters of the myoelectric signal as a measure of muscle conduction velocity. *Trans. IEEE-BME* 28:515–23.

Verroust, J., and Blinowska, A. (1981). Functioning of the ensemble of motor units of the muscle determined from global EMG signal. *Electromyogr. Clin. Neurophysiol.* 21:11–24.

Welch, P. D. (1967). The use of fast Fourier transform for the estimation of power spectra: A method based on time averaging over short, modified periodograms. *Trans. IEEE-AU* 15:70–73.

3. EMG Methods and Sample Studies

Andreassen, S., and Rosenfalck, A. (1978). Recording from a single motor unit during strong effort. *Trans. IEEE-BME* 3:236–43.

Armstrong, R. B., Marum, P., Saubert C. W., IV, Seeherman, H. J., and Taylor, C. R. (1977). Muscle fiber activity as a function of speed and gait. *J. Appl. Physiol.* 43:672–77.

Basmajian, J. V., and De Luca, C. J., eds. (1985). *Muscles alive: Their Functions Revealed by Electromyography*, ed. 5. Baltimore: Williams & Wilkins. [Standard Text for clinical EMG.]

——— (1980). Electromyography—dynamic gross anatomy: Review. *Am. J. Anat.* 159:245–60.

Cohen, A. H., and Gans, C. (1975). Muscle activity in rat locomotion: Movement analysis and electromyography of the flexors and extensors of the elbow. *J. Morphol.* 146:177–96.

Cundall, D., and Gans, C. (1979). Feeding in water snakes: An electromyographic study. *J. Exp. Zool.* 209:189–208.

Desmedt, J. E., and Godaux, E. (1977). Ballistic contractions in man: Characteristic recruitment pattern of single motor units of the tibialis anterior muscle. *J. Physiol.* 264:673–93.

DeTroyer, A., Sampson, M., Sigrist S., and Macklem, P. T. (1981). The diaphragm: Two muscles. *Science* 213:237–38.

Engel, W. K. (1967). Focal myopathic changes produced by electromyographic and hypodermic needles. *Arch. Neurol.* 16:509–11. [Discusses histological evidence of damage from intramuscular electrodes.]

Freund, H.-J. (1983). Motor unit and muscle activity in voluntary motor control. *Physiol. Rev.* 63:387–436.

Gans, C., de Vree, F., and Carrier, D. C. (1985).

Usage pattern of the complex masticatory muscles in the shingleback lizard, *Trachydosaurus rugosus*: A model for muscle placement. *Am. J. Anat.* 173:219–40. [See also Wineski and Gans (1984), section IA, which illustrates the anatomy.]

Gans, C., and Gorniak, G. C. (1980). Electromyograms are repeatable: Precautions and limitations. *Science* 210:795–97.

——— (1982a). Protrusion of the tongue in marine toads (*Bufo marinus*). *Am. J. Anat.* 163:195–222.

——— (1982b). How does the toad flip its tongue? Test of two hypotheses. *Science* 216:1335–37.

Gath, I., and Stålberg, E. V. (1976). Techniques for improving the selectivity of electromyographic recordings. *Trans. IEEE-BME* 23:467–72.

Gesteland, R. C., Howland, B., Lettvin, J. Y., and Pitts, W. H. (1959). Comments on microelectrodes. *Proc. Inst. Radio Eng.* 47:1856–62.

Goslow, G. E., Jr., Seeherman, H. J., Taylor, C. R., McCutchin, M. N., and Heglund, N. C. (1981). Electrical activity and relative length changes of dog limb muscle as a function of speed and gait. *J. Exp. Biol.* 94:15–42.

Gruner, J. A. (1976). A receptacle and plug for chronic recording from subcutaneous electrodes. *Physiol. Behav.* 17:361–63. [Describes percutaneous connector design.]

Guld, C., Rosenfalck, A., and Willison, R. G. (1969). Report of the committee on EMG instrumentation. In *Technical Factors in Recording Electrical Activity of Muscle and Nerve in Man.* Amsterdam: Elsevier.

——— (1970). Technical factors in recording electrical activity of muscle and nerve in man. *Electroencephalogr. Clin. Neurophysiol.* 28:399–413. [Presents standards for EMG measurement in humans.]

Hannerz, J. (1974). An electrode for recording single motor unit activity during strong muscle contractions. *Electroencephalogr. Clin. Neurophysiol.* 37:179–81.

Jungers, W. L., and Stern, J. T., Jr. (1980). Telemetered electromyography of forelimb muscle chains in gibbons (*Hylobates lar.*) Science 208:617–19.

Manns, A., and Spreng, M. (1977). EMG amplitude and frequency at different muscular elongations under constant masticatory force or EMG activity. *Acta Physiol. Lat. Am.* 27:259–71.

Miller, H. A., and Harrison, D. C., eds. (1974). *Biomedical Electrode Technology.* New York: Academic Press.

Mills, K. R. (1982). Power spectral analysis of electromyogram and compound muscle action potential during muscle fatigue and recovery. *J. Physiol.* 326:401–9. [Evaluates frequency domain analysis of EMG.]

Patla, A. E., Hudgins, B. S., Parker, P. A., and Scott, R. N. (1982). Myoelectric signal as a quantitative measure of muscle mechanical output. *Med. Biol. Eng. Comp.* 20:319–28.

Pierce, D. S., and Wagman, I. H. (1964). A method of recording from single muscle fibers of motor units in human skeletal muscle. *J. Appl. Physiol.* 19:366–68.

Soderberg, G. L., and Dostal, W. F. (1978). Electromyographic study of three parts of the gluteus medius muscle during functional activities. *Phys. Ther.* 58:691–96.

Stålberg, E. (1976). *Single Fiber Electromyography.* Skovlunde, Denmark: Disa Electronik A/S Information Department, pp. 1–20.

Stern, J. T., Jr., Wells, J. P., Jungers, W. L., and Vangor, A. K. (1980). An electromyographic study of serratus anterior in atelines and *Alouatta*: Implications for hominoid evolution. *Am. J. Phys. Anthropol.* 52:323–34.

Thompson, L. L. (1981). *The Electromyographer's Handbook.* Boston: Little, Brown.

Weijs, W. A., and van der Wielen-Drent, T. K. (1982). Sarcomere length and EMG activity in some jaw muscles of the rabbit. *Acta Anat.* 113:178–88. [Attempts correlation of architecture, kinematics, and EMG.]

Whiting, W. C., Gregor, R. J., Roy, R. R., and Edgerton, V. R. (1984). A technique for estimating mechanical work of individual muscles in the cat during treadmill locomotion. *J. Biomech.* 17:685–94.

4. Kinematic Transducers

Clark, B., Gans, C., and Rosenberg, H. I. (1978).

Air flow in snake ventilation. *Respir. Physiol.* 32:207–12. [Presents sample application of flowmeter.]

Hoffer, J. A., and Loeb, G. E. (1980). Implantable electrical and mechanical interfaces with nerve and muscle. *Ann. Biomed. Eng.* 8:351–60.

Loeb, G. E., Walmsley, B., and Duysens, J. (1980). Obtaining proprioceptive information from natural limbs: Implantable transducers vs. somatosensory neuron recordings. In *Physical Sensors for Biomedical Application*. Proceedings of Workshop on Solid State Physical Sensors for Biomedical Application, edited by M. E. Neuman et al. Boca Raton, Fla.: CRC Press.

Prochazka, V. J., Tate, K., Westerman, R. A., and Ziccone, S. P. (1974). Remote monitoring of muscle length and EMG in unrestrained cats. *Electroencephalogr. Clin. Neurophysiol.* 37:649–53. [Describes two-channel homemade telemetry system.]

Sherif, M. H., Gregor, R. J., Liv, L. M., Roy, R. R., and Hager, C. L. (1983). Correlation of myoelectric activity and muscle force during selected cat treadmill locomotion. *J. Biomech.* 16:691–701.

Walmsley, B., Hodgson, J. A., and Burke, R. E. (1978). Forces produced by medial gastrocnemius and soleus muscles during locomotion in freely moving cats. *J. Neurophysiol.* 41:1203–16. [Describes design and use of implanted tendon strain gauge.]

5. Kinesiological Methods

Carlsöö, S. (1972). *How Man Moves: Kinesiological Studies and Methods*, translated by W. P. Michael. New York: Crane, Russak.

Chapin, J. K., Loeb, G. E., and Woodward, D. J. (1980). A simple technique for determination of footfall patterns of animals during treadmill locomotion. *J. Neurosci. Methods* 2:97–102.

International Society for Optical Engineering. *Congress on High Speed Photography and Photomics*. Bellingham, Wash.: International Society for Optical Engineering. P. O. Box 10, Bellingham, WA 98227.

International Television. New York: Ziff-Davis, 1 Park Ave., N.Y., N.Y. 10016 [Complimentary journal.].

Loeb, G. E. (1986). Spinal Programs for Locomotion. *Prog. Brain Res.* 64.

Photomethods. New York: Ziff-Davis, 1 Park Ave., N.Y., N.Y. 10016 [Complimentary journal.]

Prochazka, A. (1984). Chronic techniques for studying neurophysiology of movement in cats. In *Methods for Neuronal Recording in Conscious Animals*. Proceedings of the 1984 International Brain Research Organization, edited by R. Lemon, pp. 113–28. [Reviews recent techniques for implanted electrodes and transducers.]

6. Anatomical Techniques

Abrahams, V. C., Richmond, F. J., and Keane, J. (1984). Projections from C2 and C3 nerves supplying muscles and skin of the cat neck: A study using transganglonic transport of horseradish peroxidase. *J. Comp. Neurol.* 230:142–54.

Adams, J. C. (1981). Heavy metal intensification of DAB-based HRP reaction product. *J. Histochem. Cytochem.* 29:775.

Aronson, C. E., ed. (1983). *Veterinary Pharmaceuticals and Biologicals*. Edwardsville, Kans.: Veterinary Medicine.

Bock, W. J., and Shear, C. R. (1972). A staining method for gross dissection of vertebrate muscles. *Anat. Anz.* 130:222–27. [Presents iodine stain.]

Brooke, M. H., and Kaiser, K. K. (1970a). Three "myosin adenosine triphosphatase" systems: The nature of their pH lability and sulfhydryl dependence. *J. Histochem. Cytochem.* 18:670–72.

——— (1970b). Muscle fiber types: How many and what kind? *Arch. Neurol.* 23:369–79.

Chayen, J., Bitensky, L., and Butcher, R. G. (1973). *Practical Histochemistry*. New York: John Wiley & Sons.

Chayen, J., Bitensky, L., Butcher, R. G., and Poulter, L. W., eds. (1969). *A Guide to Practical Histochemistry*. Philadelphia: J. B. Lippincott.

Cleworth, D. R., and Edman, K. A. P. (1972). Changes in sarcomere length during isometric tension development in frog skeletal muscle. *J. Physiol.* 227:1–17.

Cuello, A. C., ed. (1984). *Immunohistochemistry.* New York: John Wiley & Sons.

De Castro, F. (1925). Technique pour la coloration du système nerveux quand il est pourvu de ses étuis osseux. *Trab. Lab. Invest. Biol. Univ. Madrid* 23:427–47.

Drury, R. A. G., and Wallington, E. A. (1980). *Carleton's Histological Technique*, ed. 5. Oxford: Oxford University Press.

Dubowitz, V. and Neville, H. E. (1973). *Muscle Biopsy: A Modern Approach*. London, Philadelphia: W. B. Saunders Co.

Gauthier, G. F., and Lowey, S. (1979). Distribution of myosin isoenzymes among skeletal muscle fiber types. *J. Cell Biol.* 81:1025.

Goshgarian, H. G. (1977). A rapid silver impregnation for central and peripheral nerve fibers in paraffin and frozen sections. *Exp. Neurol.* 57:296–301.

Griffin, G., Watkins, L. R., and Mayer, D. J. (1979). HRP pellets and slow-release gels: Two new techniques for greater localization and sensitivity. *Brain Res.* 168:595–601.

Hanker, J. S., Yates, P. E., Metz, C. B., and Rustoni, A. 1977. New specific, sensitive and non-carcinogenic reagent for demonstration of horseradish-peroxidase. *J. Histochem.* 9(6):789–92.

Hotchkiss, R. D. (1948). A microchemical reaction resulting in the staining of polysaccharide structures in fixed tissue preparations. *Arch. Biochem.* 16:131–41.

——— (1973). The analysis of chemical components of cells and tissues: Reaction for polysaccharides. In *Practical Histochemistry*, edited by J. Chayen, L. Bitensky, and R. G. Butcher. New York: John Wiley & Sons, pp. 70–80. [Discusses PAS method.]

Humphrey, C. D., and Pittman, F. E. (1974). A simple methylene blue–azure II–basic fuchsin stain for epoxy-embedded tissue sections. *Stain Technol.* 49:9–14.

Karnovsky, M. J. (1964). The localization of cholinesterase activity in rat cardiac muscle by electron microscopy. *J. Cell Biol.* 23:217–32.

Karnovsky, M. J., and Roots, L. (1964). A "direct-coloring" thiocholine method for cholinesterases. *J. Histochem. Cytochem.* 12:219–21.

Leung, A. F. (1983). Light diffractometry for determining the sarcomere length of striated muscle: An evaluation. *J. Mus. Res. Cell Mot.* 1:473–484.

Lillie, R. D., ed. (1965). Schiff's sulfurous acid leucofuchsin reagent. In *Histopathologic Technic and Practical Histochemistry*, ed. 3. New York: McGraw-Hill, pp. 269–72.

Lillie, R. D. and Fullmer, H. M., eds. (1976). *Histopathologic Technic and Practical Histochemistry.* Fourth Edition. New York: McGraw-Hill.

Mesulam, M.-M. (1978). Tetramethyl benzidine for horseradish peroxidase neurohistochemistry: A non-carcinogenic blue reaction-product with superior sensitivity for visualizing neural afferents and efferents. *J. Histochem. Cytochem.* 26:106–17.

——— (1982). *Tracing Neural Connections with Horseradish Peroxidase* ISBRO Handbook Series. Chichester: John Wiley & Sons, Ltd.

Mesulam, M.-M., Hegarty, E., Barbas, H., Carsons, K. A., Gower, E. C., Knapp, A. G., Moss, M. B., and Mufson, E. J. (1980). Additional factors influencing sensitivity in the tetramethyl benzidine method for horseradish peroxidase neurohistochemistry. *J. Histochem. Cytochem.* 28:1255–59.

Moody-Corbett, F., and Cohen, M. W. (1981). Localization of cholinesterase at sites of high acetylcholine receptor density on embryonic amphibian muscle cells cultured without nerve. *J. Neurosci.* 1:596–605.

Nachlas, M. M., Tsou, K.-C., de Souza, E., Cheng, C.-S., and Seligman, A. M. (1957). Cytochemical demonstration of succinic dehydrogenase by the use of a new *p*-nitrophenyl substituted ditetrazole. *J. Histochem. Cytochem.* 5:420–36.

Niles, N. R. (1969). Enzyme histochemistry: Method for calcium-activated ATPase activity. In *A Guide to Practical Histochemistry*, edited by J. Chayen, L. Bitensky, R. G. Butcher, and L. W. Poulter. Philadelphia: J. B. Lippincott, pp. 131–32.

Niles, N. R., Bitensky, L., Chayen, J., Cunningham, G. J., and Braimbridge, M. V. (1964). The value of histochemistry in the analysis of myocardial dysfunction. *Lancet* 1:963–65.

Padykula, H. A., and Herman, E. (1955a). Factors affecting the activity of adenosine triphosphatase and other phosphatases as measured by histochemical techniques. *J. Histochem. Cytochem.* 3:161–69.

——— (1955b). The specificity of the histochemical method for adenosine triphosphatase. *J. Histochem. Cytochem.* 3:170–95.

Palmgren, A. (1960). Specific silver staining of nerve fibers. I. Technique for vertebrates. *Acta Zool. Scand.* 41:239–65.

Riley, D. A., and Berger, A. J. (1979). A regional histochemical and electromyographic analysis of the cat respiratory diaphragm. *Exp. Neurol.* 66:636–49.

Romeis, B. (1948). *Mikroskopische Technik.* Munich: Leibniz Verlag.

Rüdel, R., and Zite-Ferenczy, F. (1979). Interpretation of light diffraction by cross-striated muscle as Bragg reflexion of light by the lattice of contractile proteins. *J. Physiol.* 290:317–30.

Scheibel, M. E., and Scheibel, A. B. (1956). Histological localization of microelectrode placement in brain by ferrocyanide and silver staining. *Stain Technol.* 31:1–5.

Swash, M., and Fox, K. P. (1972). Techniques for the demonstration of human muscle spindle innervation in neuromuscular disease. *J. Neurol. Sci.* 15:291–92.

Tsuji, S., and Tobin-Gros, C. (1980). A simple silver nitrate impregnation of nerve fibers with preservation of acetylcholinesterase activity at the motor end-plate. *Experientia* 36:1317–19.

Williams, M. A., ed. (1983). *Autoradiography and Immunocytochemistry.* Amsterdam: North-Holland.

Yeh, Y., Baskin, R. J., Lieber, R. L., and Roos, K. P. (1980). Theory of light diffraction by single skeletal muscle fibers. *J. Biophys.* 29:509–22.

Zacks, S. I. (1973). *The Motor Endplate,* rev. ed. Huntington, N.Y.: R. E. Krieger.

7. Biomaterials

Geddes, L. A. (1972). *Electrodes and the Measurement of Bioelectric Events.* New York: Wiley-Interscience. [Handy reference to many of the different types of electrodes described in the literature. Not restricted to electromyography.]

Loeb, G. E., Bak, M. J., Salcman, M., and Schmidt, E. M. (1977). Parylene as a chronically stable, reproducible microelectrode insulator. *Trans. IEEE-BME* 24:121–28.

Loeb, G. E., McHardy, J., Keliher, E. M. and Brummer, S. B. (1983). Neural prosthesis. In *Biocompatibility in Clinical Practice,* vol. 2, edited by D. F. Williams. Boca Raton, Fla.: CRC Press. [Reviews failure modes of conductors and dielectrics and techniques for prevention.]

Loeb, G. E., Walker, A. E., Uematsu, S., and Konigsmark, B. W. (1977). Histological reaction to various conductive and dielectric films chronically implanted in the subdural space. *J. Biomed. Mater. Res.* 11:195–210.

Ödman, S. (1980). On biomechanical electrode technology with special reference to long-term properties and movement-induced noise in surface electrodes. Dissertation, Department of Biomedical Engineering, Linköping University, Linköping, Sweden.

8. General Electronics and Graphics

Analog Devices (1984). *Data Acquisition Databook,* vol. 1, *Integrated Circuits;* vol. 2, *Modules and Subsystems.* Norwood, Mass.: Analog Devices.

Brown, P. B., Franz, G. N., and Moraff, H. (1982). *Electronics for the Modern Scientist.* New York: Elsevier.

Computer Graphics News. New York: Scherago Assoc. 10th Floor, Reader's Services, 1515 Broadway, New York, N.Y. 10109-0163.

De Vree, F., and Gans, C. (1976). Mastication in pygmy goats, *Capra hircus. Ann. Soc. Zool. Belg.* 105:255–306. [Application of polar coordinate graphs.]

Gans, C. (1960). Studies on amphisbaenids (Amphisbaenia: Reptilia). I. A taxonomic revision of the Trogonophinae and a functional interpretation of the amphisbaenid adaptive pattern. *Bull. Am. Mus. Nat. Hist.* 119:129–204. [Provides sample illustration methods.]

Graeme, J. G., Tobey, G. E., and Huelsman, L. P. (1971). *Operational Amplifiers: Design and Applications.* New York: McGraw-Hill.

Ko, W. H., and Neuman, M. R. (1967). Implant biotelemetry and microelectronics. *Science* 156:351–60. [Discusses principles of telemetry equipment design.]

The New Handbook of Time Code Formats (1979). Datum Inc. Timing Division, 1363 State

College Ave., Anaheim, CA 92806.

PC-Week. P. O. Box 5920, Cherry Hill, N.J. [Guide to new computer products, mainly IBM compatible.]

Philbrick/Nexus Research (1969). *Applications Manual for Operational Amplifiers: A Library of Practical Feedback Circuits*. Dedham, Mass.: Teledyne.

Sass, D. J. (1977). Time, date, and event code generator to identify physiologic events simultaneously recorded on magnetic tape and oscillographic paper. *Trans. IEEE-BME* 24:425–29. [Describes a homemade time-code generator.]

Schmid, C. F. (1983). *Statistical Graphics*. New York: Wiley-Interscience.

Scientific Computing and Automation: Technology for the Laboratory. Dover N.J.: Gordon Publications.

Steiner, G. (1982). *Tierzeichnungen in Kurzeln*. Stuttgart: Gustav Fischer Verlag, pp. 1–94. [Illustrates how to sketch animals with minimal lines.]

Tufte, E. R. (1983). *The Visual Display of Quantitative Information*. Cheshire, Conn.: Graphics Press. [Excellent treatment of theory and applications.]

Author Index

Subject Index

Page numbers set in boldface indicate where term is defined.